De Gruyter Series in Nonlinear Analysis and Applications 16

Martin Väth

Topological Analysis

From the Basics to the Triple Degree
for Nonlinear Fredholm Inclusions

De Gruyter

Mathematics Subject Classification 2010: Primary: 47H11, 47H05, 47H08; Secondary: 54C60, 54C55, 54C20, 47H10, 55M20, 54E35, 46T05, 47A53, 54F45, 54H25, 46T20, 54C15, 54C25, 54C10, 54E50, 55M15, 55M10, 55M25, 55P05.

ISBN 978-3-11-027722-7
e-ISBN 978-3-11- 027733-3
ISSN 0941-813X

Library of Congress Cataloging-in-Publication Data

A CIP catalog record for this book has been applied for at the Library of Congress.

Bibliographic information published by the Deutsche Nationalbibliothek

The Deutsche Nationalbibliothek lists this publication in the Deutsche Nationalbibliografie; detailed bibliographic data are available in the internet at http://dnb.dnb.de.

Typesetting: Da-TeX Gerd Blumenstein, Leipzig, www.da-tex.de
Printing and binding: Hubert & Co. GmbH & Co. KG, Göttingen
∞ Printed on acid-free paper

Printed in Germany

www.degruyter.com

Preface

Topological methods are among the most important theoretical tools in analysis, useful for many questions arising in analysis, concerning finite-dimensional problems or also infinite-dimensional problems like differential or integral equations. One particularly important tool is degree theory which was originally developed by Leray and Schauder for equations of a rather specific form and which is meanwhile available for a much larger class of equations, even those which involve noncompact situations or multivalued maps.

This monograph aims to give a self-contained introduction into the whole field: Requiring essentially only basic knowledge of elementary calculus and linear algebra, it provides all required background from topology, analysis, linear and nonlinear functional analysis, and multivalued maps, containing even basic topics like separation axioms, inverse and implicit function theorems, the Hahn-Banach theorem, Banach manifolds, or the most important concepts of continuity of multivalued maps. Thus, it can be used as additional material in basic courses on such topics. The main intention, however, is to provide also additional information on some fine points which are usually not discussed in such introductory courses.

The selection of the topics is mainly motivated by the requirements for degree theory which is presented in various variants, starting from the elementary Brouwer degree (in Euclidean spaces and on manifolds) with several of its famous classical consequences, up to a general degree theory for function triples which applies for a large class of problems in a natural manner. Although it has been known to specialists that, in principle, such a general degree theory must exist, this is probably the first monograph in which the corresponding theory is developed in detail.

Berlin, March 2012 — Martin Väth

Contents

Chapter 1

Introduction

This monograph has several aims. For one thing, it is meant as an introduction to topology, functional analysis, and analysis for the advanced reader. Secondly, the aim is to develop a degree theory for function triples which unifies and extends most known degree theories.

The monograph aims to be self-contained, and so many chapters could even serve as a basis for courses on the covered topics; essentially, only knowledge in basic calculus and of linear algebra (matrices, determinants, eigenvalues, etc.) is assumed. However, some experience with the topics is useful for the reading, since the order in which the topics are presented corresponds to the mathematically logical order. For instance, topological spaces and continuity properties of multivalued maps in topological spaces are studied from the very beginning; metric spaces and single-valued maps are considered as special cases and consequently discussed much later, with emphasis on what *more* can be said in such cases. Thus, roughly speaking, we start from the abstract and turn to the more concrete case only later on in order to avoid unnecessary (from the mathematical point of view) repetitions. For this reason, the monograph cannot substitute e.g. a first course in analysis which should better be arranged from a didactical instead of a logical point of view. Nevertheless, most results from analysis courses are included in this monograph.

By this approach of coming from the abstract, we are partially able to give much more elegant and shorter proofs of "standard" results of topology. For instance, important statements like the connection between closed and proper maps (Corollary 2.107) or closedness of projections under some compactness assumptions (Corollary 2.112) which are rather cumbersome to prove in an elementary way come out rather simply by using results for multivalued maps in an appropriate manner.

Even for many "standard" results, some new approaches are presented. For instance, for the classical implicit and inverse function theorems, we separate clearly the differentiability assertions from the existence and uniqueness assertions into separate theorems (which have rather different hypotheses). This is not only a didactical advantage but also has the practical advantage that we obtain an implicit function theorem that even works for functions of two variables which are not C^1 in both variables but only in one variable. Such functions occur naturally in the later sections about degree theory as "Fredholm homotopies", and for these homotopies the corresponding generalization of the implicit function theorem is needed in some crucial places.

There also occur some really new results in the first sections like an extension theorem for continuous functions which not only includes the famous extension theorems of Dugundji (or of Ma in the multivalued case) but also easily implies the existence of Schauder projections.

The selection of the topics in the first sections is mainly inspired by the application to degree theory in the later sections. For instance, for topological spaces not all known separation axioms (T_0, T_1, T_2, T_{2a} etc.) are introduced but only those which actually play a role in connection with degree theory in the later sections. However, these are already quite a lot, including the less known axioms T_5 and T_6 for which it is hard to find references on basic properties that are therefore proved in this monograph.

Similarly, also the selection of topics about ANR spaces and dimension theory is based on the needs in degree theory in the later sections.

The notion of degree theory needs a more verbose explanation: In the simplest case, the degree of a continuous function $F: \overline{\Omega} \to X$ with an open set $\Omega \subseteq X$ is "something like" the number of solutions $x \in \Omega$ of the equation

$$F(x) = y.$$

By "something like", we mean that the solutions are counted with a certain multiplicity which corresponds to the geometric behavior of F in a neighborhood of the solution. In the case of a holomorphic function $F: \mathbb{C} \to \mathbb{C}$ this multiplicity is the usual multiplicity known from function theory, but in general the multiplicity can also be negative.

The important property is that this number turns out to be invariant under certain homotopies, that is, it does not change if the function F and the point y are "moved" in a certain continuous manner, as long as during this movement there don't arise solutions on the boundary $\partial\Omega$. This means that one can transform the equation by such a homotopy into a possibly much simpler equation. Degree theory can then be used by just considering this simpler equation to prove assertions about the original equation.

For instance, one can use this idea to prove the famous Brouwer fixed point theorem which states in its simplest form that any continuous map $f: \overline{\Omega} \to \overline{\Omega}$ has a fixed point if Ω denotes the open unit ball in $\mathbb{R}^n$. This result is not easy to prove in an elementary way, but using degree theory one can consider the homotopy $F_t(x) := x - tf(x)$ $(0 \leq t \leq 1)$ which relates the fixed point equation $x = f(x)$ (that is, $F_1(x) = 0$) with the trivial equation $x = 0$ (that is, $F_0(x) = 0$). Using this idea, one obtains Brouwer's fixed point theorem from elementary properties of the degree. This and some other classical applications of degree theory which are perhaps more surprising are presented in detail in Section 9.6.

It should be clear that degree theory implies much more powerful results than just such classical examples, and in fact a mere list of results which are nowadays obtained by degree theory for dynamical systems or partial differential equations would probably be longer than this whole monograph. However, the class of equations for which the classical degree theory is applicable is quite limited: As we have noted above, we must be able to write the equation with a map $f:\overline{\Omega} \to X$ with open $\Omega \subseteq X$; actually even $X = \mathbb{R}^n$ is required (the classical Brouwer degree) or that X is a Banach space and that $\mathrm{id}_X - F$ is a compact map. The latter is the classical Leray–Schauder degree, and classical applications of the latter, like the famous Schauder fixed point theorem, which is an infinite-dimensional analogue of the Brouwer fixed point theorem, are presented in Section 13.3. The reason why this is presented only so late is that we obtain the Leray–Schauder degree as a special case of a much more general degree theory.

In fact, several attempts had been made to generalize degree theory such that it can treat a richer class of equations, and in this monograph, we will present and unify a lot of these extensions. For instance, we will develop a degree theory which can handle certain equations of the type

$$F(x) = \varphi(x) \tag{1.1}$$

where $F, \varphi: \Omega \to Y$ with Ω not necessarily being a subset of Y, a degree theory for inclusions with multivalued maps, a degree theory where we relax the compactness hypothesis of the Leray–Schauder degree, and, most importantly, we will actually combine all these extensions. The most general degree theory which we finally obtain is able to treat problems of the type

$$F(x) \in \varphi(\Phi(x)), \tag{1.2}$$

where, roughly speaking,

(a) F is a nonlinear Fredholm operator of index 0, that is, F is continuously differentiable with all derivatives being Fredholm maps of index 0.

(b) Φ is a multivalued upper semicontinuous map with acyclic values $\Phi(x)$.

(c) φ is continuous.

(d) The composition $\varphi \circ \Phi$ is "more compact than F is proper".

Of course, the hypotheses will be explained and made precise in the monograph; they are mentioned here only to give an impression of what type of requirements are involved.

For the reader who dislikes multivalued maps, let us mention that one can equivalently replace Φ by the inverse of a certain single-valued map p, but the above formulation is more convenient from the viewpoint of applications.

Thus, the degree theory which we finally obtain is actually a degree theory which is appropriate for a triple of functions (F, Φ, φ) (or (F, p, φ) for the equivalent formulation with single-valued functions). In fact, the degree remains stable under homotopies for each of these functions, that is, roughly speaking, when we "move" all three functions simultaneously in a certain admissible manner. Thus, this degree in fact can be considered as a (function) triple degree.

We point out that problems like (1.2) occur naturally in connection with dynamical systems or partial differential equations. Consider, for instance, an initial value problem in a possibly infinite-dimensional Banach space

$$x'(t) = f(t, x(t)), \qquad x(0) = x_0.$$

It is known that, even if f fails to be differentiable, under some natural hypotheses the map Φ which associates to x_0 the set of all solutions x of this problem is upper semicontinuous with acyclic values. Hence, a large class of related problems like, for instance, the nonlinear boundary value problem

$$x'(t) = f(t, x(t)), \qquad F(x(0)) = G(x(T)),$$

can be cast in the above form (1.2) with φ being the composition of G with the map $x \mapsto x(T)$. However, such applications are still subject of future research and only sketched in Section 14.6.

Degree theory for equations of the form (1.1) for nonlinear Fredholm maps F and compact φ have been developed by P. Benevieri and M. Furi [17], [19]. Although rather recently it was observed by the author that any degree theory for the problem (1.1) extends to a degree theory for function triples, that is, for the problem (1.2), this extension was never carried out in detail for the particular Benevieri–Furi degree. Indeed, this extension procedure consists of three steps (treating the finite-dimensional [140], the compact infinite-dimensional [142], and the noncompact case [141]), and each of these steps is rather technical. Thus, although it was known, in principle, that a corresponding degree theory exists, this monograph is the first publication in which this program is carried out in detail and where the degree and its precise properties are developed.

It should be mentioned that in [146] (see also [98], [121]) also a degree theory for (1.2) is developed, using a rather different approach by approximation methods. However, the notion of orientation of Fredholm maps used in [146] is rather different from the orientation used for the Benevieri–Furi degree so that it is not straightforward to compare the two theories. Moreover, many of our results (especially most of our uniqueness results and perhaps also the homotopy invariance of the degree for Fredholm homotopies which are C^1 only in the last variable) can probably not be accessed by the approach from [146].

Now it is time that we point out an important issue which occurs in connection with nonlinear Fredholm operators but also in connection with degree theory on finite-dimensional manifolds: the notion of orientation.

In fact, a notion of orientation for Fredholm maps is the crucial point for the development of a degree theory for (1.1): P. Benevieri and M. Furi succeeded in [17], [18] to develop a definition of orientation by which it is possible to reduce the degree theory to a degree theory between finite-dimensional oriented manifolds.

In this monograph, we actually carry this one step further and develop a more general notion of orientation for Fredholm maps on Banach bundles: In this more general setting, it is perhaps even simpler to understand the basic properties of the orientation than in the special (but rather involved) case of tangent bundles studied in [17], [18]. In particular, we obtain results parallel to those from [18] in this more general setting.

The main advantage of this more general setting is that it can be applied in the finite-dimensional situation to orient continuous maps which are not necessarily C^1 as for the Benevieri–Furi approach. We thus obtain also a corresponding degree theory for continuous maps between finite-dimensional manifolds with the corresponding (rather simple) notion of orientation.

In this connection, it should be pointed out that the study of manifolds is crucial for the Benevieri–Furi approach: Even if one is interested only in the degree theory with Fredholm operators in Banach spaces, one needs for this approach as an auxiliary step a corresponding degree theory on *manifolds* (not only on normed spaces). Consequently, although it is only considered as a tool and not as a main aim of this monograph, also an introduction to Banach manifolds and to related notions is included.

This monograph was written in the framework of a DFG project (Az. VA 206/2-1). Financial support by the DFG is gratefully acknowledged.

The author would like to thank P. Benevieri and W. Kryszewski for answering several questions and sending otherwise unpublished notes about their results.

Part I

Topology and Multivalued Maps

Chapter 2

Multivalued Maps

2.1 Notations for Multivalued Maps and Axioms

2.1.1 Notations

Throughout this monograph, we use standard notations for sets, elements, functions, unions, and so on. For multivalued maps, there is not always a consensus about the notations in literature, so we specify the details here.

Let X and Y be sets. We call a map from X into the powerset of Y a *multivalued map* and use the notation $\Phi\colon X \multimap Y$. For such a map, we define its *domain of definition* as

$$\operatorname{dom}(\Phi) := \{x \in X : \Phi(x) \neq \varnothing\}.$$

A *selection* of Φ is a single-valued map $f\colon X \to Y$ satisfying $f(x) \in \Phi(x)$ for all $x \in X$. For the case that $\Phi(x)$ has exactly one element for each $x \in X$, that is, if there is exactly one selection of Φ, we call Φ *single-valued*, and in a slight misuse of notation, we use then the same symbol for Φ and its selection $\Phi\colon X \to Y$. In the same sense, we consider single-valued maps as special cases of multivalued maps. For example, the symbol id_X will denote the single-valued *identity* map of a set X, defined by $\mathrm{id}_X(x) := x$ for all $x \in X$, but the symbol will also denote the multivalued map $\mathrm{id}_X\colon X \multimap X$, defined by $\mathrm{id}_X(x) := \{x\}$.

For $\Phi\colon X \multimap Y$, we define the *image* of a set $M \subseteq X$ as

$$\Phi(M) := \bigcup_{x \in M} \Phi(x),$$

and the *small* and *large counterimage* of a set M as

$$\Phi^-(M) := \{x \in X : \Phi(x) \subseteq M\},$$

and

$$\Phi^+(M) := \{x \in X : \Phi(x) \cap M \neq \varnothing\},$$

respectively. It will be convenient to use the above definitions for arbitrary sets M, that is, we do *not* necessarily require that $M \subseteq Y$. For $\Phi\colon X \multimap Y$, we define $\Phi^{-1}\colon Y \multimap X$ by

$$\Phi^{-1}(y) := \Phi^+(\{y\}) = \{x \in X : y \in \Phi(x)\}.$$

In particular, if $f: X \to Y$ is single-valued, then by our tacit identification of single-valued and multivalued maps, we have $f^{-1}: Y \multimap X$; in this case

$$f^{-1}(y) = \{x \in X : f(x) = y\} = f^{+}(\{y\}) = f^{-}(\{y\}).$$

If $\Phi^{-1}: Y \multimap X$ is single-valued (by our tacit identification of single-valued and multivalued maps this means that $\Phi^{-1}: Y \to X$), then we call Φ *invertible*. Clearly, a single-valued function $f: X \to Y$ is invertible if and only if it is one-to-one and onto, and in this case $f^{-1}: Y \to X$ is the usual inverse.

For $\Phi: X \multimap Y$, we define the *graph* of Φ as the set

$$\operatorname{graph}(\Phi) := \{(x, y) \in X \times Y : y \in \Phi(x)\}.$$

If X and Y are topological spaces, we will equip $\operatorname{graph}(\Phi)$ with the topology inherited from the product topology of $X \times Y$ (see Section 2.2).

Proposition 2.1. *For $\Phi: X \multimap Y$ we have*

$$\operatorname{graph}(\Phi^{-1}) = \{(y, x) : (x, y) \in \operatorname{graph}(\Phi)\}.$$

In particular, $(\Phi^{-1})^{-1} = \Phi$, $\operatorname{dom}(\Phi^{-1}) = \Phi(X)$, $\operatorname{dom}(\Phi) = \Phi^{-1}(Y)$.

Proof. For fixed $y \in Y$ the set $\Phi^{-1}(y)$ consists of all $x \in X$ with $y \in \Phi(x)$, that is, with $(x, y) \in \Phi$. □

A *fixed point* of a multivalued map Φ is a point $x \in \operatorname{dom}(\Phi)$ satisfying $x \in \Phi(x)$, that is, $(x, x) \in \operatorname{graph}(\Phi)$. If f is single-valued, the fixed points are correspondingly the points x satisfying $x = f(x)$.

We use the standard notation $\Phi|_M$ to denote the *restriction* of a (single- or multivalued) map $\Phi: X \multimap Y$ to a subset $M \subseteq X$, that is, $\Phi|_M: M \multimap Y$ is defined by $\Phi|_M(x) := \Phi(x)$ for all $x \in M$. The *composition* of two maps $\Phi: X \multimap Y$ and $\Psi: Y \multimap Z$ is the map $\Psi \circ \Phi: X \multimap Z$, defined by

$$(\Psi \circ \Phi)(x) := \Psi\bigl(\Phi(x)\bigr).$$

For $\Phi: X \multimap X$, we define the n-th power of Φ in the obvious way by induction

$$\Phi^0 := \mathrm{id}_X, \qquad \Phi^n := \Phi \circ \Phi^{n-1}.$$

The *product* of two multivalued maps $\Phi: X \multimap Y$ and $\Psi: X \multimap Z$ is the multivalued map $\Phi \times \Psi: X \multimap Y \times Z$, defined by

$$(\Phi \times \Psi)(x) := \Phi(x) \times \Psi(x).$$

Similarly, if $\Phi\colon X \multimap Y$ and $\Psi\colon W \multimap Z$ then the multivalued map $\Phi \otimes \Psi\colon X \times W \multimap Y \times Z$ is defined by

$$(\Phi \otimes \Psi)(x, y) := \Phi(x) \times \Psi(y).$$

If $\Phi, \Psi\colon X \multimap Y$ satisfy

$$\Phi(x) \subseteq \Psi(x) \quad \text{for all } x \in X,$$

we write more briefly $\Phi \subseteq \Psi$. Note that for maps $\Phi, \Psi\colon X \multimap Y$ we have

$$\Phi \subseteq \Psi \iff \operatorname{graph}(\Phi) \subseteq \operatorname{graph}(\Psi),$$

so that our notation cannot lead to any confusion when the convention is used that maps are identified with their graph. However, we will not require the latter convention.

For later usage, we point out the following calculation rules for $\Phi\colon X \multimap Y$. The proof of these rules is straightforward. The first rule explains the names *small* and *large* counterimages for $\Phi^-(M)$ and $\Phi^+(M)$:

$$\begin{aligned}
&\Phi^-(M) \cap \operatorname{dom}(\Phi) \subseteq \Phi^+(M) \subseteq \operatorname{dom}(\Phi) \\
&\Phi\big(\Phi^-(M)\big) \subseteq M \cap \Phi\big(\Phi^+(M)\big) && (2.1)\\
&\Phi^+\big(\Phi(M)\big) \supseteq M \cap \operatorname{dom}(\Phi) \\
&M \cap \Phi(X) \subseteq \Phi\big(\Phi^+(M)\big) \\
&\bigcap_{i\in I} \Phi(M_i) \supseteq \Phi\Big(\bigcap_{i\in I} M_i\Big) \quad \text{if } M_i \subseteq X && (2.2)\\
&\bigcup_{i\in I} \Phi(M_i) = \Phi\Big(\bigcup_{i\in I} M_i\Big) \quad \text{if } M_i \subseteq X \\
&\bigcup_{i\in I} \Phi^-(M_i) \subseteq \Phi^-\Big(\bigcup_{i\in I} M_i\Big) && (2.3)\\
&\Phi^+(Y \setminus M) = X \setminus \Phi^-(M) && (2.4)
\end{aligned}$$

2.1.2 Axioms

Proposition 2.2. (AC). *Every multivalued map* $\Phi\colon X \multimap Y$ *with* $\operatorname{dom}(\Phi) = X$ *has a selection.*

Proof. This is one of many equivalent formulations of the axiom of choice.

Note that the point of the assertion is that there is a *function* $f\colon X \to Y$ satisfying $f(x) \in \Phi(x)$ for every $x \in X$. In particular, $\operatorname{graph}(f)$ and $f(X)$ are *sets*, even if there is not necessarily an "explicit formula" available for f. □

We denote the axiom of choice briefly by AC, and we use throughout the symbol (AC) in the formulation of a result to indicate that we use the axiom of choice to prove the result.

Without such an indication, the results obtained in this monograph can be proved within Zermelo–Fraenkel's set theory (ZF) together with the so-called axiom of dependent choices (DC) which we formulate in the subsequent Theorem 2.3.

In particular, the subsequent Theorem 2.3 implies that the so-called countable axiom of choice (which is the special case $X = \mathbb{N}$ of Proposition 2.2) follows from DC, and so we will usually use this countable axiom of choice without special mentioning:

(AC_ω) **(Countable axiom of choice).** Every multivalued map $\Phi\colon \mathbb{N} \multimap X$ with $\operatorname{dom}(\Phi) = \mathbb{N}$ has a selection.

We include in the subsequent theorem also a typical example of an important analytical result (Baire's category theorem) which is meant to demonstrate for which sort of reasoning the axiom DC is used, typically. This example is particularly interesting since it turns out to be equivalent to DC.

We will recall the terminology used in the formulation and proof concerning Baire's category theorem later on; it is meant here as an example and can be skipped at a first reading.

Theorem 2.3. *In* ZF *the following statements are equivalent.*

(DC) **(Axiom of dependent choices).** *For every multivalued map $\Phi\colon X \multimap X$ with $\operatorname{dom}(\Phi) = X \neq \varnothing$ there is a function $f\colon \mathbb{N} \to X$ with $f(n+1) \in \Phi(f(n))$.*

(DC′) **(Nonautonomous choices with history and start).** *For every family of multivalued maps $\Phi_n\colon X_1 \times \cdots \times X_n \multimap X_{n+1}$ and every $x_0 \in X_1$ with the property that $\Phi_n(x_1, \ldots, x_n) \neq \varnothing$ whenever $x_1 = x_0$ and $x_{k+1} \in \Phi_k(x_1, \ldots, x_k)$ $(k = 1, \ldots, n-1)$ there is a function $f\colon \mathbb{N} \to \bigcup_{n=1}^\infty X_n$ satisfying $f(1) = x_0$ and $f(n+1) \in \Phi_n(f(1), \ldots, f(n))$ for all $n \in \mathbb{N}$.*

(BC) **(Baire's category theorem).** *If X is a complete metric space and $N_n \subseteq X$ $(n \in \mathbb{N})$ is a family of closed sets without interior points then $\bigcup_{n=1}^\infty N_n$ is without interior points.*

Moreover, the implications

$$\text{AC} \implies \text{DC} \implies \text{AC}_\omega \tag{2.5}$$

hold in ZF, *and none of it can be reversed.*

Similarly as for Proposition 2.2, the crucial point of the assertions DC and DC′ is that f in these assertion is a *function*, in particular, $\mathrm{graph}(f)$ is a *set*, even if there is no "explicit formula" available for the values $f(n)$.

Proof. If AC holds then any map $\Phi: X \multimap X$ with $\mathrm{dom}(\Phi) = X \neq \varnothing$ has a selection $F: X \to X$ by Proposition 2.2. Hence, fixing some $x_0 \in X$, we can inductively define a function $f: \mathbb{N} \to X$ by $f(1) := x_0$ and $f(n+1) := F(f(n))$ $(n \in \mathbb{N})$. This proves the first implication in (2.5).

To see that DC implies DC′, we let Y_n $(n \in \mathbb{N})$ denote the family of all functions $g: \{1, \dots, n\} \to X_1 \cup \dots \cup X_n$ satisfying $g(1) = x_0$ and $g(k+1) \in \Phi_k(g(1), \dots, g(k))$ for all $k = 1, \dots, n-1$. We let $Y = \bigcup_{n=1}^{\infty} Y_n$, and consider the multivalued map $\Psi: Y \multimap Y$, where $\Psi(h)$ is defined for $h \in Y_n$ as the family of all $g \in Y_{n+1}$ satisfying $g|_{\{1,\dots,n\}} = h$. By (DC), there is a function $F: \mathbb{N} \to Y$ satisfying $F(n+1) \in \Psi(F(n))$ for all $n \in \mathbb{N}$. Put $f_n := F(n)$ $(n \in \mathbb{N})$. There is some index N with $f_1 \in Y_N$. Then $f_{n+1} \subset Y_{N+n}$ for all n. Hence $f(n) := f_k(n)$ is defined for $k \geq \max\{n+1-N, 1\}$, and this value $f(n)$ is independent of k, because $f_{j+1} \in \Psi(f_j)$ $(j \in \mathbb{N})$. Since $f|_{\{1,\dots,n\}} = f_{n+1-N} \in Y_n$ for all $n \geq N$, the definition of Y_n implies that f has all required properties.

If DC′ holds then we obtain AC_ω and DC for the particular choice $X_n := X$ for all n and $\Phi_n(x) := \Phi(n)$ $(x \in X^n)$ or $\Phi_n(x, y) := \Phi(y)$ $(x \in X^{n-1}, y \in X)$, respectively.

If BC is false, then $M_0 \subseteq \bigcup_{n=1}^{\infty} N_n$ for some nonempty open set $M_0 \subseteq X$, and N_n has no interior point for every n. By the latter, we can inductively choose $x_n \in M_{n-1} \setminus N_n$ and, since the set $M_{n-1} \setminus N_n$ is open and thus x_n is an interior point of this set, an open ball $M_n \subseteq X$ of radius less than $1/n$ around x_n with $\overline{M}_n \subseteq M_{n-1} \setminus N_n$. By DC′, we can assume that $n \mapsto (x_n, M_n)$ is a map, that is, a sequence. Since $\{x_n, x_{n+1}, \dots\}$ is contained in the ball M_n of radius less than $1/n$, we find that x_n form a Cauchy sequence which thus converges to some

$$x \in \bigcap_{n=1}^{\infty} \overline{M}_n \subseteq \bigcap_{n=1}^{\infty} (M_{n-1} \setminus N_n).$$

Since $M_0 \supseteq M_1 \supseteq \dots$, we obtain $x \in M_0$ and $x \notin N_n$ for all n which contradicts $M_0 \subseteq \bigcup_{n=1}^{\infty} N_n$.

Assuming BC, we now use the idea from [20] to prove DC. Thus, let $\Phi: X \multimap X$ satisfy $\mathrm{dom}(\Phi) = X \neq \varnothing$. The space $Y = X^{\mathbb{N}} = \prod_{n=1}^{\infty} X$ of all maps $f: \mathbb{N} \to X$ becomes a complete metric space if it is equipped with the metric

$$d(f, g) := \sum_{\substack{n=1 \\ f(n) \neq g(n)}}^{\infty} 2^{-n}.$$

The proof of this assertion is straightforward and skipped. (Actually, we will prove this assertion in a more general context in Corollary 3.67 (observe Remark 3.68), since the metric is in fact just the product metric (3.28) when d_n denotes the discrete metric

$$d_n(x, y) = \begin{cases} 1 & \text{if } x \neq y, \\ 0 & \text{if } x = y, \end{cases}$$

on $X_n := X$.)

For $n \in \mathbb{N}$, we let O_n denote the family of all $f \in Y$ with the property that for each $k \leq n$ there is some $\ell > k$ with $f(\ell) \in \Phi(f(k))$.

Then each O_n is open. Indeed, for each $f \in O_n$ there is some L such that for each $k \leq n$ there is some $\ell \geq k$ with $f(\ell) \in \Phi(f(k))$ and $\ell \leq L$. Then O_n contains all $g \in Y$ with $d(f, g) < 2^{-L}$, because $f(j) = g(j)$ for all $j = 1, \ldots, L$.

Hence, $N_n := Y \setminus O_n$ is closed. Moreover, for every $f \in N_n$ and every $\varepsilon > 0$, there is some $g \in O_n$ with $d(f, g) < \varepsilon$. Indeed, we fix some integer $K \geq n$ with $2^{-K} < \varepsilon$, and for $k \leq K$, we put $g(k) := f(k)$ and choose $g(K+k) \in \Phi(f(k))$; for $k > 2K$, we put $g(k) := x_0$ where $x_0 \in X$ is arbitrary. Then $g \in O_n$ by construction, and $d(f, g) \leq \sum_{k>K} 2^{-k} = 2^{-K} < \varepsilon$.

We thus have shown that N_n has no interior points. We conclude from BC that $Y \neq \bigcup_{n=1}^{\infty} N_n$, that is, there is some $g \in \bigcap_{n=1}^{\infty} O_n$. Hence, for every $k \in \mathbb{N}$ there is some smallest $\ell = \ell(k) \in \mathbb{N}$ with $g(\ell) \in \Phi(g(k))$. Then $\ell \colon \mathbb{N} \to \mathbb{N}$, and we can define $f \colon \mathbb{N} \to X$ inductively by $f(1) := g(1)$ and $f(k+1) := g(\ell(k))$. Then $f(n+1) \in \Phi(f(n))$ holds for all $n \in \mathbb{N}$.

The proofs that none of the implications in (2.5) can be reversed in ZF require to use Cohen's forcing method which is beyond the scope of this monograph. We refer to [81] for these proofs. □

The above application of DC in Baire's category theorem is typical for many results from analysis which use an inductive "construction". It turns out that practically all "standard" results of analysis (including Lebesgue integration theory) can be carried out in ZF + DC. As the above equivalence shows, one can often not omit DC. However, results which can be obtained with DC (instead of AC) might be considered as "approximately constructive". For these reasons, we will use DC tacitly throughout, but mark those results explicitly which we can prove only with AC.

Several applications where AC cannot be avoided refer to partial orders. Recall that a relation $\leq$ on a set X is called a *partial order* if it is

(a) *reflexive*, that is $x \leq x$ holds for all $x \in X$,

(b) *transitive*, that is $x \le y \le z$ implies $x \le z$, and

(c) *antisymmetric*, that is $x \le y$ and $y \le x$ implies $x = y$.

A *chain* in such a set is a subset $C \subseteq X$ which is *totally ordered*, that is, for every $x, y \in C$ one of the relations $x \le y$ or $y \le x$ holds. Recall that an element $y \in X$ is an *upper bound* (*lower bound*) for $M \subseteq X$ if $x \le y$ ($y \le x$) for every $x \in M$. If additionally $y \in M$ then y is a *maximum* (*minimum*) of M. The antisymmetry implies that each subset of M has at most one maximum (minimum), denoted by $\max M$ ($\min M$). The *supremum* of $M \subseteq X$ is the minimum of the set of upper bounds of M and is denoted by $\sup M$. Similarly, the *infimum* of $M \subseteq X$ is the maximum of the set of lower bounds of M and is denoted by $\inf M$.

Theorem 2.4 (Bourbaki–Witt). *Let $X \neq \varnothing$ be partially ordered such that every nonempty chain in X has a supremum. If $f: X \to X$ satisfies $x \le f(x)$ for all $x \in X$ then f has a fixed point.*

Proof. Fix $x_0 \in X$. Let $\mathcal{A}$ denote the system of all $A \subseteq X$ satisfying $x_0 \in A$, $f(A) \subseteq A$, and such that if $C \subseteq A$ is a chain then $\sup C \in A$. Then $X \in \mathcal{A}$, and so we can define $A_0 := \bigcap \mathcal{A}$.

Note that $A_0 \in \mathcal{A}$. Indeed, $x_0 \in A_0$ by definition, and $f(A_0)$ is contained in each $A \in \mathcal{A}$ and thus in A_0. Similarly, for each chain $C \subseteq A_0$ the element $\sup C$ is contained in each $A \in \mathcal{A}$ and thus in A_0.

If we can show that A_0 is a chain, we put $x_* := \sup A_0 = \max A_0$. Then $f(x_*) \in A_0$ implies $x_* \ge f(x_*)$, and so x_* is a fixed point of f.

To see that A_0 is a chain, we show first that $x_0 = \min A_0$. Indeed, $A_1 := \{x \in A_0 : x_0 \le x\}$ belongs to $\mathcal{A}$, and so $A_0 \subseteq A_1 \subseteq A_0$ implies $A_0 = A_1$.

Let B denote the set of all $x \in A_0$ such that for each $y \in A_0 \setminus \{x\}$ with $y \le x$ we have $f(y) \le x$.

For $x \in B$, let B_x denote the set of all $y \in A_0$ satisfying $y \le x$ or $f(x) \le y$. We show that $B_x \in \mathcal{A}$. Indeed, since $x \in A_0$ we have $x_0 = \min A_0 \le x$, and so $x_0 \in B_x$. If $y \in B_x$ then we have in case $f(x) \le y$ that $f(x) \le y \le f(y)$ and in case $y \le x$ in view of $x \in B$ either $f(y) \le x$ or $y = x$ (hence $f(x) \le f(y)$). In all cases, we have $f(y) \in B_x$. Hence, we have shown $f(B_x) \subseteq B_x$. Finally, if $C \subseteq B_x$ is a chain, we have to show that $z := \sup C \in A_0$ belongs to B_x. This is clear by definition of the supremum if x is an upper bound for C. Otherwise, there is some $y \in C$ with $f(x) \le y$. Then $f(x) \le y \le z$ and thus also $z \in B_x$.

Since $B_x \in \mathcal{A}$, we have $A_0 \subseteq B_x$, and so every $x \in B$ has the property that every $y \in A_0$ satisfies $y \le x$ or $f(x) \le y$.

We show now that $B \in \mathcal{A}$. Indeed, from $x_0 = \min A_0$ we obtain that there is no $y \in A_0 \setminus \{x_0\}$ with $y \le x$, and so $x_0 \in B$. If $x \in B$ then for every $y \in A_0$ we have, by what we just proved, $y \le x$ or $f(x) \le y$. Hence, if $y \in A_0 \setminus \{f(x)\}$

and $y \le f(x)$ then $f(x) \not\le y$, and so we must have $y \le x$. Thus, $x \in B$ implies $f(y) \le x \le f(x)$ or $y = x$. In both cases, it follows that $f(y) \le f(x)$, hence $f(x) \in B$, and so we have shown $f(B) \subseteq B$. Finally, if $C \subseteq B$ is a chain, we have to show that $z := \sup C \in A_0$ belongs to B. Thus, let $y \in A_0 \setminus \{z\}$ satisfy $y \le z$. Since $z = \sup C$, it follows that y is not an upper bound for C, and so there is some $x \in C$ such that $x \not\le y$. Since $x \le f(x)$, we have $f(x) \not\le y$, and so $x \in B$ implies $y \le x$ by what we have shown above. From $x \in B$, we obtain now $f(y) \le x \le z$. Hence, $z \in B$, as required.

Since $B \in \mathcal{A}$, we have $A_0 \subseteq B$. Hence, if $x, y \in A_0$ we have $x \in B$. By what we have shown above, we have $y \le x$ or $x \le f(x) \le y$, and so A_0 is a chain. □

Recall that a chain $C \subseteq X$ in a partially ordered set X is *maximal* if $C \subseteq C_0 \subseteq X$ implies that $C_0 = C$ or that C_0 is not a chain. An element $x \in X$ is called *maximal* (*minimal*) if $x \not\le y$ ($y \not\le x$) for every $y \in X \setminus \{x\}$.

Theorem 2.5. *In* ZF *the following statements are equivalent:*

(a) AC.

(b) **(Hausdorff's Maximality Theorem; Kuratowski's Lemma).** *Every partially ordered set contains a maximal chain.*

(c) **(Zorn's Lemma).** *If every chain in a partially ordered set X has a supremum then X has a maximal element.*

Proof. It is trivial that Hausdorff's maximality theorem implies Zorn's lemma. Indeed, let C be a maximal chain in a partially ordered set X. If $x := \sup C$ exists then it is a maximal element of X, since if there is some $y \in X \setminus \{x\}$ with $x \le y$ then $y \notin C$, and so the chain $C \cup \{y\}$ contradicts the maximality of C.

We show now that Hausdorff's maximality theorem follows from each of the other statements. Let $X \neq \emptyset$ be partially ordered, and X_0 be the set of all chains in X, ordered by inclusion. Then $X_0 \neq \emptyset$, and each chain in X_0 has a supremum, namely the union of the chain. Every maximal element in X_0 is a maximal chain in X, so we are to show that X_0 has a maximal element.

If Zorn's lemma holds then X_0 has a maximal element, and we are done. If AC holds, we consider the multivalued map $\Phi\colon X_0 \multimap X_0$, defined by

$$\Phi(C) := \{C_0 \in X_0 \setminus \{C\} : C \subseteq C_0\}.$$

If X_0 has no maximal element then $\operatorname{dom}(\Phi) = X_0$, and so AC (Proposition 2.2) implies that Φ has a single-valued selection f. Theorem 2.4 implies that f has a fixed point which is then a maximal element of X_0.

For the remaining implication, suppose that Hausdorff's maximality theorem holds, and that $\Phi\colon X \multimap Y$ satisfies $\operatorname{dom}(\Phi) = X$. Let X_0 denote the set of all single-valued functions from a subset of X into Y which are selections of Φ. Then X_0 is partially ordered by inclusion of the graph of these functions. If C is a maximal chain in X_0 then $\bigcup C$ is the graph of a selection $f: M \to Y$ of $\Phi|_M$. Assume by contradiction that $M \neq X$. Then there is some $x \in X \setminus M$. We extend f to $M \cup \{x\}$ by letting $f(x)$ denote some element of $\Phi(x)$. Then $C \cup \{f\}$ is a chain in X_0, contradicting the maximality of C. □

We will usually use Theorem 2.5 in the form of Hausdorff's maximality theorem. As a simple first example of this type, we show the well-known fact that every vector space has a Hamel basis.

Recall that a *Hamel basis* of a vector space X over some field $\mathbb{K}$ is a family of linearly independent $e_i \in X$ $(i \in I)$ such that X is the linear hull of all e_i $(i \in I)$.

Corollary 2.6. (AC). *Every vector space X has a Hamel basis.*

We will see in Proposition 6.15 that AC is crucial here.

Proof. Let $\mathcal{B}$ denote the family of all linearly independent subsets of X. Then $\mathcal{B}$ is partially ordered by inclusion (that is, $B_1 \preceq B_2 \iff B_1 \subseteq B_2$), and by Hausdorff's maximality theorem, $\mathcal{B}$ contains a maximal chain $\mathcal{C}$.

Since $\mathcal{C}$ is a chain, it follows that $C = \bigcup \mathcal{C}$ is linearly independent. Indeed, if $e_1, \dots, e_n \in C$ and $\lambda_1, \dots, \lambda_n \in \mathbb{K}$ satisfy $\lambda_1 e_1 + \dots + \lambda_n e_n = 0$, there are $C_k \in \mathcal{C}$ with $e_k \in C_k$ for $k = 1, \dots, n$. Since the finitely many sets C_k are pairwise contained in each other, some of these sets, say C_1, contains all other C_k. In particular, we have $e_1, \dots, e_n \in C_1$. Since C_1 is linearly independent, it follows that $\lambda_1 = \dots = \lambda_n = 0$, and so also C is linearly independent.

The maximality of $\mathcal{C}$ implies that the linear hull of C is all of X. Indeed, otherwise there is some $e \in X$ which is linearly independent from C, and then $\mathcal{C}$ together with $C \cup \{e\} \in \mathcal{B}$ would be a strictly larger chain than $\mathcal{C}$. □

2.2 Topological Notations and Basic Results

For clarity, let us first fix some topological conventions which sometimes vary slightly in literature.

By a topological space, we mean a space X and an associated family of *open* subsets with the property that unions and finite intersections of open sets are open, and X and $\varnothing$ are open. The *closed* sets are by definition the complements of open sets.

Unless we explicitly say so, we do not require that neighborhoods are open. Moreover, we also define neighborhoods of sets in the obvious way:

Definition 2.7. Let X be a topological space, and $A \subseteq X$. A set $M \subseteq X$ is a *neighborhood* of A if there is an open set $U \subseteq X$ with $A \subseteq U \subseteq M$. For the case $A = \{x\}$, we call M a *neighborhood* of $x \in X$.

We recall the following simple fact:

Proposition 2.8. *A set $M \subseteq X$ is open if and only if it is a neighborhood of each of its points.*

Proof. If M is a neighborhood of each of its points, then M is the union of all open sets $U \subseteq X$ satisfying $U \subseteq M$. The converse is trivial. □

Unless we say something else, subsets of topological spaces are always endowed with the inherited topology, that is:

Definition 2.9. If X is a topological space and $Y \subseteq X$, then the open sets $M \subseteq Y$ are those of the form $M = U \cap Y$ with an open set $U \subseteq X$.

It is well-known that one can equivalently replace "open" by "closed" in this definition.

The *interior*, *closure*, and *boundary* of a set $M \subseteq X$ will be denoted by $\mathring{M}$, $\overline{M}$, and ∂M, respectively. If X is unclear, we also write $\partial_X M$ for the boundary of M in X. Recall that the closure of M is the smallest closed set containing M, that is, it consists of all points $x \in M$ with the property that every neighborhood of x intersects M. The interior of M is the largest open set contained in M, and the boundary is the difference between these two sets, that is

$$\partial M = \overline{M} \setminus \mathring{M}.$$

Proposition 2.10. *Let $M \subseteq Y \subseteq X$. If M is open (closed) in X then M is open (closed) in Y. The converse holds if Y is open (closed) in X. The closure of M in Y is $\overline{M} \cap Y$ where $\overline{M}$ denotes the closure of M in X.*

Proof. The first assertions follow immediately from the definition, and the last assertion follows from the first, since $\overline{M}$ is the intersection of all closed in X sets containing M. □

The following generalization of the last assertion for the case that M is not necessarily a subset of Y will be useful in Section 14.4.

Proposition 2.11. *Let $M, Y \subseteq X$. If there is $N \subseteq X$ such that $N \setminus Y$ is closed (in X) and contains $M \setminus Y$ then*

$$\overline{\overline{M} \cap Y} = \overline{M \cap Y}.$$

All closures are understood with respect to X.

Proof. Putting $C := N \setminus Y$, we have $M \subseteq (M \cap Y) \cup C$, and so $\overline{M} \subseteq \overline{(M \cap Y)} \cup \overline{C}$. Since C is closed, we have $C = \overline{C}$, and so

$$\overline{M} \cap Y \subseteq (\overline{M \cap Y}) \cup (C \cap Y) = \overline{M \cap Y}.$$

Hence, $\overline{\overline{M} \cap Y} \subseteq \overline{M \cap Y}$, and the converse inclusion is trivial. □

Occasionally, the following characterization of closed sets turns out to be useful:

Proposition 2.12. *A set $M \subseteq X$ is closed if and only if every $x \in \overline{M}$ has a neighborhood $U \subseteq X$ such that $M \cap U$ is closed in U.*

Proof. If M is closed then the definition of the inherited topology implies that $M \cap U$ is closed in U, even for every set $U \subseteq X$. Conversely, if M fails to be closed then there is some $x \in \overline{M}$ with $x \notin M$. For every neighborhood $U \subseteq X$ of x the following holds: Every neighborhood $V \subseteq U$ of x in U is also a neighborhood of x in X and thus contains some $y \in M$. Since $y \in M \cap V = (M \cap U) \cap V$, we conclude that $x \in U \setminus M$ belongs to the closure of $M \cap U$ in U, and so $M \cap U$ fails to be closed in U. □

Recall that a map $f : X \to Y$ is said to be *continuous at* $x \in X$ if for each neighborhood $V \subseteq Y$ of $f(x)$ the set $f^{-1}(V)$ is a neighborhood of x, that is, if there is a neighborhood $U \subseteq X$ of x with $f(U) \subseteq V$. A map $f : X \to Y$ is *continuous* if it is continuous at each $x \in X$. It is well-known that this holds if and only if $f^{-1}(M)$ is open (closed) for every open (closed) set $M \subseteq Y$. The set of continuous functions from X into Y is denoted by $C(X, Y)$.

A *homeomorphism* is a continuous map $f : X \to Y$ between topological spaces with a continuous inverse $f^{-1} : Y \to X$. If such an f exists then X and Y are called *homeomorphic*.

We recall the definition of connectedness and path-connectedness.

Definition 2.13. A topological space X is *connected* if it is not the union of two disjoint open nonempty subsets. It is *path-connected* if for each two points $x_1, x_2 \in X$ there exists a *path* in X connecting them, that is, a continuous map $f : [0, 1] \to X$ with $f(0) = x_1$ and $f(1) = x_2$.

In situations like these, the definition refers also to subsets of topological spaces: We equip the subset with the inherited topology.

Proposition 2.14. *Every path-connected space is connected. In particular, intervals in $\mathbb{R}$ are connected.*

Proof. Assume by contradiction that $X = A_1 \cup A_2$ is path-connected although $A_1, A_2 \subseteq X$ are open and nonempty. There is a continuous $f\colon [0,1] \to X$ with $f(0) \in A_1$ and $f(1) \in A_2$. Let $t_0 \in [0,1]$ denote the supremum of all $t \in [0,1]$ with $f(t) \in A$. Then $f(t_0) \in A_i$ for $i = 1$ or $i = 2$. However, then $f(U) \subseteq A_i$ for a neighborhood $U_i \subseteq [0,1]$ of t_0 which is a contradiction to the definition of t_0. □

Recall that the converse to Proposition 2.14 does not hold: The "topologist's sine curve" (the closure of graph of the map $x \mapsto \sin(1/x)$ for $x > 0$ in $\mathbb{R}^2$) is an example of a space which is connected but not path-connected. To see that this space is connected, we observe that the graph of the map $x \mapsto \sin(1/x)$ $(x > 0)$ is path-connected, hence connected, and apply the following observation:

Lemma 2.15. *If $M \subseteq X$ is connected and $M \subseteq N \subseteq \overline{M}$ then N is connected.*

Proof. If $N = A_1 \cup A_2$ with two disjoint open (in N) nonempty sets A_i, then $M = B_1 \cup B_2$ with the disjoint open (in M) sets $B_i := M \cap A_i$. We obtain a contradiction if we can show that $B_i \neq \varnothing$ for $i = 1, 2$. However, if $B_i = \varnothing$ then $A_i \subseteq N \setminus M$. Since A_i is open in N there is an open set $U \subseteq X$ with $A_i = U \cap N$. Choose $x \in A_i$. Then U is a neighborhood of x which is disjoint from M, a contradiction to $x \in \overline{M}$. □

Definition 2.16. The *component* of a point $x \in X$ in X is the union of all connected sets containing x. The *path-component* of x in X is the union of all path-connected sets containing x.

Proposition 2.17. *The components are connected, and the path-components are path-connected. Being in the same (path-)component of a space X is an equivalence relation on X, that is, each space divides into its components (and into its path-components). The components are closed.*

Proof. Let Y_i $(i \in I)$ denote the family of all path-connected subsets of X containing x, and let Y be their union. If $x_1, x_2 \in Y$ then $x_k \in Y_{i_k}$ and so x_k can be connected with x by some path in $Y_{i_k} \subseteq Y$. The obvious "concatenation" of the corresponding paths defines a path from x_1 to x_2 in Y; hence Y is path-connected.

Let X_i $(i \in I)$ denote the family of all connected subsets of X containing x, and let U be their union. Assume that $U = A_1 \cup A_2$ with open (in U) disjoint nonempty $A_1, A_2 \subseteq U$. Without loss of generality, assume $x \in A_1$. Fix some $y \in A_2$ and some $i \in I$ with $y \in X_i$. Then $B_k := A_k \cap X_i$ are nonempty disjoint open subsets of X_i with $X_i = B_1 \cup B_2$ which is a contradiction. Hence, U is connected. Since also $\overline{U}$ is connected by Lemma 2.15, we have $X_i = \overline{U}$ for some $i \in I$ and thus $\overline{U} \subseteq U$. Hence, U is closed.

For $x \in X$, we let now $K(x)$ denote the (path-)component of x. By what we have shown, each $K(x)$ is (path-)connected. If $K(x_1) \cap K(x_2) \neq \varnothing$, say $x_3 \in K(x_1) \cap K(x_2)$, we thus obtain first $K(x_3) \supseteq K(x_i)$ $(i = 1, 2)$ and then $K(x_i) \supseteq K(x_3)$ $(i = 1, 2)$ and thus even $K(x_i) = K(x_3)$ $(i = 1, 2)$. □

Corollary 2.18. *Let $X = A_1 \cup A_2$ with disjoint open sets $A_i \subseteq X$ $(i = 1, 2)$. Then any component of X is contained in either A_1 or in A_2.*

Proof. Let C be a component of X. Then C is connected by Proposition 2.17. However, $B_i := C \cap A_i$ are open in C and disjoint with $C = B_1 \cup B_2$. Thus, $B_i = \varnothing$ for some i. □

Recall that a function $\Phi\colon X \multimap Y$ is called *locally constant* if each $x \in X$ has a neighborhood $U \subseteq X$ such that $\Phi|_U$ is constant (that is, independent of the argument).

Proposition 2.19. *Let $\Phi\colon X \multimap Y$ be locally constant. Then $\Phi|_C$ is constant for any component C of X.*

Proof. We fix $x_0 \in C$ and put $A_1 := \{x \in X : \Phi(x) = \Phi(x_0)\}$ and $A_2 := \{x \in X : \Phi(x) \neq \Phi(x_0)\}$. Then $X = A_1 \cup A_2$, and A_1 and A_2 are disjoint and open since Φ is locally constant. Hence, the assertion follows from Corollary 2.18. □

We recall the following important property of continuous functions:

Proposition 2.20. *If $f\colon X \to Y$ is continuous and $C \subseteq X$ is connected then $f(C)$ is connected.*

Proof. Otherwise $f(C) = A_1 \cup A_2$ with nonempty disjoint open sets $A_1, A_2 \subseteq f(C)$. There are open sets $B_i \subseteq Y$ with $A_i = B_i \cap f(C)$ $(i = 1, 2)$. Then $C_i := C \cap f^{-1}(B_i)$ are nonempty, disjoint and open in C with $C = C_1 \cup C_2$, contradicting the connectedness of C. □

Definition 2.21. Let X be a topological space, and $x \in X$. A *basis of neighborhoods* of x is a family $\mathscr{U}$ of neighborhoods of x such that every neighborhood of x is contained in some set from $\mathscr{U}$.

Definition 2.22. A topological space X is *locally connected* or *locally path-connected* if each point has a neighborhood basis consisting of connected or path-connected sets, respectively.

Proposition 2.23. *X is locally connected if and only if for each open $U \subseteq X$ the components of U are open in X. Moreover, the following statements are equivalent:*

(a) *X is locally path-connected.*

(b) *For each open $U \subseteq X$ the path-components of U are open in X.*

(c) *X is locally connected and each connected open $U \subseteq X$ is path-connected.*

(d) *The components of each open $U \subseteq X$ are open and coincide with its path-components.*

Proof. Suppose that the (path-)components of open subsets of X are open in X. If $x \in X$ and $U \subseteq X$ is an open neighborhood of x then the (path-)component of U containing x is open in X and thus a neighborhood of x contained in U. Hence, X is locally (path-)connected.

Conversely, let X be locally (path-)connected, and $U \subseteq X$ be open. Let M be a component of U. Each $x \in M$ has a (path-)connected neighborhood $V \subseteq X$ with $V \subseteq U$. Proposition 2.17 implies that M is the (path-)component of x and thus $V \subseteq M$. Hence, Proposition 2.8 implies that M is open in X.

We thus have shown the first assertion and the equivalence of (a) and (b).

Suppose now (b). Let $U \subseteq X$ be open and connected. By hypothesis, the path-components of U are open. Hence, if $A_1 \subseteq U$ is a nonempty path-component of U then A_1 is open, and $A_2 := U \setminus A_1$ is the union of the path-components disjoint from A_1 and thus also open. Since U is connected, it follows that $A_2 = \varnothing$, that is, $U = A_1$ is path-connected. Hence, (c) holds.

If (c) holds and $U \subseteq X$ is open, recall that we have already shown that the components of U are open. Hence, by hypothesis, the components are path-connected. In view of Proposition 2.14, it follows that the components are the path-components. Thus, (d) holds. Finally, (d) evidently implies (b). □

Definition 2.24. A family $\mathscr{O}$ of subsets of a topological space X is a *cover* of $M \subseteq X$ if $\bigcup \mathscr{O} = M$. If additionally each $O \in \mathscr{O}$ is open (closed), $\mathscr{O}$ is called an *open cover* (*closed cover*).

A *locally finite cover* of X is a cover $\mathcal{O}$ of X with the property that each $x \in X$ has a neighborhood which intersects at most finitely many elements of $\mathcal{O}$.

A cover $\mathcal{O}$ of X is a *refinement* of a cover $\mathcal{U}$ of X if for each $O \in \mathcal{O}$ there is some $U \in \mathcal{U}$ with $O \subseteq U$.

We use throughout the convention that by a *countable* set we mean a set which is either finite or countably infinite.

Definition 2.25. A subset of a topological space $M \subseteq X$ is *compact* if every open cover $\mathcal{O}$ of M has a finite subcover. It is *Lindelöf* if every open cover $\mathcal{O}$ of M has a countable subcover.

A topological space X is *(countably) paracompact* if every (countable) open cover $\mathcal{U}$ of X has a locally finite open refinement.

Proposition 2.26. *Every compact space X is paracompact.*

Proof. Any finite subcover of an open cover $\mathcal{U}$ is a locally finite open refinement of $\mathcal{U}$. □

We recall the most important properties of compact spaces:

Definition 2.27. A nonempty family A_i $(i \in I)$ has the *finite intersection property* if $\bigcap_{i \in I_0} A_i \neq \emptyset$ for each nonempty finite set $I_0 \subseteq I$.

Proposition 2.28. *X is compact if and only if every family of closed subsets $A_i \subseteq X$ $(i \in I)$ with the finite intersection property satisfies $\bigcap_{i \in I} A_i \neq \emptyset$.*

Proof. The family $O_i := X \setminus A_i$ $(i \in I)$ consists of open sets, and $\bigcap_{i \in I_0} A_i = \emptyset$ if and only if $\{O_i : i \in I_0\}$ is a cover of X. □

Proposition 2.29. *$M \subseteq X$ is compact if and only if M is a compact space with the inherited topology. If X is compact and $M \subseteq X$ is closed then M is compact.*

Proof. Let $M \subseteq X$ be compact, and $\mathcal{O}$ be an open in M cover of M. Let $\mathcal{O}_0$ be the family of all open set $O \subseteq X$ with $O \cap M \in \mathcal{O}$. Since M is compact, there is a finite subcover $\mathcal{O}_1 \subseteq \mathcal{O}_0$ for M, and then the sets $O \cap M$ with $O \in \mathcal{O}_1$ constitute a finite subcover of $\mathcal{O}$ for M.

Conversely, let $M \subseteq X$ be compact with respect to the inherited topology. Now if $\mathcal{O}$ is an open in X cover of M then the family of all $O \cap M$ with $O \in \mathcal{O}$ is an open in M cover of M and thus has a finite subcover $O_k \cap M$ $(k = 1, \dots, n)$ with $O_k \in \mathcal{O}$. Then $O_1, \dots, O_n \in \mathcal{O}$ form a finite open subcover of M.

The second assertion follows from the first and Proposition 2.28, since subsets of M are closed if and only if they are closed in X (Proposition 2.10). □

There is an alternative proof of the second assertion of Proposition 2.29 which has the advantage that the same idea can be used for paracompact spaces:

Proposition 2.30. *If X is (countably) paracompact and $M \subseteq X$ is closed then M is (countably) paracompact.*

Proof. Let $\mathcal{O}$ be a (countable) open in M cover of M. Let $\mathcal{O}_0$ be the family of all sets of the form $O \cup (X \setminus M)$ with $O \in \mathcal{O}$. Then $\mathcal{O}_0$ is an open cover of X and thus has a locally finite open refinement $\mathcal{U}$. Then, in the space M, the family $\{U \cap M : U \in \mathcal{U}\}$ is a locally finite open refinement of $\mathcal{O}$. □

Definition 2.31. A set $M \subseteq X$ is *relatively compact in X* if it is contained in a compact subset of X.

We point out that our definition differs slightly from the definition used in some text books: In many text books, it is required that $\overline{M}$ be compact. In general, this is equivalent only if X is a Hausdorff space (Corollary 2.47).

Proposition 2.32. *Let $M \subseteq X$ be relatively compact. Then every subset of M is relatively compact, and every subset of M which is closed in X is compact.*

Proof. The first assertion is an immediate consequence of the definition, and the second follows from Proposition 2.29. □

2.3 Separation Axioms

Concerning separation axioms of topological spaces, literature is not completely unique: In some references T_3 or T_4 spaces are by definition supposed to be T_1. We will not assume this. In order to avoid any misunderstanding, we formulate exactly the definitions which we will use.

We only list those axiom that will actually play a role for us later. Since we have no other use for the axiom T_1, we use the Hausdorff property instead in the definition of "(completely) regular" and "(perfectly) normal" spaces which is unusual but easily seen to be equivalent in this context.

Definition 2.33. Let X be a topological space.

(a) X is *Hausdorff* if each two different points of X have disjoint neighborhoods.

(b) X is T_3 if for each closed subset $A \subseteq X$ and each $x \in X \setminus A$ the set A and the point x have disjoint neighborhoods.

(c) X is T_{3a} if for each closed subset $A \subseteq X$ and each $x \in X \setminus A$ there is $f \in C(X, [0,1])$ with $f(x) = 0$ and $f(A) = \{1\}$.

(d) X is T_4 if each two disjoint closed subsets of X have disjoint neighborhoods.

(e) X is T_5 if each subset of X is T_4.

(f) X is T_6 if for each two disjoint closed subsets $A, B \subseteq X$ there is $f \in C(X, [0,1])$ with $f^{-1}(0) = A$ and $f^{-1}(1) = B$.

A *regular space* is a T_3 Hausdorff space, a *completely regular space* is a T_{3a} Hausdorff space, and a *normal space* is a T_4 Hausdorff space. A *completely normal space* is a T_5 Hausdorff space. A *perfectly normal space* is a T_6 Hausdorff space.

Remark 2.34. It is equivalent to require that the neighborhoods in the above definition are open.

The class of T_5 and T_6 spaces is perhaps not so well-known, but these separation axioms will appear rather naturally as hypotheses in Proposition 4.40, Theorem 12.22, and in dimension theory (Section 5.2). Moreover, T_5 spaces will play an important role in connection with the degree for noncompact function triples (Section 11.4, Remark 14.39, Theorem 14.49 and its consequences).

Recall that $A, B \subseteq X$ are *separated* in X if $\overline{A} \cap B = \emptyset$ and $A \cap \overline{B} = \emptyset$. In several text books some of the following equivalent characterizations is used as the definition of T_5 spaces.

Proposition 2.35. *For a topological space X the following statements are equivalent:*

(a) *X is T_5.*

(b) *All open subsets of X are T_4.*

(c) *Any two separated subsets $A, B \subseteq X$ have disjoint neighborhoods.*

Proof. Suppose that (c) holds. If $M \subseteq X$ and if $A, B \subseteq M$ are disjoint and closed in M then they are separated in X. Hence, A and B have disjoint open neighborhoods $U, V \subseteq X$. Then $U \cap M$ and $V \cap M$ are disjoint open neighborhoods of A and B in M. Thus, (a) holds.

Trivially, (a) implies (b). Suppose now that (b) holds and that $A, B \subseteq X$ are separated. The open subset $M := X \setminus (\overline{A} \cap \overline{B})$ is T_4. Since $A_0 := M \cap \overline{A}$ and $B_0 := M \cap \overline{B}$ are closed in M and disjoint, they have disjoint open neighborhoods

$U, V \subseteq M$. Since M is open, U, V are also open in X, and we have $A \subseteq A_0 \subseteq U$ and $B \subseteq B_0 \subseteq V$. □

As a rule of thumb, the number n in "T_n" indicates how restrictive the corresponding separation axiom is. More precisely, this holds with only one exception, and this exception does not exist for Hausdorff spaces.

Theorem 2.36. *For every topological space X we have the chains of implications*

$$T_6 \Rightarrow T_5 \Rightarrow T_4, \qquad T_{3a} \Rightarrow T_3.$$

With the Hausdorff property, we have even

perfectly normal ⇒ completely normal ⇒ normal ⇒
completely regular ⇒ regular ⇒ Hausdorff

Proof. The implication $T_5 \Rightarrow T_4$ is trivial, and the implication $T_{3a} \Rightarrow T_3$ (and also $T_6 \Rightarrow T_4$) follows by considering the disjoint neighborhoods $f^{-1}([0, 1/2))$ and $f^{-1}((1/2, 1])$.

We show now the implication $T_6 \Rightarrow T_5$. Thus, let X be T_6. By Proposition 2.35, we have to show that any separated sets $A, B \subseteq X$ have disjoint neighborhoods. Since X is T_6, there are $f, g \in C(X, [0, 1])$ with $f^{-1}(0) = \overline{B}$, $g^{-1}(0) = \overline{A}$, $f^{-1}(1) = g^{-1}(1) = \varnothing$. For $n = 1, 2, \dots$, we put

$$U_n := f^{-1}((\tfrac{1}{n}, 1]) \setminus g^{-1}([\tfrac{1}{n}, 1]),$$
$$V_n := g^{-1}((\tfrac{1}{n}, 1]) \setminus f^{-1}([\tfrac{1}{n}, 1]).$$

Then U_n and V_n and thus also $U := \bigcup_{n=1}^{\infty} U_n$ and $V := \bigcup_{n=1}^{\infty} V_n$ are open. For $k \le n$, we have $U_k \cap V_n \subseteq f^{-1}((\frac{1}{n}, 1]) \cap V_n = \varnothing$ and $V_k \cap U_n \subseteq g^{-1}((\frac{1}{n}, 1]) \cap U_n = \varnothing$. Hence $U_k \cap V_n = \varnothing$ for all $k, n = 1, 2, \dots$ which implies $U \cap V = \varnothing$. We have $A \subseteq U$ since for each $x \in A$ we have $x \notin \overline{B}$ and thus $f(x) > 0 = g(x)$; for $n \ge 1/f(x)$, we thus have $x \in U_n \subseteq U$. An analogous argument shows that $B \subseteq V$. Hence, U and V are the required disjoint neighborhoods of A and B.

It remains to show that normal spaces are completely regular. This follows from Urysohn's lemma which is discussed below. □

Remark 2.37. We point out that the only nontrivial assertions of Theorem 2.36 are the two implications $T_6 \Rightarrow T_5$ and "normal⇒completely regular". In particular, the implication $T_6 \Rightarrow T_4$ is easy to obtain, as we have observed in the above proof.

The famous lemma of Urysohn implies immediately that any normal space is completely regular (put $B := \{x\}$ and consider the function $1 - f$):

Lemma 2.38 (Urysohn). *A space X is T_4 if and only if for each two disjoint closed subsets $A, B \subseteq X$ there is a continuous function $f: X \to [0,1]$ with $f(A) = \{0\}$ and $f(B) = \{1\}$.*

We postpone the proof of Lemma 2.38 for a moment.

It is well-known that topological vector spaces are T_{3a}. Unfortunately, they are not T_4, in general. For this reason, the class of T_{3a} spaces is of a particular interest.

Also for T_{3a} spaces there is some form of Urysohn's lemma which is less known but rather simple to prove:

Lemma 2.39 (Urysohn for T_{3a} spaces). *If X is T_{3a} then for each two disjoint subsets $A, B \subseteq X$, one being closed and the other being compact, there is $f \in C(X, [0,1])$ with $f(A) = \{0\}$ and $f(B) = \{1\}$.*

Proof. Assume that A is compact and B is closed. Let $\mathcal{O}$ be the family of all sets of the form $g^{-1}([-1,0))$ where $g \in C(X, [-1,1])$ and with $g(A) = \{1\}$. Since X is T_{3a}, the set $\mathcal{O}$ is an open covering of A. By the compactness of A, we thus find finitely many continuous functions $g_1, \dots, g_n \in C(X, [-1,1])$ with $g_k(B) = \{1\}$ such that $h(x) := \min\{g_1(x), \dots, g_n(x)\} < 0$ for every $x \in A$. Hence, $f(x) := \max\{0, h(x)\}$ has the required properties. □

Using Urysohn's lemma, we can give an alternative characterization of T_6 spaces.

Recall that a subset M of a topological space X is called a G_δ if it is the intersection of countably many open sets $U_n \subseteq X$.

Proposition 2.40. *Let X be T_4 (T_{3a}) and $A \subseteq X$ be closed (compact). Then the following assertions are equivalent:*

(a) *A is a G_δ.*

(b) *There is $f \in C(X, [0,1])$ with $f^{-1}(0) = A$.*

(c) *For each closed $B \subseteq X$ with $A \cap B = \varnothing$ there is $f \in C(X, [0,1])$ with $f^{-1}(0) = A$ and $f(B) = \{1\}$.*

Proof. The implication (c)⇒(b) follows with the choice $B = \varnothing$. For the proof of the implication (b)⇒(a), suppose that (b) holds. Then $U_n := f^{-1}([0, 1/n))$ are open neighborhoods of A. For each $x \in X \setminus A$, we have $f(x) > 0$, and so there is some n with $f(x) > 1/n$, hence $x \notin U_n$. It follows that $A = \bigcap_{n=1}^{\infty} U_n$ is a G_δ.

To prove (a)⇒(c), let A be the intersection of open sets $U_n \subseteq X$ ($n = 1, 2, \dots$). Let $B \subseteq X$ be closed with $A \cap B = \varnothing$. For each $n = 1, 2, \dots$, the closed (compact) set A and the closed set $B_n := B \cup (X \setminus U_n)$ are disjoint, and so Lemma 2.38 (or Lemma 2.39) implies that there is $f_n \in C(X,[0,1])$ with $f_n(A)=\{0\}$ and $f_n(B_n)=\{1\}$. Then $f(x):=\sum_{n=1}^{\infty} 2^{-n} f_n(x)$ is the uniform limit of a sequence of continuous functions and thus belongs to $C(X,[0,1])$. By construction, we have $f(A)=\{0\}$. For every $x \in X \setminus A$ there is some n with $x \notin U_n$, and so $x \in B_n$ implies $f_n(x)=1$ and thus $f(x) \geq 2^{-n} > 0$. Hence, $f^{-1}(0)=A$. Finally, for each $x \in B$, we have $x \in B_n$ $(n=1,2,\dots)$, hence $f(x)=\sum_{n=1}^{\infty} 2^{-n}=1$, and so $f(B)=\{1\}$. Thus, (c) holds. □

Corollary 2.41. *A space X is T_6 if and only if X is T_4 and each closed subset is a G_δ.*

Proof. If $A \subseteq X$ is closed and X is T_6, then Proposition 2.40 implies that A is a G_δ. Moreover, X is T_4 by Remark 2.37.

Conversely, suppose that X is T_4 and that $A, B \subseteq X$ are closed and disjoint G_δ-sets. By Proposition 2.40, there are functions $F, G \in C(X,[0,1])$ with $F^{-1}(0)=A$, $G^{-1}(0)=B$, and $F(B)=G(A)=\{1\}$. Then the function $f:=(F+1-G)/2 \in C(X,[0,1])$ satisfies

$$f^{-1}(0)=\{x : F(x)=G(x)-1\}=\{x : F(x)=0 \text{ and } G(x)=1\}=A$$

and

$$f^{-1}(1)=\{x : F(x)=G(x)+1\}=\{x : F(x)=1 \text{ and } G(x)=0\}=B.$$

Thus, the given condition is sufficient for T_6. □

All separation properties except for T_4 carry over to subsets:

Theorem 2.42. *If X is Hausdorff, T_3, T_{3a}, T_5, or T_6 then every subset $M \subseteq X$ has the same property. Hence, if X is regular, completely regular, completely normal, or perfectly normal then every subset $M \subseteq X$ has the same property.*

Proof. The proof for Hausdorff spaces and T_5 spaces is trivial, and for T_3 and T_{3a} spaces, it follows from the fact that if $A_0 \subseteq M$ is closed in M then there is a closed set $A \subseteq X$ with $A_0 = A \cap M$. Hence, if $x \in M \setminus A_0$ then also $x \in X \setminus A$, and so the T_3 or T_{3a} property of X implies the corresponding property of M.

Let X be T_6 and $M \subseteq X$. Since Theorem 2.36 implies that X is T_5, the subset M is T_4. By Corollary 2.41, it suffices to show that every closed (in M) subset $A \subseteq M$ is a G_δ in M. Letting $\overline{A}$ denote the closure of A in X, we have by Proposition 2.10 that $A=\overline{A} \cap M$. Since X is T_6, Corollary 2.41 implies that $\overline{A}$ is a G_δ in X. Hence, $\overline{A}$ is the intersection of a sequence of open (in X) subsets $U_n \subseteq X$. Then $V_n := U_n \cap M$ are open in M, and their intersection is $\overline{A} \cap M = A$. Thus, A is a G_δ in M, as required. □

Remark 2.43. Theorem 2.42 for T_6 spaces implies the nontrivial implication $T_6 \Rightarrow T_5$ of Theorem 2.36. Indeed, if X is T_6 then Theorem 2.42 implies that any $M \subseteq X$ is T_6 and thus T_4 by Remark 2.37. The latter means that X is T_5.

Note, however, that we have *used* the implication $T_6 \Rightarrow T_5$ to prove Theorem 2.42 for T_6 spaces so that we have actually shown that Theorem 2.42 for T_6 spaces is equivalent to the implication $T_6 \Rightarrow T_5$ of Theorem 2.36.

For T_4 spaces, there cannot be an analogue of Theorem 2.42: Every T_4 space which fails to be T_5 is a counterexample by definition of T_5 spaces. (A particular example of such a space is the so-called Tychonoff plank [132] which is even Hausdorff and compact.) However, a weak form of Theorem 2.42 holds also for T_4 spaces:

Proposition 2.44. *If X is T_4 then every closed subset is T_4.*

Proof. If X is T_4, $M \subseteq X$ is closed and $A, B \subseteq M$ are closed and disjoint then A, B are closed in X by Proposition 2.10. Hence, there are disjoint open neighborhoods $U, V \subseteq X$ of A and B, respectively. Then $U \cap M$ and $V \cap M$ are disjoint open in M neighborhoods of A and B, respectively. □

We recall that in the definition of Hausdorff and T_3 spaces the single point may be replaced by a compact set. We formulate the well-known proof to make clear that it can be carried out without referring to the axiom of choice; not even the countable axiom of choice is needed.

Proposition 2.45. *In a Hausdorff space every compact subset is closed. If X is Hausdorff then each two disjoint compact subsets have disjoint neighborhoods. If X is T_3 and $A, B \subseteq X$ are disjoint, one of them being closed and the other compact, then A and B have disjoint neighborhoods.*

Proof. Let X be Hausdorff, $A \subseteq X$ be compact, and $x \in X \setminus A$. Let $\mathcal{O}$ denote the family of all open sets in X which are disjoint from some neighborhood of x. Since X is Hausdorff, $\mathcal{O}$ is an open cover of A, and so a finite subcover of sets $O_1, \dots, O_n \in \mathcal{O}$ suffices. There are corresponding neighborhood $U_1, \dots, U_n \subseteq X$ of x such that $O_i \cap U_i = \varnothing$ for $i = 1, \dots, n$. Then $\bigcup_{i=1}^n O_i$ and $\bigcap_{i=1}^n U_i$ are disjoint neighborhoods of A and x, respectively. In particular, $x \notin \overline{A}$, and so A is closed.

Now if $B \subseteq X$ is compact and disjoint from A, we have just seen that the set $\mathcal{U}$ of all open sets with the property that there is a disjoint neighborhood of A constitutes an open cover of B; the same holds if X is T_3 and A is just closed.

There is finite subcover of sets $V_1, \dots, V_m \in \mathscr{U}$ and corresponding neighborhoods $W_1, \dots, W_m \subseteq X$ of B with $V_i \cap W_i = \varnothing$ $(i = 1, \dots, m)$. Then $\bigcup_{i=1}^m V_i$ and $\bigcap_{i=1}^m W_i$ are disjoint neighborhoods of A and B, respectively. □

Corollary 2.46. *Every compact Hausdorff space X is normal.*

Proof. By Proposition 2.29 closed subsets of X are compact, and so the assertion follows from Proposition 2.45. □

Corollary 2.47. *A subset M of a Hausdorff space X is relatively compact in X if and only if $\overline{M}$ is compact.*

Proof. If $M \subseteq K$ for some compact set $K \subseteq X$ then K is closed by Proposition 2.45, and so $\overline{M} \subseteq K$. Proposition 2.29 then implies that $\overline{M}$ is compact. The converse implication is clear from the definition. □

Usually, the T_3 or T_4 property is applied in the following form:

Corollary 2.48. *The following statements for a topological space X are equivalent:*

(a) *X is T_3 (T_4).*

(b) *For every open $U \subseteq X$ and every compact (closed) $A \subseteq X$ with $A \subseteq U$ there is an open $V \subseteq X$ with*

$$A \subseteq V \subseteq \overline{V} \subseteq U.$$

In other words: A has a closed neighborhood which is contained in U.

All topological notions are understood with respect to the space X; in particular, $\overline{V}$ denotes the closure in X.

Proof. By Proposition 2.45 (or by the definition of T_4 spaces), the sets A and $X \setminus U$ have disjoint open neighborhoods $V, W \subseteq X$ respectively. Since $X \setminus W$ is closed, it contains $\overline{V}$, and so $\overline{V} \subseteq X \setminus U$ has the required property.

Conversely, if $A, B \subseteq X$ are disjoint, B is closed and $A = \{x\}$ (or A is closed) then (b) with $U := X \setminus B$ implies that there is an open $V \subseteq X$ with $A \subseteq V$ such that the disjoint open set $V_0 := X \setminus \overline{V}$ contains B. Hence, we have (a). □

The following consequence will be used several times in the proof of the sum theorem of dimension theory.

Corollary 2.49. *If X is T_4 then each two disjoint closed subsets $A, B \subseteq X$ have disjoint closed neighborhoods, that is, there are open sets $U, V \subseteq X$ with $A \subseteq U$, $B \subseteq V$ and $\overline{U} \cap \overline{V} = \varnothing$.*

Proof. By Corollary 2.48, there is some open set $U \subseteq X$ with $A \subseteq U \subseteq \overline{U} \subseteq X \setminus B$ and then some open set $V \subseteq X$ with $B \subseteq V \subseteq \overline{V} \subseteq X \setminus \overline{U}$. □

Recall that X is *locally compact* if every point has a compact neighborhood.

Corollary 2.50. *Every locally compact Hausdorff space X is regular.*

Proof. Let $A \subseteq X$ be closed and $x \in X \setminus A$. By hypothesis, there is a compact neighborhood $N \subseteq X$ of x. Then $A \cap N$ is closed in N and thus compact by Proposition 2.29. Proposition 2.45 implies that $A \cap N$ and x have disjoint open neighborhoods $U, V \subseteq X$, respectively. Since N is closed by Proposition 2.29, we obtain that $U \cup (X \setminus N)$ and $V \cap N$ are disjoint neighborhoods of x and A, respectively. □

Now we return to Urysohn's lemma. Actually, this Lemma 2.38 is a special case of the Tietze extension theorem which we formulate and prove now.

Since the formulation of that result requires the notion of a convex hull, let us first recall the latter in a more general context for later usage. A subset M of a real or complex vector space X is called *convex* if for each $x, y \in M$ the "line segment joining x and y" is contained in M, that is, if for each $x, y \in M$ and each $\lambda \in [0, 1]$, we have $\lambda x + (1 - \lambda) y = y + \lambda(x - y) \in M$.

Proposition 2.51. *Let $\mathcal{M}$ be a nonempty family of convex subsets of a real or complex vector space X.*

Then $\bigcap \mathcal{M}$ is convex. Moreover, if $\mathcal{M}$ is directed upwards with respect to set inclusion, that is, if for each $M, N \in \mathcal{M}$ there is some $K \in \mathcal{M}$ with $M \cup N \subseteq K$, then also $\bigcup \mathcal{M}$ is convex.

In particular, if $M_1 \subseteq M_2 \subseteq \cdots \subseteq X$ are convex then also $\bigcup_{n=1}^{\infty} M_n$ is convex.

Proof. If $x, y \in \bigcap \mathcal{M}$ and $\lambda \in [0, 1]$ then the point $\lambda x + (1 - \lambda) y$ belongs to any $M \in \mathcal{M}$ and thus to $\mathcal{M}$. For the second assertion, assume that $\mathcal{M}$ is directed upwards. Hence, for each $x, y \in \bigcup \mathcal{M}$ there is some $M \in \mathcal{M}$ with $x, y \in M$. Since M is convex, we find for each $\lambda \in [0, 1]$ that the point $\lambda x + (1 - \lambda) y$ belongs to M and thus to $\bigcup \mathcal{M}$. □

If M is a subset of a real or complex vector space X, we use the symbol $\operatorname{conv} M$ to denote the *convex hull* of M, that is, the intersection of all convex subsets containing M. By Proposition 2.51, $\operatorname{conv} M$ is convex and thus the smallest convex subset of X containing M. A more explicit description of $\operatorname{conv} M$ is the following.

Proposition 2.52. *Let X be a vector space over $\mathbb{R}$ or $\mathbb{C}$. Then* $\operatorname{conv} M = \bigcup_{n=1}^{\infty} \operatorname{conv}_n(M)$ *with*

$$\operatorname{conv}_n(M) := \Big\{ \sum_{k=1}^{n} \lambda_k x_k : x_k \in M, \lambda_k \geq 0, \sum_{k=1}^{n} \lambda_k = 1 \Big\}.$$

Proof. We show first that $M_0 := \bigcup_{n=1}^{\infty} \operatorname{conv}_n(M)$ is convex. Thus, let $x, y \in M_0$, without loss of generality $x = \sum_{k=1}^{n} \lambda_k x_k$ and $y = \sum_{k=n+1}^{N} \lambda_k x_k$ with $x_k \in M$, $\lambda_k \geq 0$, $\sum_{k=1}^{n} \lambda_k = \sum_{k=n+1}^{N} \lambda_k = 1$. For $t \in [0,1]$, we put

$$\mu_k(t) = \begin{cases} t\lambda_k & \text{if } k \leq n, \\ (1-t)\lambda_k & \text{if } k > n. \end{cases}$$

Then $\mu_k(t) \geq 0$ and $\sum_{k=1}^{N} \mu_k(t) = t \cdot 1 + (1-t) \cdot 1 = 1$, and so $tx + (1-t)y = \sum_{k=1}^{N} \mu_k(t) x_k \in M_0$. Hence, M_0 is convex, and so $\operatorname{conv} M \subseteq M_0$.

Conversely, we show by induction on n that any convex set $N \subseteq X$ containing M satisfies $\operatorname{conv}_n(M) \subseteq N$. For $n = 1$, this is trivial, since $\operatorname{conv}_1(M) = M$. Suppose now $x \in \operatorname{conv}_{n+1}(M)$, say $x = (\sum_{k=1}^{n} \lambda_k x_k) + ty$ with $x_k, y \in M$, $\lambda_k, t \geq 0$ and $s := \sum_{k=1}^{n} \lambda_k = 1 - t$. In case $s = 0$, we have $x = y \in M \subseteq N$, and in case $s > 0$ we have $x_s := \sum_{k=1}^{n} s^{-1} \lambda_k x_k \in \operatorname{conv}_n(M)$. The induction hypothesis thus implies $x_s \in N$, and since $y \in N$, we obtain by the convexity of N that $x = s x_s + ty = (1-t) x_s + ty \in N$, as required. □

We will use the above results later on. The special case $M \subseteq X := \mathbb{R}$ which we need now is actually much simpler: In this case, $\operatorname{conv} M$ is the smallest interval (closed, open, or half-open) containing M.

Theorem 2.53 (Tietze–Urysohn). *Let X be a T_4 space, $A \subseteq X$ be closed, and $f \in C(A, \mathbb{R})$. Then f has an extension to $F \in C(X, \mathbb{R})$ with $F(X) \subseteq \operatorname{conv} f(A)$.*

Historically, Lemma 2.38 was used as a tool to prove Theorem 2.53. The original proof of the latter also used rather different methods (uniformly convergent series etc). We use an approach from [103] to obtain both results simultaneously by a method similarly to the original proof of Lemma 2.38.

Proof of Lemma 2.38 *and Theorem* 2.53. Let $f \in C(A, \mathbb{R})$. We assume first that $f(A) \subseteq [0,1]$. We put $Q := \mathbb{Q} \cap [0,1]$. For $r \in Q$, we will construct closed sets $B_r \subseteq X$ such that $B_1 = X$, and

$$\begin{aligned} B_r \cap A = f^{-1}([0,r]) \quad &\text{for all } r \in Q, \\ B_r \subseteq \mathring{B}_s \text{ for every } r, s \in Q, r < s. \end{aligned} \tag{2.6}$$

To this aim, we let $\Delta := \{(r,s) \in Q \times Q : r < s\}$. Note that for $(r,s) \in \Delta$ the set $A_r := f^{-1}([0,r])$ is closed in A and thus closed in X by Proposition 2.10. Analogously, $U_s := X \setminus f^{-1}([s,1])$ is open in X, and $A_r \subseteq U_s$. We show first that for all $(r,s) \in \Delta$ there are open sets $V_{r,s} \subseteq X$ satisfying

$$\begin{aligned} A_r \subseteq V_{r,s} \subseteq \overline{V}_{r,s} \subseteq U_s \quad &\text{for all } (r,s) \in \Delta, \\ \overline{V}_{\rho,\sigma} \subseteq V_{r,s} \quad \text{for all } (\rho,\sigma), (r,s) \in \Delta &\text{ with } \rho < r \text{ and } \sigma < s. \end{aligned} \tag{2.7}$$

To see that such sets exist, we let (r_n, s_n) be an enumeration of the elements of Δ and assume that $V_{r,s}$ is already defined and satisfies (2.7) for all $(r,s) \in \Delta_n := \{(r_k, s_k) : k < n\}$. Since Δ_n is finite, the sets

$$C_n := A_{r_n} \cup \bigcup_{\substack{(\rho,\sigma)\in\Delta_n \\ \rho < r_n,\, \sigma < s_n}} \overline{V}_{\rho,\sigma}$$

and

$$O_n := U_{s_n} \cap \bigcap_{\substack{(r,s)\in\Delta_n \\ r_n < r,\, s_n < s}} V_{r,s}$$

are closed and open, respectively. Since $C_n \subseteq O_n$, we obtain from Corollary 2.48 that there is an open set $V_{r_n,s_n} \subseteq X$ satisfying $C_n \subseteq V_{r_n,s_n} \subseteq \overline{V}_{r_n,s_n} \subseteq O_n$. Hence, we have defined $V_{r,s}$ satisfying (2.7) for all $(r,s) \in \Delta_{n+1}$. The existence of open $V_{r,s} \subseteq X$ satisfying (2.7) for all $(r,s) \in \Delta$ now follows by induction (and by the principle of dependent choices).

For $r \in Q$ with $r < 1$, we put now $B_r := \bigcap_{(r,s)\in\Delta} \overline{V}_{r,s}$, and $B_1 := X$. These sets are closed and satisfy (2.6). Indeed, $A_r \subseteq V_{r,s} \subseteq \overline{V}_{r,s}$ holds for all $(r,s) \in \Delta$, and so $A_r \subseteq B_r \cap A$. Conversely,

$$B_r \cap A \subseteq \bigcap_{(r,s)\in\Delta} (U_s \cap A) = \bigcap_{\substack{s\in Q \\ s>r}} f^{-1}([0,s)) = f^{-1}([0,r]),$$

and for $(\rho,\sigma) \in \Delta$ there is some $r \in \mathbb{Q}$ with $\rho < r < \sigma$, and so

$$B_\rho \subseteq \overline{V}_{\rho,r} \subseteq V_{r,\sigma}.$$

Since $V_{r,\sigma}$ is open and contained in $V_{\sigma,s}$ for each $s \in Q$ with $s > \sigma$, we have $V_{r,\sigma} \subseteq \mathring{B}_\sigma$ and thus have shown that $B_\rho \subseteq \mathring{B}_\sigma$. Hence, (2.6) is established.

Now we observe that (2.6) implies $B_s \subseteq B_r$ and $\mathring{B}_s \subseteq \mathring{B}_r$ for $s \le r$ $(s, r \in Q)$. Hence, for any $x \in X = B_1$, the sets

$$I(x) := \{s \in Q : x \in B_s\} \quad \text{and} \quad J(x) := \{s \in Q : x \in \mathring{B}_s\}$$

are intervals in Q containing 1, that is, putting $I^c(x) := Q \setminus I(x)$ and $J^c(x) := Q \setminus J(x)$, we have

$$F_1(x) := \inf I(x) = \sup I^c(x) \quad \text{and} \quad F_2(x) := \inf J(x) = \sup J^c(x),$$

when we use the convention $\sup \varnothing := 1$. Since (2.6) implies $J(x) \subseteq I(x)$, we have $F_1(x) \leq F_2(x)$. On the other hand, for any $r \in J^c(x)$ and any $s \in Q$ with $s < r$, we have $s \in I^c(x)$ by (2.6) and thus $F_1(x) \geq F_2(x)$, whence we have

$$F(x) := F_1(x) = F_2(x) \qquad \text{for all } x \in X.$$

Then $F\colon X \to [0,1]$ is continuous at every $x_0 \in X$. To see this, we note that for $\varepsilon > 0$ there are $r \in J(x_0)$ and $s \in I^c(x_0)$ with $r - \varepsilon < F(x_0) < s + \varepsilon$. Then $U := \mathring{B}_r \setminus B_s$ is a neighborhood of x_0. For all $x \in U$ we have $r \in J(x)$ and $s \in I^c(x)$ and thus $s \leq F(x) \leq r$. We obtain $|F(x) - F(x_0)| < \varepsilon$ for all $x \in U$, and so F is continuous at x_0.

For $x \in A$ we have by (2.6) that

$$I(x) = \{s \in Q : x \in f^{-1}([0,s])\} = \{s \in Q : f(x) \leq s\}.$$

Since $\overline{Q} = [0,1] \supseteq f(A)$, we obtain $F(x) = \inf I(x) = f(x)$.

We thus have shown that every continuous function $f\colon A \to [0,1]$ has a continuous extension $F\colon X \to [0,1]$. This already implies Urysohn's Lemma 2.38 (define $f\colon A \cup B \to [0,1]$ by $f|_A = 0$ and $f|_B = 1$). Now if $f(A)$ is disjoint from a set $M \subseteq \{0,1\}$, then A and $F^{-1}(M)$ are disjoint closed subsets of X, and so we find by Urysohn's lemma a continuous function $\lambda\colon X \to [0,1]$ with $\lambda(A) = \{0\}$ and $\lambda(F^{-1}(M)) = \{1\}$. The continuous function

$$\widetilde{F}(x) := F(x) - \lambda(x)\Big(F(x) - \frac{1}{2}\Big)$$

then satisfies $\widetilde{F}|_A = F|_A = f$ and

$$\widetilde{F}(X) \subseteq \operatorname{conv}\Big(\big(F(X) \setminus M\big) \cup \Big\{\frac{1}{2}\Big\}\Big) \subseteq [0,1] \setminus M.$$

Hence, we have shown that if I is any of the intervals $I_1 := [0,1]$, $I_2 := (0,1]$, $I_3 := [0,1)$, $I_4 := (0,1)$ and if $f\colon A \to I$ is continuous then f has a continuous extension $\widetilde{F}\colon X \to I$.

As the last step, we consider a general continuous map $f\colon A \to \mathbb{R}$. Then $\operatorname{conv} f(A)$ is an interval, and so there is a homeomorphism $h\colon \operatorname{conv} f(A) \to I$ for some $I \in \{I_1, I_2, I_3, I_4\}$. We have shown that $h \circ f$ has a continuous extension $\widetilde{F}\colon X \to I$, and then $h^{-1} \circ \widetilde{F}\colon X \to \operatorname{conv} f(A)$ is the required continuous extension of f. □

Remark 2.54. In [123], there was a shortcut suggested to obtain sets B_r satisfying (2.6) faster (with Q replaced by the dyadic numbers in $[0,1]$). Unfortunately, this shortcut seems to have a problem: With the notation from [123], it is not clear why $A(2^{-n-1}(2i+1))$ should be contained in the interior (with respect to X!) of $X(2^{-n}(i+1))$.

An important property of Hausdorff spaces is the uniqueness of limits and extensions.

Recall that a sequence $x_n \in X$ is said to be *convergent* to $x \in X$, if each neighborhood of x contains x_n for all except at most finitely many n. We write in this case $x = \lim_{n\to\infty} x_n$ or, more briefly, $x_n \to x$.

Lemma 2.55. *Let Y be Hausdorff, and X be a topological space.*

(a) *If $x_n \in Y$ satisfies $x_n \to x$ and $x_n \to y$ then $x = y$.*

(b) *If $F, G \in C(X,Y)$ are such that there is some $M \subseteq X$ with $F|_M = G|_M$ and $\overline{M} = X$ then $F = G$.*

Proof. Otherwise, there are disjoint neighborhoods $O_1, O_2 \subseteq Y$ of x and y respectively or $F(z)$ and $G(z)$ respectively for some $z \in X$. However, $x_n \in O_1 \cap O_2$ for all except finitely many n or $F(u) = G(u) \in O_1 \cap O_2$ for all $u \in M$ in a neighborhood of z. □

Definition 2.56. A *basis of the topology* of X is a family $\mathcal{O}$ of open sets in X with the property that every nonempty open set in X is the union of sets from $\mathcal{O}$.

A *subbasis of the topology* of X is a family $\mathcal{O}$ of open sets in X with the property that the family of all finite intersections of elements from $\mathcal{O}$ together with X constitutes a basis of the topology.

Proposition 2.57. *Let $\mathcal{O}$ be a family of subsets of X. Then there is a unique topology with $\mathcal{O}$ as a subbasis. A nonempty set $O \neq X$ is open with respect to this topology if and only if it is a union of finite intersections of elements from $\mathcal{O}$.*

Proof. The uniqueness is clear: Since $\mathcal{O}$ is a subbasis, the open sets must be as described in the assertion. For the existence, one just has to verify that the family $\mathcal{T}$ of open sets defined in this way is indeed closed under unions and finite intersections. Only the latter is nontrivial. By induction, it suffices to show that the intersection of each two sets from $\mathcal{T}$ belongs to $\mathcal{T}$. Thus, assume that $O_k \in \mathcal{T}$ $(k = 1,2)$. By hypothesis, O_k is the union of sets $F_{k,i}$ $(i \in I_k)$ where $F_{k,i}$ is the finite intersection of elements from $\mathcal{O}$. Then

$$O_1 \cap O_2 = \Big(\bigcup_{i\in I_1} F_{1,i}\Big) \cap \Big(\bigcup_{j\in I_2} F_{2,j}\Big) = \bigcup_{(i,j)\in I_1\cap I_2} (F_{1,i} \cap F_{2,j})$$

and the fact that each $F_{1,i} \cap F_{2,j}$ is a finite intersection of elements from $\mathcal{O}$ implies that $O_1 \cap O_2 \in \mathcal{T}$. □

Theorem 2.58 (Alexander's Subbasis Theorem). (AC). *Let X have a subbasis $\mathcal{O}$ of open sets such that every cover of X by elements from $\mathcal{O}$ has a finite subcover. Then X is compact.*

Proof. Suppose by contradiction that X is not compact. Let $\mathcal{C}$ denote the system of all open covers of X without finite subcover. Then $\mathcal{C}$ is partially ordered by inclusion, and by Hausdorff's maximality theorem, there is a maximal chain $\mathcal{C}_0 \subseteq \mathcal{C}$. Then $\mathcal{U} = \bigcup \mathcal{C}_0$ is an open cover of X. We show first that $\mathcal{U}$ has no finite subcover. Indeed, if $\mathcal{U}$ would have a finite subcover $\mathcal{V}$ then each element of $\mathcal{V}$ would be contained in some element of $\mathcal{C}_0$. Since $\mathcal{C}_0$ is a chain and $\mathcal{V}$ is finite, $\mathcal{V}$ would be a subset of some element from $\mathcal{C}_0 \subseteq \mathcal{C}$, contradicting the definition of $\mathcal{C}$.

In particular, the family $\mathcal{O}_0 = \mathcal{O} \cap \mathcal{U}$. cannot be a cover of X, since otherwise it would have a finite subcover by hypothesis. Hence, there is some $x \in X$ with $x \notin \bigcup \mathcal{O}_0$. There is some $U \in \mathcal{U}$ with $x \in U$. Since $\mathcal{O}$ is a subbasis, there are $O_1, \dots, O_n \in \mathcal{O}$ with $x \in O_1 \cap \dots \cap O_n \subseteq U$. For each $k = 1, \dots, n$, the set O_k is not contained in $\mathcal{U}$, since otherwise $x \in O_k \in \mathcal{O}_0$. It follows that $\mathcal{U} \cup \{O_k\}$ has a finite subcover, since otherwise it would belong to $\mathcal{C}$ and contradict the maximality of $\mathcal{C}_0$. Hence, for each $k = 1, \dots, n$, there is a finite $\mathcal{U}_k \subseteq \mathcal{U}$ such that $X = O_k \cup \bigcup \mathcal{U}_k$. Let $\mathcal{U}_0 = \mathcal{U}_1 \cup \dots \cup \mathcal{U}_n$. Then $\mathcal{U}_0 \cup \{U\}$ is a finite subset of $\mathcal{U}$ and thus cannot cover X. But by construction, for every element $y \in X \setminus U$ there is some k with $y \notin O_k$ and thus $y \in \bigcup \mathcal{U}_k \subseteq \bigcup \mathcal{U}_0$. □

In some situations, we make use of countability axioms:

Definition 2.59. X is *first countable* if every point $x_0 \in X$ has a countable basis of neighborhoods.

A topological space X is *second countable* if it has a countable basis of the topology.

Clearly, each second countable space is first countable, but the converse need not hold.

Recall that $f: X \to Y$ is called *sequentially continuous at $x \in X$* if $x_n \to x$ implies $f(x_n) \to f(x)$. We call f *sequentially continuous* if f is sequentially continuous at every $x \in X$.

Proposition 2.60. *If $f: X \to Y$ is continuous at $x \in X$ then f is sequentially continuous at $x \in X$. The converse holds if x has a countable basis of neighborhoods. In particular, if X is first countable, the continuous functions $f: X \to Y$ are exactly the sequentially continuous functions.*

Proof. If f is continuous at x, $V \subseteq Y$ is a neighborhood of f, and $x_n \to x$ then $f^{-1}(V)$ is a neighborhood of x and thus contains x_n for all except finitely many n. Hence, $f(x_n) \to f(x)$.

Conversely, suppose that x has a countable basis $U_1, U_2, \dots \subseteq U$ of neighborhoods and that $x_n \to x$ implies $f(x_n) \to f(x)$. Let $V \subseteq Y$ be a neighborhood of $f(x)$. If $f^{-1}(V)$ is not a neighborhood of x then there is for each n some $x_n \in U_1 \cap \dots \cap U_n$ with $f(x_n) \notin V$. It follows that $x_n \to x$ and thus by hypothesis $f(x_n) \to f(x)$ which is a contradiction. □

Proposition 2.61. *Every second countable space is Lindelöf.*

Proof. Let $\mathscr{U}$ be an open cover of a space X which has the countable basis $\mathscr{O}$ of the topology. For each $O \in \mathscr{O}$ choose some $U \in \mathscr{U}$ with $O \subseteq U$, if such a set U exists. The collection $\mathscr{U}_0$ of these U is a countable subcover of $\mathscr{U}$. Indeed, for every $x \in X$ there are $V \in \mathscr{U}$ with $x \in V$ and $O \in \mathscr{O}$ with $x \in O \subseteq V$. Hence, there is some $U \in \mathscr{U}_0$ with $O \subseteq U$, and so $x \in U$. □

If X and Y are topological spaces, we understand $X \times Y$ equipped with the product topology, that is, it has as a basis the sets of the form $U \times V$ where $U \subseteq X$ and $V \subseteq Y$ are open.

Proposition 2.62. *Let $K \subseteq X \times Y$ be compact. If $A \subseteq X$ is closed then*

$$K_A := \{y \in Y : (A \times \{y\}) \cap K \neq \varnothing\}$$

is compact. In particular, if $\{x\}$ is closed in X then $\{y : (x, y) \in K\}$ is compact.

Proof. By Proposition 2.29, we have to show that K_A is compact in Y. Thus, let $\mathscr{O}$ be an open in Y cover of K_A. Then the sets $U_O := X \times O$ $(O \in \mathscr{O})$ together with $U := (X \setminus A) \times Y$ constitute an open cover of K, since every $(x, y) \in K \setminus U$ satisfies $x \in A$ and thus $y \in K_A$. By the compactness, there are finitely many $O_1, \dots, O_n \in \mathscr{O}$ with $K \subseteq U \cup U_{O_1} \cup \dots \cup U_{O_n}$. For every $y \in K_A$ there is $x \in A$ with $(x, y) \in K$. Since $(x, y) \in K \setminus U$, we have $(x, y) \in U_{O_1} \cup \dots \cup U_{O_n}$ and thus $y \in O_1 \cup \dots \cup O_n$. Hence, $O_1, \dots, O_n \in \mathscr{O}$ constitute a finite subcover of K_A. □

We will frequently use Tychonoff's theorem for finite products which in contrast to Tychonoff's theorem for infinite products does not require the axiom of choice.

Theorem 2.63 (Tychonoff). *$X \times Y$ is compact (Lindelöf) if one of the spaces X and Y is compact (Lindelöf) and the other compact.*

Proof. Without loss of generality, let Y be compact. Let $\mathcal{O}$ be an open cover of $X \times Y$. Let $\mathcal{U}$ denote the family of all open sets $U \subseteq X$ which have the property that $U \times Y$ is covered by finitely many sets from $\mathcal{O}$. We show first that $\mathcal{U}$ is an open cover of X.

Thus, let $x \in X$, and let $\mathcal{V}_x$ denote the family of all open sets $V \subseteq Y$ which have the property that there is some open neighborhood $U \subseteq X$ of x and some $O \in \mathcal{O}$ with $U \times V \subseteq O$. Note that for each $y \in Y$ there is some $O \in \mathcal{O}$ with $(x, y) \in O$ and by definition of the product topology some open neighborhoods $U \subseteq X$ of x and $V \subseteq Y$ of y with $U \times V \subseteq O$. Hence, $\mathcal{V}_x$ is an open cover of Y. Since Y is compact, Y is covered by finitely many $V_1, \dots, V_n \in \mathcal{V}_x$. Let $U_1, \dots, U_n \subseteq U$ and $O_1, \dots, O_n \in \mathcal{O}$ be corresponding sets satisfying $x \in U_k$ and $U_k \times V_k \subseteq O_k$ $(k = 1, \dots, n)$. Then $U := U_1 \cap \dots \cap U_n$ is an open neighborhood of x with the property that

$$U \times Y \subseteq U \times (V_1 \cup \dots \cup V_n) \subseteq O_1 \cup \dots \cup O_n,$$

and so $U \in \mathcal{U}$. Hence, $\mathcal{U}$ is an open cover of X.

Since X is compact (Lindelöf) there is a finite (countable) $\mathcal{U}_0 \subseteq \mathcal{U}$ which covers X. For each $U \in \mathcal{U}_0$ there is a finite set $\mathcal{O}_U \subseteq \mathcal{O}$ with $U \times Y \subseteq \bigcup \mathcal{O}_U$. Then $\bigcup_{U \in \mathcal{U}_0} \mathcal{O}_U$ is a finite (countable) subcover of $X \times Y$, and so $X \times Y$ is compact (Lindelöf). □

In general, if X_i $(i \in I)$ is a family of topological spaces, the product $X := \prod_{i \in I} X_i$ is equipped with the topology which has as basis the sets $\prod_{i \in I} O_i$ where $O_i \subseteq X_i$ is open for every $i \in I$ and $O_i = X_i$ for all but finitely many $i \in I$.

Proposition 2.64. *Let X_i $(i \in I)$ be a family of connected spaces. Then $X := \prod_{i \in I} X_i$ is connected.*

Proof. Suppose by contradiction that $X = A \cup B$ with nonempty disjoint open sets $A, B \subseteq X$. Let $x = (x_i)_{i \in I} \in A$. Since B is open and nonempty, it contains an element of the basis of the topology as a subset, and so there is some $y = (y_i)_{i \in I} \in B$ with $x_i = y_i$ for all but finitely many $i \in I$, say $x_i \neq y_i$ if and only if $i \in J = \{i_1, \dots, i_n\}$. For $k \in \{0, \dots, n\}$, we put $z_k := (x_{k,i})_{i \in I}$ where

$$z_{k,i} := \begin{cases} x_i & \text{if } i \in \{i_j : j \leq k\}, \\ y_i & \text{if } i \in \{i_j : j > k\}, \\ x_i = y_i & \text{if } i \in I \setminus J. \end{cases}$$

Since $z_0 = y \in B$ and $z_n = x \in A$, there is a smallest $k \in \{1, \dots, n\}$ with $z_k \in A$, and then $z_{k-1} \in B$.

Let $A_0, B_0 \subseteq X_{i_k}$ denote the set of all $w \in X_{i_k}$ such that $(w_i)_{i \in I}$ belongs to A or B, respectively, if $w_i = z_i$ for $i \neq i_k$ and $w_{i_k} = w$. Then A_0 and B_0 are nonempty, since $z_{k,i_k} \in A_0$ and $z_{k-1,i_k} \in B_0$. On the other hand, since A and B are disjoint with $X = A \cup B$, it follows that A_0 and B_0 are disjoint with $X_{i_k} = A_0 \cup B_0$. The definition of the product topology implies that A_0 and B_0 are both open, contradicting the connectedness of X_{i_k}. □

Proposition 2.65. *A function $f: Y \to \prod_{i \in I} X_i$, $f(y) = (f_i(y))_{i \in I}$, is continuous at $y_0 \in Y$ if and only if each $f_i: Y \to X_i$ $(i \in I)$ is continuous at y_0.*

Proof. Suppose that each f_i is continuous at y_0. Let $U \subseteq X := \prod_{i \in I} X_i$ be a neighborhood of $f(y_0)$. Then there are open sets $O_i \subseteq X$ $(i \in I)$ with $O_i \neq X$ only for those i from a finite set $I_0 \subseteq I$ such that $f(y_0) \in \prod_{i \in I} O_i \subseteq U$. Then $V := \bigcap_{i \in I_0} f_i^{-1}(O_i)$ is a neighborhood of y_0 with $f(V) \subseteq U$, and so f is continuous at y_0.

Conversely, if f is continuous at y_0 and $O_{i_0} \subseteq X_{i_0}$ is an open neighborhood of $f_{i_0}(y_0)$ for some $i_0 \in I$ then we put $O_i := X_i$ for $i \in I \setminus \{i_0\}$. Then $U := \prod_{i \in I} O_i$ defines an open neighborhood of $f(y_0)$, and so there is some neighborhood $V \subseteq Y$ of y_0 with $f(V) \subseteq U$, hence $f_{i_0}(V) \subseteq O_{i_0}$. Thus, f_{i_0} is continuous at y_0. □

Corollary 2.66. (AC). *If all X_i $(i \in I)$ are path-connected then so is $X = \prod_{i \in I} X_i$.*

Proof. Let X_i $(i \in I)$ be path-connected, and $x = (x_i)_{i \in I}, y = (y_i)_{i \in I} \in X$. For each $i \in I$, there are continuous $\gamma_i: [0, 1] \to X_i$ with $\gamma_i(0) = x_i$ and $\gamma_i(1) = y_i$. Then Proposition 2.65 implies that $\gamma(t) := (\gamma_i(t))_{i \in I}$ defines a continuous map into X with $\gamma(0) = x$ and $\gamma(1) = y$. □

Corollary 2.67 (Tietze–Urysohn). (AC). *If $I_i \subseteq \mathbb{R}$ $(i \in I)$ is a family of nonempty intervals, X is a T_4 space, and $A \subseteq X$ is closed then every continuous map $f: A \to \prod_{i \in I} I_i$ has an extension to a continuous map $f: X \to \prod_{i \in I} I_i$.*

Proof. By Proposition 2.65, We have $f(y) = (f_i(y))_{i \in I}$ $(y \in Y)$ with continuous maps $f_i: A \to I_i$. By the Tietze–Urysohn extension theorem (Theorem 2.53), every f_i has a continuous extension $f_i: X \to I_i$, and so $f(y) := (f_i(y))_{i \in I}$ defines the required extension by Proposition 2.65. □

Remark 2.68. If I is finite (or countable) then AC is not needed (or the countable axiom of choice suffices) for the proof of Corollaries 2.66 and 2.67. Indeed, AC is only needed in the proof to ensure that γ_i $(i \in I)$ or the extensions f_i $(i \in I)$ form a family of functions.

Corollary 2.69. *If X and Y are locally (path-)connected then $X \times Y$ is locally (path-)connected.*

Proof. For $(x, y) \in X \times Y$ let $\mathcal{U}$ and $\mathcal{V}$ be (path-)connected neighborhood bases of $x \in X$ and $y \in Y$. Then the family of all sets $U \times V$ with $U \in \mathcal{U}$ and $V \in \mathcal{V}$ is a neighborhood basis of $(x, y) \in X \times Y$. Proposition 2.64 or Corollary 2.66 (and Remark 2.68) imply that the sets $U \times V$ are (path-)connected. □

The important role of products of intervals is that they can serve as a class of universal spaces of T_{3a} spaces with the property of Corollary 2.67:

Proposition 2.70. *Let X be completely regular. Then there is a homeomorphism of X onto a subset of the product $[0,1]^I := \prod_{i \in I}[0,1]$ with $I := C(X, [0,1])$.*

Proof. We show that a required homeomorphism is the map $g: X \to [0,1]^I$ which associates to each $x \in X$ the element $g(x) = (f(x))_{f \in I}$. Since each f is continuous, Proposition 2.65 implies that g is continuous. For every $x, y \in X$ with $x \neq y$ the Hausdorff property implies that $A = \{y\}$ is closed, and so there is some $f \in I$ with $f(x) = 0$ and $f(y) = 1$. Hence, $g(x) \neq g(y)$. Thus g is a continuous one-to-one map onto a subset $Y \subseteq [0,1]^I$. It remains to show that $g^{-1}: Y \to X$ is continuous. Thus, let $y_0 \in Y = g(X)$, and let $V \subseteq X$ be a neighborhood of $x_0 = g^{-1}(y_0)$. Since X is T_{3a} there is a continuous function $f_0: X \to [0,1]$ satisfying $f_0(x_0) = 0$ and $f_0(X \setminus V) = \{1\}$. Let $U := \prod_{f \in I} O_f$ where $O_f = [0,1]$ for $f \neq f_0$ and $O_{f_0} = [0,1)$. Then U is open in $[0,1]^I$, and so $U_0 := U \cap Y$ is open in Y. The open set U_0 consists of all $y = (f(x))_{f \in I}$ with $x = g^{-1}(y) \in f_0^{-1}([0,1))$. Hence, $y_0 \in U_0$ and $g^{-1}(U_0) \subseteq f_0^{-1}([0,1)) \subseteq V$, and so g^{-1} is continuous at y_0. □

Also for infinite products, the result of Theorem 2.63 holds, and so the universal space of Proposition 2.70 is actually compact:

Theorem 2.71 (Tychonoff). (AC). *If X_i $(i \in I)$ are compact then $\prod_{i \in I} X_i$ is compact.*

Proof. For $i \in I$, we define a map $m_i \colon X_i \multimap X := \prod_{j \in I} X_j$ by $m_i(x) := \{(x_j)_{j \in I} : x_i = x\}$. By definition of the topology, the sets of the form $m_i(O)$ with $i \in I$ and open $O \subseteq X_i$ form a subbasis of the topology and cover X. By Alexander's subbasis theorem (Theorem 2.58), it suffices to show that every open cover $\mathcal{O}$ of X consisting of sets of such a form has a finite subcover.

For $i \in I$, let $\mathcal{O}_i$ denote the family of all open sets $O \subseteq X_i$ with $m_i(O) \in \mathcal{O}$. Assume first that there is some $i \in I$ such that $\mathcal{O}_i$ is a cover of X_i. Then $\mathcal{O}_i$ has a finite subcover $O_1, \ldots, O_n \subseteq X_i$. Hence, $m_i(O_k) \in \mathcal{O}$ $(k = 1, \ldots, n)$ is a finite subcover of X, and we are done.

In the other case, we have for each $i \in I$ some $x_i \in X \setminus \bigcup \mathcal{O}_i$. Then $(x_i)_{i \in I} \in X \setminus \mathcal{O}$, contradicting the fact that $\mathcal{O}$ is a cover of X. (In the previous sentence, we used AC a second time.) □

It is known that the statement of Theorem 2.71 is actually equivalent to AC, see [85] (and a minor correction of the latter in [100]).

For completeness, we also provide proofs for the following well-known facts which we will need.

Proposition 2.72. *Let X be paracompact. If X is Hausdorff or T_3 then X is T_4. In particular, every paracompact Hausdorff space is normal.*

Proof. Let $A, B \subseteq X$ be closed and disjoint. Let $\mathcal{U}_{A,B}$ denote the family of all open sets with the property that they are disjoint from a neighborhood of B. We will show that A and B have disjoint neighborhoods if we assume in addition that $\mathcal{U}_{A,B}$ is a cover of A.

This implies the assertion if X is T_3, since in this case $\mathcal{U}_{A,B}$ is a cover of A. However, it also implies that X is T_3 if X is Hausdorff, since if B consists of a single point $x \in X \setminus A$, the set $\mathcal{U}_{A,\{x\}}$ is a cover of A.

Thus, assume that $\mathcal{U}_{A,B}$ is a cover of A. Then $\mathcal{U}_0 := \mathcal{U}_{A,B} \cup \{X \setminus A\}$ is an open cover of X and thus has a locally finite open refinement $\mathcal{O}$. Let $\mathcal{O}_0$ denote the family of all $O \in \mathcal{O}$ which intersect A. Since $\mathcal{O}$ is a cover of X, it follows that $U := \bigcup \mathcal{O}_0$ is an open neighborhood of A. We claim that U and $X \setminus \overline{U}$ are disjoint open neighborhoods of A and B, respectively.

It remains to show that $B \cap \overline{U} = \varnothing$. Thus, let $x \in B$. Since $\mathcal{O}$ is locally finite there is a neighborhood $V \subseteq X$ of x which intersects only finitely many elements from $\mathcal{O}$ and thus at most finitely many elements $O_1, \ldots, O_n$ from $\mathcal{O}_0 \subseteq \mathcal{O}$. Since $\mathcal{O}$ is a refinement of $\mathcal{U}_0$ we can for each $k = 1, \ldots, n$ argue as follows: Since O_k is not contained in $X \setminus A$ there is some $U_k \in \mathcal{U}_{A,B}$ with $O_k \subseteq U_k$. By definition of $\mathscr{U}_{A,B}$ there is a neighborhood $V_k \subseteq X$ of B which is disjoint from U_k and thus disjoint from O_k. Hence, $V \cap V_1 \cap \cdots \cap V_n$ is a neighborhood of $x \in B$ which is disjoint from each O_k $(k = 1, \ldots, n)$ and thus disjoint from $U = \bigcup \mathscr{O}_0$. Hence, $x \notin \overline{U}$. □

Theorem 2.73. (AC). *$X \times Y$ is paracompact if one of the spaces X and Y is paracompact and the other compact.*

Proof. Without loss of generality, let X be paracompact and Y be compact. Let $\mathscr{O}$ be an open cover of $X \times Y$. Let $\mathscr{B}$ be a basis for the topology of Y. We define a multivalued map Φ from the family of open subsets of X into the family $\mathscr{B}_0$ of finite subsets of $\mathscr{B}$ as follows. For open $U \subseteq X$, let $\Phi(U)$ be the system of all finite open covers $\mathcal{V}_U \subseteq \mathscr{B}$ of Y with the property that for each $V \in \mathcal{V}_U$ the set $U \times V$ is contained in a set from $\mathscr{O}$.

Then $\operatorname{dom}(\Phi)$ is an open cover of X. Indeed, for $x \in X$ let $\mathcal{V}_x$ denote the family of all $V \in \mathscr{B}$ with the property that there are an open neighborhood $U \subseteq X$ of x and $O \in \mathscr{O}$ with $U \times V \subseteq O$. Then $\mathcal{V}_x$ is an open cover of Y since for each $y \in Y$ there is some $O \in \mathscr{O}$ with $(x, y) \in O$, and thus there is an open set $U \subseteq X$ and $V \in \mathscr{B}$ with $(x, y) \in U \times V \subseteq O$. Since Y is compact, there is a finite subcover by sets $V_1, \ldots, V_n \in \mathcal{V}_x$. There are corresponding open neighborhoods $U_1, \ldots, U_n \subseteq U$ of x such that $U_k \times V_k$ is for each $k = 1, \ldots, n$ contained in a set from $\mathscr{O}$. Then $U := U_1 \cap \cdots \cap U_n$ is an open neighborhood of x, and $\mathcal{V}_U := \{V_1, \ldots, V_n\}$ is a finite open cover of Y, and for each $V_k \in \mathcal{V}_U$ the set $U \times V_k$ is contained in a set from $\mathscr{O}$. Hence, $x \in U \in \operatorname{dom}(\Phi)$, and so $\operatorname{dom}(\Phi)$ is an open cover of X.

Since X is paracompact, there is a locally finite open refinement $\mathscr{U}_0$ of $\operatorname{dom}(\Phi)$. Note that the definition of Φ implies that every open subset of a set from $\operatorname{dom}(\Phi)$ belongs to $\operatorname{dom}(\Phi)$. Hence, $\mathscr{U}_0 \subseteq \operatorname{dom}(\Phi)$. By the axiom of choice, the multivalued map $\Phi|_{\mathscr{U}_0}$ has a selection $U \mapsto \mathcal{V}_U$. Put $\mathscr{O}_U := \{U \times V : V \in \mathcal{V}_U\}$. Then $\mathscr{O}_0 := \bigcup_{U \in \mathscr{U}_0} \mathscr{O}_U$ is a locally finite open refinement of $\mathscr{O}$. To see that $\mathscr{O}_0$ is locally finite, note that for each $(x, y) \in X \times Y$ there is a neighborhood $U_x \subseteq X$ of x which intersects at most finitely many sets $U_1, \ldots, U_n \in \mathscr{U}_0$. Then $U_x \times Y$ is a neighborhood of (x, y) which intersects only finitely many sets from $\mathscr{O}_0$, namely the finitely many sets from $\mathscr{O}_{U_1} \cup \cdots \cup \mathscr{O}_{U_n}$. □

Remark 2.74. The countable axiom of choice suffices for Theorem 2.73 if the paracompact space in the product is Lindelöf.

Indeed, in this case, we can assume that the set $\mathscr{U}_0$ in the proof of Theorem 2.73 is countable, and thus $\Phi|_{\mathscr{U}_0}$ has a selection.

Remark 2.75. Theorem 2.73 does not even require the countable axiom of choice if the compact space in the product is second countable.

Indeed, in this case we can assume that the set $\mathscr{B}$ and thus also $\mathscr{B}_0$ in the proof of Theorem 2.73 is countable, say $\mathscr{B}_0 = \{\mathcal{V}_1, \mathcal{V}_2, \ldots\}$. To define a selection $U \mapsto \mathcal{V}_U$ of $\Phi|_{\mathscr{U}_0}$, we can thus put $\mathcal{V}_U := \mathcal{V}_{n(U)}$ where $n(U) := \min\{n \in \mathbb{N} : \mathcal{V}_n \in \Phi(U)\}$.

Corollary 2.76. *Let X be paracompact. If X is T_3 or Hausdorff then* $[0, 1] \times X$ *is paracompact, T_3, and T_4 (or even normal, respectively).*

Proof. Since $[0, 1]$ is second countable and compact, Remark 2.75 implies that $[0, 1] \times X$ is paracompact. Since X is T_3 or Hausdorff, also $[0, 1] \times X$ is T_3 or Hausdorff. Proposition 2.72 thus implies that $[0, 1] \times X$ is T_4. □

We point out that our proof of Corollary 2.76 did not even require the countable axiom of choice.

Corollary 2.76 is almost the most general result by which the normality of $[0, 1] \times X$ can be obtained. In fact, only a very slight refinement is possible: Instead of requiring paracompactness, one can require countable paracompactness and normality. But this is already the best which can be done, since Dowker has shown in [40] that for every topological space X the following assertions are equivalent:

(a) $[0, 1] \times X$ is normal.

(b) $I \times X$ is normal for every infinite compact metric space I.

(c) X is normal and countably paracompact.

Historically, Dowker had conjectured that normal spaces satisfy these properties automatically. It took quite a while before the first counterexamples to this conjecture have been found, and the first constructions required spaces of extremely large cardinality [128]. Such counterexamples are called *Dowker spaces* in literature.

2.4 Upper Semicontinuous Multivalued Maps

When one speaks about multivalued maps, one usually means multivalued maps between topological spaces which satisfy some sort of continuity. Classically, one distinguishes two notions of continuity for multivalued maps: upper semicontinuity and lower semicontinuity, but there are also some other variants. A lot about such maps can be found in [9], [10], [21]–[24], [71], [102], but we will recall in this section all required results (and provide some new).

Degree theory of multivalued maps is mainly related with the notion of upper semicontinuity:

Definition 2.77. Let X and Y be topological spaces, and $x_0 \in X$. Then $\Phi\colon X \multimap Y$ is *upper semicontinuous at* x_0 if for each open neighborhood $V \subseteq Y$ of $\Phi(x_0)$ the set $\Phi^-(V)$ is a neighborhood of x_0, that is, there is an open neighborhood $U \subseteq X$ of x_0 with $\Phi(U) \subseteq V$. If Φ is upper semicontinuous at every $x_0 \in X$, we call Φ *upper semicontinuous.*

For completeness, and since we will need it in the context of orientation of nonlinear Fredholm maps, we also recall the dual notion:

Definition 2.78. Let X and Y be topological spaces, and $x_0 \in X$. Then $\Phi\colon X \multimap Y$ is *lower semicontinuous at* x_0 if $\Phi(x_0) = \varnothing$ or if for each $y \in \Phi(x_0)$ and each open neighborhood $V \subseteq Y$ of y the set $\Phi^+(V)$ is a neighborhood of x_0. If Φ is lower semicontinuous at every $x_0 \in X$, we call Φ *lower semicontinuous.*

Remark 2.79. In Definition 2.77 and 2.78, it is equivalent to consider neighborhoods U and/or V which are not necessarily open.

In the single-valued case, both notions are equivalent to continuity:

Proposition 2.80. *For $f\colon X \to Y$ the following statements are equivalent:*

(a) *f is continuous at $x_0 \in X$.*

(b) *$f\colon X \multimap Y$ is upper semicontinuous at x_0.*

(c) *$f\colon X \multimap Y$ is lower semicontinuous at x_0.*

Proof. For $V \subseteq Y$, we have $f^-(V) = f^{-1}(V) = f^+(V)$. □

Intuitively, for an upper semicontinuous map the values in a neighborhood of x_0 must not "explode" (compared to the value at x_0), while for a lower semicontinuous map the values in a neighborhood of x_0 must not "implode". This is best explained by some simple examples with $X = Y = \mathbb{R}$ at $x_0 = 0$:

Example 2.81. The function $\Phi\colon \mathbb{R} \multimap \mathbb{R}$,

$$\Phi(x) := \begin{cases} M_1 & \text{if } x \neq 0, \\ M_0 & \text{if } x = 0 \end{cases}$$

is upper semicontinuous if and only if $M_1 \subseteq M_0$ and lower semicontinuous if and only if $M_0 \subseteq \overline{M}_1$.

In this (and many other) examples the "duality" between upper and lower semicontinuity occurs only for functions with closed values. In fact, in most applications where upper semicontinuity occurs "naturally", one has to do with functions with closed values.

Example 2.82. The function $\Phi: \mathbb{R} \multimap \mathbb{R}$,

$$\Phi(x) := \begin{cases} \{+x, -x\} & \text{if } x \neq 0, \\ M & \text{if } x = 0 \end{cases}$$

is upper semicontinuous if and only if $0 \in M$ and lower semicontinuous if and only if $M \subseteq \{0\}$.

Example 2.83. The extension $\Phi: \mathbb{R} \multimap \mathbb{R}$ of the signum function

$$\Phi(x) := \begin{cases} \{+1\} & \text{if } x > 0, \\ \{-1\} & \text{if } x < 0, \\ M & \text{if } x = 0 \end{cases}$$

is upper semicontinuous if and only if $\{+1, -1\} \subseteq M$, and is lower semicontinuous at 0 if and only if $M = \varnothing$. Note that the restriction of Φ to $(-\infty, 0]$ and $[0, \infty)$ is lower semicontinuous if and only if $M \subseteq \{-1\}$ or $M \subseteq \{1\}$, respectively.

In the particular case that $M \supseteq \{+1, -1\}$ is a compact interval, the values $\Phi(x)$ of the function from Example 2.83 are nonempty, compact and acyclic for all $x \in X$ (cf. Definition 4.55). Because of these properties, this function is in a sense the prototypical example for a function which should be treated by our degree theory.

In fact, the following result implies that this and similar functions satisfy a multivalued form of the intermediate value theorem. Degree theory can in a sense be considered as a formulation of such an intermediate value theorem in a more general topological setting: We will obtain the intermediate value theorem (of single-valued functions) as a special case of degree theory in Remark 9.87.

Theorem 2.84 (Intermediate Value). *Let $\Phi: [a, b] \multimap [-\infty, \infty]$ be upper semicontinuous and such that $\Phi(x)$ is a nonempty interval for every $x \in [a, b]$. Then for each $y_1 \in \Phi(a)$, $y_2 \in \Phi(b)$ the set $\Phi([a, b])$ contains all values between y_1 and y_2.*

Proof. Without loss of generality, we assume $a < b$ and $y_1 < y_2$. We have to show that $\Phi([a, b])$ contains every $y \in (y_1, y_2)$. To this end, we put

$M := \Phi^-((y, \infty])$ and $x_0 := \inf M$ (in case $M = \emptyset$, we put $x_0 := b$). We claim that $y \in \Phi(x_0)$.

Assume on the one hand by contradiction that $\Phi(x_0) \subseteq V_1 := (y, \infty]$. Since $y > y_1 \in \Phi(a)$, we must have $x_0 > a$, and the upper semicontinuity of Φ at x_0 would imply that $M = \Phi^-(V_1)$ is a neighborhood of x_0. In particular, there is some $\varepsilon > 0$ such that M contains the point $x_0 - \varepsilon > a$, contradicting the fact that x_0 is a lower bound of M.

Assume on the other hand by contradiction that $\Phi(x_0) \subseteq V_2 := [-\infty, y)$. Since $y < y_2 \in \Phi(b)$, we must have $x_0 < b$, and since Φ is upper semicontinuous at x_0, there is in particular some $\varepsilon > 0$ such that for all $x \in [x_0, x_0 + \varepsilon] \subseteq [a, b)$ the inclusion $\Phi(x) \subseteq V_2$ holds. Since $\Phi(x) \neq \emptyset$, we obtain $[x_0, x_0+\varepsilon] \cap M = \emptyset$. Since x_0 is a lower bound for M, it follows that also $x_0 + \varepsilon < b$ is a lower bound for M, contradicting the maximality of $x_0 = \inf M$.

We thus have seen that $\Phi(x_0)$ contains at least one point from $[-\infty, y]$ and at least one point from $[y, \infty]$. Since $\Phi(x_0)$ is an interval, we conclude $y \in \Phi(x_0)$. □

The fact that lower semicontinuous functions cannot "implode" is reflected by the following observation:

Proposition 2.85. *Let $M \subseteq X$ and $\Phi: X \multimap Y$ be lower semicontinuous in each point from $\overline{M} \setminus M$. Then*

$$\Phi(\overline{M}) \subseteq \overline{\Phi(M)}.$$

Proof. For $x_0 \in M$, we have $\Phi(x_0) \subseteq \Phi(M) \subseteq \overline{\Phi(M)}$. Thus, let $x_0 \in \overline{M} \setminus M$ and $y \in \Phi(x_0)$. For each open neighborhood $V \subseteq Y$ of y, the set $U := \Phi^+(V)$ is a neighborhood of x_0. Since $x_0 \in \overline{M}$, there is some $x \in U \cap M$. Hence, $V \cap \Phi(x) \neq \emptyset$ and thus $V \cap \Phi(M) \neq \emptyset$. It follows that $y \in \overline{\Phi(M)}$. □

An analogous assertion to Proposition 2.85 for upper semicontinuous maps does not hold as can easily be seen with $X = \mathbb{R}$ and $M = [-1, 1] \setminus \{0\}$ in each of the Examples 2.81, 2.82, and 2.83.

Remark 2.86. The latter has an important practical implication: In the single-valued case, it is more or less only a matter of taste whether one develops degree theory for continuous functions $\varphi: \overline{\Omega} \to Y$ or for continuous functions $\varphi: \Omega \to Y$ where Ω is an open subset of some space X. However, in the case of multivalued functions $\Phi: \overline{\Omega} \multimap Y$, the upper semicontinuity gives no control of Φ on the boundary $\partial\Omega$. Hence, in this case, one obtains a much more natural theory if one considers upper semicontinuous functions $\Phi: \Omega \multimap Y$. For this reason, we will always consider degree theories for functions defined on Ω (and not on $\overline{\Omega}$). We come back to this remark in Chapter 14.

Unfortunately, the terms "upper semicontinuous" and "lower semicontinuous" are historically used also in the context of single-valued maps with values in $[-\infty, \infty]$ with a different meaning:

Definition 2.87. Let X be a topological space, $f: X \to [-\infty, \infty]$ and $x_0 \in X$. Then f is *upper semicontinuous at* x_0 if for each $\eta > f(x_0)$ there is a neighborhood $U \subseteq X$ of x_0 with $f(U) \subseteq [-\infty, \eta)$. The function f is *lower semicontinuous at* x_0 if for each $\eta < f(x_0)$ there is a neighborhood $U \subseteq X$ of x_0 with $f(U) \subseteq (\eta, \infty]$.

We call f *upper semicontinuous* (*lower semicontinuous*) if f is upper (lower) semicontinuous at every point of X.

Formally, the meaning of "upper semicontinuous" and "lower semicontinuous" for single- and multivalued maps is different and so some confusion might arise by our convention to interpret single-valued maps as multivalued maps, although we hope that the meaning will always be clear from the context. Nevertheless, the following two results show that there is a relation between these notions.

Proposition 2.88. *For $f: X \to [-\infty, \infty]$ the following statements are equivalent:*

(a) *f is upper/lower semicontinuous at x_0.*

(b) *The multivalued map $x \mapsto [-\infty, f(x)]$ is upper/lower semicontinuous at x_0, or $f(x_0) = \infty$.*

(c) *The multivalued map $x \mapsto [f(x), \infty]$ is lower/upper semicontinuous at x_0, or $f(x_0) = -\infty$.*

Moreover, the cases $f(x_0) = \pm\infty$ have to be distinguished only for the lower semicontinuous multivalued maps.

Proof. Clear from the definitions. □

Proposition 2.89. *Let $\Phi: X \multimap [-\infty, \infty]$ be upper/lower semicontinuous at x_0 with nonempty values.*

(a) *If $\max \Phi(x_0)$ exists then $f(x) := \sup \Phi(x)$ is upper/lower semicontinuous at x_0.*

(b) *If $\min \Phi(x_0)$ exists then $f(x) := \inf \Phi(x)$ is lower/upper semicontinuous at x_0.*

Proof. Clear from the definitions. □

The definition of upper/lower semicontinuity does not change if we enlarge or shrink the space Y, and it is preserved under restrictions of X:

Proposition 2.90. *$\Phi: X \multimap Y$ is upper/lower semicontinuous at x_0 if and only if for the restricted image space $Y_0 := \Phi(X)$ the map $\Phi: X \multimap Y_0$ is upper/lower semicontinuous at x_0.*

If Φ is upper/lower semicontinuous at x_0, then for any subset $X_0 \subseteq X$ with $x_0 \in X_0$ the restriction $\Phi|_{X_0}: X_0 \multimap Y$ is upper/lower semicontinuous at x_0 for every $X_0 \subseteq X$ with $x_0 \in X_0$.

Proof. Clear from the definition of the inherited topology. □

Since we do not require (as many other authors) for $\Phi: X \multimap Y$ that $\Phi(x) \neq \varnothing$ for all $x \in X$, that is, we do not require $\operatorname{dom}(\Phi) = X$, let us point out that this is "almost" a consequence of upper semicontinuity:

Proposition 2.91. *Let $\Phi: X \multimap Y$ be upper semicontinuous at $x_0 \in \overline{\operatorname{dom}(\Phi)}$. Then $x_0 \in \operatorname{dom}(\Phi)$.*

In particular, if Φ is upper semicontinuous on X then $\operatorname{dom}(\Phi)$ is closed in X: $M \subseteq \operatorname{dom}(\Phi)$ implies $\overline{M} \subseteq \operatorname{dom}(\Phi)$.

Proof. If $x_0 \notin \operatorname{dom}(\Phi)$ then $V = \varnothing$ is a neighborhood of $\Phi(x_0) = \varnothing$. Hence, there is a neighborhood $U \subseteq X$ of x_0 with $\Phi(U) \subseteq V = \varnothing$, that is $U \cap \operatorname{dom}(\Phi) = \varnothing$, and so $x \notin \overline{\operatorname{dom}(\Phi)}$. □

Proposition 2.92. *For $\Phi: X \multimap Y$ the following statements are equivalent:*

(a) *Φ is upper semicontinuous.*

(b) *$\Phi^-(V)$ is open for every open $V \subseteq Y$.*

(c) *$\Phi^+(C)$ is closed for every closed $C \subseteq Y$.*

Also the following statements are equivalent:

(a) *Φ is lower semicontinuous.*

(b) *$\Phi^+(V)$ is open for every open $V \subseteq Y$.*

(c) *$\Phi^-(C)$ is closed for every closed $C \subseteq Y$.*

Proof. If Φ is upper (lower) semicontinuous, and $V \subseteq Y$ is open, then $\Phi^-(V)$ ($\Phi^+(V)$) is a neighborhood of each $x_0 \in \Phi^-(V)$ ($x_0 \in \Phi^+(V)$) and thus open by Proposition 2.8. The converse implication is obvious, and the equivalence of the last two assertions follows from (2.4) with $M = V$. □

As an almost immediate corollary, we obtain a multivalued version of the so-called glueing lemma:

Lemma 2.93 (Multivalued Glueing). *Let $A_1, \dots, A_n \subseteq X$ be closed with $X = A_1 \cup \dots \cup A_n$, and $\Phi_k \colon A_k \multimap Y$ be compatible in the sense that $\Phi_k(x) = \Phi_j(x)$ whenever $x \in A_j \cap A_k$ $(j \neq k)$. Then there is a unique map $\Phi \colon X \multimap Y$ with $\Phi|_{A_k} = \Phi_k$. Moreover, Φ is upper/lower semicontinuous if and only if all Φ_k are upper/lower semicontinuous.*

Proof. The unique existence of Φ is clear by the compatibility condition. If Φ is upper/lower semicontinuous then all Φ_k are upper/lower semicontinuous by Proposition 2.90. We show the converse.

For a closed set $C \subseteq Y$, we put $M := \Phi^+(C)$ (or $M := \Phi^-(C)$ for the lower semicontinuous case) and $M_k := \Phi_k^+(C)$ (or $M_k := \Phi_k^-(C)$, respectively) for $k = 1, \dots, n$. By Proposition 2.92, each M_k is closed in A_k and thus closed in X by Proposition 2.10. Hence, $M = M_1 \cup \dots \cup M_n$ is closed in X. Since $C \subseteq Y$ was an arbitrary closed set, the assertion follows from Proposition 2.92. □

Upper and lower semicontinuity works well with compositions:

Proposition 2.94. *If $\Phi \colon X \to Y$ is upper/lower semicontinuous on x_0 and $\Psi \colon Y \to Z$ is upper/lower semicontinuous at each point of $\Phi(x_0)$ then $\Psi \circ \Phi$ is upper/lower semicontinuous at x_0.*

Proof. We prove first the upper semicontinuous case. If $W \subseteq Z$ is an open neighborhood of $\Psi(\Phi(x_0))$ then the union V of all open sets $V_0 \subseteq Y$ with $\Psi(V_0) \subseteq W$ is an open neighborhood of $\Phi(x_0)$. Hence, there is some open neighborhood $U \subseteq X$ of x_0 with $\Phi(U) \subseteq V$. Then $\Psi(\Phi(U)) \subseteq \Psi(V) \subseteq W$.

To prove the lower semicontinuous case, let $W \subseteq Z$ be an open neighborhood of some $z \in \Psi\big(\Phi(x_0)\big)$. There is some $y \in \Phi(x_0)$ with $z \in \Psi(y)$. Then $V := \Psi^+(W)$ is a neighborhood of y, and so $U := \Phi^+(V)$ is a neighborhood of x_0 with $U \subseteq (\Psi \circ \Phi)^+(W)$. □

An analogous result for the Cartesian product is trivial in the lower semicontinuous case:

Proposition 2.95. *If $\Phi \colon X \multimap Y$ and $\Psi \colon X \multimap Z$ are lower semicontinuous at $x_0 \in X$, then $\Phi \times \Psi \colon X \multimap Y \times Z$ is lower semicontinuous at $x_0 \in X$.*

Proof. Let $M \subseteq Y \times Z$ be an open neighborhood of some $(y_0, z_0) \in (\Phi \times \Psi)(x_0)$. By definition of the product topology, there are open sets $V \subseteq Y$ and $W \subseteq Z$ with $(y_0, z_0) \in V \times W \subseteq M$. Then $U := \Phi^+(V) \cap \Psi^+(W)$ is a neighborhood of x_0, and for all $x \in U$ there is $y \in \Phi(x) \cap V$ and $z \in \Psi(x) \cap W$. Then $(y, z) \in (\Phi \times \Psi)(x) \cap M$, and so $U \subseteq (\Phi \times \Psi)^+(M)$. □

A corresponding result for the Cartesian product of upper semicontinuous maps does not hold (in contrast to some claims in text books like e.g. [71, (14.8)]), even if one of the maps in this product is single-valued, as the following example shows.

Example 2.96. Let $\Phi\colon \mathbb{R} \multimap \mathbb{R}$ be defined by $\Phi(x) := (-1, 1)$. Then the map $\mathrm{id}_{\mathbb{R}} \times \Phi\colon \mathbb{R} \multimap \mathbb{R} \times \mathbb{R}$ fails to be upper semicontinuous at $x_0 = 0$. Indeed, the Euclidean open unit disk $V \subseteq \mathbb{R}^2 \cong \mathbb{R} \times \mathbb{R}$ contains $(\mathrm{id}_{\mathbb{R}} \times \Phi)(0)$, but there is no neighborhood $U \subseteq \mathbb{R}$ of 0 with $(\mathrm{id}_{\mathbb{R}} \times \Phi)(U) \subseteq V$.

Example 2.96 might appear a bit artificial, since the value $\Phi(x_0)$ is open: Upper semicontinuity might appear not a very natural condition for such maps, since the set V in Definition 2.77 need not be strictly larger than $\Phi(x_0)$. However, one can give examples similar to Example 2.96 where $\Phi(x_0)$ is closed but unbounded.

Example 2.97. Let $\Phi\colon \mathbb{R} \multimap \mathbb{R}$ be defined by $\Phi(x) = \mathbb{N}$. Then $\mathrm{id}_{\mathbb{R}} \times \Phi\colon \mathbb{R} \multimap \mathbb{R} \times \mathbb{R}$ fails to be upper semicontinuous. To see this, we note that $V := \{(x, y) \in \mathbb{R} \times \mathbb{R} : |xy| < 1\}$ is open with $\Phi(0) \subseteq V$, but there is no neighborhood $U \subseteq \mathbb{R}$ of 0 with $(\mathrm{id}_{\mathbb{R}} \times \Phi)(0) \subseteq V$.

If we go to infinite-dimensional spaces, it can even be arranged that $\Phi(x_0)$ is closed and bounded.

Example 2.98. Let X be an infinite-dimensional normed space. It follows by Riesz Lemma (which we will recall in Lemma 3.31) that there is a bounded countable set $M = \{e_1, e_2, \dots\} \subseteq X$ such that $\|e_n - e_m\| \geq 1$ for all $m < n$. Then the map $\Phi\colon \mathbb{R} \multimap X$, defined by $\Phi(x) := M$, assumes closed bounded values, but $\mathrm{id}_{\mathbb{R}} \times \Phi\colon \mathbb{R} \multimap \mathbb{R} \times X$ fails to be upper semicontinuous at 0. To see this, let V be the union of all open balls in $\mathbb{R} \times X$ with center $(0, e_n)$ and radius $1/n$ for $n = 2, 3, \dots$

Then V is open with $\Phi(0) \subseteq V$, but there is no neighborhood $U \subseteq \mathbb{R}$ of 0 with $(\mathrm{id}_{\mathbb{R}} \times \Phi)(U) \subseteq V$.

The previous examples suggest that it is actually the lack of compactness of the map Φ which makes these examples possible: If we assume that the values of both maps are compact, we can prove the upper semicontinuity of products of these maps. The proof is somewhat technical, but the result turns out to be surprisingly powerful: Actually, we will use only the special case $\Phi = \mathrm{id}_X$, but already this special case will provide us some elegant (and to the author's knowledge new) proofs for some facts from elementary topology (Corollary 2.112

and Corollary 2.119 for single-valued maps) which are mathematical folklore, but whose elementary proofs would be rather cumbersome.

Proposition 2.99. *If $\Phi\colon X \multimap Y$ and $\Psi\colon X \multimap Z$ are upper semicontinuous at $x_0 \in X$ and $\Phi(x_0)$ and $\Psi(x_0)$ are compact, then $\Phi \times \Psi\colon X \multimap Y \times Z$ is upper semicontinuous at x_0.*

Proof. Let $M \subseteq Y \times Z$ be an open neighborhood of $C := \Phi(x_0) \times \Psi(x_0)$. There is a family of open set $V_i \subseteq Y$ and $W_i \subseteq Z$ with $V_i \times W_i \subseteq M$ $(i \in I)$ and $C \subseteq \bigcup_{i \in I}(V_i \times W_i)$. By the compactness of C (Tychonoff's Theorem 2.63), we can assume that I is finite. For each $y \in \Phi(x_0)$ let V^y denote the intersection of all V_i containing y, and for each $z \in \Psi(x_0)$, let W^z denote the intersection of all W_i containing z. We denote the union of all sets V^y $(y \in \Phi(x_0))$ and W^z $(z \in \Psi(x_0))$ by V and W, respectively. For every $v \in V$ and $w \in W$ there are $y \in \Phi(x_0)$ and $z \in \Psi(x_0)$ with $v \in V^y$ and $w \in W^z$. Since $(y, z) \in C$, there is some $i \in I$ with $(y, z) \in V_i \times W_i$, and so $V^y \subseteq V_i$ and $W^z \subseteq W_i$. Hence, $(v, w) \in V_i \times W_i \subseteq M$. Since $(v, w) \in V \times W$ was arbitrary, we have shown that $V \times W \subseteq M$. There is a neighborhood $U \subseteq X$ of x_0 with $\Phi(U) \subseteq V$ and $\Psi(V) \subseteq W$, and so $(\Phi \times \Psi)(U) \subseteq V \times W \subseteq M$. □

Also the proof of the following result is somewhat cumbersome, at least, if one compares it to the much simpler proof in the single-valued case. However, this is not so surprising since we will also see that it provides a surprisingly easy proof for a nontrivial fact about single-valued maps (Corollary 2.107).

Proposition 2.100. *If $\Phi\colon X \multimap Y$ is upper semicontinuous with $\Phi(x)$ being compact for every $x \in X$, then $\Phi(C)$ is compact for every compact $C \subseteq X$.*

Proof. Let $\mathcal{O}$ be an open cover of $\Phi(C)$, and let $\mathcal{O}_0$ be the set of all finite unions of sets from $\mathcal{O}$. Then $\mathcal{U}_0 := \{\Phi^-(O) : O \in \mathcal{O}_0\}$ is an open cover of C. Indeed, for any $x \in C$ the compactness of $\Phi(x)$ implies that there is some $O \in \mathcal{O}_0$ which contains $\Phi(x)$, hence, $x \in \Phi^-(O)$. Since C is compact, there are finitely many $O_1, \dots, O_n \in \mathcal{O}_0$ with

$$C \subseteq \Phi^-(O_1) \cup \cdots \cup \Phi^-(O_n) \subseteq \Phi^-(O_1 \cup \cdots \cup O_n),$$

where the last inclusion follows from (2.3). The set $O := O_1 \cup \cdots \cup O_n$ thus is a union of finitely many sets from $\mathcal{O}$ and satisfies $\Phi(C) \subseteq O$ by (2.1). □

Corollary 2.101. *Let $K \subseteq \prod_{i \in I} X_i$ be compact and/or connected. Then there is a family of compact and/or connected sets $K_i \subseteq X_i$ $(i \in I)$ with $K \subseteq \prod_{i \in I} K_i$.*

More precisely, one can choose K_i as the set of all $x \in X_i$ with the property that there is some $(x_j)_{j \in I} \in X$ with $x_i = x$.

Proof. The maps $p_i((x_i)_{i \in I}) := x_i$ are continuous from $\prod_{i \in I} X_i$ into X_i by definition of the product topology. Hence, $K_i = p_i(K)$ have the required property by Proposition 2.100 and/or Proposition 2.20. □

Corollary 2.102. *Let $f: X \to [-\infty, \infty]$ be upper (lower) semicontinuous. Then $\max f(C)$ ($\min f(C)$) exists for any compact $C \subseteq X$.*

Proof. By Proposition 2.88 the map $\Phi(x) := [-\infty, f(x)]$ ($\Phi(x) := [f(x), \infty]$) is upper semicontinuous with compact values. Hence, for any compact set $C \subseteq X$ the set $\Phi(C) \subseteq [-\infty, \infty]$ is compact and thus has a maximum (minimum). □

2.5 Closed and Proper Maps

Definition 2.103. A multivalued map $\Phi: X \multimap Y$ is called *closed* if for every closed set $C \subseteq X$ the image $\Phi(C)$ is closed in Y.

Proposition 2.104. *The map $\Phi: X \multimap Y$ is upper semicontinuous if and only if $\Phi^{-1}: Y \multimap X$ is closed.*

Proof. For every $C \subseteq Y$, we have $\Phi^{+}(C) = \Phi^{-1}(C)$. Hence, the claim follows from Proposition 2.92. □

Definition 2.105. A multivalued map $\Phi: X \multimap Y$ is called *proper* if $\Phi^{-1}(C)$ is compact for every compact $C \subseteq Y$.

Corollary 2.106. *For $\Phi: X \multimap Y$ and $\Psi = \Phi^{-1}: Y \multimap X$ (that is $\Phi = \Psi^{-1}$) the following statements are equivalent:*

(a) *Φ is upper semicontinuous, and $\Phi(x)$ is compact for every $x \in X$.*

(b) *Ψ is closed, and $\Psi^{-1}(x)$ is compact for every $x \in X$.*

(c) *Ψ is closed and proper.*

Proof. In view of $\Psi^{-1} = \Phi$, the equivalence of the first two statements follows from Proposition 2.104. If these assertions are satisfied, then Proposition 2.100 implies that $\Psi^{-1}(C) = \Phi(C)$ is compact for compact $C \subseteq X$, and so Ψ is proper. Conversely, if Ψ is proper then the compactness of $\{x\}$ implies that $\Psi^{-1}(x)$ is compact. □

We point out that our proof of Corollary 2.106 essentially only used Proposition 2.100. To the author's knowledge, this is a new (and perhaps the shortest possible) proof of the following well-known statement about single-valued functions whose elementary proof is not so short:

Corollary 2.107. *A closed function $p: X \to Y$ is proper if and only if $p^{-1}(x)$ is compact for every $x \in X$.*

Proof. This is a special case of the last equivalence of Corollary 2.106 with $\Psi :=$ p and $\Phi := p^{-1}$. □

Let us mention some sort of converse of Corollary 2.107 for upper semicontinuous maps in particular spaces.

Recall that a topological space X is called *compactly generated*, if for a set $M \subseteq X$ the property that $M \cap C$ is closed in C for every compact set $C \subseteq X$ implies that M is closed in X.

Proposition 2.108. *Every first countable space X is compactly generated.*

Proof. If $M \subseteq X$ fails to be closed in X, then there is some $x \in \overline{M} \setminus M$. Let $U_1, U_2, \dots \subseteq X$ constitute a countable neighborhood basis of x. For each n, we choose some $x_n \in M \cap U_1 \cap \dots \cap U_n$. Then $C := \{x_1, x_2, \dots\} \cup \{x\}$ is compact, and $M \cap C = C \setminus \{x\}$ is not closed in C. □

Proposition 2.109. *Let Y be a compactly generated Hausdorff space, and let $\Phi: X \multimap Y$ be proper and upper semicontinuous and such that $\Phi(x)$ is compact for every $x \in X$. Then Φ is closed.*

Proof. Let $A \subseteq X$ be closed. We have to show that $\Phi(A)$ is closed. Since Y is compactly generated, it suffices to show that for each compact set $C \subseteq Y$ the intersection $M := \Phi(A) \cap C$ is closed in C. Since Φ is proper, the set $\Phi^{-1}(C) = \Phi^{+}(C)$ is compact, and so also its intersection $A_C := A \cap \Phi^{+}(C)$ with the closed set A is compact (Proposition 2.29). Since Φ is upper semicontinuous, Proposition 2.100 implies that $\Phi(A_C)$ is compact and thus closed, since Y is Hausdorff (Proposition 2.45). Hence, $\Phi(A_C) \cap C = M$ is closed in C. □

Corollary 2.110. *Let Y be a compactly generated Hausdorff space. Then every continuous proper function $p: X \to Y$ is closed.*

Proof. In view of Proposition 2.80, this follows from Proposition 2.109 with $\Phi(x) = \{p(x)\}$. □

We note that if X and Y are both metric spaces, Corollary 2.110 is well-known and can be simpler proved by observing that preimages of convergent sequences have convergent subsequences. However, the weaker variant if only Y is metrizable (or even just compactly generated) seems to be less known. The multivalued variant (Proposition 2.109) is perhaps even new.

Now we come to the first application of Proposition 2.99 which turns out to be crucial for our definition of the degree. Recall that a single-valued map of topological spaces is called *perfect* if it is continuous, proper, closed, and onto.

Theorem 2.111. *Let $\Phi\colon X \multimap Y$ be upper semicontinuous and such that $\Phi(x)$ is compact for every $x \in X$. Let the map $p\colon \operatorname{graph}(\Phi) \to X$ be defined by $p(x,y) := x$. Then $p^{-1}(x) = \{x\} \times \Phi(x)$ for every $x \in X$, and p is continuous, closed, and proper. The map p is perfect if and only if $\operatorname{dom}(\Phi) = X$.*

Proof. For every $x_0 \in X$, we have

$$\begin{aligned} p^{-1}(x_0) &= \{(x,y) \in \operatorname{graph}(\Phi) : x = x_0\} = \{(x_0,y) : y \in \Phi(x_0)\} \\ &= \{x_0\} \times \Phi(x_0). \end{aligned}$$

Hence, $\Psi := p^{-1}\colon X \multimap \operatorname{graph}(\Phi)$ is the same map as $\operatorname{id}_X \times \Phi\colon X \multimap X \times Y$. Since $\operatorname{id}_X \times \Phi$ is upper semicontinuous by Proposition 2.99, also Ψ is upper semicontinuous by Proposition 2.90. Since $\Psi(x) = \{x\} \times \Phi(x)$ is compact for every $x \in X$, we obtain from Corollary 2.106 that p is closed and proper. The continuity of p follows from the definition of the product topology. Finally, $p(\operatorname{graph}(\Phi)) = \operatorname{dom}(\Phi)$ implies that p is onto if and only if $\operatorname{dom}(\Phi) = X$. □

Our proof of Theorem 2.111 is based on Proposition 2.99 and thus simplifies the argument from the related result in [4, Proposition (4.10)] (whose proof is only so short because a cumbersome elementary argument was omitted). It is somewhat surprising that we do not have to assume for our proof that X or Y are Hausdorff spaces. In particular, $\operatorname{graph}(\Phi)$ need not be closed under our hypotheses (we come to the closedness of $\operatorname{graph}(\Phi)$ in a moment).

We point out that the multivalued Theorem 2.111 (which we obtained as a special case of Proposition 2.99) provides a short proof for the following well-known result whose elementary proof would be somewhat cumbersome:

Corollary 2.112. *For $i = 1,2$, let $p_i\colon X_1 \times X_2 \to X_i$ be defined by $p_i(x_1,x_2) := x_i$. Let $C \subseteq X_1 \times X_2$ be closed.*

(a) *If $p_2(C)$ is relatively compact in X_2 then $p_1(C)$ is closed in X_1.*

(b) *If $p_1(C)$ is relatively compact in X_1 then $p_2(C)$ is closed in X_2.*

Proof. For symmetry reasons, it suffices to show the first assertion. We apply Theorem 2.111 with $X := X_1$, $Y \subseteq X_2$ being a compact subset containing $p_2(C)$, and $\Phi(x) := Y$. Theorem 2.111 implies that $p = p_1|_{X\times Y}$ is closed. Since $p_2(C) \subseteq Y$ implies that $C \subseteq X \times Y$, and thus C is closed in $X \times Y$, it follows that $p(C) = p_1(C)$ is closed in $X = X_1$. □

Corollary 2.113. *Let $A \subseteq X$ be closed, and I be a compact space. Then for every neighborhood $U \subseteq I \times X$ of $I \times A$ there is an open neighborhood $V \subseteq X$ of A with $I \times V \subseteq U$.*

Proof. We can assume that U is open. Applying Corollary 2.112 with $X_1 := I$, $X_2 := X$, $C := (I \times X) \setminus U$, we find that

$$V := X \setminus p_2(C) = \{x \in X : (I \times \{x\}) \cap C = \varnothing\} = \{x \in X : I \times \{x\} \subseteq U\}$$

is open. Hence, $A \subseteq V$ and $I \times V \subseteq U$ □

2.6 Coincidence Point Sets and Closed Graphs

Now we come to some properties of multivalued maps which require some separation axioms. For the case that we have *two* upper semicontinuous maps, the following observation about the coincidence point set

$$\operatorname{coin}_X(\Phi, \Psi) := \{x \in X : \Phi(x) \cap \Psi(x) \neq \varnothing\}$$

is crucial.

Proposition 2.114. *Let $\Phi, \Psi \colon X \multimap Y$ be upper semicontinuous, and let $x_0 \in X$ be such that $\Phi(x_0)$ and $\Psi(x_0)$ are disjoint. Assume in addition one of the following:*

(a) *Y is Hausdorff, and the sets $\Phi(x_0)$ and $\Psi(x_0)$ are compact.*

(b) *Y is T_3, one of the sets $\Phi(x_0)$ and $\Psi(x_0)$ is compact and the other closed.*

(c) *Y is T_4, and the sets $\Phi(x_0)$ and $\Psi(x_0)$ are closed.*

Then there is a neighborhood $U \subseteq X$ of x_0 such that $\Phi(U)$ and $\Psi(U)$ have disjoint neighborhoods in Y. In particular, U is disjoint from $\operatorname{coin}_X(\Phi, \Psi)$.

Proof. In view of Proposition 2.45, the hypothesis implies that there are disjoint open neighborhoods $V_1, V_2 \subseteq X$ of $\Phi(x_0)$ and $\Psi(x_0)$, respectively. Hence, $U := \Phi^-(V_1) \cap \Phi^-(V_2)$ is a neighborhood of x_0 with $\Phi(U) \subseteq V_1$ and $\Psi(U) \subseteq V_2$. □

Corollary 2.115. *Suppose that the hypotheses of Proposition* 2.114 *hold for every $x_0 \in X \setminus \operatorname{coin}_X(\Phi, \Psi)$. Then $\operatorname{coin}_X(\Phi, \Psi)$ is closed in X.*

Proof. By Proposition 2.114, the complement of $\operatorname{coin}_X(\Phi, \Psi)$ is a neighborhood for each of its points and thus open (Proposition 2.8). □

Corollary 2.116. *Let $\Phi: X \multimap Y$ be upper semicontinuous at $x \in X$. Assume in addition one of the following:*

(a) *Y is Hausdorff and $\Phi(x)$ is compact.*

(b) *Y is T_3 and $\Phi(x)$ is closed.*

Then (the closure being understood in $X \times Y$)

$$\overline{\operatorname{graph}(\Phi)} \cap (\{x\} \times Y) = \operatorname{graph}(\Phi) \cap (\{x\} \times Y). \tag{2.8}$$

Proof. Let $y \in Y \setminus \Phi(x)$. Proposition 2.114 with the constant function $\Psi \equiv \{y\}$ implies that there is a neighborhood $U \subseteq X$ of x_0 such that $\Phi(U)$ and y have disjoint neighborhoods, and so $(x, y) \notin \overline{\operatorname{graph}(\Phi)}$. □

Corollary 2.117. *Suppose that Y is Hausdorff or T_3. Let $\Phi: X \multimap Y$ be upper semicontinuous with $\Phi(x)$ being closed (even compact if Y is not T_3) for every $x \in X$. Then $\operatorname{graph}(\Phi)$ is closed in $X \times Y$.*

Proof. Clear by Corollary 2.116. □

Our second application of Proposition 2.99 consists in showing that the conclusion of Proposition 2.114 concerning $\operatorname{coin}_X(\Phi, \Psi)$ holds even if we do *not* assume that Φ is upper semicontinuous, but if we merrily assume that the conclusion of Corollary 2.116 holds.

Proposition 2.118. *Let $\Phi, \Psi: X \multimap Y$ and $x_0 \in X$ satisfy*

(a) *$\overline{\operatorname{graph}(\Phi)} \cap (\{x_0\} \times \Psi(x_0)) = \varnothing$, the closure being understood in $X \times Y$.*

(b) *Ψ is upper semicontinuous at x_0, and $\Psi(x_0)$ is compact.*

Then there is a neighborhood $U \subseteq X$ of x_0 which is disjoint from $\operatorname{coin}_X(\Phi, \Psi)$.

Proof. By Proposition 2.99, the map $\operatorname{id}_X \times \Psi: X \multimap X \times Y$ is upper semicontinuous at x_0. Since $V := (X \times Y) \setminus \overline{\operatorname{graph}(\Phi)}$ is open in $X \times Y$ and $(\operatorname{id}_X \times \Psi)(x_0) \subseteq V$ by hypothesis, there is a neighborhood $U \subseteq X$ of x_0 with $(\operatorname{id}_X \times \Psi)(U) \subseteq V$, that is, $\Phi(x) \cap \Psi(x) = \varnothing$ for all $x \in U$. □

Corollary 2.119. *Let $\Phi, \Psi: X \multimap Y$ be such that $\operatorname{graph}(\Phi)$ is closed in $X \times Y$, and that for every x in the complement of $\operatorname{coin}_X(\Phi, \Psi)$ the map Ψ is upper semicontinuous at x with compact $\Psi(x)$. Then $\operatorname{coin}_X(\Phi, \Psi)$ is closed in X.*

Proof. Proposition 2.118 implies that the complement of the set $\mathrm{coin}_X(\Phi, \Psi)$ is open in X. □

We point out that the requirement in Corollary 2.119 that $\mathrm{graph}(\Phi)$ be closed is a much weaker requirement than the upper semicontinuity of Φ. The reader familiar with weak convergence in Banach spaces will see this immediately from the following example:

Example 2.120. (AC). Let X be a metric space, Y a normed space, and $F\colon X \to Y$ be demicontinuous, that is, $x_n \to x$ implies that $F(x_n)$ converges weakly to $F(x)$. Then $\mathrm{graph}(F)$ is closed in $X \times Y$. Indeed, if $(x, y) \in \overline{\mathrm{graph}(F)}$, there is a sequence $x_n \in X$ with $x_n \to x$ and $F(x_n) \to y$. Then $F(x_n)$ converges weakly to y and to $F(x)$, and so the classical Hahn–Banach extension theorem (we will prove the corresponding assertion in Corollary 6.25) implies $y = F(x)$, that is, $(x, y) \in \mathrm{graph}(F)$.

Under additional compactness hypotheses, Corollary 2.117 has a strong converse as we will show now.

Definition 2.121. Let $\Phi\colon X \multimap Y$.

(a) Φ is *compact into* Y if $\Phi(X)$ is relatively compact in Y.

(b) Φ is *locally compact at* x_0 *into* Y, if there is a neighborhood $U \subseteq X$ of x_0 such that $\Phi|_U\colon U \to Y$ is compact, that is, $\Phi(U)$ is contained in a compact subset of Y.

(c) Φ is *locally compact into* Y if it is locally compact at each $x_0 \in X$.

Proposition 2.122. *Let $Y \subseteq Z$ be closed and $\Phi\colon X \multimap Y$. Then Φ is compact into Y if and only if Φ is compact into Z.*

Proof. If $\Phi(X) \subseteq Y$ is contained in a compact subset $K \subseteq Z$ then $\Phi(X)$ is also contained in the set $K \cap Y \subseteq Y$ which by Proposition 2.29 is compact. Conversely, every compact $K \subseteq Y$ is also compact in Z by Proposition 2.29. □

Proposition 2.123. *Let $\Phi\colon X \multimap Y$ and $x \in X$ satisfy* (2.8), *the closure being understood in $X \times Y$. If Φ is locally compact at x in Y then Φ is upper semicontinuous at x.*

Proof. Let $V \subseteq Y$ be a neighborhood of $\Phi(x)$. Assume by contradiction that for each neighborhood $U \subseteq X$ of x the set $\Phi(U) \setminus V$ is nonempty. By hypothesis there is a compact set $Y_0 \subseteq Y$ and a neighborhood $U_0 \subseteq X$ of x with $\Phi(U_0) \subseteq Y_0$. Let $\mathscr{U}$ denote the family of all neighborhoods $U \subseteq X$ of x which satisfy $U \subseteq$

U_0. Proposition 2.10 implies that $A_U := \overline{\Phi(U)} \setminus V$ is closed in Y_0. The family of all sets A_U ($U \in \mathscr{U}$) has the finite intersection property, since if $\mathscr{U}_0 \subseteq \mathscr{U}$ is finite then $U := \bigcap \mathscr{U}_0$ belongs to $\mathscr{U}$ and satisfies by hypothesis and (2.2)

$$\varnothing \neq \Phi(U) \setminus V \subseteq \bigcap_{U \in \mathscr{U}_0} \Phi(U) \setminus V \subseteq \bigcap_{U \in \mathscr{U}_0} A_U.$$

Proposition 2.28 thus implies that there is some $y \in \bigcap_{U \in \mathscr{U}} A_U$. Then $(x, y) \in \overline{\text{graph}(\Phi)}$. Indeed, for any neighborhood $W \subseteq X \times Y$ of (x, y) there are $U \in \mathscr{U}$ and a neighborhood $V_0 \subseteq Y$ of y with $U \times V_0 \subseteq W$. Since $y \in A_U \subseteq \overline{\Phi(U)}$, there is $\hat{y} \in V_0 \cap \Phi(U)$ and thus some $\hat{x} \in U$ with

$$(\hat{x}, \hat{y}) \in \text{graph}(\Phi) \cap (U \times V_0) \subseteq \text{graph}(\Phi) \cap W.$$

Hence $(x, y) \in \overline{\text{graph}(\Phi)}$ which by (2.8) implies $(x, y) \in \text{graph}(\Phi)$, that is, $y \in \Phi(x) \subseteq V$. On the other hand, $y \in \bigcap_{U \in \mathscr{U}} A_U$ implies $y \notin V$ which is a contradiction. □

Corollary 2.124. *Let $\Phi\colon X \multimap Y$ be locally compact. If* graph(Φ) *is closed in $X \times Y$ then Φ is upper semicontinuous.*

Example 2.120 shows that even for single-valued functions it is not possible to drop the hypothesis that Φ be locally compact.

Proposition 2.125. *Let $\Phi\colon X \multimap Y$ be locally compact. Then each compact set $K \subseteq X$ has an open neighborhood $U \subseteq X$ such that $\Phi(U)$ is relatively compact in Y.*

Proof. Let $\mathscr{U}$ denote the system of all open sets $U \subseteq X$ such that $\Phi(U)$ is relatively compact in Y. By hypothesis, $\mathscr{U}$ is an open cover of K. Since K is compact, it is covered by finitely many $U_1, \dots, U_n \in \mathscr{U}$. Then $U := U_1 \cup \cdots \cup U_n$ is an open neighborhood of K, and

$$\Phi(U) = \Phi(U_1) \cup \cdots \cup \Phi(U_n)$$

is relatively compact in Y. □

Chapter 3

Metric Spaces

3.1 Notations and Basic Results for Metric Spaces

Recall that a function $d\colon X \times X \to [0,\infty)$ is called a *metric* if it satisfies:

(a) $d(x,y) = 0 \iff x = y$ for all $x, y \in X$.

(b) $d(x,y) = d(y,x)$ for all $x, y \in X$ (*symmetry*).

(c) $d(x,y) \le d(x,z) + d(z,y)$ for all $x, y, z \in X$ (*triangle inequality*).

The couple (X, d) is called a *metric space*. Notationally, we usually just write X and mean by d the corresponding metric. We define for nonempty $M \subseteq X$ the *diameter*

$$\operatorname{diam} M := \sup\{d(x,y) : x, y \in M\}$$

and the *distance function* $\operatorname{dist}(\cdot, M)\colon X \to [0,\infty)$ by

$$\operatorname{dist}(x, M) := \inf\{d(x,y) : y \in M\}.$$

Moreover, for nonempty $N, M \subseteq X$, we put

$$\operatorname{dist}(N, M) := \inf\{d(x,y) : x \in N, y \in M\} = \inf\{\operatorname{dist}(x, M) : x \in N\}.$$

In case $M = \varnothing$, it is convenient to define $\operatorname{diam} M = 0$ and $\operatorname{dist}(x, M) = \operatorname{dist}(N, M) = \operatorname{dist}(M, N) = \infty$ (also if $N = \varnothing$).

For $x_0 \in M \subseteq X$ and $r \in [0, \infty]$, we define the *open* and *closed ball* and *sphere* (with radius r and center x_0), and the *ball-neighborhood* of M (with radius r) by

$$\begin{aligned}
B_r(x_0) &:= \{x \in X : d(x, x_0) < r\},\\
K_r(x_0) &:= \{x \in X : d(x, x_0) \le r\},\\
S_r(x_0) &:= \{x \in X : d(x, x_0) = r\},\\
B_r(M) &:= \bigcup_{x \in M} B_r(x) = \{x \in X : \operatorname{dist}(x, M) < r\},
\end{aligned}$$

respectively. An *ε-net* for $M \subseteq X$ is a set $N \subseteq X$ satisfying $M \subseteq B_\varepsilon(N)$.

Each metric on X induces a topology which has $B_r(x)$ ($r > 0$, $x \in \mathbb{R}$) as the basis of open sets. We tacitly understand this topology when we speak about a metric space. A topological space is *metrizable* if there is a metric which induces the given topology.

Note that if X and Y are metric spaces then $f: X \to Y$ is continuous at x_0 with respect to the induced topologies if and only if for each $\varepsilon > 0$ there is some $\delta > 0$ with $f(B_\delta(x_0)) \subseteq B_\varepsilon(f(x_0))$.

Similarly, we have $x_n \to x$ if and only if for each $\varepsilon > 0$ we have $x_n \in B_\varepsilon(x)$ for all but at most finitely many n.

We recall the so-called *inverse triangle inequality*:

Proposition 3.1. *If X is a metric space then*

$$|d(x,y) - d(w,z)| \le d(x,w) + d(y,z) \qquad \text{for all } x,y,w,z \in X.$$

Proof. By the triangle inequality $d(x,y) \le d(x,w) + d(w,z) + d(y,z)$ and $d(w,z) \le d(w,x) + d(x,y) + d(y,z)$ which together implies the claim in the cases $d(x,y) \ge d(w,z)$ or $d(x,y) \le d(w,z)$, respectively. □

The inverse triangle inequality implies in particular that $d: X \times X \to \mathbb{R}$ is continuous.

Corollary 3.2. *All open balls are open, all closed balls and spheres are closed.*

Proof. Since $f := d(\cdot, x_0)$ is continuous, the assertion follows from $B_r(x_0) = f^{-1}((-\infty, r))$, $K_r(x_0) = f^{-1}([0,r])$ and $S_r(x_0) = f^{-1}(\{r\})$. □

Easy examples show that it need not be the case that $K_r(x_0) = \overline{B_r(x_0)}$. However, this is the case for normed spaces. Recall that X is a (real or complex) *normed space* if it is a vector space (over $\mathbb{K} = \mathbb{R}$ or $\mathbb{K} = \mathbb{C}$) and if it is equipped with a *norm*, that is a map $\|\cdot\|: X \to [0,\infty]$ satisfying the well-known properties

(a) $\|x\| = 0 \iff x = 0$ $(x \in X)$.

(b) $\|\lambda x\| = |\lambda| \|x\|$ for all $\lambda \in \mathbb{K}$, $x \in X$.

(c) $\|x + y\| \le \|x\| + \|y\|$ for all $x, y \in X$.

We equip a normed space with the metric $d(x,y) = \|x - y\|$. The inverse triangle inequality (Proposition 3.1 with $y = z = 0$) becomes for normed spaces

$$|\|x\| - \|y\|| \le \|x - y\|.$$

In particular, $\|\cdot\|: X \to \mathbb{R}$ is continuous.

Proposition 3.3. *If X is a normed space and $r \in (0,\infty)$ then $B_r(x_0)$ is the interior of $K_r(x_0)$, $K_r(x_0) = \overline{B_r(x_0)}$, $S_r(x_0) = \partial B_r(x_0) = \partial K_r(x_0)$, and $B_r(x_0)$ and $K_r(x_0)$ are convex.*

Proof. If $x \in S_r(x_0)$ and $t \in (0, \infty)$ then $x_t := x_0 + t(x - x_0)$ belongs to $B_r(x_0)$ in case $t < r$ and does not belong to $K_r(x_0)$ in case $t > r$. Since $B_r(x_0)$ is open and $K_r(x_0)$ is closed, this implies the first two assertions and then also the third. The last assertion follows from the estimate

$$d(tx + (1-t)y, x_0) = \|t(x - x_0) + (1-t)(y - x_0)\| \le t d(x, x_0) + (1-t) d(y - x_0)$$

for $t \in [0, 1]$. □

Definition 3.4. Let X and Y be metric spaces, $f: X \to Y$, and $\omega: [0, \infty) \to [0, \infty]$ be continuous at 0 with $\omega(0) = 0$. Then ω is called a *modulus of continuity at* x_0 *for* f if

$$d(f(x), f(x_0)) \le \omega(d(x, x_0)) \qquad \text{for all } x \in X.$$

If there is $L \in [0, \infty)$ such that $\omega(t) = Lt$ is a modulus of continuity at every $x_0 \in X$, we call f *Lipschitz* with *Lipschitz constant* L. Similarly, if there is $L \in [0, \infty)$ and $\alpha \in (0, 1)$ such that $\omega(t) = Lt^\alpha$ is a modulus of continuity at every $x_0 \in X$, we call f *Hölder* (of exponent α) with *Hölder constant* L.

Proposition 3.5. *Let X and Y be metric spaces, and $f: X \to Y$. Then f has a modulus of continuity at $x_0 \in X$ if and only if f is continuous at $x_0 \in X$.*

Proof. Let ω be a modulus of continuity of f at x_0. For each $\varepsilon > 0$ there is $\delta > 0$ such that $\omega(t) < \varepsilon$ for all $t \in [0, \delta)$. Then $f(B_\delta(x_0)) \subseteq B_\varepsilon(f(x_0))$.

Conversely, if f is continuous at x_0 then

$$\omega(t) := \sup_{x \in K_t(x_0)} d(f(x), f(x_0))$$

is a modulus of continuity for f at x_0. Indeed, $\omega(0) = 0$, and for each $\varepsilon > 0$ there is $\delta > 0$ with $f(B_\delta(x_0)) \subseteq B_\varepsilon(f(x_0))$, that is, for all $x \in B_\delta(x_0)$ we have $d(f(x), f(x_0)) < \varepsilon$. Hence, $\omega([0, \delta)) \subseteq [0, \varepsilon]$, and so ω is continuous at 0. □

Corollary 3.6. *Every Lipschitz or Hölder function is continuous.*

Proof. Every Lipschitz or Hölder function has a modulus of continuity at every point. So the assertion follows from Proposition 3.5. □

The following definition of a Cauchy sequence is not the one which is usually found in text books on analysis. However, we will show in a moment that it is equivalent. Moreover, it is much more convenient to verify and actually is even much more natural since the relation with convergent sequences is immediately clear: Convergent sequences are exactly those Cauchy sequences for which x in the following definition can be chosen independently of ε.

Definition 3.7. A sequence x_n in a metric space X is a *Cauchy sequence* if for each $\varepsilon > 0$ there is some $x \in X$ such that the relation $d(x_n, x) < \varepsilon$ holds for all except finitely many n.

Recall that a metric space is *complete* if each Cauchy sequence converges. We collect the well-known properties in the following lemma.

Lemma 3.8. *Let X be a metric space.*

(a) *Each convergent sequence in X is a Cauchy sequence. Conversely, a Cauchy sequence converges if and only if it has a convergent subsequence.*

(b) *A sequence $x_n \in X$ is a Cauchy sequence if and only if for each $\varepsilon > 0$ there is some N with $d(x_n, x_m) \le \varepsilon$ for all $n, m \ge N$.*

(c) *If $x_n \to x$ and $x_n \to y$ then $x = y$.*

(d) *If $M \subseteq X$ is complete then M is closed, and the converse holds if X is complete.*

Proof. If $x_n \to x$ then one can just choose the limit x in the definition of a Cauchy sequence. Conversely, if $x_n \in X$ is a Cauchy sequence with a convergent subsequence $x_{n_k} \to x$ then $x_n \to x$. Indeed, for each $\varepsilon > 0$ there is $y \in X$ and $N \in \mathbb{N}$ such that $d(x_n, y) < \varepsilon$ for all $n \ge N$ and some k with $n_k \ge N$ and $d(x_{n_k}, x) < \varepsilon$. Then we have for all $n \ge N$ that

$$d(x_n, x) \le d(x_n, y) + d(y, x_{n_k}) + d(x_{n_k}, y) < 3\varepsilon.$$

If x_n is a Cauchy sequence and $\varepsilon > 0$ then there is $x \in X$ and $N \in \mathbb{N}$ with $d(x_n, x) \le \varepsilon$ for all $n \ge N$. Hence,

$$d(x_n, x_m) \le d(x_n, x) + d(x, x_m) \le 2\varepsilon$$

for all $n \ge N$. Conversely, if for each $\varepsilon > 0$ there is some N with $d(x_n, x_m) \le \varepsilon$ for all $n, m \ge N$ then the choice $x = x_N$ shows that x_n is a Cauchy sequence.

Assertion (c) follows from $d(x, y) \le d(x, x_n) + d(x_n, y) \to 0$.

If M is complete and $x \in \overline{M}$ then for each n there is some $x_n \in M \cap K_{1/n}(x)$, in particular $x_n \to x$. Then x_n is a Cauchy sequence by (a) and thus convergent to some $y \in M$. By (c), we have $x = y \in M$. Conversely, if X is complete, $M \subseteq X$ is closed, and $x_n \in M$ is a Cauchy sequence then $x_n \to x$ for some $x \in X$. Since M is closed it follows that $x \in M$, and so M is complete. □

For later usage, we recall Banach's fixed point theorem.

Theorem 3.9 (Banach Fixed Point). *Let $X \neq \varnothing$ be a complete metric space, and $f\colon X \to X$ be Lipschitz with constant $L < 1$. Then f has a fixed point in X.*

Proof. We choose $x_0 \in X$ and define inductively $x_n := f(x_{n-1})$. Using the shortcut $d_n := d(x_{n+1}, x_n)$, we have $d_n = d(f(x_n), f(x_{n-1})) \leq L d_{n-1}$, and thus an induction shows $d_n \leq L^n d_0$ for all $n \geq 0$. For $0 < m \leq n$, we obtain

$$d(x_n, x_m) \leq \sum_{k=m}^{n-1} d_k \leq \sum_{k=m}^{\infty} L^k d_0 = \frac{L^m}{1-L} d_0.$$

By Lemma 3.8(a), it follows that x_n is a Cauchy sequence in X and thus convergent to some $x \in X$. The continuity of f implies $x_n = f(x_{n-1}) \to f(x)$, and so Lemma 3.8(c) implies $f(x) = x$. □

Actually, the fixed point of Theorem 3.9 is unique. We prove this in the following generalization which is more convenient for most applications. Theorem 3.9 follows from the following result in the special case $r = \infty$.

Theorem 3.10 (Banach Fixed Point on Balls). *Let X be a complete metric space, $x_0 \in X$ and $r \in [0, \infty]$, and let $f\colon K_r(x_0) \to X$ be Lipschitz with constant $L < 1$. If $\rho := (1-L)^{-1} d(f(x_0), x_0) \leq r$ then f has exactly one fixed point, and this fixed point belongs to $K_\rho(x_0)$.*

Proof. For $x \in K_\rho(x_0)$, we have

$$d(f(x), x_0) \leq d(f(x), f(x_0)) + d(f(x_0), x_0) \leq L\rho + (1-L)\rho = \rho.$$

Hence, $f\colon K_\rho(x_0) \to K_\rho(x_0)$. Since $K_\rho(x_0)$ is closed and thus a complete metric space by Lemma 3.8(d), the existence of a fixed point in $K_\rho(x_0)$ follows from Theorem 3.10. Concerning the uniqueness, observe that if x and y are two fixed points of f then $d(x, y) = d(f(x), f(y)) \leq L d(x, y)$ implies that $d(x, y) = 0$, hence $x = y$. □

Lemma 3.11. *Let X be a metric space, and $\mathcal{F}$ be a nonempty family of functions $f\colon X \to \mathbb{R}$ and such that there is a modulus of continuity ω at x_0 for every $f \in \mathcal{F}$ with ω being independent of $f \in \mathcal{F}$ and finite on $[0, r)$. Suppose that $S(x) := \sup_{f \in \mathcal{F}} f(x)$ is finite for $x = x_0$. Then $S(x)$ is finite in $B_r(x_0)$, and $S\colon B_r(x_0) \to \mathbb{R}$ is continuous at x_0 and has the modulus of continuity ω at x_0. An analogous result holds with* "inf" *instead of* "sup".

Proof. Let $S(x_0) < \infty$ and $x \in B_r(x_0)$. For every $f \in \mathcal{F}$, we have

$$f(x) \le f(x_0) + \omega(d(x, x_0)) \le S(x_0) + \omega(d(x, x_0)).$$

Taking the supremum over all $f \in \mathcal{F}$, we find $S(x) < \infty$, hence $S(x) \in \mathbb{R}$, and

$$S(x) - S(x_0) \le \omega\big(d(x, x_0)\big).$$

Conversely, taking the supremum over all $f \in \mathcal{F}$ in

$$f(x_0) \le f(x) + \omega\big(d(x, x_0)\big) \le S(x) + \omega\big(d(x, x_0)\big),$$

we find

$$S(x_0) - S(x) \le \omega\big(d(x, x_0)\big).$$

For the last assertion, note that $\inf_{f \in \mathcal{F}} f(x) = -S(x)$ where S is the function corresponding to the family $\{-f : f \in \mathcal{F}\}$. □

Proposition 3.12. *For every metric space X and nonempty $M \subseteq X$ the function $f := \operatorname{dist}(\cdot, M)\colon X \to [0, \infty)$ is Lipschitz with constant 1. We have $f = \operatorname{dist}(\cdot, \overline{M})$, and $f^{-1}(0) = \overline{M}$.*

Proof. Put $\mathcal{F} := \{d(\cdot, y) : y \in M\}$. Since each $f \in \mathcal{F}$ is Lipschitz with constant 1, Lemma 3.11 implies that $S(x) := \operatorname{dist}(x, M) = \inf_{f \in \mathcal{F}} f(x)$ is Lipschitz with constant 1.

For each $x \in X$ there is a sequence $x_n \in \overline{M}$ with $d(x, x_n) \to \operatorname{dist}(x, \overline{M})$. Choose $y_n \in M$ with $d(x_n, y_n) \le 1/n$. Then $d(x, y_n) \le d(x, x_n) + d(x_n, y_n) \to \operatorname{dist}(x, \overline{M})$, and so $f(x) \le \operatorname{dist}(x, \overline{M})$. The converse inequality is obvious. We have $f(x) = 0$ if and only if there is a sequence $x_n \in M$ with $d(x, x_n) \to 0$ which means $x \in \overline{M}$. □

We obtain that metric spaces satisfy all separation properties introduced in Section 2.3:

Corollary 3.13 (Urysohn's Lemma for metric spaces). *Every metric space X is perfectly normal, that is, X is Hausdorff and T_6. More precisely, there is a map which associates to each pair (A, B) of closed disjoint subsets $A, B \subseteq X$ some $f_{(A,B)} \in C(X, [0, 1])$ satisfying $f_{(A,B)}^{-1}(0) = A$ and $f_{(A,B)}^{-1}(1) = B$.*

Note that without AC, the described property is slightly stronger than T_6, since in general T_6 spaces it is not clear without AC, whether there is a selection $(A, B) \mapsto f_{(A,B)}$.

Proof. For the Hausdorff property, we note that $B_r(x) \cap B_r(y) = \emptyset$ for $r < d(x,y)/2$. The continuous function

$$f_{(A,B)}(x) := \frac{\operatorname{dist}(x,A)}{\operatorname{dist}(x,A) + \operatorname{dist}(x,B)}$$

(in case $A = \emptyset$ and/or $B = \emptyset$, we replace in this formula $\operatorname{dist}(x,A)$ and/or $\operatorname{dist}(x,B)$ by 1) has the required properties by Proposition 3.12. □

Corollary 3.14. *Let A and B be disjoint nonempty subsets of a metric space. If A is compact and B is closed then there is some $x \in A$ with* $\operatorname{dist}(x,B) = \operatorname{dist}(A,B) > 0$.

Proof. The continuous function $\operatorname{dist}(\cdot\,,B)|_A\colon A \to (0,\infty)$ attains its minimum on the compact set A. □

Proposition 3.15. *Let K be a compact space, X be a metric space, and let $\Phi, \Psi\colon K \multimap X$ be upper semicontinuous and such that for each $x \in K$ the sets $\Phi(x)$ and $\Psi(x)$ are nonempty, compact and disjoint. Then*

$$\min_{x\in K} \operatorname{dist}\bigl(\Phi(x), \Psi(x)\bigr) > 0.$$

Proof. Since $\Phi \times \Psi$ is upper semicontinuous by Proposition 2.99, we obtain by the continuity of d and Proposition 2.94 that $F := d \circ (\Phi \times \Psi)$, that is $F\colon K \multimap (0,\infty]$, $F(x) := d \circ (\Phi \times \Psi) = d(\Phi(x), \Psi(x))$ is upper semicontinuous. Then $f(x) := \min F(x) > 0$ exists by Corollary 3.14, and Proposition 2.88 implies that f is lower semicontinuous. Corollary 2.102 implies that $\min f(K) > 0$ exists. □

Recall that a set $M_0 \subseteq M \subseteq X$ is called *dense* in M if $M \subseteq \overline{M}_0$; a set $M \subseteq X$ is *separable* if it has a countable dense subset.

Proposition 3.16. *Every metric space X is first countable. It is second countable if and only if it is separable.*

Proof. The sets $B_{1/n}(x_0)$ $(n \in \mathbb{N})$ constitute a countable neighborhood basis for x_0. If $\mathcal{O}$ is a countable basis of the topology of X, we choose from each nonempty element of $\mathcal{O}$ some point. The countable collection C of these points is dense in X, since every neighborhood intersects C. Conversely, suppose that $C = \{x_1, x_2, \ldots\}$ is dense in X. Then the family $\mathcal{O}$ of balls $B_r(x_n)$ with $n \in \mathbb{N}$ and $r \in \mathbb{Q} \cap (0,\infty)$ is countable and dense in X. To see this, we note that every open set $O \subseteq X$ is the union of all elements from $\mathcal{O}$ contained in O. Indeed, for every $x \in O$ there is some $\varepsilon > 0$ with $B_{2\varepsilon}(x) \subseteq O$. We choose some $r \in \mathbb{Q} \cap (0,\varepsilon)$ and some n with $d(x,x_n) < r$; then $x \in B_r(x_n) \subseteq B_{2\varepsilon}(x) \subseteq O$. □

Corollary 3.17. *If X is a separable metric or second countable space then every subset $M \subseteq X$ has the same property*

Proof. For second countable spaces, the assertion is obvious: If $\mathcal{O}$ is a countable basis of the topology of X then the family of all $O \cap M$ $(O \in \mathcal{O})$ is a countable basis of the topology of M. For separable metric spaces, the assertion follows from Proposition 3.16. □

Definition 3.18. Let X be a topological space and Y be a metric space. A map $\Phi\colon X \multimap Y$ is *upper semicontinuous in the uniform sense at* $x \in X$ if for each $\varepsilon > 0$ there is some neighborhood $U \subseteq X$ of x with $\Phi(U) \subseteq B_\varepsilon(\Phi(x))$. If this property holds for every $x \in X$, we call Φ *upper semicontinuous in the uniform sense.*

Proposition 3.19. *Let X be a topological space, Y be a metric space, and Φ: $X \multimap Y$. If Φ is upper semicontinuous at $x \in X$, then Φ is upper semicontinuous in the uniform sense at x. The converse holds if $\Phi(x)$ is compact.*

Proof. For the first claim, it suffices to note that $B_\varepsilon(\Phi(x))$ is a neighborhood of $\Phi(x)$. For the converse implication, note that if $V \subseteq Y$ is an open neighborhood of the compact set $\Phi(x)$ then $\varepsilon := \operatorname{dist}(\Phi(x), X \setminus V) > 0$ by Corollary 3.14, and so $B_\varepsilon(\Phi(x)) \subseteq V$. Hence, if U is as in Definition 3.18, we have $\Phi(U) \subseteq V$. □

Recall that the inclusion $\Phi(\overline{C}) \subseteq \overline{\Phi(C)}$ of Proposition 2.85 fails in general for maps which are upper semicontinuous. Hence, it may be somewhat surprising that in metric spaces, this relation holds by an *appropriate* choice of a countable dense subset C (if $\operatorname{dom}(\Phi)$ is separable). Actually, this is the case even if Φ is just upper semicontinuous in the uniform sense as we show now.

Proposition 3.20. *Let X and Y be metric spaces and $\Phi\colon X \multimap Y$ be upper semicontinuous in the uniform sense. If $M \subseteq X$ is separable then there is a countable set $C \subseteq M$ with $\Phi(M) \subseteq \overline{\Phi(C)}$.*

Proof. Replacing X by M, we may assume without loss of generality that $M = X$. For $n \in \mathbb{N}$, let $\mathcal{O}_n$ be the family of all $B_\delta(x) \subseteq X$ with the property that $\Phi(B_\delta(x)) \subseteq B_{1/n}(\Phi(x))$. Since Φ is upper semicontinuous in the uniform sense, $\mathcal{O}_n$ is an open cover of X. Since X is Lindelöf by Propositions 2.61 and 3.16, $\mathcal{O}_n$ has a countable subcover $B_{\delta_{n,k}}(x_{n,k})$ $(k \in \mathbb{N})$. Then the family C of all $x_{n,k}$ has the required property. Indeed, let $y \in \Phi(x)$ and $\varepsilon > 0$ be arbitrary. Choose some $n \in \mathbb{N}$ with $1/n < \varepsilon$. There is some k with $x \in B_{\delta_{n,k}}(x_{n,k})$. Then

$$y \in \Phi\big(B_{\delta_{n,k}}(x_{n,k})\big) \subseteq B_{1/n}\big(\Phi(x_{n,k})\big)$$

implies that there is some $y_0 \in \Phi(x_{n,k}) \subseteq \Phi(C)$ with $d(y, y_0) < 1/n < \varepsilon$. □

Corollary 3.21. *Let X and Y be metric spaces and $\Phi\colon X \multimap Y$ be upper semi-continuous in the uniform sense. If $M \subseteq X$ is separable and $\Phi(x)$ is separable for every $x \in M$ then $\Phi(M)$ is separable.*

Proof. By Proposition 3.20 there is a countable $C \subseteq M$ with $\Phi(M) \subseteq \overline{\Phi(C)}$. By hypothesis there is a countable $A \subseteq \Phi(C)$ with $\Phi(C) \subseteq \overline{A}$. Then $A \subseteq \Phi(M)$ is countable with $\Phi(M) \subseteq \overline{A}$. □

3.2 Three Measures of Noncompactness

Definition 3.22. Let X be a metric space, and $M \subseteq X$. The *Hausdorff measure of noncompactness* of M in X is

$$\chi_X(M) := \inf\{\varepsilon \in [0,\infty] : M \text{ has a finite } \varepsilon\text{-net } N \subseteq X\}.$$

The *Kuratowski measure of noncompactness* of M is

$$\alpha(M) := \inf\{\varepsilon \in [0,\infty] : M \text{ has a finite cover by sets with diameter} \le \varepsilon\}.$$

The *Istrăţescu measure of noncompactness* of M is

$$\beta(M) := \sup\{\varepsilon \in [0,\infty] : \text{There are } x_1, x_2, \dots \in M \text{ with } d(x_n, x_m) \ge \varepsilon \text{ for all } n \ne m\}.$$

Note that $\alpha(M)$ is in fact independent of X since the sets in the cover can be chosen to be subsets of M.

We first collect some well-known estimates for the above quantities in the following Proposition 3.23. Only the last of the assertions of Proposition 3.23 seems to be less known. The proof of that last assertion is surprisingly involved, but we will need that assertion several times.

Proposition 3.23. *For $M \subseteq X_0 \subseteq X$, we have*

$$\chi_X(M) \le \chi_{X_0}(M) \le \chi_M(M) \le \beta(M) \le \alpha(M) \le 2\chi_X(M), \tag{3.1}$$

and for each $r > \chi_X(M)$

$$\chi_X(M) = \chi_{B_r(M)}(M). \tag{3.2}$$

Moreover,

$$\begin{aligned}\chi_X(M) = \inf\{\delta \in [0,\infty] : &\textit{There is an } \varepsilon\textit{-net } N \subseteq X \textit{ for } M \\ &\textit{with } \chi_X(N) + \varepsilon \le \delta\},\end{aligned} \tag{3.3}$$

$$\begin{aligned}\alpha(M) = \inf\{\delta \in [0,\infty] : &\textit{There is an } \varepsilon\textit{-net } N \subseteq X \textit{ for } M \\ &\textit{with } \alpha(N) + 2\varepsilon \le \delta\},\end{aligned} \tag{3.4}$$

$$\begin{aligned}\beta(M) = \inf\{\delta \in [0,\infty] : &\textit{There is an } \varepsilon\textit{-net } N \subseteq X \textit{ for } M \\ &\textit{with } \beta(N) + 2\varepsilon \le \delta\},\end{aligned} \tag{3.5}$$

and for each sequence $x_n \in M$ *and each* $\varepsilon > 0$ *there is a subsequence satisfying*

$$d(x_{n_j}, x_{n_k}) < \beta(M) + \varepsilon \quad \textit{for all } j, k \in \mathbb{N}. \tag{3.6}$$

Proof. The first two inequalities of (3.1) are obvious, and the last follows from (3.4) with finite sets $N \subseteq X$.

To prove $\chi_M(M) \le \beta(M)$, we note that $0 < \varepsilon < \chi_M(M)$ implies that there is no finite ε-net for M in M. Hence, we can inductively define a sequence $x_n \in M$ with $d(x_n, x_m) \ge \varepsilon$ for all $m < n$, and so $\beta(M) \ge \varepsilon$.

To prove $\beta(M) \le \alpha(M)$, we note that $0 < \varepsilon < \beta_M(M)$ implies that there is a sequence $x_n \in M$ with $d(x_n, x_m) \ge \varepsilon$ with $n \ne m$. Now for each finitely many sets $M_1, \dots, M_n \subseteq X$ with $M \subseteq M_1 \cup \dots \cup M_n$, at least one of these sets M_k must contain two elements of the sequence x_n and thus satisfy $\operatorname{diam} M_k \ge \varepsilon$. Hence, $\alpha(M) \ge \varepsilon$.

Now we prove (3.2). Thus, let $r > \chi_X(M)$. For each $\rho > \chi_X(M)$ with $\rho < r$, there is a finite ρ-net $N \subseteq X$ for X. Then also $N_0 := N \cap B_r(M) \subseteq B_r(M)$ is a finite ρ-net for M, since for every $x \in M$ there is some $x_0 \in N$ with $d(x, x_0) < \rho < r$. Hence, $x_0 \in N_0$ with $d(x, x_0) < \rho$. This shows $\chi_{B_r(M)}(M) \le \rho$, and so $\chi_{B_r(M)}(M) \le \chi_X(M)$. The converse inequality follows from (3.1), and so (3.2) is established.

Note that "$\ge$" in (3.3), (3.4), and (3.5) follows for the choice $N := M$. For the converse inequalities, let $N \subseteq X$ be an ε-net for M, and we have to show that $\varepsilon_0 > \chi_X(N)$ implies $\chi_X(M) \le \varepsilon_0 + \varepsilon$, $\varepsilon_0 > \alpha(N)$ implies $\alpha(M) \le \varepsilon_0 + 2\varepsilon$, and $\varepsilon_0 > \beta(N)$ implies $\beta(M) \le \varepsilon_0 + 2\varepsilon$.

For the first of these implications, we choose an ε_0-net $N_0 \subseteq X$ for N, and then the triangle inequality implies that N_0 is an $(\varepsilon_0 + \varepsilon)$-net for M. For the second of these implications, we choose finitely many $N_k \subseteq X$ with $N \subseteq N_1 \cup \dots \cup N_n$

and $\operatorname{diam} N_k \le \varepsilon_0$, and then $M_k := B_\varepsilon(N_k)$ cover M and satisfy $\operatorname{diam} M_k \le \operatorname{diam} N_k + 2\varepsilon \le \varepsilon_0 + 2\varepsilon$. For the last of these implications, suppose by contradiction that there is a sequence $x_n \in M$ satisfying $d(x_n, x_m) \ge \varepsilon_0 + 2\varepsilon$ for all $n \ne m$. There are $y_n \in N$ with $d(y_n, x_n) < \varepsilon$, and then Proposition 3.1 implies $d(y_n, y_m) \ge d(x_n, x_m) - 2\varepsilon \ge \varepsilon_0$ for all $n \ne m$ which gives the contradiction $\beta(N) \ge \varepsilon_0$.

For the last assertion, we can assume that $r := \beta(M) + \varepsilon$ is finite and show first that for each sequence $y_n \in M$ there is some n_0 such that $B_r(y_{n_0})$ contains y_n for infinitely many n. Indeed, if such an n_0 would not exist, then we could inductively define a family of infinite subsets $N_k \subseteq \mathbb{N}$ as follows: We put $N_1 = \mathbb{N}$, and if N_k is already defined, we put $n_k := \min N_k$ and observe that by assumption $y_n \in B_r(y_{n_k})$ only for finitely many n. Hence, the set $N_{k+1} \subseteq N_k$ of all $n \in N_k$ satisfying $d(y_n, y_{n_k}) \ge r$ is infinite. Now the infinite sequence $x_k := y_{n_k}$ has for $m > k$ the property that $x_m \in N_m \subseteq N_{k+1}$ and thus $d(x_m, x_k) = d(x_m, y_{n_k}) \ge r$. Thus, the sequence x_k shows that $\beta(M) \ge r$ which is a contradiction.

Now if $x_n \in M$ is any sequence, we define inductively infinite sets $N_k \subseteq \mathbb{N}$ and $n_k \in N_k$ as follows: We put $N_1 := \mathbb{N}$. If N_k is already defined, we consider the subsequence x_n $(n \in N_k)$. By what we have shown above there is some smallest index $n_k \in N_k$ such that $x_n \in B_r(x_{n_k})$ for infinitely many $n \in N_k$. Let N_{k+1} denote the set of all such n with $n > n_k$. Now the sequence n_k has the required property, for if $j > k$ then $n_j \in N_j \subseteq N_{k+1}$ is larger than n_k and satisfies $x_{n_j} \in B_r(x_{n_k})$, that is, $d(x_{n_j}, x_{n_k}) < r$. □

The reason for the terminology "measure of noncompactness" will be explained in a moment. We recall the famous Hausdorff criterion for compactness:

Theorem 3.24 (Hausdorff). *For a subset M of a metric space X the following statements are equivalent:*

(a) *M is relatively compact in X:*

(b) *$\overline{M}$ is compact.*

(c) *Every sequence in M has a subsequence convergent in X.*

(d) *$\overline{M}$ is complete and $\chi_X(M) = 0$.*

Proof. Since X is Hausdorff, the equivalence of the first two statements follows from Corollary 2.47. If $\overline{M}$ is compact and $x_n \in M$ is a sequence, then the family of closed subsets $A_n := \overline{\{x_n, x_{n+1}, \dots\}} \subseteq \overline{M}$ $(n \in \mathbb{N})$ has the finite intersection property. Proposition 2.28 thus implies that there is some $x \in \bigcup_{n=1}^{\infty} A_n$. We put $n_0 := 0$ and choose inductively $n_k > n_{k-1}$ $(k \in \mathbb{N})$ with $d(x, x_{n_k}) < 1/k$. Then x_{n_k} is a convergent subsequence of x_n, and so (c) holds.

Now suppose that (c) holds. If $x_n \in \overline{M}$ is a Cauchy sequence, we choose $y_n \in M$ with $d(x_n, y_n) < 1/n$ and a convergent subsequence $y_{n_k} \to x$. Then $d(x_{n_k}, x) \leq d(y_{n_k}, x) + 1/n_k \to 0$ implies $x_{n_k} \to x$. By Lemma 3.8(a), we obtain $x_n \to x$. Hence, $\overline{M}$ is complete. Moreover, $\chi_X(M) > 0$ would imply $\beta(M) > 0$ by (3.1), and for $0 < \varepsilon < \beta(M)$, we could find a sequence $x_n \in M$ with $d(x_n, x_m) \geq \varepsilon$ for $n \neq m$. This sequence has no Cauchy subsequence and thus no convergent subsequence by Lemma 3.8(a).

It remains to show that (d) implies the compactness of $\overline{M}$. Assume by contradiction that (d) holds but there is an open cover $\mathcal{O}$ of $\overline{M}$ without a finite subcover. Let $N_n \subseteq X$ be a finite $1/n$-net for M. We define now inductively closed sets $C_n \subseteq \overline{M}$ such that C_n is not covered by finitely many sets from $\mathcal{O}$: We put $C_1 := \overline{M}$, and if C_n is already defined, we note that the family $C_{n,x} := K_{2/n}(x) \cap C_n$ $(x \in N_n)$ is a finite cover of C_n, and so there is some $x_n \in N_n$ such that $C_{n+1} := C_{n,x_n}$ is not covered by finitely many sets from $\mathcal{O}$.

In particular, $C_n \neq \varnothing$, and so there is a sequence $y_n \in C_n$ $(n \in \mathbb{N})$. Since $C_1 \supseteq C_2 \supseteq \cdots$ and $\operatorname{diam} C_n \leq 4/n$, it follows that $y_n \in \overline{M}$ is a Cauchy sequence and thus convergent to some $x \in \overline{M}$. Since the sets C_n are closed, we have $x \in C_n$ for every $n \in \mathbb{N}$. There is some $O \in \mathcal{O}$ with $x \in O$, and since O is open and $\operatorname{diam} C_n \leq 4/n$, we have $C_n \subseteq O$ for all sufficiently large $n \in \mathbb{N}$. These C_n are even covered by a single set $O \in \mathcal{O}$, contradicting our choice of C_n. □

Recall that the *completion* of a metric space X is a complete metric space $\overline{X}$ in which X is dense and such that the metric on X is the restriction of the metric of $\overline{X}$. It is well-known (and we will give a convenient proof in Corollary 3.64) that every metric space has a completion. It is also well-known that one can speak about "the" completion of a metric space by the following simple observation:

Proposition 3.25. *The completion of a metric space is unique up to a natural isometry, that is, if $\overline{X}$ and Y are two completions of X then* id_X *extends (uniquely) to an isometric bijection i of $\overline{X}$ onto Y.*

Proof. For $x \in \overline{X}$ there is a sequence $x_n \in X$ with $x_n \to x$ in $\overline{X}$. Then x_n is a Cauchy sequence in Y, and so Lemma 3.8 implies that x_n is convergent in Y to a unique $y \in Y$. If $\hat{x} \in \overline{X}$ and $\hat{x}_n \in X$ satisfies $\hat{x}_n \to \hat{x}$ then also $\hat{x}_n$ is convergent in Y to a unique $\hat{y} \in Y$. The continuity of the metric d in $\overline{X}$ and the metric d_Y in Y implies $d(x_n, \hat{x}_n) \to d(x, \hat{x})$ and $d(x_n, \hat{x}_n) = d_Y(x_n, \hat{x}_n) \to d_Y(y, \hat{y})$. Hence, $d(x, \hat{x}) = d_Y(y, \hat{y})$. In particular, for $x = \hat{x}$ we have $y = \hat{y}$, and so y is actually independent of the particular choice of x_n. Hence, we can define a map $i: \overline{X} \to Y$ by $x \mapsto y$. By what we have shown, we have $d(x, \hat{x}) = d_Y(i(x), i(\hat{x}))$, that is, d is an isometry and in particular one-to-one. To see that i

is onto, we note that as in the beginning of the proof we find for each $y \in Y$ a sequence $x_n \in X$ with $x_n \to y$ in Y which then also converges in $\overline{X}$ to some $x \in \overline{X}$, and so $y = i(x)$. □

Using the notion of completion of a metric space, we can now explain the notion "measure of noncompactness". We collect also some other basic properties in the following result.

Proposition 3.26. *Let X be a metric space, and $\gamma \in \{\chi_X, \alpha, \beta\}$. Then we have for all $M, N \subseteq X$:*

(a) $\gamma(M) = 0$ *if and only if the completion of M is compact.*

(b) *If M is relatively compact in X then $\gamma(M) = 0$. The converse holds if $\overline{M}$ is complete.*

(c) $\gamma(M) = 0$ *implies that M is separable.*

(d) $\gamma(M) < \infty$ *if and only if* $\operatorname{diam} M < \infty$.

(e) $\gamma(M) \le \gamma(N)$ *if* $M \subseteq N \subseteq X$.

(f) $\gamma(M \cup N) = \max\{\gamma(M), \gamma(N)\}$.

(g) $\gamma(\overline{M}) = \gamma(M)$.

Proof. For the first assertion we note that by (3.1) the property $\gamma(M) = 0$ is actually independent of X. Hence, considering the completion of X if necessary, we can assume without loss of generality that X is complete. Then the completion of M is just $\overline{M}$, and so the first assertion follows from the second which in turn is contained in Theorem 3.24. If $M \subseteq X$ satisfies $\gamma(M) = 0$ then $\chi_M(M) = 0$ by (3.1), and this trivially implies that M is separable. The proof of the other assertions is straightforward and left to the reader. □

Corollary 3.27. *If X is a metric space and $M \subseteq X$ is relatively compact then M is separable.*

Proof. If M is relatively compact then $\gamma(M) = 0$ and thus M is separable by Proposition 3.26. □

It is well-known and not hard to see that Lipschitz maps with constant L increase measures of noncompactness at most by the factor L. The following result generalizes this fact even for the case of functions of two variables. In the case $\gamma = \beta$, this is not so trivial as it might appear at a first glance. In fact, we need the tricky last assertion of Proposition 3.23 for the proof in this case.

Proposition 3.28. *Let X, Y and Z be metric spaces and $f\colon X \times Y \to Z$ be such that $f(\cdot, y)$ and $f(x, \cdot)$ are Lipschitz with constant L_1 or L_2 for every $y \in Y$ or $x \in X$, respectively. Then we have for every $M \subseteq X$ and $N \subseteq Y$ the estimate*

$$\gamma\big(f(M \times N)\big) \le L_1\gamma(M) + L_2\gamma(N),$$

where $\gamma \in \{\chi, \alpha, \beta\}$ (in the respective spaces X, Y, or Z).

Note that the assertion depends only on the metric on X, Y, and Z, not on a metric on the product space $X \times Y$.

Proof. For $\varepsilon_1 > \gamma(M)$ and $\varepsilon_2 > \gamma(N)$, put $c := L_1\varepsilon_1 + L_2\varepsilon_2$. We are to show that $\gamma(f(M \times N)) \le c := L_1\varepsilon_1 + L_2\varepsilon_2$. To this end, we note first that

$$d\big(f(x,y), f(x_0,y_0)\big) \le c \quad \text{if } d(x,x_0) \le \varepsilon_1 \text{ and } d(y,y_0) \le \varepsilon_2. \tag{3.7}$$

Indeed, (3.7) follows from $d(f(x,y), f(x_0,y)) \le L_1\varepsilon_1$, $d(f(x_0,y), f(x_0, y_0)) \le L_2\varepsilon_2$, and the triangle inequality.

In case $\gamma = \chi$, there is a finite ε_1-net $N_1 \subseteq X$ for M and a finite ε_2-net $N_2 \subseteq Y$ for N. Then (3.7) implies that $f(N_1 \times N_2)$ is a finite c-net for $f(M \times N)$.

In the case $\gamma = \alpha$, the sets M and N are covered by finitely many sets $M_1, \dots, M_n \subseteq X$ and $N_1, \dots, N_m \subseteq Y$ with $\operatorname{diam} M_k \le \varepsilon_1$ and $\operatorname{diam} N_j \le \varepsilon_2$, respectively. Then the finitely many sets $D_{k,j} := f(M_k \times N_j)$ cover $f(M \times N)$, and satisfy $\operatorname{diam} D_{k,j} \le c$ by (3.7).

In case $\gamma = \beta$, we assume by contradiction that there is a sequence $z_n \in f(M \times N)$ with $d(z_n, z_m) > c$ $(n \ne m)$. Then $z_n = f(x_n, y_n)$ with $(x_n, y_n) \in M \times N$. Passing to a subsequence, we can assume by (3.6) that $d(x_n, x_m) \le \varepsilon_1$ for all $n, m \in \mathbb{N}$. Since $\varepsilon_2 > \beta(N)$, there are $n \ne m$ with $d(y_n, y_m) \le \varepsilon_2$. For this choice, we obtain by (3.7) the contradiction $d(z_n, z_m) \le c$. □

Corollary 3.29. *Let X be a normed space over $K = \mathbb{R}$ or $\mathbb{K} = \mathbb{C}$, and $M \subseteq X$. If $\Lambda \subseteq \mathbb{K}$ is bounded by $r \in [0, \infty)$ then $\gamma(\Lambda M) \le r\gamma(M)$ for all $\gamma \in \{\chi_X, \alpha, \beta\}$.*

Proof. We can assume that $\gamma(M) < \infty$, and so M is bounded by Proposition 3.26. If $M_0 \subseteq X$ is bounded with $M \subseteq M_0$, then $f\colon \Lambda \times M_0 \to X$, defined by $f(t,x) := tx$, is Lipschitz with respect to x with constant $L_2 = r$ and Lipschitz with respect to t with some constant L_1. Since $\gamma(\Lambda) = 0$, we obtain from Proposition 3.28 that $\gamma(\Lambda M) \le r\gamma(M)$ for $\gamma \in \{\alpha, \beta\}$ and $\chi_X(\Lambda M) \le r\chi_{M_0}(M)$. Choosing $M_0 := B_R(M)$ with $R > \chi_X(M)$, we have $r\chi_{M_0}(M) = r\chi_X(M)$ by (3.2). □

Proposition 3.30. *Let X be a normed space over $\mathbb{K} = \mathbb{R}$ or $\mathbb{K} = \mathbb{C}$, and $\gamma \in \{\chi_X, \alpha, \beta\}$. Then we have for every $M, N \subseteq X$ and $\lambda \in \mathbb{K}$:*

(a) $\gamma(\lambda M) = |\lambda|\gamma(M)$.

(b) $\gamma(M+N) \le \gamma(M) + \gamma(N)$.

Proof. Corollary 3.29 with $\Lambda := \{\lambda\}$ implies $\gamma(\lambda M) \le |\lambda|\gamma(M)$. If $\lambda \ne 0$, we apply this inequality with λ^{-1} in place of λ and obtain the converse inequality

$$|\lambda|\gamma(M) = |\lambda|\gamma\big(\lambda^{-1}(\lambda M)\big) \le |\lambda||\lambda^{-1}|\gamma(\lambda M) = \gamma(\lambda M).$$

For the second assertion, we are to show that for all $\varepsilon > \gamma(M)$ and $\varepsilon_0 > \gamma(N)$, we have $\gamma(M+N) \le \varepsilon + \varepsilon_0$.

In the case $\gamma = \chi_X$, there is a finite ε-net $N_0 \subseteq X$ for M and a finite ε_0-net $N_1 \subseteq X$ for M. Then $N_0 + N_1$ is a finite $(\varepsilon + \varepsilon_0)$-net for M, and so $\chi_X(M+N) \le \varepsilon + \varepsilon_0$.

In the case $\gamma = \alpha$, note that M is covered by sets $M_1, \dots, M_n \subseteq X$ with $\operatorname{diam} M_k < \varepsilon$ $(k = 1, \dots, n)$ and N is covered by sets $N_1, \dots, N_m \subseteq X$ with $\operatorname{diam} N_j < \varepsilon_0$ $(j = 1, \dots, m)$. Then the finitely many sets $M_k + N_j$ cover $M + N$ and satisfy

$$\operatorname{diam}(M_k + N_j) \le \operatorname{diam} M_k + \operatorname{diam} N_j \le \varepsilon + \varepsilon_0,$$

and so $\alpha(M+N) \le \varepsilon + \varepsilon_0$.

In the case $\gamma = \beta$, assume by contradiction $\beta(M+N) > \varepsilon+\varepsilon_0$, that is, there is a sequence $z_n \in M + N$ with $\|z_n - z_m\| \ge \varepsilon + \varepsilon_0$ $(n \ne m)$. There are $x_n \in M$ and $y_n \in N$ with $z_n = x_n + y_n$. Since $\varepsilon > \beta(M)$ we can assume by (3.6), passing to a subsequence if necessary, that $\|x_n - x_m\| < \varepsilon$ for all $n, m \in \mathbb{N}$. Since $\varepsilon_0 > \beta(M)$, there are $n \ne m$ with $\|y_n - y_m\| < \varepsilon_0$. For this choice, we find

$$\|z_n - z_m\| \le \|x_n - x_m\| + \|y_n - y_m\| < \varepsilon + \varepsilon_0,$$

contradicting our choice of z_n. □

Lemma 3.31 (Riesz). *Let X be normed and $U \subseteq X$ be a linear subspace. Assume $\overline{U} \ne X$ (or $\dim U < \infty$ and $U \ne X$). Then for each $\varepsilon > 0$ (or $\varepsilon = 0$) there is $x \in X$ with $\|x\| = 1$ and $\operatorname{dist}(x, U) \ge 1 - \varepsilon$.*

Proof. In case $\dim U < \infty$ we will show in Proposition 3.59 that U is closed, that is $\overline{U} = U \ne X$. Hence, in both cases there is $x_0 \in X \setminus \overline{U}$, and so $\delta := \operatorname{dist}(x_0, U) > 0$. There is some $u_0 \in U$ with $\|x_0 - u_0\| \le \delta/(1-\varepsilon)$ (in case $\dim U < \infty$ and $\varepsilon = 0$, we used here Corollary 3.14 and that we will also show in Proposition 3.59 that closed bounded subsets of U are compact.) Then $x := \|x_0 - u_0\|^{-1}(x_0 - u_0)$ has the required property. Indeed, for any $u \in U$, we have in view of $u_1 := u_0 + \|x_0 - u_0\|u \in U$ that $\|x - u\| = \|x_0 - u_0\|^{-1}\|x - u_1\| \ge \|x_0 - u_0\|^{-1}\delta \ge 1 - \varepsilon$. □

Proposition 3.32. *Let X be normed and infinite-dimensional, and $S = S_1(0)$, $K = K_1(0)$. Then the following holds:*

(a) $\beta(S) \geq 1$ *even in the strong sense that there is a sequence $x_n \in S$ with $\|x_n - x_m\| \geq 1$ for $n \neq m$.*

(b) $\chi_X(S) = \chi_X(K) = \chi_K(K) = 1$.

(c) $\alpha(S) = \alpha(K) = 2$.

(d) $\sqrt{2} \leq \chi_S(S) \leq \beta(S) \leq 2$ *if X is an inner product space, that is, if the norm satisfies the parallelogram identity*

$$2(\|x\|^2 + \|y\|^2) = \|x+y\|^2 + \|x-y\|^2 \quad \textit{for all } x, y \in X.$$

Proof. By Lemma 3.31, we can inductively choose x_n with $\|x_n\| = 1$ and $\operatorname{dist}(x_n, U_n) \geq 1$ where U_n is the linear hull of $x_1, \ldots, x_{n-1}$. Then $\|x_n - x_m\| \geq 1$ for all $n > m$ and thus for all $n \neq m$.

Since $\{0\} \subseteq K$ is a finite $(1+\varepsilon)$-net for K for every $\varepsilon > 0$, we have by (3.1) that $\chi_X(S) \leq \chi_X(K) \leq \chi_K(K) \leq 1$. For the converse inequality, assume that $N \subseteq X$ is a finite ε-net for S for some $\varepsilon \in (0,1)$. Let U denote the linear hull of N. Then Lemma 3.31 implies that there is some $x \in S$ with $\operatorname{dist}(x, N) \geq \operatorname{dist}(x, U) \geq 1 > \varepsilon$, a contradiction.

Since $\operatorname{diam} K \leq 2$, we have in view of (3.1) that $\alpha(S) \leq \alpha(K) \leq 2$. Conversely, suppose that S can be decomposed into finitely many sets $M_1, \ldots, M_n$ of diameter less than 2. Let U be an n-dimensional subspace and $S := U \cap S_1(0)$. Then $\overline{M_k} \cap U$ $(k = 1, \ldots, n)$ are closed sets which cover the unit sphere of an n-dimensional space. By the famous Ljusternik–Schnirel'man theorem (we will give a proof in Theorem 9.89(c)), at least one of these sets $\overline{M_k} \cap U$ must contain an antipodal pair and thus has diameter at least 2. Hence, also $\operatorname{diam} M_k \geq 2$ for the corresponding k which is a contradiction.

By (3.1), we obtain $\chi_S(S) \leq \beta(S) \leq \alpha(S) \leq 2$. Conversely, suppose that $\{x_1, \ldots, x_n\}$ is a finite ε-net. Then $S \subseteq M_1 \cup \cdots \cup M_n$ with the closed sets $M_k := S \cap K_\varepsilon(x_k)$. As above, we can use the Ljusternik–Schnirel'man theorem to find that some M_k must contain an antipodal pair, that is, there is some k and some $y \in S$ with $y, -y \in K_\varepsilon(x_k)$. Since $2(\|x_k\|^2 + \|y\|^2) = 4$, the parallelogram identity implies that $\|x_k - y\|^2 \geq 2$ or $\|x_k - (-y)\|^2 \geq 2$. In both cases, we obtain $\varepsilon \geq \sqrt{2}$. Hence, $\chi_S(S) \geq \sqrt{2}$. □

For countable sets, we will occasionally use the following result:

Lemma 3.33. *Let X be a metric space, $C = \{x_1, x_2, \ldots\} \subseteq X$, and $D = \{y_1, y_2, \ldots\} \subseteq X$. If $d(x_n, y_n) \to 0$ then $\gamma(C) = \gamma(D)$ for $\gamma \in \{\chi_X, \alpha, \beta\}$.*

Proof. Put $C_n := \{x_1, \dots, x_n\}$. Let $\varepsilon > 0$. For sufficiently large n, the set $D \cup C_n$ is an ε-net for C, and so (3.3), (3.5) and (3.4) implies $\gamma(C) \le \gamma(D \cup C_n) + 2\varepsilon$. Since Proposition 3.26 shows that $\gamma(D \cup C_n) = \max\{\gamma(D), \gamma(C_n)\} = \gamma(D)$, we thus have shown $\gamma(C) \le \gamma(D) + 2\varepsilon$. Since $\varepsilon > 0$ was arbitrary, we obtain $\gamma(C) \le \gamma(D)$. For symmetry reasons, this implies also $\gamma(D) \le \gamma(C)$. □

3.3 Condensing Maps

Definition 3.34. Let X be a set. A function γ which associates to each $M \subseteq X$ an element from $[0, \infty]$ is called a *set function* on X. A set function γ_X on X is called *monotone* if

$$M \subseteq N \subseteq X \implies \gamma_X(M) \le \gamma_X(N).$$

Practically all set functions considered in applications are monotone. We are mainly interested in the set functions χ_X, α, and β which are all monotone by Proposition 3.26.

Example 3.35. The so-called "inner Hausdorff measure of noncompactness" $\gamma_X(M) := \chi_M(M)$ fails to be monotone if X is an infinite-dimensional inner product space. Indeed, for $S = S_1(0)$ and $K = K_1(0)$ we have $S \subseteq K$, but Proposition 3.32 implies $\gamma_X(S) \ge \sqrt{2} > 1 = \gamma_X(K)$.

Definition 3.36. Let γ_X and γ_Y be set functions on X and Y. Let $L \in [0, \infty]$. Then $\Phi\colon X \multimap Y$ is called $(L, {}^{\gamma_Y}_{\gamma_X})$*-bounded* if

$$\gamma_Y(\Phi(M)) \le L\gamma_X(M) \qquad \text{for all } M \subseteq X, \tag{3.8}$$

or if $L = \infty$. In the special case $\gamma_X(M) = \infty$, we consider (3.8) as satisfied, even if $L = 0$. The smallest such constant $L \in [0, \infty]$ is denoted by $[\Phi]^{\gamma_Y}_{\gamma_X}$.

The smallest such constant exists, because if L_0 is the infimum of all constants $L \in [0, \infty]$ satisfying (3.8) (or $L = \infty$), then (3.8) holds also for $L = L_0$ (or $L_0 = \infty$). The most important examples of maps with this property are the following two:

Proposition 3.37. *Let Y be a metric space, $\gamma_Y \in \{\chi_Y, \alpha, \beta\}$ and $\Phi\colon X \multimap Y$. If Φ is compact then $[\Phi]^{\gamma_Y}_{\gamma_X} = 0$, and the converse holds if Y is complete and $\gamma_X(X) > 0$.*

Proof. This follows from Proposition 3.26. □

Proposition 3.38. *If X and Y are metric spaces and $F: X \to Y$ is Lipschitz with constant L then $[F]_{\chi_X}^{\chi_Y} \le L$, $[F]_\alpha^\alpha \le L$, and $[F]_\beta^\beta \le L$.*

Proof. This is a special case of Proposition 3.28. □

Proposition 3.38 shows the main disadvantage of the Hausdorff measure of noncompactness: While α and β do not depend on the space, and so one also has a corresponding estimate if $F: X_0 \to Y$ with $X_0 \subseteq X$, this is not the case for χ_X. For χ_X, one really must require that F is defined (and Lipschitz) on the whole space X as the following example shows:

Example 3.39. Let X be an infinite-dimensional inner product space, $Y := S_1(0)$, and $F = \mathrm{id}_Y$. Then F is Lipschitz with constant 1, but $[F]_{\chi_X}^{\chi_Y} \ge \sqrt{2}$, since $\chi_X(Y) = 1 < \sqrt{2} \le \chi_Y(Y)$ by Proposition 3.32.

Even if one cannot obtain good Lipschitz constants for a map F, one can often obtain much better estimates for the quantity $[F]_{\gamma_X}^{\gamma_Y}$. Here is an example which is a key to many further examples.

Lemma 3.40. *If X is a normed space, $r \in [0, \infty)$, and $\nu_r: X \setminus \{0\} \to S_r(0)$ is defined by $\nu_r(x) := rx/\|x\|$ then*

$$[\nu_r|_{X \setminus B_\rho(0)}]_\gamma^\gamma \le r/\rho,$$

for all $\gamma \in \{\chi_X, \alpha, \beta\}$ and $\rho > 0$. Moreover, we have equality unless $\dim X < \infty$.

Proof. Let $M \subseteq X \setminus B_\rho(0)$. Then $\rho_r(M) \subseteq [0, r/\rho]M$ implies by Proposition 3.26 and Corollary 3.29 that $\gamma(\rho_r(M)) \le \gamma([0, r/\rho]M) \le (r/\rho)\gamma(M)$. For $M = S_\rho(0)$, we have equality by Proposition 3.30. □

Proposition 3.41. *In a normed space X the radial retraction $\rho_r: X \to K_r(0)$, defined by*

$$\rho_r(x) := \begin{cases} x & \text{if } x \in K_r(0), \\ r\dfrac{x}{\|x\|} & \text{otherwise,} \end{cases}$$

satisfies $[\rho]_\gamma^\gamma \le 1$ for $\gamma \in \{\chi_X, \alpha, \beta\}$. Moreover, we have equality unless $\dim X < \infty$.

Proof. We use the notation of Lemma 3.40. For any $M \subseteq X$, we have $\rho_r(M) = M_1 \cup M_2$ with $M_1 = M \cap K_r(0)$ and $M_2 = \nu_r(M \setminus K_r(0))$. Hence, Proposition 3.26 and Lemma 3.40 imply

$$\gamma\big(\rho_r(M)\big) = \max\{\gamma(M_1), \gamma(M_2)\} \le \gamma(M).$$

For $M = K_r(0)$, we have equality. □

Note that in most spaces the function ρ_r fails to be Lipschitz with constant 1. Of particular importance is the case that the inequality in (3.8) is strict.

Definition 3.42. If the inequality in (3.8) is even strict in case $\gamma_X(\Phi(M)) > 0$ then Φ is called *$(L, \frac{\gamma_Y}{\gamma_X})$-condensing*.

The reader be warned that the case $\gamma_X(M) = \infty$ often has to be considered separately in the above definition so that *not* every $(L_0, \frac{\gamma_Y}{\gamma_X})$-bounded map Φ is automatically $(L, \frac{\gamma_Y}{\gamma_X})$-condensing for $L < L_0$.

More precisely, we have the following result:

Proposition 3.43. *Let γ_X and γ_Y be monotone set functions on X and Y. Suppose that $\Phi: X \multimap Y$ is $(L_0, \frac{\gamma_Y}{\gamma_X})$-bounded and $0 \le L < L_0$. Then the following statements are equivalent:*

(a) *Φ is $(L, \frac{\gamma_Y}{\gamma_X})$-condensing.*

(b) *$\gamma_X(X) < \infty$ or $\gamma_Y(\Phi(X)) < \infty$.*

Proof. By the monotonicity, we have $\gamma_Y(\Phi(M)) \le \gamma_Y(\Phi(X))$ and $\gamma_X(M) \le \gamma_X(X)$ for every $M \subseteq X$. Hence, if the second case holds and $\gamma_X(M) = \infty$ then $\gamma_Y(\Phi(M)) < \infty$. □

In literature, also the opposite inequality to (3.8) plays an important role. Since we will only need the opposite inequality for single-valued maps, it turns out that we do not need a separate definition for this case, since we can deal with multivalued inverse maps:

Proposition 3.44. *Let γ_X and γ_Y be set functions on X and Y, and $F: X \to Y$. If γ_Y is monotone then*

$$\gamma_Y(F(M)) \ge L\gamma_X(M) \qquad \text{for all } M \subseteq X \tag{3.9}$$

implies that $F^{-1}: Y \multimap X$ is $(L^{-1}, \frac{\gamma_X}{\gamma_Y})$-bounded. If γ_X is monotone, we have the converse implication.

In case $L > 0$ we also have an analogous assertion for strict inequalities: If γ_Y is monotone then strict inequality in (3.9) for $\gamma_X(M) > 0$ implies that F^{-1} is $(L^{-1}, \frac{\gamma_X}{\gamma_Y})$-condensing. If γ_X is monotone, the converse implication holds.

Proof. F^{-1} is $(K, \frac{\gamma_X}{\gamma_Y})$-bounded if and only if

$$\gamma_X\big(F^{-1}(N)\big) \leq K\gamma_Y(N) \quad \text{for all } N \subseteq Y, \tag{3.10}$$

when we consider the inequality as satisfied for $\gamma_Y(N) = \infty$ or $K = \infty$. Moreover, F^{-1} is $(K, \frac{\gamma_X}{\gamma_Y})$-condensing if and only if the inequality is even strict for positive left-hand side.

If this holds and $M \subseteq X$, we put $N := F(M)$ in (3.10). Since $M \subseteq F^{-1}(N)$, we find for monotone γ_X that

$$\gamma_X(M) \leq \gamma_X\big(F^{-1}(N)\big) \leq K\gamma_Y\big(F(M)\big),$$

the last inequality being strict if $\gamma_X(M) > 0$ and F^{-1} is $(K, \frac{\gamma_X}{\gamma_Y})$-condensing. Hence, (3.9) holds with $L = K^{-1}$, with strict inequality if $K < \infty$, $\gamma_X(M) > 0$, and F^{-1} is $(K, \frac{\gamma_X}{\gamma_Y})$-condensing.

Conversely, if (3.9) holds and $N \subseteq Y$, we put $M := F^{-1}(N)$. Then $F(M) \subseteq N$, and if γ_Y is monotone, we thus obtain

$$\gamma_Y(N) \geq \gamma_Y\big(F(M)\big) \geq L\gamma_X\big(F^{-1}(N)\big).$$

Hence, (3.10) holds with $K = L^{-1}$. In case of strict inequality and $L > 0$ also the inequality in (3.10) is strict. □

If the operator Φ involves integration of vector functions, estimates like (3.8) or (3.9) are usually hard to obtain: One can often obtain such estimates only for countable subsets, see e.g. [76], [109] or [138, §11 and §12]. For this reason, we introduce "countable" variants of Definition 3.36.

Definition 3.45. Let γ_X and γ_Y be set functions on X or Y, respectively. For $L \in [0, \infty]$, a map $\Phi\colon X \multimap Y$ is called $(L, \frac{\gamma_Y}{\gamma_X^c})$-*bounded* or $(L, \frac{\gamma_Y^c}{\gamma_X^c})$-*bounded* if (3.8) holds for all countable sets $M \subseteq X$ or if

$$\gamma_Y(C) \leq L\gamma_X(M) \quad \text{for all countable } M \subseteq X \text{ and countable } C \subseteq \Phi(M), \tag{3.11}$$

respectively. We denote the corresponding minimal $L \in [0, \infty]$ by $[\Phi]_{\gamma_X^c}^{\gamma_Y}$ and $[\Phi]_{\gamma_X^c}^{\gamma_Y^c}$, respectively, and also define $(L, \frac{\gamma_Y}{\gamma_X^c})$-*condensing* and $(L, \frac{\gamma_Y^c}{\gamma_X^c})$-*condensing* maps by requiring in addition that the corresponding inequalities be strict if the left-hand side is positive.

The notation "c" in γ_X^c and γ_Y^c should indicate that we consider only countable subsets of X and Y, respectively.

The reader might have observed that one case seems to be missing: This is the case which one obtains by dropping the hypothesis that M be countable in (3.11). However, this case actually gives nothing new:

Proposition 3.46. *If γ_X is monotone then* (3.11) *is equivalent to*

$$\gamma(C) \leq L\gamma(M) \quad \textit{for all } M \subseteq X \textit{ and countable } C \subseteq \Phi(M). \tag{3.12}$$

Proof. If $C \subseteq \Phi(M)$ is countable for some $M \subseteq X$, then there is a countable $M_0 \subseteq M$ with $C \subseteq \Phi(M_0)$. Hence, if L satisfies (3.11) then $\gamma_Y(C) \leq L\gamma_X(M_0) \leq L\gamma_X(M)$, and so (3.12) holds. The converse implication is trivial. □

Proposition 3.47. (a) *Each $(L, \frac{\gamma_Y}{\gamma_X})$-bounded map is $(L, \frac{\gamma_Y}{\gamma_X^c})$-bounded. If γ_Y is monotone then each $(L, \frac{\gamma_Y}{\gamma_X^c})$-bounded map is $(L, \frac{\gamma_Y^c}{\gamma_X^c})$-bounded.*

(b) *Each $(L, \frac{\gamma_Y}{\gamma_X})$-condensing map is $(L, \frac{\gamma_Y}{\gamma_X^c})$-condensing. If γ_Y is monotone then each $(L, \frac{\gamma_Y}{\gamma_X^c})$-condensing map is $(L, \frac{\gamma_Y^c}{\gamma_X^c})$-condensing.*

In particular,

$$[\Phi]_{\gamma_X^c}^{\gamma_Y^c} \leq [\Phi]_{\gamma_X^c}^{\gamma_Y} \leq [\Phi]_{\gamma_X}^{\gamma_Y}, \tag{3.13}$$

where we require for the first inequality that γ_Y is monotone.

Proof. If (3.12) holds then the monotonicity of γ_Y implies for each countable $C \subseteq \Phi(M)$ that $\gamma_X(M) \leq L\gamma_Y(C) \leq L\gamma_Y(\Phi(M))$. The other assertions are shown similarly. □

Proposition 3.48. *Let γ_X, γ_Y, and γ_Y be set functions on X, Y, and Z, respectively. Then for any $\Phi\colon X \multimap Y$ and $\Psi\colon Y \multimap Z$:*

$$[\Psi \circ \Phi]_{\gamma_X}^{\gamma_Z} \leq [\Psi|_{\Phi(X)}]_{\gamma_Y}^{\gamma_Z} [\Phi]_{\gamma_X}^{\gamma_Y}$$

$$[\Psi \circ \Phi]_{\gamma_X^c}^{\gamma_Z} \leq [\Psi|_{\Phi(X)}]_{\gamma_Y}^{\gamma_Z} [\Phi]_{\gamma_X^c}^{\gamma_Y^c}$$

$$[\Psi \circ \Phi]_{\gamma_X^c}^{\gamma_Z^c} \leq [\Psi|_{\Phi(X)}]_{\gamma_Y^c}^{\gamma_Z^c} [\Phi]_{\gamma_Y^c}^{\gamma_X^c}$$

Proof. The proof is straightforward from the definitions. □

For our particular three measures of noncompactness, we can compare all these quantities up to the factor 2:

Proposition 3.49. *Let X and Y be metric spaces, and $\Phi\colon X \multimap Y$. Then we have for all monotone set functions γ_X and γ_Y on X and Y that*

$$[\Phi]_{\gamma_X}^{\chi_Y} \le [\Phi]_{\gamma_X}^{\beta} \le [\Phi]_{\gamma_X}^{\alpha} \le 2[\Phi]_{\gamma_X}^{\chi_Y} \tag{3.14}$$

$$[\Phi]_{\alpha}^{\gamma_Y} \le [\Phi]_{\beta}^{\gamma_Y} \le [\Phi]_{\chi_X}^{\gamma_Y} \le 2[\Phi]_{\alpha}^{\gamma_Y} \tag{3.15}$$

$$[\Phi]_{\gamma_X^c}^{\chi_Y} \le [\Phi]_{\gamma_X^c}^{\beta} \le [\Phi]_{\gamma_X^c}^{\alpha} \le 2[\Phi]_{\gamma_X^c}^{\chi_Y} \tag{3.16}$$

$$[\Phi]_{\alpha^c}^{\gamma_Y} \le [\Phi]_{\beta^c}^{\gamma_Y} \le [\Phi]_{\chi_X^c}^{\gamma_Y} \le 2[\Phi]_{\alpha^c}^{\gamma_Y} \tag{3.17}$$

$$[\Phi]_{\gamma_X^c}^{\chi_Y^c} \le [\Phi]_{\gamma_X^c}^{\beta^c} \le [\Phi]_{\gamma_X^c}^{\alpha^c} \le 2[\Phi]_{\gamma_X^c}^{\chi_Y^c} \tag{3.18}$$

$$[\Phi]_{\alpha^c}^{\gamma_Y^c} \le [\Phi]_{\beta^c}^{\gamma_Y^c} \le [\Phi]_{\chi_X^c}^{\gamma_Y^c} \le 2[\Phi]_{\alpha^c}^{\gamma_Y^c} \tag{3.19}$$

Suppose in addition that $\gamma_Y \in \{\chi_Y, \alpha, \beta\}$. Then

$$[\Phi]_{\gamma_X^c}^{\gamma_Y^c} \le [\Phi]_{\gamma_X^c}^{\gamma_Y} \le [\Phi]_{\gamma_X}^{\gamma_Y} \le 2[\Phi]_{\gamma_X^c}^{\gamma_Y^c} \tag{3.20}$$

and moreover:

(a) *If $\Phi(X)$ is separable or $\gamma_Y = \beta$ then*

$$[\Phi]_{\gamma_X^c}^{\gamma_Y^c} = [\Phi]_{\gamma_X^c}^{\gamma_Y} = [\Phi]_{\gamma_X}^{\gamma_Y}. \tag{3.21}$$

(b) *If Φ is upper semicontinuous in the uniform sense and X is separable then*

$$[\Phi]_{\gamma_X^c}^{\gamma_Y} = [\Phi]_{\gamma_X}^{\gamma_Y}.$$

If additionally $\Phi(x)$ is separable for every $x \in X$ then (3.21) *holds.*

(c) *Either all of the quantities $[\Phi]_{\gamma_X}^{\gamma_Y}$, $[\Phi]_{\gamma_X^c}^{\gamma_Y}$, $[\Phi]_{\gamma_X^c}^{\gamma_Y^c}$ with $\gamma_X \in \{\chi_X, \alpha, \beta\}$ and $\gamma_Y \in \{\chi_Y, \alpha, \beta\}$ are finite or none, and if they are finite then* $\operatorname{diam} \Phi(M) < \infty$ *for each $M \subseteq X$ with* $\operatorname{diam} M < \infty$.

Proof. The inequalities (3.14)–(3.19) follow immediately from (3.1), and the first two inequalities of (3.20) follow from (3.13). The latter implies that, in order to prove (3.21), we have to prove only $[\Phi]_{\gamma_X^c}^{\gamma_Y^c} \le [\Phi]_{\gamma_X}^{\gamma_Y}$.

If $\Phi(X)$ is separable and $M \subseteq X$ then $\Phi(M)$ is separable by Corollary 3.17, and so there is a countable $C \subseteq \Phi(M)$ with $\Phi(M) \subseteq \overline{C}$. By Propositions 3.26 and 3.46, we obtain

$$\gamma_Y\bigl(\Phi(M)\bigr) = \gamma_Y(C) \le [\Phi]_{\gamma_X^c}^{\gamma_Y^c}\gamma(M).$$

If $\gamma_Y = \beta$ and $M \subseteq X$, put $\beta_0 := \beta(\Phi(M))$. Assume first $\beta_0 > 0$. For each $c < \beta_0$ there is a sequence $x_n \in \Phi(M)$ with $d(x_n, x_m) \geq c$ $(n \neq m)$. Putting $C := \{x_1, x_2, \ldots\}$, we have $\beta(C) \geq c$. By Propositions 3.26 and 3.46, we obtain

$$c \leq \beta(C) \leq [\Phi]_{\gamma_X^c}^{\beta^c} \gamma(M).$$

Since $c < \beta_0$ was arbitrary, we obtain (also in case $\beta_0 = 0$) that

$$\beta\big(\Phi(M)\big) = \beta_0 \leq [\Phi]_{\gamma_X^c}^{\beta^c} \gamma(M).$$

Hence, also in this case $[\Phi]_{\gamma_X^c}^{\gamma_Y^C} \leq [\Phi]_{\gamma_X}^{\gamma_Y}$. We thus have established (a) in all cases. Combining (3.21) in the special case $\gamma_X = \beta$ with (3.14) and (3.18), we obtain the last inequality of (3.20).

To prove (b), let $M \subseteq X$. Corollary 3.17 and Proposition 3.20 imply that there is a countable $M_0 \subseteq M$ with $\Phi(M) \subseteq \overline{\Phi(M_0)}$. Hence, Proposition 3.26 implies

$$\gamma_Y\big(\Phi(M)\big) \leq \gamma_Y\big(\Phi(M_0)\big) \leq [\Phi]_{\gamma_X^c}^{\gamma_Y} \gamma_X(M_0) \leq [\Phi]_{\gamma_X^c}^{\gamma_Y} \gamma_X(M).$$

This proves $[\Phi]_{\gamma_X}^{\gamma_Y} \leq [\Phi]_{\gamma_X^c}^{\gamma_Y}$, and the converse inequality comes from (3.20). If additionally $\Phi(x)$ is separable for every $x \in X$ then $\Phi(X)$ is separable by Corollary 3.21, and so (a) applies.

Finally, the first assertion of (c) follows by combining the previous inequalities, and in case $[\Phi]_{\gamma_X}^{\gamma_Y} < \infty$, the second assertion follows from Proposition 3.26. □

For our three particular measures of noncompactness, we have also a natural compatibility with algebraic operations:

Proposition 3.50. *Let X and Y be normed spaces over $\mathbb{K} = \mathbb{R}$ or $\mathbb{K} = \mathbb{C}$. Then for all $\Phi, \Psi\colon X \multimap Y$, $\lambda \in \mathbb{K}$, and $\gamma_Y \in \{\chi_Y, \alpha, \beta\}$ and any set function γ_X on X there holds:*

(a) $[\Phi + \Psi]_{\gamma_X}^{\gamma_Y} \leq [\Phi]_{\gamma_X}^{\gamma_Y} + [\Psi]_{\gamma_X}^{\gamma_Y}$.

(b) $[\Phi + \Psi]_{\gamma_X^c}^{\gamma_Y} \leq [\Phi]_{\gamma_X^c}^{\gamma_Y} + [\Psi]_{\gamma_X^c}^{\gamma_Y}$.

(c) $[\Phi + \Psi]_{\gamma_X^c}^{\gamma_Y^c} \leq [\Phi]_{\gamma_X^c}^{\gamma_Y^c} + [\Psi]_{\gamma_X^c}^{\gamma_Y^c}$.

(d) $[\lambda\Phi]_{\gamma_X}^{\gamma_Y} = |\lambda|[\Phi]_{\gamma_X}^{\gamma_Y}$.

(e) $[\lambda\Phi]_{\gamma_X^c}^{\gamma_Y} = |\lambda|[\Phi]_{\gamma_X^c}^{\gamma_Y}$.

(f) $[\lambda\Phi]_{\gamma_X^c}^{\gamma_Y^c} = |\lambda|[\Phi]_{\gamma_X^c}^{\gamma_Y^c}$.

Proof. The assertions follow straightforwardly from Proposition 3.30. □

Recall that a map $A\colon X \multimap Y$ between real or complex vector spaces X and Y is *positively homogeneous* if $A(\lambda x) = \lambda A(x)$ for all $\lambda \geq 0$.

Theorem 3.51. *Let X and Y be normed spaces. For positively homogeneous A: $X \multimap Y$, $\gamma_X \in \{\chi_X, \alpha, \beta\}$, and $\gamma_Y \in \{\chi_X, \alpha, \beta\}$ we have*

$$[A]_{\gamma_X}^{\gamma_Y} = [A|_{S_r(0)}]_{\gamma_X}^{\gamma_Y}$$
$$[A]_{\gamma_X^c}^{\gamma_Y} = [A|_{S_r(0)}]_{\gamma_X^c}^{\gamma_Y}$$
$$[A]_{\gamma_X^c}^{\gamma_Y^c} = [A|_{S_r(0)}]_{\gamma_X^c}^{\gamma_Y^c}$$

for all $r \in (0, \infty)$. If some of these quantities is finite then A is upper semicontinuous at 0.

We point out that an analogue of Theorem 3.51 for condensing maps is unknown, in general, even in many simple cases. We refer to the discussion in [87] for some particular cases.

Proof. We have to show "$\leq$", and to this end, we can assume that r is fixed and that the respective right-hand side $S = [A|_{S_r(0)}]_{\gamma_X}^{\gamma_Y}$, $S = [A|_{S_r(0)}]_{\gamma_X^c}^{\gamma_Y}$, or $S = [A|_{S_r(0)}]_{\gamma_X^c}^{\gamma_Y^c}$ is finite. Proposition 3.49(c) implies in particular that $A(S_r(0))$ is bounded by some $N \in (0, \infty)$. Since A is positively homogeneous, it follows that $A(S_\rho(0))$ is bounded by ρN, and so

$$A\big(K_\rho(0)\big) \subseteq K_{\rho N}(0) \quad \text{for all } \rho \geq 0. \tag{3.22}$$

This implies that A is upper semicontinuous at 0. Let now $M \subseteq X$ be arbitrary or countable, respectively. Since the inequalities (3.8) or (3.11) are trivial if $\gamma_X(M) = \infty$, we can assume by Proposition 3.26 that $M \subseteq B_R(0)$ for some $R \in (0, \infty)$. Let also $C = A(M)$ be $C \subseteq A(M)$ be countable, respectively.

For $\varepsilon \in (0, 1)$, put $r_n := \varepsilon(1 + \varepsilon)^{n-1}$. Fix some n with $r_n > R$. We cut M into finitely many slices $M_k := \{x \in M : \|x\| \in (r_{k-1}, r_k]\}$ $(k = 1, \dots, n)$ and $M_0 := M \cap K_\varepsilon(0)$. We also divide C correspondingly into disjoint sets $C_0, \dots, C_n$ such that $C_k \subseteq A(M_k)$ $(k = 0, \dots, n)$. For $k = 1, \dots, n$, we put $I_k := [r_{k-1}, r_k]$. Since A is positively homogeneous, we find for $\nu(x) := x/\|x\|$ in view of

$$C_k \subseteq A\big(I_k \nu(M_k)\big)$$

that

$$C_k \subseteq I_k A_k \quad \text{with appropriate } A_k \subseteq A\big(\nu(M_k)\big).$$

If C_k is countable, we can also choose A_k countable in this inclusion. Hence, we obtain for $k = 1, \dots, n$ by Proposition 3.26, Corollary 3.29, and Lemma 3.40 that

$$\begin{aligned}\gamma_Y(C_k) &\leq \gamma_Y(I_k A_k) \leq r_k \gamma_Y(A_k) \leq r_k S \gamma_X(\nu(M_k)) \\ &\leq \frac{r_k}{r_{k-1}} S \gamma_X(M_k) \leq (1+\varepsilon) S \gamma_X(M).\end{aligned}$$

Moreover, since (3.22) implies $C_0 \subseteq A(K_\varepsilon(0)) \subseteq K_\varepsilon(0)$, we find by (3.3), (3.4), or (3.5) that $\gamma_Y(C_0) \leq 2\varepsilon$. Hence, Proposition 3.26 implies

$$\gamma_Y(C) = \max\{\gamma_Y(C_k) : k = 0, \dots, n\} \leq \max\{(1+\varepsilon) S \gamma_X(M), 2\varepsilon\}.$$

Since $\varepsilon > 0$ was arbitrary, we obtain $\gamma(C) \leq S \gamma_X(M)$, and so the assertion follows. □

Recall that a *Banach space* is a complete normed space.

Corollary 3.52. *For positively homogeneous $A\colon X \multimap Y$ in normed spaces X and Y the following statements are equivalent.*

(a) $\gamma(A(S_r(0))) = 0$ *for some* $r > 0$ *and some* $\gamma \in \{\chi_X, \alpha, \beta\}$.

(b) $\gamma(A(M)) = 0$ *for all bounded* $M \subseteq X$ *and all* $\gamma \in \{\chi_X, \alpha, \beta\}$.

(c) *(If Y is a Banach space.) A is locally compact.*

In this case, A is upper semicontinuous at 0.

Proof. The assertion follows from Theorem 3.51 and Proposition 3.26. □

Recall that a map $A\colon X \to Y$ between vector spaces over $\mathbb{K}$ is *additive* if $A(x + y) = A(x) + A(y)$ for all $x, y \in X$. It is *linear* if additionally $A(\lambda x) = \lambda A(x)$ for all $x \in X$ and all $\lambda \in \mathbb{K}$.

Proposition 3.53. *Let X and Y be normed spaces. If $A\colon X \multimap Y$ is positively homogeneous and upper semicontinuous at* 0 *then there is a finite constant $C \geq 0$ with*

$$\|y\|_Y \leq C \|x\|_X \quad \text{for all } x \in X,\ y \in A(x). \tag{3.23}$$

If $A\colon X \to Y$ is linear with (3.23), *that is*

$$\|A(x)\| \leq C \|x\| \quad \text{for all } x \in X,$$

then A is Lipschitz continuous on X with constant C.

Proof. Since $V = K_1(0) \subseteq Y$ is a neighborhood of $A(0)$, the set $A^-(V)$ is a neighborhood of 0 and thus contains $K_r(0) \subseteq X$ for some $r > 0$. If $x \neq 0$ and $y \in A(x)$, we put $\lambda := r/\|x\|_X$. Then we have $\lambda x \in K_r(0)$ and thus $\lambda y \in A(\lambda x) \subseteq K_1(0)$ which implies $\|y\|_Y \leq \lambda^{-1} = \|x\|_X/r$. Hence, (3.23) holds with $C := 1/r$. Conversely, if $A\colon X \to Y$ is linear with (3.23) then $\|A(x) - A(y)\| = \|A(x - y)\| \leq C\|x - y\|$ implies that A is Lipschitz with constant C. □

3.4 Convexity

By a *topological vector space* we mean a vector space X over $\mathbb{K} = \mathbb{R}$ or $\mathbb{K} = \mathbb{C}$, endowed with a topology such that addition $+\colon X \times X \to X$ and scalar multiplication $\cdot\colon \mathbb{K} \times X \to X$ are continuous operations. A *locally convex space* is a topological vector space with the property that some (hence: every) point of the space has a neighborhood basis consisting of convex sets. Proposition 3.3 implies that every normed space is a locally convex space.

For a subset M of a topological vector space X, we define $\overline{\operatorname{conv}}\, M$ as the intersection of all closed convex subsets of X containing M. Then $\overline{\operatorname{conv}}\, M$ is closed and convex by Proposition 2.51 and thus the smallest closed convex subset of X containing M.

Proposition 3.54. *We have $\overline{\operatorname{conv}}\, M \supseteq \overline{\operatorname{conv} M}$, and equality holds in locally convex spaces.*

Proof. Since $\overline{\operatorname{conv}}\, M$ is closed, the inclusion follows from $\operatorname{conv} M \subseteq \overline{\operatorname{conv}}\, M$. To prove the converse inclusion, it suffices to show that $M_0 := \overline{\operatorname{conv} M}$ is convex. Thus, let $x, y \in M_0$, and $z = tx + (1 - t)y$ for some $t \in [0, 1]$. Since X is locally convex, it suffices to show that for every convex neighborhood $U \subseteq X$ of 0 there is some $z_0 \in \operatorname{conv} M$ with $z_0 \in z + U$. There are $x_0, y_0 \in \operatorname{conv} M$ with $x_0 \in x + U$ and $y_0 \in y + U$. Then the convexity of U implies that $z_0 := tx_0 + (1 - t)y_0 \in z + U$. □

Proposition 3.55. *Suppose that $\overline{\operatorname{conv}}\, M$ is a metrizable subset of a locally convex space. Then for $A \subseteq \overline{\operatorname{conv}}\, M$ the following statements are equivalent:*

(a) *A is separable.*

(b) *There is a countable $C \subseteq M$ with $A \subseteq \overline{\operatorname{conv}}\, C$.*

Proof. Assume first that $A = \overline{\{x_1, x_2, \ldots\}}$. For each $n = 1, 2, \ldots$, we have $x_n \in \overline{\operatorname{conv}}\, M$, and so Proposition 3.54 implies that there is a sequence $y_{n,k} \in \operatorname{conv} M$ with $y_{n,k} \to x_n$ as $k \to \infty$. With the notation of Proposition 2.52, we find for each n and k some number $m(n,k)$ with $y_{n,k} \in \operatorname{conv}_{m(n,k)}(M)$. Hence, there is a set $C_{n,k} \subseteq M$ consisting of $m(n,k)$ points with $y_{n,k} \in \operatorname{conv}_{m(n,k)} C_{n,k} \subseteq \operatorname{conv} C_{n,k}$. If C denotes the union of all sets $C_{n,k}$, we thus have $y_{n,k} \in \operatorname{conv} C$ for all n and k, and so Proposition 3.54 implies $x_n \in \overline{\operatorname{conv} C} = \overline{\operatorname{conv}} C$ for all n. It follows that $C \subseteq M$ is countable with $A \subseteq \overline{\operatorname{conv}} C$.

Conversely, suppose that there is a countable $C \subseteq M$ with $A \subseteq \overline{\operatorname{conv}} C$. For $n = 1, 2, \ldots$, let C_n denote the set of all points $\sum_{k=1}^{n} \lambda_k x_k$ with rational $\lambda_k \geq 0$, $\sum_{k=1}^{n} \lambda_k = 1$, and $x_k \in M$. Then C_n is countable, and with the notation of Proposition 2.52, we have $\overline{C}_n = \overline{\operatorname{conv}_n(C)}$ and thus $\operatorname{conv} C \subseteq \bigcup_{n=1}^{\infty} \overline{C_n}$. Let A_0 denote the union of the countably many countable sets C_n. Then $\operatorname{conv} C \subseteq \overline{A}_0$, and so Proposition 3.54 implies that $A \subseteq \overline{A}_0$. Corollary 3.17 thus implies that A is separable. □

An *isomorphism* between topological vector spaces X and Y is a linear homeomorphism of X onto Y. If such an isomorphism exists, we call X and Y *isomorphic*.

Proposition 3.56. *Let X and Y be normed spaces. A linear $J: X \to Y$ is an isomorphism of X onto the subspace $J(X) \subseteq Y$ if and only if*

$$c\|x\|_X \leq \|Jx\|_Y \leq C\|x\|_X \quad \textit{for all } x \in X. \tag{3.24}$$

If this is the case and X is a Banach space then $J(X)$ is a Banach space and thus closed in Y.

Proof. If J is a homeomorphism onto $J(X)$ then $J^{-1}: J(X) \to X$ and $J: X \to Y$ are continuous which implies the two inequalities in (3.24) by Proposition 3.53. Conversely, (3.24) holds then

$$c\|x - \widetilde{x}\|_X \leq \|Jx - J\widetilde{x}\|_Y \tag{3.25}$$

implies that J is one-to-one, that is, $J^{-1}: J(X) \to X$, and the continuity of J and J^{-1} follows by Proposition 3.53. If this holds and X is a Banach space, let $y_n = Jx_n$ be a Cauchy sequence in $J(X)$. Then (3.25) implies that x_n is a Cauchy sequence in X and thus convergent to some $x \in X$. The continuity of J implies $y_n \to Jx$, and so $J(X)$ is complete. □

In the following assertion, we equip $\mathbb{K}^n$ with the *sum norm*

$$\|(x_1, \ldots, x_n)\| := |x_1| + \cdots + |x_n|,$$

but we will see in a moment that we could have chosen also every other norm.

Proposition 3.57. *Let X be a vector space over $\mathbb{K} = \mathbb{R}$ or $\mathbb{K} = \mathbb{C}$ of finite dimension n. Then there is a map $J: \mathbb{K}^n \to X$ which is an isomorphism whenever X is equipped with a topology which turns it into a topological Hausdorff vector space.*

Proof. If $e_1, \dots, e_n$ is a basis of X, we define J by $J(x_1, \dots, x_n) := x_1 e_1 + \dots + x_n e_n$. Then J is linear, one-to-one and onto, and we have to show that J is a homeomorphism if X is a topological Hausdorff vector space. The continuity of J is obvious. We are to show that $J(B_r(x_0))$ is a neighborhood of $y_0 := J(x_0)$ for each $x_0 \in \mathbb{K}^n$ and each $r > 0$.

To this end, we put $S := S_r(x_0) \subseteq \mathbb{K}^n$. Since S is bounded, we have clearly $\chi_X(S) = 0$. Hence S is relatively compact by Theorem 3.24. Since S is closed and thus complete, it is actually compact. Proposition 2.100 implies that $J(S)$ is compact and thus closed by Proposition 2.45. Since $y_0 \notin J(S)$, the set $U := X \setminus J(S)$ is an open neighborhood of y_0. Since X is a topological vector space, there is a neighborhood $V \subseteq X$ of 0 and $\varepsilon > 0$ with $y_0 + sV \subseteq U$ for all $s \in [0, \varepsilon]$.

We are to show that $J(B_r(x_0))$ contains $W := y_0 + \varepsilon V$ which implies the claim since W is a neighborhood for $y_0 = J(x_0)$. Thus, let $y_1 \in W$, that is, $y_1 = y_0 + \varepsilon v$ for some $v \in V$. Put $x_1 := J^{-1}(y_1)$, and $x_t := x_0 + t(x_1 - x_0)$, $y_t := J(x_t)$ for $0 < t < 1$. Then $y_t = y_0 + t\varepsilon v \in U$ and thus $x_t \notin S$ for all $t \in [0, 1]$. It follows that $x_1 \in B_r(x_0)$ since otherwise there would be a convex combination x_t of x_0 and x_1 which lies on S. Hence, $y_1 = J(x_1) \in J(B_r(x_0))$. □

Corollary 3.58. *All topological Hausdorff vector spaces of finite dimension n over the same field are isomorphic. Their topology is induced by a norm, and all norms induce this same topology. Any two norms $\|\cdot\|$ and $\|\cdot\|^*$ on a finite-dimensional (real or complex) vector space X are equivalent in the sense that there are constants $0 < c \le C < \infty$ with*

$$c\|x\| \le \|x\|^* \le C\|x\| \quad \text{for all } x \in X, \tag{3.26}$$

and X is a Banach space with that norm.

Proof. If X and Y are topological Hausdorff vector spaces of finite dimension n, let J_X, J_Y denote the corresponding maps from Proposition 3.57. Then $J_Y J_X^{-1}: X \to Y$ is an isomorphism. We obtain a norm on X by $\|x\| := \|J_X^{-1} x\|$.

If $\|\cdot\|^*$ is any norm on $Y := X$, then $J_Y J_X^{-1} = J_X J_X^{-1} = \mathrm{id}_X$ (note that $J_X = J_Y$ does not depend on the topology) is an isomorphism. Hence, $\mathrm{id}_X: X \to Y = X$ is an isomorphism which means that the topologies on X and $Y = X$ (the latter induced by $\|\cdot\|^*$) are the same. If $\|\cdot\|$ is any norm on X, then $\mathrm{id}_X: X \to Y = X$

is an isomorphism, and so (3.26) follows from (3.24). Since $\mathbb{K}^n$ is complete and $J_X: \mathbb{K}^n \to X$ is an isomorphism, X is a Banach space by Proposition 3.56. □

Proposition 3.59 (Heine–Borel). *Let X denote a topological Hausdorff vector space. Then each finite-dimensional subspace $Y \subseteq X$ is closed, and each subset $M \subseteq Y$ with* $\operatorname{diam} M < \infty$ *for some norm on Y is relatively compact in Y and in X.*

Proof. In view of Corollary 3.58, we can assume that Y is equipped with the norm $\|y\| := \|J^{-1}x\|$ where $J: \mathbb{K}^n \to Y$ is the isomorphism of Proposition 3.57.

We prove first the last assertion. Now if $M \subseteq Y$ has finite diameter then $M_0 := J^{-1}(M) \subseteq \mathbb{K}^n$ is bounded and thus also its closure $K \subseteq \mathbb{K}^n$ is bounded. We thus have obviously $\chi_X(K) = 0$, and so K is compact by Theorem 3.24. Hence, $J(K) \subseteq Y$ is compact by Proposition 2.100 and contains M.

For the proof of the first assertion, assume that $x \in \overline{Y}$. By definition of the inherited topology, there is an open set $U \subseteq X$ with $U \cap Y = B_1(0)$ (open unit ball in Y). Let $\mathcal{A}$ denote the family of all sets of the form $A := Y \cap \overline{(x + V) \cap Y}$, where $V \subseteq U$ is a neighborhood of 0. Note that A is the closure of $(x + V) \cap Y$ in Y by Proposition 2.10 and thus closed in Y and contained in the set $K_1(0) \subseteq Y$ which by the second assertion is compact. Moreover $x \in \overline{Y}$ implies that $\mathcal{A}$ has the finite intersection property. Proposition 2.28 thus implies that there is some $y \in Y$ which belongs to every element of $\mathcal{A}$. We must have $x = y \in Y$, since otherwise there would exist a neighborhood $V \subseteq U$ of 0 such that $x+V$ and $y+V$ are disjoint which would imply that $y \notin \overline{x + V}$ (closure in X), contradicting the fact that $A := Y \cap \overline{(x + V) \cap Y}$ belongs to $\mathcal{A}$ and thus contains y and satisfies $A \subseteq \overline{x + V}$. □

Now we are in a position to define the general notion of a measure of noncompactness:

Definition 3.60. A set function γ on a topological vector space X is a *measure of noncompactness* if

$$\gamma(M) = \gamma(\overline{\operatorname{conv}} M) \quad \text{for all } M \subseteq X.$$

A measure of noncompactness is *regular* if

$$\gamma(M) = 0 \iff M \text{ is relatively compact in } X.$$

We note that in literature also measures of noncompactness are considered which assume their values not only in $[0, \infty]$ but also in partially ordered sets like e.g. in $[0, \infty]^I$ for some index set I. The latter is important if one wants to deal with locally convex spaces, since for each seminorm, one can define a corresponding measure of noncompactness with values in $[0,\infty]$, and so the family of these measures of noncompactness assumes values in $[0,\infty]^I$ (where I is a family of seminorms). However, in this monograph, we are mainly interested in Banach spaces, and so we do not consider this more general definition, although many results would carry over to this case.

Theorem 3.61. *Let X be a normed space. Then χ_X, α, and β are measures of noncompactness on X. They are regular if X is a Banach space.*

Proof. The last assertion follows from Proposition 3.26. In view of Proposition 3.26, it suffices to show that $\gamma(\operatorname{conv} M) \leq \gamma(M)$ for $\gamma \in \{\chi_X, \alpha, \beta\}$.

For $\gamma = \chi_X$, let $\rho > \chi_X(M)$. There is a finite ρ-net $N \subseteq X$ for M. We claim that $\operatorname{conv} N$ is a ρ-net for $\operatorname{conv} M$. Indeed, by Proposition 2.52, any $x \in \operatorname{conv} M$ can be written in the form $x = \sum_{k=1}^n \lambda_k x_k$ with $n \in \mathbb{N}$, $x_k \in M$, $\lambda_k \geq 0$, and $\lambda_1 + \cdots + \lambda_n = 1$.

There are $y_k \in N$ with $\|x_k - y_k\| < \varepsilon$, and so $y := \sum_{k=1}^n \lambda_k y_k$ belongs to $\operatorname{conv} N$ (Proposition 2.52) and satisfies

$$\|x - y\| = \|\sum_{k=1}^n \lambda_k (x_k - y_k)\| \leq \sum_{k=1}^n \lambda_k \|x_k - y_k\| < \sum_{k=1}^n \lambda_k \varepsilon = \varepsilon.$$

Hence, the estimate $\chi_X(\operatorname{conv} M) \leq \rho$ now follows from (3.3), if we can show that $\chi_X(\operatorname{conv} N) = 0$. Since Proposition 2.52 implies that $\operatorname{conv} N$ is a bounded subset of a finite-dimensional subspace of Y, the latter follows from Proposition 3.59 (or alternatively from the assertion $\alpha(\operatorname{conv} N) = \alpha(N) = 0$ which we will prove now).

For $\gamma = \alpha$, let $\rho > \alpha(M)$ be arbitrary. There are sets $M_1, \ldots, M_n \subseteq X$ with $M \subseteq M_1 \cup \cdots \cup M_n$ and $\operatorname{diam} M_k \leq \rho$ ($k = 1, \ldots, n$). Without loss of generality, we can assume that $M_k \neq \varnothing$. There is some $r > 0$ with $M_k \subseteq B_r(0)$ ($k = 1, \ldots, n$). We denote by σ_n the set of all $(\lambda_1, \ldots, \lambda_n) \in [0,1]^n$ with $\lambda_1 + \cdots + \lambda_n = 1$. Then σ_n is a bounded subset of $\mathbb{R}^n$ and thus satisfies $\chi_{\mathbb{R}^n}(\sigma_n) = 0$. Hence, $\chi_{\sigma_n}(\sigma_n) = 0$ by (3.1), and so for every $\varepsilon > 0$ there is a finite ε-net $\sigma_{n,\varepsilon} \subseteq \sigma_n$. We define now

$$N_\varepsilon := \bigcup_{(\lambda_1,\ldots,\lambda_n) \in \sigma_{n,\varepsilon}} (\lambda_1 \operatorname{conv} M_1 + \cdots + \lambda_n \operatorname{conv} M_n).$$

We claim that N_ε is a finite $nr\varepsilon$-net for $\operatorname{conv} M$. Indeed, by Proposition 2.52, any $\widetilde{x} \in \operatorname{conv} M$ can be written in the form $\widetilde{x} = \mu_1 x_1 + \cdots + \mu_m x_m$ with $x_k \in M_1 \cup \cdots \cup M_n$, $\mu_k \geq 0$, and $\mu_1 + \cdots + \mu_m = 1$. We divide the index set

$\{1,\dots,m\}$ into disjoint sets $J_1,\dots,J_n$ such that $j \in J_k$ implies $x_j \in M_k$, and put $\widetilde{\lambda}_k := \sum_{j\in J_k} \mu_j$ and $y_k := \sum_{j\in J_k} \mu_j x_j/\widetilde{\lambda}_k$; in case $\widetilde{\lambda}_k = 0$ choose $y_k \in M_k$ arbitrary. Then Proposition 2.52 implies $y_k \in \operatorname{conv} M_k$, we have $\widetilde{s} = (\widetilde{\lambda}_1,\dots,\widetilde{\lambda}_n) \in \sigma_n$, and $\widetilde{x} = \widetilde{\lambda}_1 y_1 + \dots + \widetilde{\lambda}_n y_n$. Choosing some $s = (\lambda_1,\dots,\lambda_n) \in \sigma_{n,\varepsilon}$ with $\|s - \widetilde{s}\| < \varepsilon$, we obtain that $x := \lambda_1 y_1 + \dots + \lambda_n y_n \in N_\varepsilon$ satisfies $\|x - \widetilde{x}\| < nr\varepsilon$.

By (3.4), we obtain now $\alpha(\operatorname{conv} M) \le \alpha(N_\varepsilon) + 2nr\varepsilon$. Moreover, since $\sigma_{n,\varepsilon}$ is finite, Proposition 3.26 implies

$$\alpha(N_\varepsilon) = \max_{(\lambda_1,\dots,\lambda_n)\in\sigma_{n,\varepsilon}} \alpha(\lambda_1 \operatorname{conv} M_1 + \dots + \lambda_n \operatorname{conv} M_n).$$

Using the first two assertions of Theorem 3.61, we find

$$\alpha(\lambda_1 \operatorname{conv} M_1 + \dots + \lambda_n \operatorname{conv} M_n) \le \lambda_1 \alpha(\operatorname{conv} M_1) + \dots + \lambda_n \alpha(\operatorname{conv} M_n).$$

Since $\operatorname{diam}(\operatorname{conv} M_k) \le \operatorname{diam}(M_k) \le \rho$, we have $\alpha(\operatorname{conv} M_k) \le \rho$, and thus obtain altogether

$$\alpha(\operatorname{conv} M) \le \alpha(N_\varepsilon) + 2nr\varepsilon \le \max_{(\lambda_1,\dots,\lambda_n)\in\sigma_{n,\varepsilon}} (\lambda_1 + \dots + \lambda_n)\rho + 2nr\varepsilon = \rho + 2nr\varepsilon.$$

Since $\varepsilon > 0$ was arbitrary, we have $\alpha(\operatorname{conv} M) \le \rho$. Since $\rho > \alpha(M)$ was arbitrary, we obtain $\alpha(\operatorname{conv} M) \le \alpha(M)$.

The proof of the remaining estimate $\beta(\operatorname{conv} M) \le \beta(M)$ is very cumbersome. Since we will not need it, we just refer to the literature. One proof can be found in [2, Theorem 1.3.4]. □

Corollary 3.62 (Mazur). *Let X be normed and $M \subseteq X$. Then $\operatorname{conv} M$ is relatively compact in X if and only if M is relatively compact in X and $\overline{\operatorname{conv}}\, M$ is complete.*

Proof. In view of Proposition 3.26, the assertion follows from Theorem 3.61. □

3.5 Two Embedding Theorems for Metric Spaces

Theorem 3.63 (Arens–Eells). *Arens-Eells Let X be a metric space. Then X is isometric to a closed subset of a complex normed space and to a (not necessarily closed) subset of a complex Banach space.*

One can replace "complex" by "real", since every complex normed space is also a real normed space by just restricting the definition of multiplication to that with real scalars.

Proof. As a first step we show that X is isometric to a subset Y of a complex Banach space Z. As a second step we show that Y is a closed subset of its (complex) linear hull in Z so that this linear hull is the required space. If X is complete, we could omit the second step of course, since then also Y is complete and thus closed in Z. (In fact, in this case one could simplify the definition of M in the following proof.)

To define the Banach space Z, we let M denote the system of all nonempty finite subsets of X. (If we are only interested in the first step, we could also let $M \cong X$ denote the system of all sets $A = \{x\}$ with $x \in X$.) Let $Z := \ell_\infty(M, \mathbb{C})$ denote the Banach space of all bounded function $f\colon M \to \mathbb{C}$, equipped with the sup-norm $\|f\|_\infty := \sup_{A \in M} |f(A)|$. The isometry is as follows: We fix some $x_0 \in X$ and will then associate with $x \in X$ the element $f_x \in Z$, defined by

$$f_x(A) := \operatorname{dist}(x, A) - \operatorname{dist}(x_0, A) \qquad \text{for all } A \in M.$$

Note that f_x is indeed bounded, since Proposition 3.12 implies $|f_x(A)| \le d(x, x_0)$ for all $A \in M$. Moreover, Proposition 3.12 implies for all $x, y \in X$ that

$$|f_x(A) - f_y(A)| = |\operatorname{dist}(x, A) - \operatorname{dist}(y, A)| \le d(x, y)$$

for all $A \in M$, and we have equality for the particular choice $A := \{y\} \in M$. Hence,

$$\|f_x - f_y\|_\infty = d(x, y) \quad \text{for all } x, y \in X, \tag{3.27}$$

that is, the mapping $x \mapsto f_x$ is an isometry of X onto a subset Y of Z.

It remains to show that Y is closed in its (complex) linear hull Z_0 in Z. Hence, we have to show that every $f \in Z_0 \setminus Y$ has a neighborhood which is disjoint from Y. Since $f \in Z_0$, there are $\lambda_1, \dots, \lambda_n \in \mathbb{C}$ and $x_1, \dots, x_n \in X$ with $f = \lambda_1 f_{x_1} + \dots + \lambda_n f_{x_n}$. Note also that $f_{x_0} \in Y$. The hypothesis $f \notin Y$ thus implies

$$r := \min\{\|f - f_{x_k}\|_\infty : k = 0, 1, \dots, n\} > 0.$$

We claim that $B_{r/2}(f) \cap Y = \emptyset$. Assume by contradiction that there is some $x \in X$ with $f_x \in B_{r/2}(f)$. Then

$$r \le \|f - f_{x_k}\|_\infty \le \|f - f_x\|_\infty + \|f_x - f_{x_k}\|_\infty < \frac{r}{2} + \|f_x - f_{x_k}\|_\infty$$

implies by (3.27) that

$$d(x, x_k) = \|f_x - f_{x_k}\|_\infty > \frac{r}{2} \qquad (k = 0, 1, \dots, n).$$

Hence, putting $A := \{x_0, x_1, \dots, x_n\}$, we have

$$f_x(A) - f(A) = f_x(A) - \sum_{k=1}^{n} \lambda_k f_k(A) = \operatorname{dist}(x, A) - \sum_{k=1}^{n} \lambda_k 0 = \operatorname{dist}(x, A) > \frac{r}{2}.$$

In particular, $\| f_x - f \|_\infty > r/2$, contradicting $f_x \in B_{r/2}(f)$. □

The Arens–Eells embedding theorem (even the simplified version without the technical second step) provides an easy proof of the following well-known fact without using a technical definition like equivalence classes of Cauchy sequences:

Corollary 3.64. *Every metric space has a completion.*

Proof. Let X be a metric space. By Theorem 3.63, there is a Banach space Z such that X is isometric to a subset X_0 of Z. Then $\overline{X}_0$ is a complete metric space in which X_0 is dense. Hence, when we replace in $\overline{X}_0$ the elements of X_0 by the corresponding elements of X (by the given isometry), the space obtained is the completion of X. □

Corollary 3.65. *Let X be a separable metric space. Then X is isometric to a closed subset of a separable complex normed space and to a (not necessarily closed) subset of a separable complex Banach space.*

Proof. Suppose that X is isometric to a subset M of a complex normed space Y. Since M is separable, it follows that also the linear hull Y_0 of M in Y is separable.

If additionally M is closed in Y (which we can assume by Theorem 3.63) then M is also closed in Y_0 by Proposition 2.10, and so Y_0 is the required space for the first assertion. If alternatively Y is a Banach space (which we could also assume by Theorem 3.63) then $\overline{Y}_0$ is a separable Banach space by Lemma 3.8 and thus the required space for the second assertion. □

For *separable* metric spaces, there is another important embedding theorem which we prepare now.

Recall that if X_n ($n \in \mathbb{N}$) are countably many metric spaces with respective metric d_n, then the product $X := \prod_{n=1}^{\infty} X_n$ becomes a metric space if we endow it with the metric

$$d(x, y) := \sum_{n=1}^{\infty} d_n^*(x, y), \tag{3.28}$$

where $d_n^*(x, y) := \min\{d_n(x, y), 2^{-n}\}$. Moreover, the corresponding uniform structure is that of the product of uniform spaces. Since we did not introduce uniform structures, we formulate the assertion only in terms of sequences and topologies:

Proposition 3.66. *With the above notation and* $x_k = (x_{k,n})_n, y_k = (y_{k,n})_n \in X$ *we have*

$$\lim_{k\to\infty} d(x_k, y_k) = 0 \iff \lim_{k\to\infty} d(x_{k,n}, y_{k,n}) = 0 \quad \textit{for every } n \in \mathbb{N}.$$

The metric induces the product topology on X.

If Y *is a topological space then* $f: Y \to X$, $f(y) = (f_1(y), f_2(y), \ldots)$ *is continuous if and only if each* f_n *is continuous.*

Proof. Let $d(x_{k,n}, y_{k,n}) \to 0$ for every $n \in \mathbb{N}$. Then for every $\varepsilon > 0$ there is some n_0 with $2^{-n_0} < \varepsilon$ and some k_0 with $d(x_{k,n}, y_{k,n}) < \varepsilon/n_0$ for $k \ge k_0$ and $n \le n_0$. Then $d(x_k, y_k) < \varepsilon + \sum_{k>n_0} 2^{-k} < 2\varepsilon$ for all $k \ge k_0$. Hence, $d(x_k, y_k) \to 0$. The converse implication follows from

$$\min\{d(x_{k,n}, y_{k,n}), 2^{-k}\} \le d(x_n, y_n).$$

To see that the induced topology is the product topology, note that if $O \subseteq X$ is a neighborhood of $x = (x_n)_n$ with respect to d, say $B_{2^{1-k}}(x) \subseteq O$ for some $k \in \mathbb{N}$ then O contains all $y = (y_n)_n \in X$ satisfying $d_n(y_n, x_n) < 2^{-k}/k$ for $n = 1, \ldots, k$, since $d(y, x) < \sum_{n=1}^{k} 2^{-k}/k + \sum_{n>k} 2^{-n} = 2^{1-k}$. Hence, O is a neighborhood of x with respect to the product topology. Conversely, if O is a neighborhood of $x = (x_n)_n$ with respect to the product topology, then there are $\varepsilon > 0$ and n_0 such that all $y = (y_n)_n \in X$ satisfying $d_n(y_n, x_n) < \varepsilon$ for $n < n_0$ belong to O. Since these y satisfy $d(y, x) \le r := n_0\varepsilon + 2^{-n_0}$, we have $B_r(x) \subseteq O$, and so O is a neighborhood with respect to d.

The last assertion now follows from Proposition 2.65. □

For countable products of metric spaces, we obtain Tychonoff's theorem without AC:

Corollary 3.67. *If* X_n $(n \in \mathbb{N})$ *is a family of complete (compact) metric spaces then* $X = \prod_{n=1}^{\infty} X_n$ *is a complete (compact) metric space.*

Proof. If $x_n = (x_{n,k})_k$ is a Cauchy sequence in X then

$$\min\{d(x_{n,k}, x_{m,k}), 2^{-k}\} \le d(x_n, x_m)$$

implies that for each fixed k the sequence $x_{n,k}$ is a Cauchy sequence and thus convergent to some $y_k \in X_k$. Applying Proposition 3.66, we find that x_n converges to $y = (y_k)_k$.

Similarly, if all X_k are compact and $x_n = (x_{n,k})_k$ is a sequence in X then a standard diagonal argument shows that there is a subsequence such that $x_{n,k}$ converges in X_k for every k. Indeed, for $k = 1, 2, \ldots$ one can successively choose subsequences $n_{k,m}$ (each a subsequence of the previous subsequence) such that $x_{n_{k,m},k} \in X_k$ converges to some $y_k \in X_k$ as $m \to \infty$; then the diagonal sequence $n_{m,m}$ ($m \in \mathbb{N}$) is a subsequence of $n_{k,m}$ for every k and thus $x_{n_{m,m},k} \in X_k$ converges to y_k as $m \to \infty$. Proposition 3.66 thus implies that $x_{n_{m,m}}$ converges in X to $y = (y_k)_k$. □

Remark 3.68. Concerning completeness, our argument in Corollary 3.67 did not use any form of the axiom of choice, not even AC_ω. In particular, our argument can be used for the proof of Theorem 2.3.

Lemma 3.69. *Let $H := [0,1]^{\mathbb{N}} = \prod_{n=1}^{\infty}[0,1]$ be the Hilbert cube with the product metric. For every normal second countable space X there is a homeomorphism of X onto a subset of H.*

We will see later (Corollary 3.78) that it is equivalent to replace "normal" by "regular" in Lemma 3.69.

Proof. Let $\mathcal{O}$ be a countable basis of the topology consisting of nonempty sets. Then the family $\mathcal{U}$ of all pairs (U, O) with $U, O \in \mathcal{O}$ and $\overline{U} \subseteq O$ is countable. Since X is T_4, there is a countable family of continuous functions $f_n \colon X \to [0,1]$ such that for each $(U, O) \in \mathcal{U}$ there is some n and some $t_n \in (0,1]$ with $f_n(\overline{U}) \subseteq [0, t_n)$ and $f_n(X \setminus O) \subseteq [t_n, 1]$. We claim that the required homeomorphism is given by $f \colon X \to H$ by $f(x) := (f_n(x))_{n=1}^{\infty}$. Proposition 3.66 implies that f is continuous.

We prove now that $f(x_k) \to f(x)$ implies $x_k \to x$.

Indeed, let $V \subseteq X$ be a neighborhood of x. There is $O \in \mathcal{O}$ with $x \in O \subseteq V$. Since X is T_3 there is an open $U \subseteq X$ with $x \in U$ and $\overline{U} \subseteq O$, without loss of generality $U \in \mathcal{O}$. Hence, $(U, O) \in \mathcal{U}$ and so there is some n and t_n with $f_n(x) < t_n$ and $f_n(X \setminus O) \subseteq [t_n, 1]$. Since $f(x_k) \to f(x)$, Proposition 3.66 implies $f_n(x_k) \to f_n(x) < t_n$. Hence, for all sufficiently large k we have $x_k \in O$ (since otherwise $f_n(x_k) \geq t_n$). Hence, $x_k \in V$ for all sufficiently large k, that is $x_k \to x$.

If $f(x) = f(y)$ then the particular choice $x_k := y$ implies $x_k \to x$ and thus $x = y$ since X is Hausdorff. Hence, f is one-to-one. Thus $f \colon X \to H_0$ is invertible for $H_0 := f(X)$. Moreover, we have shown that $f^{-1} \colon H_0 \to X$ is sequentially continuous and thus continuous by Proposition 2.60. □

Remark 3.70. If Y is a separable metric space in Lemma 3.69 then it can be arranged that the homeomorphism f is uniformly continuous (that is, whenever $d(x_k, y_k) \to 0$ then $d(f(x_k), f(y_k)) \to 0$). Note, however, that f^{-1} is not uniformly continuous, in general.

Indeed, if $\{y_1, y_2, \ldots\}$ is dense in Y, the proof of Proposition 3.16 shows that a basis $\mathcal{O}$ of the topology of Y consists of all balls $B_r(y_k)$ with rational $r > 0$. Hence, the functions $f_n(x) := \min\{d(x, y_n), 1\}$ have the property required in the proof of Lemma 3.69. Since each f_n is Lipschitz (with constant 1) and thus uniformly continuous, Proposition 3.66 implies that the function $f(x) = (f_n(x))_{n=1}^{\infty}$ in the proof of Lemma 3.69 is uniformly continuous.

Actually, one could even simplify the proof of Lemma 3.69 in this special case by showing elementary (without considering any basis) that the function f has the property that $f(x_k) \to f(x)$ implies $x_k \to x$. Indeed, for every $\varepsilon > 0$ there is some n with $y_n \in B_\varepsilon(x)$. Since $f_n(x_k) \to f_n(x) < \varepsilon$ by Proposition 3.66, we have for all except finitely many k that $f_n(x_k) < \varepsilon$ and thus $d(x_k, x) \leq d(x_k, y_n) + d(y_n, x) < 2\varepsilon$. Hence, $x_k \to x$.

Lemma 3.71. *If X is a metric space and $X_0 \subseteq X$ is homeomorphic to a complete metric space, then X_0 is a G_δ in X.*

Proof. Let $h\colon X_0 \to Y$ be a homeomorphism onto a complete metric space Y. For $x \in \overline{X}_0$, put $h_r(x) := \operatorname{diam} h(B_r(x) \cap X_0)$. If $h_r(x) < 1/n$ then $h_{r/2}(B_{r/2}(x) \cap \overline{X}_0) \subseteq [0, 1/n)$. Hence, the sets

$$A_n := \{x \in \overline{X}_0 : h_r(x) \geq 1/n \text{ for all } r > 0\}$$

are closed in $\overline{X}_0$ and thus closed in X, and so $X_1 := \overline{X}_0 \setminus \bigcup_{n=1}^{\infty} A_n$ is a G_δ, since it is the intersection of the open sets $B_{1/n}(X_0) \setminus A_n$. We are to show that $X_1 = X_0$. If $x \in X_0$, then the continuity of h at x implies $x \notin A_n$ for all n, and so $x \in X_1$. Hence, $X_0 \subseteq X_1$.

It remains to prove the converse inclusion. Thus, let $x \in X_1$. Choose a sequence $x_n \in X_0$ with $x_n \to x$. We show first that $h(x_n)$ is a Cauchy sequence. Indeed, for each $\varepsilon > 0$, there is some $k \in \mathbb{N}$ with $1/k < \varepsilon$. Since $x \notin A_k$, we find some $r_k > 0$ with $h_{r_k}(x) < 1/k < \varepsilon$. Choose some n_0 with $x_n \in B_{r_k}(x)$ for $n \geq n_0$. The definition of h_{r_k} thus implies $d(h(x_n), h(x_m)) < \varepsilon$ for $n, m \geq n_0$.

Since Y is complete, we thus have $h(x_n) \to y$ for some $y \in Y$. The continuity of h^{-1} at y implies $x_n \to h^{-1}(y)$. Since $x_n \to x$, we have $x = h^{-1}(y) \in h^{-1}(Y) = X_0$. In particular, $x \in X_0$, and so we have proved $X_1 \subseteq X_0$. □

The proof of the following Lemma 3.72 slightly simplifies that of [96, §24 IX]:

Lemma 3.72. *Let X be a metric space and $M \subseteq X$ be a G_δ subset of X. Then there is a homeomorphism of X onto a closed subset of $X \times [0,1]^{\mathbb{N}}$ which maps M onto a closed subset of $X \times (0,1]^{\mathbb{N}}$.*

Proof. Let $M = \bigcap_{n=1}^{\infty} U_n$ with open sets $U_n \subseteq X$. Since $X \setminus U_n$ is closed, Corollary 3.13 implies that there are $f_n \in C(X, [0,1])$ satisfying $f_n^{-1}(0) = X \setminus U_n$ (and $f_n^{-1}(1) = \varnothing$). Proposition 3.66 implies that we obtain a continuous function $f: X \to X \times [0,1]^{\mathbb{N}}$ by putting

$$f(x) := \big(x, f_1(x), f_2(x), \dots\big).$$

Let $p: X \times [0,1]^{\mathbb{N}} \to X$ be defined by $p(x,y) := x$ for all $x \in X$ and all $y \in [0,1]^{\mathbb{N}}$. Proposition 3.66 implies that p is continuous, and the definition of f implies $p \circ f = \mathrm{id}_X$. Hence, putting $X_0 := f(X)$, we find that $f: X \to X_0$ is invertible with $f^{-1} = p|_{X_0}: X_0 \to X$. Since $p|_{X_0}$ is continuous, it follows that $f: X \to X_0$ is a homeomorphism. To see that X_0 is closed in $Y := X \times [0,1]^{\mathbb{N}}$, let $y \in \overline{X}_0$, that is, there is a sequence $x_n \in X$ with $f(x_n) \to y$. The continuity of p at y implies $x_n = p(f(x_n)) \to p(y)$, and so the continuity of f at $x := p(y)$ implies $f(x_n) \to f(x)$. Hence, $y = f(x) \in f(X) = X_0$.

It remains to show that $f(M)$ is a closed subset of $Y_0 := X \times (0,1]^{\mathbb{N}}$. Since $f_n(x) \in (0,1]$ if and only if $x \in U_n$, we have $f(x) \in Y_0$ if and only if $x \in \bigcap_{n=1}^{\infty} U_n = M$. Hence, $f(M) = f(X) \cap Y_0 = X_0 \cap Y_0$ is a closed subset of Y_0, since X_0 is closed in Y (Proposition 2.10). □

Lemma 3.73. *For any sequence of metric spaces X_n and corresponding homeomorphic metric spaces Y_n the spaces $X = \prod_{n=1}^{\infty} X_n$ and $Y = \prod_{n=1}^{\infty} Y_n$ are homeomorphic. Moreover, if X_n and Y_n are independent of n then also $X \times Y$ is homeomorphic to X and Y.*

Proof. Let $f_n: X_n \to Y_n$ be a homeomorphism onto Y_n. Then Proposition 3.66 implies that homeomorphisms of X onto Y or $X \times Y$, respectively, are given as follows: To each sequence $x_n \in X$ we associate the sequence $f_n(x_n) \in Y_n$ or the sequence $(x_{2n}, f_n(x_{2n+1})) \in X_n \times Y_n$, respectively. □

Combining the previous preparations, we obtain now the announced embedding theorem which was perhaps first observed in [106] (for the case $X = \mathbb{R}^{\mathbb{N}}$).

Theorem 3.74 (Embedding into Open Hilbert Cube). *Let $X = \prod_{n=1}^{\infty} I_n$ where each $I_n \subseteq \mathbb{R}$ is a noncompact interval. Then every separable (complete) metric space is homeomorphic to a (closed) subset of X.*

Proof. We claim first that the assertion is true for the case that $I_n = (0,1]$ for all $n \in \mathbb{N}$. By Lemma 3.73, the spaces $[0,1]^{\mathbb{N}}$ and $[1/2,1]^{\mathbb{N}}$ are homeomorphic. Hence, if Y is a separable metric space, then Lemma 3.69 implies that Y is homeomorphic to a subset of $[0,1]^{\mathbb{N}}$ and thus also homeomorphic to a subset X_0 of $[1/2,1]^{\mathbb{N}} \subseteq X = (0,1]^{\mathbb{N}}$. Thus, the claim is proved if Y is incomplete.

If Y is complete, then Lemma 3.72 implies that X_0 is even a G_δ subset of X. In this case, Lemma 3.72 implies that X_0 is homeomorphic to a closed subset X_1 of $X \times (0,1]^{\mathbb{N}} = X \times X$. By Lemma 3.73 there is a homeomorphism f of $X \times X$ onto X. Then $X_2 = f(X_1)$ is closed in $f(X \times X) = X$, and composing the corresponding homeomorphisms, we find that Y is homeomorphic to X_2.

Now we prove the general case: Since each I_n is an open or half-open interval, there are homeomorphisms f_n of $(0,1]$ onto closed (in I_n) subsets $J_n \subseteq I_n$. Lemma 3.73 implies that there is a homeomorphism f of $(0,1]^{\mathbb{N}}$ onto $X_0 = \prod_{n=1}^{\infty} J_n$. Since J_n is closed in I_n, Proposition 3.66 implies that X_0 is a closed subset of $X = \prod_{n=1}^{\infty} I_n$. By what we have shown, we find for each separable (complete) metric space Y a homeomorphism h if X onto a (closed) subset $Z \subseteq (0,1]^{\mathbb{N}}$. Then $f \circ h$ is a homeomorphism onto the (closed in X_0 and thus closed in X) set $f(Z) \subseteq X_0$. □

3.6 Some Old and New Extension Theorems for Metric Spaces

We recall Stone's famous theorem:

Theorem 3.75 (Stone). (AC). *Every metric space is paracompact.*

Proof. An elementary proof can be found e.g. in [127]. □

We point out that AC cannot be avoided for Theorem 3.75, in principle [69].

However, using only the countable axiom of choice, we can prove the following variant which was originally proved by Morita [110]. Using ideas from Michael [105], we give a simple proof of a stronger statement.

Theorem 3.76 (Morita). *Let X be a T_3 Lindelöf space. Then X is paracompact. More precisely, every open cover $\mathcal{U}$ of X has a refinement to a locally finite countable open cover $\mathcal{O}$ of X.*

Proof. We show first that every open cover $\mathcal{U}_0$ of X has a locally finite countable refinement to a cover $\mathcal{A}$ of X consisting of closed sets.

Indeed, since $\mathcal{U}_0$ is an open cover for X, we find for every $x \in X$ some $U \in \mathcal{U}_0$ with $x \in U$. Since X is T_3, we find by Corollary 2.48 an open set $O \subseteq X$ with $x \in O \subseteq \overline{O} \subseteq U$. Hence, the family $\mathcal{O}_0$ of all open sets $O \subseteq X$ satisfying $\overline{O} \subseteq U$ for some $U \in \mathcal{U}_0$ constitute an open cover of X. Since X is Lindelöf, $\mathcal{O}_0$ has a countable subcover consisting of sets $O_1, O_2, \dots \in \mathcal{O}_0$. Then the sets $A_n := \overline{O}_n \setminus \bigcup_{k<n} O_k$ are closed. Since for each $x \in X$ there is a smallest n

with $x \in O_n$ and thus $x \in A_n$, we find that the sets A_n constitute a cover $\mathcal{A}$ of X. In view of $A_n \subseteq \overline{O}_n$, this cover is a refinement of $\mathcal{U}_0$. Moreover, this cover is locally finite, since for any $x \in X$ there is some n with $x \in O_n$, and so O_n is a neighborhood of x which intersects at most the finitely many sets $A_1, \dots, A_n$ from $\mathcal{A}$.

To prove the assertion of the theorem, we note that we have just proved that there is a refinement $\mathcal{A} = \{A_1, A_2, \dots\}$ of $\mathcal{U}$ which is a locally finite countable cover of X. Let $\mathcal{U}_0$ denote the system of all open sets which intersect only finitely many elements from $\mathcal{A}$. Since $\mathcal{A}$ is locally finite, it follows that $\mathcal{U}_0$ is an open cover of X. Using once more what we have proved, we find a refinement $\mathcal{A}_0$ of $\mathcal{U}_0$ which is a locally finite countable cover of X consisting of closed sets.

For each n, we choose some $U_n \in \mathcal{U}$ with $A_n \subseteq U_n$, and let

$$B_n := \bigcup \{A \in \mathcal{A}_0 : A \cap A_n = \emptyset\}.$$

Then B_n is closed. Indeed, every $x \in X$ has a neighborhood $O \subseteq X$ which intersects only finitely many elements from $\mathcal{A}_0$. Hence, $O \cap B_n$ is a finite union of sets of the form $O \cap A$ with $A \in \mathcal{A}_0$ and thus closed in O. Thus, B_n is closed by Proposition 2.12.

Hence, $V_n := U_n \setminus B_n$ are open sets, and the definition of B_n implies $A_n \subseteq V_n$. Since $\mathcal{A}$ is a cover of X, it follows that $\mathcal{O} := \{V_1, V_2, \dots\}$ is also a cover of X. In view of $V_n \subseteq U_n \in \mathcal{U}$, this cover is a refinement of $\mathcal{U}$. It remains to show that $\mathcal{O}$ is locally finite.

To see this, we note first that each $A \in \mathcal{A}_0$ is contained in a set from $\mathcal{U}_0$ which by definition of $\mathcal{U}_0$ intersects at most finitely many of the sets A_n ($n \in \mathbb{N}$). Note also that for $A \in \mathcal{A}_0$ the relation $A \cap A_n = \emptyset$ implies $A \subseteq B_n$ and thus $A \cap V_n = \emptyset$. Hence, every $A \in \mathcal{A}_0$ intersects at most finitely sets from $\mathcal{O}$.

Every $x \in X$ has a neighborhood U such that $\mathcal{A}_U := \{A \in \mathcal{A}_0 : A \cap U \neq \emptyset\}$ is finite. Since $\mathcal{A}_0$ is a cover of U, it follows that $\mathcal{A}_U$ is a cover of U. Since we have just noted that each of the finitely elements from $\mathcal{A}_U$ intersects at most finitely elements from $\mathcal{O}$ and since $\mathcal{A}_U$ is finite, there are at most finitely elements from $\mathcal{O}$ which intersect some element from $\mathcal{A}_U$. Since $\mathcal{A}_U$ is a cover of U, only these finitely many elements can intersect U. □

Corollary 3.77 (Stone for Separable Spaces). *Let X be a separable metric space. Then every open cover $\mathcal{U}$ of X has a refinement to a locally finite countable open cover $\mathcal{O}$ of X.*

Proof. Note that Corollary 3.13 (and Theorem 2.36) implies that X is regular. Moreover, X is second countable by Proposition 3.16 and thus Lindelöf by Proposition 2.61. Hence, Theorem 3.76 applies. □

Corollary 3.78. *Every regular Lindelöf space is normal. In particular, every second countable regular space is normal.*

Proof. Theorem 3.76 implies that every regular Lindelöf space is paracompact and thus normal by Proposition 2.72. The second assertion follows from Proposition 2.61. □

In view of Lemma 3.69 we obtain the following side result:

Theorem 3.79 (Urysohn Metrization). *Every second countable regular space is metrizable.*

Proof. Let X be second countable and regular. Corollary 3.78 implies that X is normal, and so Lemma 3.69 implies that there is a homeomorphism f of X onto a subset H_0 of the Hilbert cube. If d_H denotes the metric on the Hilbert cube, a metric on X is thus given by $d(x, y) := d_H(f(x), f(y))$. Since f is a homeomorphism and, by construction, an isometry, it follows that this metric generates the original topology of X. □

The following result is an important refinement of Theorem 3.76.

Theorem 3.80. *Let X be a regular Lindelöf space. Then for every open cover $\mathscr{U}$ of X there are countably many open sets $U_n, O_n \subseteq X$ with $X = \bigcup_{n=1}^{\infty} U_n$, $\overline{U}_n \subseteq O_n$ for all n, and such that the family $\overline{O}_n$ $(n \in \mathbb{N})$ is a locally finite refinement of $\mathscr{U}$.*

Proof. By Theorem 3.76 there is a locally finite countable open cover V_n $(n \in \mathbb{N})$ of X which refines $\mathscr{U}$. Note that X is normal by Corollary 3.78. By the subsequent Lemma 3.81 there are open sets $U_n \subseteq X$ with $\overline{U}_n \subseteq V_n$ and $X = \bigcup_{n=1}^{\infty} U_n$. Since X is T_4, Corollary 2.48 implies that there are open neighborhoods $O_n \subseteq X$ of $\overline{U}_n$ with $\overline{O}_n \subseteq V_n$. Since V_n $(n \in \mathbb{N})$ is locally finite, it follows that $\overline{O}_n$ $(n \in \mathbb{N})$ is locally finite. Moreover, for every $n \in \mathbb{N}$ there is some $U \in \mathscr{U}$ with $V_n \subseteq U$ and thus $\overline{O}_n \subseteq U$. Hence, $\overline{O}_n$ $(n \in \mathbb{N})$ is a locally finite refinement of $\mathscr{U}$. □

In the above proof of Theorem 3.80, we needed the following lemma.

Lemma 3.81. *Let X be a T_4 space, and U_n $(n \in \mathbb{N})$ a locally finite countable open cover of X. Then there is a countable open cover O_n $(n \in \mathbb{N})$ of X with $\overline{O}_n \subseteq U_n$ for all n.*

Proof. We show first by induction the there are open sets $O_n \subseteq X$ such that $\overline{O}_n \subseteq U_n$ and

$$X = \bigcup_{k \leq n} O_n \cup \bigcup_{k>n} U_k. \tag{3.29}$$

Indeed, if O_k is already defined for all $k < n$ then $V_n := \bigcup_{k<n} O_k \cup \bigcup_{k>n} U_k$ is an open set with $U_n \cup V_n = X$. Hence, $X \setminus V_n \subseteq U_n$. By Corollary 2.48 there is an open neighborhood $O_n \subseteq X$ of $X \setminus V_n$ with $\overline{O}_n \subseteq U_n$. Since $X \setminus V_n \subseteq O_n$, we have (3.29).

To see that $\mathscr{O} = \{O_1, O_2, \ldots\}$ is a cover of X, let $x \in X$. Since the cover $\mathscr{U}$ is locally finite, there is a maximal n with $x \in U_n$. Then (3.29) implies $x \in O_1 \cup \cdots \cup O_n$. □

We recall that the important role of paracompact spaces is that they are related with partitions of unity.

Definition 3.82. A *partition of unity* subordinate to an open cover $\mathscr{U}$ of a space X is a family of continuous functions $\lambda_i \colon X \to [0,1]$ $(i \in I)$ such that the family $\mathscr{A}$ of sets

$$\operatorname{supp} \lambda_i := \overline{\{x \in X : \lambda_i(x) \neq 0\}} \qquad (i \in I)$$

is a locally finite cover of X which is a refinement of $\mathscr{U}$, and

$$\sum_{i \in I} \lambda_i(x) = 1 \quad \text{for every } x \in X.$$

Note that all except finitely many terms of the sum vanish since $\mathscr{A}$ is locally finite.

Corollary 3.83. *Let X be a regular Lindelöf space or a second countable metrizable space. Then every open cover $\mathscr{U}$ of X has a subordinate countable partition of unity.*

Proof. Actually in both cases X is a regular Lindelöf space (recall Corollary 3.13 and Propositions 2.61 and 3.16). Hence, Corollary 3.78 implies that X is normal.

If $\mathscr{U}$ is an open cover of X, let $U_n, O_n \subseteq X$ be as in Theorem 3.80. Since X is normal there are continuous functions $f_n \colon X \to [0,1]$ satisfying $f_n(\overline{O}_n) = \{1\}$ and $f_n(X \setminus V_n) = \{0\}$. Put $S_n := \{x \in X : f_n(x) \neq 0\}$. Then $O_n \subseteq S_n$ and $\overline{S}_n \subseteq U_n$ imply that the family S_n $(n \in \mathbb{N})$ is a cover of X, and that $\overline{S}_n$ $(n \in \mathbb{N})$ is locally finite. It follows that $g(x) := \sum_{n=1}^{\infty} f_n(x)$ satisfies $g(x) \neq 0$ for all $x \in X$ and that g is continuous at every $x \in X$, since x has a neighborhood which intersects only finitely many S_n. Hence, $\lambda_n(x) := f_n(x)/g(x)$ is the required countable subordinate partition of unity. □

Now we show that any function (not necessarily continuous) defined on a subset M of a metric space with values in a topological vector space "matches almost" to a map which is continuous outside of $\overline{M}$. Although the proof has many similarities with the proof of Dugundji's famous extension theorem, this seems to be a new and much more general result: We will obtain sharpenings of Dugundji's extension theorem and also sharpenings of Ma's extension theorem for multivalued functions as special cases. The sharpenings are actually so general that we can obtain results like the existence of so-called Schauder projections (which cannot directly be deduced from the classical Dugundji extension theorem) as a trivial special case in Proposition 13.8.

The key property is the following result. For the classical extension theorems, the given map f in the subsequent result is continuous. It is rather surprising (and important for us to obtain Ma's extension theorem) that the result holds even for discontinuous $f: M \to Y$: The result states that there is a continuous map defined on the complement of $\overline{M}$ which up to certain convex combinations is "very close" to the given function f, especially near the boundary of M. Moreover, the map which associates to f such an extension is actually a linear operator and in a sense independent of the image space. This statement about the map will lead to Dugundji's result about "simultaneous extension" of continuous maps.

Theorem 3.84 (Continuous Matching of Discontinuous Maps). (AC). *Let (X, d) be a metric space, $M \subseteq X$ nonempty, and $L: X \setminus \overline{M} \to (1, \infty]$ be lower semicontinuous. Then there is an operator $\mathcal{M}_{L,M,X}$ which associates to each topological vector space Y over $\mathbb{R}$ or $\mathbb{C}$ and each function $f: M \to Y$ (not necessarily continuous) a function $F = \mathcal{M}_{L,M,X}(f, Y)$ where $F: X \setminus \overline{M} \to Y$ is continuous and satisfies*

$$F(x) \in \operatorname{conv}\{f(y) : y \in M,\ d(x, y) \le L(x) \operatorname{dist}(x, M)\}. \tag{3.30}$$

The maps $\mathcal{M}_{L,M,X}(\cdot, Y)$ are linear; if $Y_0 \subseteq Y$ as sets, then $\mathcal{M}_{L,M,X}(f, Y_0) = \mathcal{M}_{L,M,X}(f, Y)$.

By "$Y_0 \subseteq Y$ as sets" we mean that the topology on Y_0 need not necessarily be the topology inherited from Y.

Proof. Putting $X_0 := X \setminus \overline{M}$, we can define a lower semicontinuous function $\varepsilon: X_0 \to (0, 1/2]$ by

$$\varepsilon(x) := \begin{cases} \frac{L(x)-1}{2L(x)+4} & \text{if } 1 < L(x) < \infty, \\ 1/2 & \text{if } L(x) = \infty. \end{cases}$$

For $x \in X_0$, we put

$$r(x) := \sup\{\rho \in [0,\infty] : \rho \le \varepsilon(x)\operatorname{dist}(x,M) \text{ and } \inf \varepsilon(B_\rho(x)) \ge \varepsilon(x)/2\}.$$

Since $\operatorname{dist}(x,M) = \operatorname{dist}(x,\overline{M}) > 0$ and ε is positive and lower semicontinuous at $x \in X_0$, we have $r: X_0 \to (0,\infty)$. The sets $U_x := B_{r(x)}(x)$ $(x \in X_0)$ are an open cover of X_0. Since X_0 is metrizable and thus paracompact by Stone's Theorem 3.75, there is a locally finite open refinement $\mathcal{O}$ of that cover. For each $O \in \mathcal{O}$, we choose some $x_O \in X_0$ with $O \subseteq U_{x_O}$ and some $y_O \in M$ with

$$\operatorname{dist}(x_O, y_O) \le \big(1 + \varepsilon(x_O)\big)\operatorname{dist}(x_O, M). \tag{3.31}$$

Since $1 + \varepsilon(x_O) > 1$ and $\operatorname{dist}(x_O, M) = \operatorname{dist}(x_O, \overline{M}) > 0$, a point $y_O \in M$ satisfying (3.31) indeed exists. By Corollary 3.13, there are $\lambda_O \in C(X,[0,1])$ satisfying $\lambda_O^{-1}(0) = X \setminus O$ (and $\lambda_O^{-1}(1) = \varnothing$). We claim that

$$\mathcal{M}_{L,M,X}(f,Y)(x) := \frac{\sum_{O\in\mathcal{O}} \lambda_O(x) f(y_O)}{\sum_{O\in\mathcal{O}} \lambda_O(x)}$$

has the required properties. Indeed, since each $x \in X_0$ is contained in at least one set from $\mathcal{O}$ and contains a neighborhood which intersects at most finitely many sets from $\mathcal{O}$, it follows that $F := \mathcal{M}_{L,M,X}(f)$ is defined and continuous on that neighborhood, and $F(x) \in \operatorname{conv}\{f(y_O) : x \in O \in \mathcal{O}\}$ by Proposition 2.52. Hence, the claim follows if we can show the implication

$$x \in O \in \mathcal{O} \implies d(x, y_O) \le L(x)\operatorname{dist}(x, M). \tag{3.32}$$

Thus, let $x \in O \in \mathcal{O}$. By construction, we have $x \in O \subseteq U_{x_O} = B_{r(x_O)}(x_O)$, hence $d(x, x_O) < r(x_O)$. The definition of $r(x_O)$ thus implies

$$d(x, x_O) \le \varepsilon(x_O)\operatorname{dist}(x_O, M) \quad \text{and} \quad \varepsilon(x) \ge \varepsilon(x_O)/2. \tag{3.33}$$

We obtain by (3.31) that

$$\begin{aligned} d(x, y_O) &\le d(x, x_O) + d(x_O, y_O) \le \big(1 + 2\varepsilon(x_O)\big)\operatorname{dist}(x_O, M) \\ &\le \big(1 + 4\varepsilon(x)\big)\operatorname{dist}(x_O, M). \end{aligned} \tag{3.34}$$

Since Proposition 3.12 and (3.33) imply

$$\begin{aligned} \operatorname{dist}(x_O, M) &\le d(x_O, x) + \operatorname{dist}(x, M) \\ &\le \varepsilon(x_O)\operatorname{dist}(x_O, M) + \operatorname{dist}(x, M) \\ &\le 2\varepsilon(x)\operatorname{dist}(x_O, M) + \operatorname{dist}(x, M), \end{aligned}$$

we have $(1 - 2\varepsilon(x))\operatorname{dist}(x_O, M) \leq \operatorname{dist}(x, M)$. In case $L(x) < \infty$, we have $2\varepsilon(x) < 1$ and thus obtain with (3.34)

$$d(x, y_O) \leq \frac{1 + 4\varepsilon(x)}{1 - 2\varepsilon(x)} \operatorname{dist}(x, M) = L(x) \operatorname{dist}(x, M).$$

Hence, (3.32) is established in case $L(x) < \infty$; the case $L(x) = \infty$ is trivial in view of $\operatorname{dist}(x, M) > 0$. □

Remark 3.85. We point out that the proof of Theorem 3.84 makes essential use of AC (in using Stone's theorem and in the choice of y_O for each $O \in \mathcal{O}$). However, if $X \setminus M$ is separable then AC_ω suffices: Corollary 3.77 requires only AC_ω, and by that result, $X \setminus M$ is paracompact, and the set $\mathcal{O}$ in the proof of Theorem 3.84 can be assumed to be countable, so we have to choose only countably many $y_O \in \mathcal{O}$.

We show in a moment that we obtain as a special case the following generalization of Dugundji's celebrated extension theorem [43] (see also [28, Section III.7]).

Theorem 3.86 (Dugundji's Extension Theorem). (AC). *Let X be a metric space, $A \subseteq X$ be closed, and Y be a locally convex space. Then each continuous $f: A \to Y$ has a continuous extension $F: X \to Y$ with $F(X) \subseteq \operatorname{conv} f(A)$.*

Moreover, for every lower semicontinuous function $L: X \to (1, \infty]$ and every $M \subseteq A$ with $\partial A \subseteq \overline{M}$ the extension may be chosen such that that (3.30) *holds, in particular $F(X \setminus A) \subseteq \operatorname{conv} f(M)$. Simultaneously, it may be arranged that the association $(f, Y) \mapsto F = \mathcal{M}_{L,M,X}(f, Y)$ is linear with respect to f and satisfies $\mathcal{M}_{L,M,X}(f, Y_0) = \mathcal{M}_{L,M,X}(f, Y)$ if $Y_0 \subseteq Y$ as sets.*

The importance of the linearity of the extension operation was already observed by Dugundji.

Theorem 3.86 is actually only the single-valued special case of the following result. This result generalizes [86] which in turn extends Ma's extension theorem [102, (2.1)] for upper semicontinuous maps with convex values.

Theorem 3.87. (AC). *Let X be a metric space, $A \subseteq X$ closed, Y a locally convex space, and $\Phi: A \multimap Y$ be upper semicontinuous. Assume that $\Phi(x)$ is convex, nonempty, and either compact or open for each $x \in \partial A$. Then Φ has an upper semicontinuous extension $\Phi: X \multimap Y$ with $\Phi(X) \subseteq \operatorname{conv} \Phi(A)$ and such that $\Phi|_{X \setminus A}$ is a single-valued continuous function $F: X \setminus A \to Y$.*

Moreover, for every lower semicontinuous function $L: X \to (1, \infty]$, every $M \subseteq A$ with $\partial A \subseteq \overline{M}$ and every selection $f: M \to Y$ of $\Phi|_M$, it may be arranged that (3.30) *holds, in particular $\Phi(X \setminus A) \subseteq \operatorname{conv} f(M)$, and that the association $(f, Y) \mapsto F = \mathcal{M}_{L,M,X}(f, Y)$ is linear with respect to $f|_M$ and satisfies $\mathcal{M}_{L,M,X}(f, Y_0) = \mathcal{M}_{L,M,X}(f, Y)$ if $Y_0 \subseteq Y$ as sets.*

Remark 3.88. Our proof will show that the requirement $\Phi(x) \neq \varnothing$ is actually only needed for $x \in M$. However, since $\partial A \subseteq \overline{M}$, it follows automatically that $\Phi(x) \neq \varnothing$ for all $x \in \partial A$ (Proposition 2.91).

Proof. Given L and M, we can assume without loss of generality that L assumes its values in $(1, 2]$, since we can replace L by $L_0(x) := \min\{L(x), 2\}$ if necessary. We choose $\mathcal{M}_{L,M,X}$ as in Theorem 3.84, and for $f: M \to Y$, we put $F := \mathcal{M}_{L,M,X}(f, Y)$ and $\Phi(x) := \{F(x)\}$ for $x \in X \setminus A$. We are to show that this extension is automatically upper semicontinuous at every $x_0 \in \partial A$.

Thus, let V be a neighborhood of $\Phi(x_0)$. If $\Phi(x_0)$ is compact, there is some neighborhood $U \subseteq Y$ of 0 with $V_0 := \Phi(x_0)+U \subseteq V$. Since Y is locally convex, we can assume without loss of generality that U is convex. If $\Phi(x_0)$ is open, we put instead $V_0 := \Phi(x_0)$. In both cases, $V_0 \subseteq V$ is a convex neighborhood of $\Phi(x_0)$.

Since $\Phi|_A$ is upper semicontinuous at x_0, there is $r > 0$ with $\Phi(A \cap B_r(x_0)) \subseteq V_0$. Since V_0 is convex, this implies $\operatorname{conv} f(M \cap B_r(x_0)) \subseteq V_0$. By $L(x) \leq 2$, we obtain from (3.30) that $F(B_{r/2}(x_0) \setminus A) \subseteq V_0$. □

Remark 3.89. In accordance with Remark 3.85, we point out that we used AC for Theorem 3.87 not only to apply Theorem 3.84 but also to find a selection f of $\Phi|_M$.

However, if $X \setminus A$ is separable, then there is also a countable dense $M \subseteq \partial A$, and so in this case AC_ω is sufficient for our proof of Theorem 3.87 (and thus of Theorem 3.86).

Chapter 4

Spaces Defined by Extensions, Retractions, or Homotopies

4.1 AE and ANE Spaces

Definition 4.1. A topological space Y is an AE ("absolute extensor") for a class $\mathfrak{C}$ of topological spaces if for every X from $\mathfrak{C}$, every closed subset $A \subseteq X$ and every continuous map $f: A \to Y$ there is a continuous extension $F: X \to Y$ of f.

A topological space Y is an ANE ("absolute neighborhood extensor") for a class $\mathfrak{C}$ of topological spaces if for every X from $\mathfrak{C}$, every closed subset $A \subseteq X$ and every continuous map $f: A \to Y$ there is a neighborhood $U \subseteq X$ with $A \subseteq U$ and a continuous extension $F: U \to Y$ of f.

If $\mathfrak{C}$ consists only of one space X, we call Y simply an AE (or ANE) for X.

Clearly, each AE for $\mathfrak{C}$ is an ANE for $\mathfrak{C}$.

Remark 4.2. In Definition 4.1, one can equivalently require that the neighborhood $U \subseteq X$ be open.

Dugundji's extension theorem (Theorem 3.86) can now be formulated as follows.

Theorem 4.3 (Dugundji's Extension Theorem). (AC). *Every convex subset of a locally convex space is an* AE *for the class of metric spaces.*

Remark 4.4. Without AC, we claim only that every locally convex space is an AE for the class of *separable* metric spaces, cf. Remark 3.89.

Dugundji's theorem provides us with a large class of examples of AE spaces (and thus also ANE spaces) for some class. Once we know such examples, we can construct more examples by the following observations:

Proposition 4.5. *If Y is an* ANE *for a class $\mathfrak{C}$, then every open subset $U \subseteq Y$ is also an* ANE *for $\mathfrak{C}$.*

Proof. Let X be a space from $\mathcal{C}$, $A \subseteq X$ be closed, and $f \in C(A, U)$. Then $f \in C(A, Y)$, and so there is an open neighborhood $V \subseteq X$ of A and an extension $f \in C(V, Y)$. Then $V_0 := f^{-1}(U) \subseteq V$ is an open (in V and thus in X by Proposition 2.10) neighborhood of A, and so $f|_{V_0}: V_0 \to U$ is the required extension of f. □

Proposition 4.6. *If each $Y_1, \dots, Y_n$ is* ANE (AE) *for a class $\mathcal{C}$ then also $Y := Y_1 \times \cdots \times Y_n$ is an* ANE (AE) *for $\mathcal{C}$.*

Proof. Let X be a space from $\mathcal{C}$, $A \subseteq X$ be closed, and $f \in C(A, Y)$. Then $f = (f_1, \dots, f_n)$ with $f_k \in C(A, Y_k)$. There are extensions $F_k \in C(U_k, Y_k)$ of f_k with neighborhoods $U_k \subseteq X$ of A (or $U_k = X$). Putting $U := U_1 \cap \cdots \cap U_n$, we can define the required extension $F \in C(U, Y)$ by $F := (F_1|_U, \dots, f_n|_U)$. □

Proposition 4.7. *Let $g \in C(Y, Y_0)$ be such that there is $h \in C(Y_0, Y)$ with $g \circ h = \mathrm{id}_{Y_0}$. If Y is an* ANE (AE) *for some class $\mathcal{C}$ of spaces then so is Y_0.*

Proof. Let X be a space from $\mathcal{C}$, $A \subseteq X$ be closed, and $f \in C(A, Y_0)$. Then $h \circ f: A \to Y$ is continuous and thus has a continuous extension $H: U \to Y$ with a neighborhood $U \subseteq X$ of A (or $U = X$, respectively). Then the map $F := g \circ H: U \to Y_0$ is continuous and satisfies $F|_A = g \circ h \circ f = f$. □

Corollary 4.8. *If Y and Y_0 are homeomorphic, and if Y is an* AE (ANE) *for a class $\mathcal{C}$ then also Y_0.*

Proof. Apply Proposition 4.7 with a homeomorphism $g: Y \to Y_0$ and $h = g^{-1}$. □

Definition 4.9. A *retraction* in a topological space X is a continuous map $\rho: X \to X$ with $\rho \circ \rho = \rho$. The range of a retraction is called a *retract* of X.

Definition 4.10. A *neighborhood retract* of a topological space X is a set $A \subseteq X$ with the property that it is the retract of some of its neighborhoods.

Remark 4.11. $A \subseteq X$ is a retract of X if and only if id_A has a continuous extension $\rho: X \to A$. $A \subseteq X$ is a neighborhood retract of X if and only if id_A has a continuous extension $\rho: U \to A$ for some neighborhood $U \subseteq X$ of A.

Corollary 4.12. *If Y is an* AE (ANE) *for a class $\mathcal{C}$ then so is every (neighborhood) retract $Y_0 \subseteq Y$.*

Proof. There is a retraction $\rho\colon U \to Y_0$ onto Y_0 with $U := Y$ (resp. an open neighborhood $U \subseteq Y$ of Y_0). Then U is an AE (ANE) for $\mathcal{C}$ (by Proposition 4.5), and so Proposition 4.7 with $g := \rho$ and $h := \mathrm{id}_{Y_0}$ implies that Y_0 is an AE (ANE) for $\mathcal{C}$. □

Proposition 4.13. *Every retract of a Hausdorff space X is closed in X.*

Proof. Let $\rho\colon X \to Y$ be a retraction onto $Y \subseteq X$. Applying Corollary 2.115 with $\Phi := \rho$ and $\Psi := \mathrm{id}_X$, we find that $Y = \{x \in X : \rho(x) = \mathrm{id}_X(x)\}$ is closed in X. □

Definition 4.14. A topological space X is called a LCR ("locally convex retract") if it is homeomorphic to a retract of a convex subset of a locally convex space.

A topological space X is called a LCNR ("locally convex neighborhood retract"), if it is homeomorphic to a neighborhood retract of a convex subset of a locally convex space.

Using this notion, Dugundji's extension theorem can be formulated as follows.

Theorem 4.15 (Dugundji for Retracts). (AC). *Every* LCR *(*LCNR*) is an* AE *(*ANE*) for the class of metric spaces.*

Proof. In view of Corollaries 4.8 and 4.12, the result follows from Theorem 4.3. □

Remark 4.16. Without AC, we obtain in view of Remark 4.4 that every LCR (LCNR) is an AE (ANE) for the class of separable metric spaces.

4.2 ANR and AR Spaces

Of particular importance is the following class of AE/ANE spaces:

Definition 4.17. A topological space X is an AR ("absolute retract") if it is metrizable and an AE for the class of metrizable spaces.

A topological space X is an ANR ("absolute neighborhood retract") if it is metrizable and an ANE for the class of metrizable spaces.

In literature, it is usually not distinguished between AR/ANR spaces and LCR/LCNR spaces. The reason is the following corollary of the theorems of Arens–Eells and Dugundji which also explains the name:

Theorem 4.18 (Arens–Eells for Retracts). (a) *Every metric* AR *(*ANR*) is isometric to a closed (neighborhood) retract of a normed space. In particular, every* AR *(*ANR*) is a metrizable* LCR *(*LCNR*), and the converse holds if we assume* AC.

(b) *Every separable metric space which is an* AE *(*ANE*) for the class of separable metric spaces is isometric to a closed (neighborhood) retract of a separable normed space and thus a metric* LCR *(*LCNR*). Conversely, every* LCR *(*LCNR*) is an* AE *(*ANE*) for the class of separable metric spaces.*

Proof. The Arens–Eells embedding theorem states that there is an isometry $f: X \to X_0$ onto a closed subset X_0 of a normed space Y. By Corollary 3.65, we can assume that Y is separable if X is separable. Now if X is an AE (ANE) for Y, then also X_0 has this property by Corollary 4.8, and so $\mathrm{id}_{X_0}: X_0 \to X_0$ has an extension to a continuous map $\rho: Y \to X_0$ (or $\rho: U \to Y_0$ with an open neighborhood $U \subseteq Y$ of X_0, respectively.) The converse assertion is the content of Theorem 4.15 or Remark 4.16, respectively. □

Proposition 4.19. (a) *Every open subset of an* ANR *is an* ANR.

(b) *Every finite Cartesian products of* AR *(*ANR*) spaces is an* AR *(*ANR*).*

(c) *Every (neighborhood) retract of an* AR *(*ANR*) is an* AR *(*ANR*).*

(d) *Every space which is homeomorphic to an* AR *(*ANR*) is an* AR *(*ANR*).*

(e) *If X_0 is a topological space, X is an* AR *(*ANR*), and if there are $g \in C(X, X_0)$ and $h \in C(X_0, X)$ with $g \circ h = \mathrm{id}_{X_0}$ then X_0 is an* AR *(*ANR*).*

Proof. The assertions follow immediately from Proposition 4.5, Proposition 4.6, Corollary 4.12, Corollary 4.8, or Proposition 4.7, respectively. For the last assertion, it remains to show that X_0 is metrizable. Let d be a metric on X generating the topology. Then a metric on X_0 can be defined by

$$d_h(y_1, y_2) := d(h(y_1), h(y_2)). \tag{4.1}$$

Since $g \circ h = \mathrm{id}_{X_0}$ implies that $h(x) \neq h(y)$ for $x \neq y$, a straightforward calculation shows that this is indeed a metric on X_0. This metric is compatible with the topology of X_0. To see this, let $y_0 \in X_0$. Since h and $f := d(\cdot, h(y_0)): X \to [0, \infty)$ are continuous, it follows that $f \circ h$ is continuous. Hence, for every $r > 0$ the set

$$B_h(r, y_0) := \{y \in X_0 : d_h(y, y_0) < r\} = (f \circ h)^{-1}([0, r))$$

is an open neighborhood of y_0. Conversely, if $U \subseteq X_0$ is a neighborhood of $y_0 \in X_0$, we put $x_0 := h(y_0)$. Since $g(x_0) = y_0$, the continuity of g implies that $g^{-1}(U)$ is a neighborhood of x and thus contains some ball $B_r(x_0) \subseteq X$. It follows that U contains $B_h(r, y_0)$, since for all $y \in B_h(r, y_0)$ we have with $x := h(y)$ that $d(x, x_0) = d_h(y, y_0) < r$, hence $x \in B_r(x_0) \subseteq g^{-1}(U)$, and so $y = g(x) \in U$. □

We want to find spaces which are an AE or ANE for a larger class of spaces than just metric spaces. To this end, the following notion is useful.

Definition 4.20. A metrizable space X is *topologically complete*, if it can be equipped with a complete metric generating the topology. We call a topologically complete ANR (AR) an ANR_c (AR_c).

Of course, every complete metric space is topologically complete. In particular, every compact metrizable space is topologically complete.

However, the class of topologically complete spaces is much larger than one might expect from this example:

Theorem 4.21. *Let X be a topologically complete metrizable space. Assume one of the following:*

(a) *$X_0 \subseteq X$ is open in X.*

(b) *$X_0 \subseteq X$ is closed in X.*

(c) *X_0 is a neighborhood retract of X.*

(d) *X_0 is homeomorphic to X.*

(e) *X_0 is a topological space, and there are $g \in C(X, X_0)$, $h \in C(X_0, X)$ with $g \circ h = \mathrm{id}_{X_0}$.*

Then X_0 is a topologically complete metrizable space.

Proof. Let d be a complete metric on X generating the topology. Assume first that $X_0 \subseteq X$ is closed in X. Then Lemma 3.8 implies that d is also a complete metric on X_0.

Assume next that $X_0 \subseteq X$ is open in X. In case $X_0 = X$, we have nothing to prove, and in case $X_0 \neq X$, a complete metric on X_0 is given by

$$\delta(x, y) := d(x, y) + \left| \frac{1}{\operatorname{dist}(x, X \setminus X_0)} - \frac{1}{\operatorname{dist}(y, X \setminus X_0)} \right| .$$

Indeed, a straightforward calculation shows that δ is a metric of X_0. The metrics δ and d generate the same topology on X_0: If $x_n \in X_0$ is a sequence and $x \in X$

satisfy $\delta(x_n, x) \to 0$ then clearly $d(x_n, x) \to 0$, and conversely, if $d(x_n, x) \to 0$ then

$$\operatorname{dist}(x_n, X \setminus X_0) \to c := \operatorname{dist}(x, X \setminus X_0) > 0$$

by Lemma 3.11, and so the continuity of $t \mapsto 1/t$ at c implies $\delta(x_n, x) \to 0$. Moreover, X_0 is complete with respect to the metric δ. To see this, let $x_n \in X_0$ be a Cauchy sequence with respect to δ. Then x_n is a Cauchy sequence with respect to d and thus convergent with respect to d to some $x \in X$. By our definition of a Cauchy sequence, we find for $\varepsilon := 1$ some $y \in X_0$ such that $\delta(x_n, y) < \varepsilon$ for all except finitely many n. This implies that $1/\operatorname{dist}(x_n, X \setminus X_0)$ remains bounded, and so there is some $\varepsilon > 0$ with $\operatorname{dist}(x_n, X \setminus X_0) \geq \varepsilon$ for all n. Since Proposition 3.12 implies that $\operatorname{dist}(x_n, X \setminus X_0) \to \operatorname{dist}(x, X \setminus X_0)$, we find $x \in X_0$. Since $d(x_n, x) \to 0$ and $x \in X_0$ we thus obtain by what we have proved above that $\delta(x_n, x) \to 0$.

If $X_0 \subseteq X$ is a neighborhood retract of X, that is, there is an open neighborhood $U \subseteq X$ of X_0 and a retraction $\rho: U \to X_0$ onto X_0, then by what we just proved, U is a topologically complete metrizable space. Since X_0 is closed in U (Proposition 4.13) also X_0 is a topologically complete metrizable space by the first assertion.

The case of a homeomorphism is a special case of the last assertion. To see the latter, let $g \in C(X, X_0)$ and $h \in C(X_0, X)$ satisfy $g \circ h = \mathrm{id}_{X_0}$. Consider on X_0 the metric (4.1). As we have seen in the proof of Proposition 4.19, this metric is compatible with the topology of X_0. We show that it is complete. To this end, we note that h is an isometry of X_0 (with the metric d_h) onto the subset $Y := h(X_0) \subseteq X$ (with the metric d). Moreover, $\rho := h \circ g: X \to Y$ is a retraction, because $\rho(\rho(x)) = h \circ g \circ h \circ g = h \circ g = \rho$. Since $g \circ h = \mathrm{id}_{X_0}$, it follows that $g(X) = X_0$ and thus $\rho(X) = h(X_0) = Y$. Hence, Y is a retract of X and thus closed by Proposition 4.13 and thus complete. Since $h^{-1}: Y \to X_0$ is an isometry onto X_0 and thus preserves Cauchy sequences, convergent sequences, and thus also completeness, it follows that X_0 is complete. □

Corollary 4.22. (a) *Every open subset of an* ANR_c *is an* ANR_c.

(b) *Every finite product of* AR_c *(*ANR_c*) spaces is an* AR_c *(*ANR_c*).*

(c) *Every (neighborhood) retract of an* AR_c *(*ANR_c*) is an* AR_c *(*ANR_c*).*

(d) *If X_0 is a topological space, X is an* AR_c *(*ANR_c*), and if there are $g \in C(X, X_0)$, $h \in C(X_0, X)$ with $g \circ h = \mathrm{id}_{X_0}$ then X_0 is an* AR_c *(*ANR_c*).*

Proof. The assertions follow from Proposition 4.19 and Theorem 4.21. □

Theorem 4.23 (Arens–Eells for AR_c/ANR_c). *Every* AR_c *(*ANR_c*) is homeomorphic to a closed (neighborhood) retract of a Banach space, and the converse holds if we assume* AC.

Proof. If X is an AR_c (ANR_c), equipped with a complete metric, then Theorem 3.63 implies that X is isometric to a subset X_0 of a Banach space Y. By the completeness, X_0 is closed. Since X_0 is an AE (ANE) by Corollary 4.8, the map $\mathrm{id}_{X_0}\colon X_0 \to X_0$ has an extension to a continuous map $\rho\colon Y \to X_0$ (or $\rho\colon U \to Y_0$ with an open neighborhood $U \subseteq Y$ of X_0, respectively.)

Conversely, if we assume the axiom of choice then every Banach space is an AR_c by Theorem 4.15. Hence, the claim follows from Corollary 4.22. □

We will not use the following deep result but just cite it to give the reader an intuition about the meaning of AR_c and ANR_c spaces.

Theorem 4.24. (AC). *A metrizable space is an* AR_c *(*ANR_c*) if and only if it is an* AE *(*ANE*) for the class of paracompact Hausdorff spaces.*

Proof. A deep result from [8] resp. [41] states that every Banach space is an AE for the class of so-called fully normal space resp. collectionwise normal spaces. Since paracompact Hausdorff spaces have both this property, it follows that every Banach space is an AE for the class of paracompact Hausdorff spaces. Hence, if X is an AR_c (ANR_c) then Theorem 4.23 and Corollary 4.22 imply that X is an AE (ANE) for the class of paracompact Hausdorff spaces. The converse was obtained in [106] and, independently, in [41]. □

For the case that the space is separable, we obtain a stronger statement with a much simpler proof even without AC.

Theorem 4.25. *A metrizable space is an* AE *(*ANE*) for the class of* T_4 *spaces if and only if it is separable, topologically complete, and an* AE *(*ANE*) for the class of separable metric spaces.*

Proof. Let X be a separable complete metric space which is an AE (ANE) for the class of separable metric spaces. By Theorem 3.74, X is homeomorphic to a closed subset X_0 of $Y := \mathbb{R}^{\mathbb{N}}$. Note that Y is separable since e.g. $(\mathbb{Q} \cap (0,1))^{\mathbb{N}}$ is dense in Y by Proposition 3.66. Since X_0 is an AE (ANE) for the class of separable metric spaces by Corollary 4.8, it follows that $\mathrm{id}_{X_0}\colon X_0 \to Y$ has an extension to a continuous map $\rho\colon Y \to Y$ (or $\rho\colon U \to Y$ with a neighborhood $U \subseteq Y$ of X_0, respectively). In particular, X_0 is a (neighborhood) retract of Y.

Note that Tietze's extension theorem (note Remark 2.68) implies that Y is an ANE for the class of T_4 spaces. Since X_0 is a (neighborhood) retract of Y, Corollary 4.12 thus implies that X_0 is an AE (ANE) for the class of T_4 spaces. By Corollary 4.12, it follows that also the homeomorphic space X is an AE (ANE) for the class of T_4 spaces.

The proof of the converse assertion is more involved. Since we do not need it, we just remark that it was obtained in [106] and, independently, in [41]. □

We point out that we have shown in particular that the following result holds without the (general) axiom of choice.

Corollary 4.26. *For a separable topologically complete metrizable space X the following statements are equivalent:*

(a) *X is homeomorphic to a (neighborhood) retract of a Banach space.*

(b) *X is a* LCR *(*LCNR*).*

(c) *X is an* AR *(*ANR*).*

(d) *X is an* AR_c *(*ANR_c*).*

(e) *X is an* AE *(*ANE*) for the class of separable metric spaces.*

(f) *X is an* AE *(*ANE*) for the class of T_4 spaces.*

Proof. By our assumptions on X the implications (a)⇒(b), (f)⇒(e), (f)⇒(c), and (c)⇔(d)⇒(e) are trivial. In view of Remark 4.16, we have the implication (b)⇒(e) (without AC). Theorem 4.25 provides us the implication (e)⇒(f), and the remaining implication (d)⇒(a) is the content of Theorem 4.23. □

Following Borsuk [28, IV.(6.1)], we give now a large class of examples of ANR spaces. Let us first prepare this result:

Lemma 4.27. *Let X_1, X_2 be closed subsets of a metric space X. If X_1 and X_2 are both* AE *(*ANE*) spaces for X, and if $X_1 \cap X_2$ is a (neighborhood) retract of X then $X_1 \cup X_2$ is a (neighborhood) retract of X.*

Proof. We divide X into the three disjoint sets Y_0, Y_1, and Y_2, consisting of all $x \in X$ for which $\operatorname{dist}(x, X_1)$ is equal, less, or larger than $\operatorname{dist}(x, X_2)$, respectively. Putting $X_0 := X_1 \cap X_2$, we have then $X_0 \subseteq Y_0$. Let $\rho\colon U \to X_0$ be a retraction onto X_0 with $U = X$ (or $U \subseteq X$ being an open neighborhood of X_0, respectively).

We put $U_0 := X$ (or let $U_0 \subseteq X$ be an open neighborhood of X_0 with $\overline{U}_0 \subseteq U$ which is possible by Corollary 2.48, since X_0 is closed by Proposition 4.13.) Note that U_0 is open, and $A := \overline{U}_0$ satisfies $X_0 \subseteq U_0 \subseteq A \subseteq U$.

Note that Y_0 is closed, and thus $A_0 := A \cap Y_0$ is closed. Hence, for $i = 1, 2$ the set $A_i := X_i \cup A_0$ is closed, and the functions $\rho_i : A_i \to X_i$, defined by $\rho_i|_{X_i} := \mathrm{id}_{X_i}$ and $\rho_i|_{A_0} := \rho|_{A_0}$ are continuous by the glueing lemma (Lemma 2.93): For the compatibility of the definition, note that $X_i \cap A_0 \subseteq X_i \cap Y_0 \subseteq X_0$.

Thus, for $i = 1, 2$ there are continuous extensions $\rho_i : U_i \to X_i$ with $U_i = X$ (or with $U_i \subseteq X$ being open neighborhoods of A_i, respectively.) We define now $C_i := (Y_i \cap U_i) \cup A_0$ $(i = 1, 2)$ and $C := C_1 \cup C_2$. Since $Y_1, Y_2 \subseteq X$ are open and disjoint, it follows that $C_i = (C \setminus Y_{2-i}) \cup A_0$ $(i = 1, 2)$ are closed in C, and $C_1 \cap C_2 = A_0$. Since $\rho_i|_{A_0} = \rho|_{A_0}$ $(i = 1, 2)$, the glueing lemma thus implies that we obtain a continuous function $\rho : C \to X_1 \cup X_2$ by $\rho|_{C_i} := \rho_i|_{C_i}$ $(i = 1, 2)$.

We show that this is the desired retraction. We put

$$M := (Y_1 \cap U_1) \cup (Y_2 \cap U_2) \cup (U_0 \cap U_1 \cap U_2).$$

For $i = 1, 2$ we get by $Y_0 \cap X_i = X_0 \subseteq U_0 \cap U_1 \cap U_2 \subseteq M$ that

$$X_i = X_i \cap (Y_i \cup Y_0) \subseteq (Y_i \cap U_i) \cup (Y_0 \cap X_i) \subseteq (Y_i \cup Y_0) \cap M, \tag{4.2}$$

and since $Y_0 \cap Y_i = \varnothing$ and $Y_0 \cap U_0 \subseteq Y_0 \cap A = A_0$, we have further

$$(Y_i \cup Y_0) \cap M \subseteq (Y_i \cap M) \cup (Y_0 \cap M) \subseteq (Y_i \cap U_i) \cup (Y_0 \cap U_0) \subseteq C_i. \tag{4.3}$$

In particular, $X_i \subseteq C_i$ implies that $\rho|_{X_i} = \rho_i|_{X_i} = \mathrm{id}_{X_i}$, hence ρ is a retraction of C onto $X_1 \cup X_2$. Moreover, $C = X$ (or C is a neighborhood of $X_1 \cup X_2$, respectively. Indeed, since Y_1 and Y_2 are open, the set M is open with $X_1 \cup X_2 \subseteq M$ by (4.2), and by (4.3), we have $M = (Y_1 \cup Y_2 \cup Y_0) \cap M \subseteq C_1 \cup C_2 = C$.) □

Theorem 4.28. *Let X_1, X_2 be closed in the metrizable space $X_1 \cup X_2$.*

(a) (AC). *If all three spaces X_1, X_2, and $X_1 \cap X_2$ are* AR *(*ANR*) spaces then so is $X_1 \cup X_2$.*

(b) *If all three spaces X_1, X_2, and $X_1 \cap X_2$ are separable and an* AE *(*ANE*) for the class of separable metric spaces then so is $X_1 \cup X_2$.*

Proof. By the Arens–Eells embedding theorem, we can assume that $X_1 \cup X_2$ is a closed subspace of a normed space X. If X_1 and X_2 are separable, we can assume in view of Corollary 3.65 that also X is separable. By extending $\mathrm{id}_{X_1 \cap X_2}$ to a continuous map $\rho : U \to X_1 \cap X_2$ with $U = X$ (or a neighborhood $U \subseteq X$ of $X_1 \cap X_2$), we find in particular that $X_1 \cap X_2$ is a (neighborhood) retract of X. Lemma 4.27 implies that $X_1 \cup X_2$ is a (neighborhood) retract of X and thus a metric LCR (LCNR). The assertion follows by Theorem 4.15 or Remark 4.16, respectively. □

Corollary 4.29. (AC). *Let X be a locally convex space, and $Y = Y_1 \cup \cdots \cup Y_n$ where Y_k are metrizable convex subsets of X and closed in Y. Then Y is an* ANR.

Proof. The proof is by induction on n: For $n = 1$, the claim follows from Theorem 4.15, and if the claim is proved for $n-1$, we apply Theorem 4.28 with $X_1 := Y_1 \cup \cdots \cup Y_{n-1}$ and $X_2 := Y_n$. Since $X_1 \cap X_2$ is the union of the $n-1$ convex sets $Y_k \cap Y_n$ $(k < n)$ which are closed in $X_1 \cap X_2$, the induction hypothesis implies that $X_1 \cap X_2$ is an ANR, and thus also $X_1 \cup X_2$ is an ANR by Theorem 4.28 □

Remark 4.30. Without AC, we obtain in view of Remark 4.16 and Corollary 4.26 a slightly weaker result than Corollary 4.29:

We have to assume in addition that all Y_k are separable, and either we also have to assume that Y is topologically complete, or we can conclude only that $Y = Y_1 \cup \cdots \cup Y_n$ is an ANE for the class of separable metric spaces.

4.3 Extension of Compact Maps and of Homotopies

Theorem 4.31. *Let $X = \prod_{n \in \mathbb{N}} I_n$ with nondegenerate intervals $I_n \subseteq \mathbb{R}$. Then for any compact set $K \subseteq X$ and any neighborhood $U \subseteq X$ of K the following holds:*

(a) *There is a compact* AR $Y_X \subseteq X$ *containing K.*

(b) *There is a compact* ANR Y_U *with $K \subseteq Y_U \subseteq U$.*

Remark 4.32. The proof shows that Y_X can be chosen to be homeomorphic to $[0,1]^{\mathbb{N}}$.

Proof. Assume first that $X = [0,1]^{\mathbb{N}}$, endowed with the metric (3.28). Then X is compact by Tychonoff's theorem (Corollary 3.67) and an AR by Remark 2.68. Hence, the first assertion follows with $Y_X = X$. For the second assertion, we assume without loss of generality that $U \neq X$ is open in X.

Since K is compact and disjoint from the closed set $X \setminus U$, we find by Corollary 3.14 some $r > 0$ with $B_{2r}(K) \subseteq U$. Choose some $n \in \mathbb{N}$ with $2^{-n} < r$. We equip $\mathbb{R}^n$ with thc sum-norm, i.e.

$$\|(x_1, \ldots, x_n)\| := |x_1| + \cdots + |x_n|.$$

Let $p\colon X \to \mathbb{R}^n$ and $q\colon X \to X$ denote the projections onto the first n coordinates and onto the remainder, that is, $p((x_1, x_2, \ldots)) := (x_1, \ldots, x_n)$ and $q((x_1, x_2, \ldots)) := (x_{n+1}, x_{n+2}, \ldots)$. Proposition 3.66 implies that p and q

are continuous. Clearly $h := p \times q \colon X \to \mathbb{R}^n \times X$ is invertible, and also the inverse is continuous by Proposition 3.66.

The set $K_0 := p(K)$ is compact. Hence, since the open balls $B_r(z) \subseteq \mathbb{R}^n$ ($z \in K_0$) cover K_0 we find finitely many $z_1, \dots, z_m \in K_0$ such that K_0 is contained in $Y_0 := K_r(z_1) \cup \dots \cup K_r(z_m)$. Remark 4.30 implies that Y_0 is a compact ANR. Hence, Proposition 4.19 implies that $Y_0 \times X$ and thus also $Y_U := h^{-1}(Y_0 \times X)$ are compact ANR spaces.

Since $K_0 \subseteq Y_0$, we have $Y_U = p^{-1}(Y_0) \supseteq p^{-1}(K_0) = p^{-1}(p(K)) \supseteq K$. It remains to show that $Y_U \subseteq U$. Thus, let $y = (y_1, y_2, \dots) \in Y_U$. By definition of Y_U, we find for $p(y) = (y_1, \dots, y_n)$ some $z_k \in K_0$ with $\|p(y) - z_k\| \le r$. Since $K_0 = p(K)$, there is some $x = (x_1, x_2, \dots) \in K$ with $z_k = p(x)$. Now

$$d(y,x) \le \sum_{k=1}^{n} |y_k - x_k| + \sum_{k=n+1}^{\infty} 2^{-k} = \|p(y) - z_k\| + 2^{-n} < 2r.$$

Hence, $y \in B_{2r}(K) \subseteq U$.

For the general case $X = \prod_{n=1}^{\infty} I_n$, we observe first that the projections $\pi_n \colon X \to I_n$, $\pi_n((x_1, x_2, \dots)) := x_n$, are continuous by Proposition 3.66. Hence, $\pi_n(K)$ is contained in a nondegenerate compact interval $J_n \subseteq I_n$, and so $K \subseteq X_0 := \prod_{n=1}^{\infty} J_n \subseteq X$. Lemma 3.73 implies that X_0 and $H := [0,1]^{\mathbb{N}}$ are homeomorphic. Thus, the first claim follows with $Y_X := X_0$. For the second claim, let h denote the homeomorphism of X_0 onto H. Let $U \subseteq X$ be a neighborhood of K, without loss of generality open. Then $U_0 := U \cap X_0$ is open in X_0 with $K \subseteq U_0$, and so $h(U_0)$ is open in H and contains the compact set $h(K)$. As we have shown above there is a compact ANR $Y \subseteq H$ with $h(K) \subseteq Y \subseteq h(U_0)$. Thus Proposition 4.19 implies that $Y_U := h^{-1}(Y)$ is a compact ANR with $K \subseteq Y_U \subseteq U$. □

Remark 4.33. For the case that X is a normed space, a result analogous to Theorem 4.31 was obtained in [68].

It seems to be an open problem whether the statement of Theorem 4.31 holds also if X is replaced by an AR (for the first assertion) or an ANR (for the second assertion), respectively.

In view of the Arens–Eells embedding theorem for retracts (Theorem 4.18) and the mentioned result [68], one might conjecture at a first glance that such a generalization is possible if one considers X as a subset of a normed space $\hat{X}$. However, the corresponding compact ANR (AR) $Y \subseteq \hat{X}$ is not necessarily contained in X, and the intersection $X \cap Y$ of two ANR (AR) spaces X and Y need not necessarily be an ANR (AR).

Theorem 4.31 is rather useful if one wants to extend compact continuous maps (although one can avoid to use that result in many cases, as we will see).

In connection with degree theory it turns out that it is very important to be able to extend continuous compact maps. In fact, this is more important for us than extension results for maps which are just continuous. Somewhat surprisingly, these extension results will play a particular role when we want to deal with noncompact maps, as we will see in Chapter 14 (in particular, in Section 14.2, we will need this property often).

Unfortunately, the notion of a compact map depends on the notion of a relatively compact set which in turn does not only depend on the topology but also on the considered space: A set can be relatively compact in a space Z but fail to be relatively compact in $Y \subseteq Z$. For this reason, the following notion depends on a space Z, in general:

Definition 4.34. Let Y and Z be topological spaces. Then Y is an CE_Z ("compact extensor into Z") for a class $\mathcal{C}$ of topological spaces if for every space X from $\mathcal{C}$, every closed subset $A \subseteq X$, and every continuous $f\colon A \to Y$ which is compact into Z (that is, $f(A)$ is relatively compact in Z), there is a continuous extension $F\colon X \to Y$ of f which is compact into Z.

Similarly, Y is an CNE_Z ("compact neighborhood extensor into Z") for a class $\mathcal{C}$ of topological spaces if for every space X from $\mathcal{C}$, every closed subset $A \subseteq X$, and every continuous map $f\colon A \to Y$ which is compact into Z there is a neighborhood $U \subseteq X$ of A and a continuous compact extension $F\colon U \to Y$ of f which is compact into Z.

If $\mathcal{C}$ consists only of one space X, we call Y simply an CE_Z (or CNE_Z) for X.

Although we formally do not require it, this definition is only useful if $Y \cap Z \neq \emptyset$. Moreover, although formally it is not required in Definition 4.34 that on the set $Y \cap Z$ the topologies induced by Y and Z coincide, we will typically have $Y \subseteq Z$ (with the inherited topology).

Let us first discuss the case that $Y \subseteq Z$ is closed in Z. In this case, the space Z actually plays no role:

Proposition 4.35. *A closed set $Y \subseteq Z$ is a CE_Z (CNE_Z) for a class $\mathcal{C}$ if and only if Y is a CE_Y (CNE_Y) for $\mathcal{C}$.*

Proof. The assertion follows immediately from Proposition 2.122. □

For the case that $Y \subseteq Z$ is closed in Z, the following criterion is sufficient for most practical purposes:

Theorem 4.36. *Let $Y \subseteq Z$ be closed in Z. If Y is a metrizable* LCNR *(*LCR*), then Y is an* CNE_Z *(*CE_Z*) for the class of T_4 spaces.*

Proof. By Proposition 4.35, we are to show that Y is an CNE_Y (CE_Y) for the class of T_4 spaces. Thus, let X be T_4, $A \subseteq X$ be closed, and $f \in C(A, Y)$ with $f(A)$ contained in a compact set $K \subseteq Y$. By Lemma 3.69 there is a homeomorphism h of K onto a subset $K_0 \subseteq [0,1]^{\mathbb{N}}$. Since $K_0 = h(K)$ is compact, it is closed (Proposition 2.45). In view of Remark 4.16, Y is an ANE (AE) for the separable metric space $H := [0,1]^{\mathbb{N}}$, and so $h^{-1}\colon K_0 \to K \subseteq Y$ has a continuous extension $g\colon V_0 \to Y$ where $V_0 \subseteq H$ is an open neighborhood of K_0 (or $V_0 = H$, respectively). In view of Theorem 4.31, there is a compact ANR (AR) Y_0 satisfying $K_0 \subseteq Y_0 \subseteq V_0$. Then Y_0 is an ANE (AE) for the class of T_4 spaces by Theorem 4.25, and so $h \circ f\colon A \to K_0 \subseteq Y_0$ has a continuous extension $F\colon U \to V$ with a neighborhood $U \subseteq X$ of A ($U = X$). Then $Y_1 := h^{-1}(Y_0)$ is a compact subset of Y, and $h^{-1} \circ F\colon U \to Y_1$ is the required extension for f. □

Remark 4.37. Theorem 4.31 will be used essentially in our proof of Theorem 4.54. In the above proof of Theorem 4.36, the usage of Theorem 4.31 only slightly simplified the argument, but it could have been avoided.

In fact, in the proof of Theorem 4.36, we could alternatively have observed that, since H is T_4, Corollary 2.48 implies that there is an open neighborhood $V_1 \subseteq H$ of K_0 with $Y_0 := \overline{V_1} \subseteq V_0$. Then Y_0 is compact but not necessarily an ANR. Nevertheless, the open subset set $V_1 \subseteq H$ is an ANE for the class of T_4 spaces by Proposition 4.5, and so we can find similarly as in the above proof of Theorem 4.36 a continuous extension $F\colon U \to V_1 \subseteq Y_0$ of $h \circ f$ with a neighborhood $U \subseteq X$ of A.

Corollary 4.38. *If K is an* ANR *(*AR*) then K is a* CNE_K *(*CE_K*) for the class of T_4 spaces.*

Proof. Theorem 4.18 implies that K is a metrizable LCNR (LCR). Since K is closed in K, the assertion follows from Theorem 4.36. □

The following property will turn out to be crucial for degree theory in the noncompact case.

Proposition 4.39. *Let Z be a Banach space, and $Y_1, Y_2 \subseteq Z$ closed convex subsets with $Y_1 \cap Y_2 \neq \varnothing$. Then $Y := Y_1 \cup Y_2$ is a* CE_Y *for the class of T_4 spaces.*

Proof. Let X be T_4, $A \subseteq X$ be closed, and $f \in C(A, Y)$ with $f(A)$ being relatively compact in Y. Let $y \in Y_1 \cap Y_2$. By Corollary 3.62, the sets

$K_i := \overline{\operatorname{conv}}((\{y\} \cup f(A)) \cap Y_i)$ are compact convex subsets of Y_i $(i = 1, 2)$. Since K_1, K_2, and $K_1 \cap K_2$ are convex and compact (and nonempty in view of $y \in K_1 \cap K_2$), we obtain by Corollary 4.26 and Theorem 4.28 that $K_1 \cup K_2$ is an AE for the class of separable metric spaces and thus even for the class of T_4 spaces. Hence, f has an extension $F \in C(X, K_1 \cup K_2)$ which is a required extension of f. □

Now we discuss Definition 4.34 for the case that $Y \subseteq Z$ is not necessarily closed in Z. In this case, a map $f: X \to Y$ can be compact into Z although it is not compact into Y. This is the case, for example, if Y is a non-closed bounded subset of a finite-dimensional normed space Z and $f = \mathrm{id}_Y$. Nevertheless, it may happen that Y is an CE_Z for a large class of spaces, as we will show now.

Although such results appear important for applications in degree theory, there are not many sufficient criteria known for this property. It seems that the following criteria are essentially new.

By a *homotopy* (from X into Y) we mean a continuous map h: $[0, 1] \times X \to Y$.

Proposition 4.40. *Let Y and Z be subsets of a topological space Z_0 with the inherited topology. Assume that $Y_Z := \overline{Y} \cap Z$ (closure in Z_0) is a CNE_{Y_Z} (CE_{Y_Z}) for a class $\mathfrak{C}$ of spaces. If there is a homotopy h: $[0, 1] \times Y_Z \to Y_Z$ satisfying*

$$h(0, \cdot) = \mathrm{id}_{Y_Z} \text{ and } h((0, 1] \times Y_Z) \subseteq Y, \tag{4.4}$$

then Y is a CNE_Z (CE_Z) for all T_6 spaces from $\mathfrak{C}$.

Proof. Let X be a T_6 space from the class $\mathfrak{C}$, let $A \subseteq X$ be closed, and let $f: A \to Y$ be continuous and compact into Z. Since Y_Z is closed in Z, Proposition 2.122 implies that f is compact into Y_Z. The hypothesis thus implies that f has a continuous extension $f: U \to Y_Z$ where $U \subseteq X$ is a neighborhood of A (or $U = X$, respectively), and $f(U)$ is contained in a compact subset $K \subseteq Y_Z$. Since X is T_6, there is $\lambda \in C(U, [0, 1])$ with $\lambda^{-1}(0) = A$ (and $\lambda^{-1}(1) = \varnothing$). Then $F(x) := h(\lambda(x), f(x))$ defines a required extension of f. Indeed, for $x \in A$, we have $F(x) = h(0, f(x)) = f(x)$ by (4.4), and for $x \in U \setminus A$, we have $\lambda(x) > 0$ and thus $F(x) \in Y$ by (4.4). Finally, $F(U) \subseteq h([0, 1] \times K)$, and the latter set is compact by Theorem 2.63 and Proposition 2.100, and it is contained in $Y_Z \subseteq Z$. □

In order to apply Proposition 4.40, we use the following simple observation which we will also use later in Corollaries 9.78, 13.22, and 14.52.

Lemma 4.41. *Let Y be a convex subset of a topological vector space, and $y_0 \in \mathring{Y}$. Then for every $z \in \overline{Y}$ and every $t \in (0, 1]$ we have $(1 - t)z + ty_0 \in Y$.*

Proof. Since $y_0 \in \mathring{Y}$, there is a neighborhood U of 0 in the topological vector space with $y_0 + U \subseteq Y$. In case $t = 1$, the assertion is trivial. Thus, assume $t \in (0, 1)$. Then $r := t/(1-t) \in (0, \infty)$, and so $z - rU$ is a neighborhood of $z \in \overline{Y}$ and thus must contain some $y_1 \in Y$. Then $z - y_1 \in rU$, and so $y_2 := y_0 + r^{-1}(z - y_1) \in y_0 + U \subseteq Y$. Since Y is convex, it follows that also the convex combination $ty_2 + (1-t)y_1 = ty_0 + (1-t)(z - y_1) + (1-t)y_1 = (1-t)z + ty_0$ belongs to Y. □

Summarizing the previous observations, we obtain the following result which seems to be new and generalizes a similar result from [86] in several respects.

Theorem 4.42. *Let Z_0 be a locally convex vector space, and $Y \subseteq Z \subseteq Z_0$. Suppose that Y is convex and that $Y_Z := Z \cap \overline{Y}$ is metrizable with* conv $Y_Z \subseteq Z$. *If $\mathring{Y} \neq \varnothing$ then Y is an* CE_Z *for the class of T_6 spaces.*

Proof. Proposition 3.54 implies that $\overline{Y}$ is convex. Hence, $M := \overline{Y} \cap \operatorname{conv} Y_Z$ is convex with $Y_Z \subseteq M \subseteq \overline{Y} \cap Z = Z_Y$, and so $Y_Z = M$ is actually convex and thus an LCR. Theorem 4.36 implies that Y_Z is an CE_{Y_Z} for the class of T_4 spaces. Let $y_0 \in \mathring{Y}$. Lemma 4.41 shows that the homotopy $h(t, z) := (1-t)z + ty_0$ satisfies (4.4), and so the assertion follows from Proposition 4.40. □

One of the most important properties of ANE spaces is Borsuk's famous homotopy extension property which was originally only formulated for ANR spaces [27] (see also [28, Theorem 8.1]). We provide a simple proof in a more general situation, based on Corollary 2.112 and Urysohn's lemma:

Theorem 4.43 (Homotopy Extension). *Suppose that Y is an* ANE *(*CNE_Z*) for* $[0, 1] \times X$. *Let $A \subseteq X$ be closed, and suppose that one of the following holds:*

(a) *X is T_4.*

(b) *X is T_{3a} and $A \subseteq X$ is compact.*

Let $h \in C([0, 1] \times A, Y)$ (and be compact into Z), and let $f \in C(X, Y)$ (and compact into Z) with $h(0, \cdot) = f|_A$. Then there is a continuous (and compact into Z) extension $H\colon [0, 1] \times X \to Y$ of h with $H(0, \cdot) = f$.

Proof. The set $C := (\{0\} \times X) \cup ([0, 1] \times A)$ is closed in $P := [0, 1] \times X$. We extend h continuously to $h\colon C \to Y$ by putting $h(0, \cdot) := f$. Since Y is an ANE (CNE_Z) for P, there is a continuous extension $h\colon U \to Y$ (compact into Z) where $U \subseteq P$ is a neighborhood of C. By Corollary 2.113, there is an open neighborhood $V \subseteq X$ of A with $[0, 1] \times V \subseteq U$. By Urysohn's Lemma 2.38 or

Lemma 2.39, respectively, there is $\lambda \in C(X,[0,1])$ with $\lambda(X \setminus V) = \{0\}$ and $\lambda(A) = \{1\}$. For all $t \in [0,1]$, we have

$$h_0(t,x) := (\lambda(x)t, x) \in (\{0\} \times X) \cup ([0,1] \times V) \subseteq C \cup ([0,1] \times V) \subseteq U,$$

that is, $h_0 \colon P \to U$. Hence, $H := h \circ h_0 \colon [0,1] \times X \to Y$ is defined and continuous (and compact into Z). For all $(t,x) \in [0,1] \times A$, we have $h_0(t,x) = (t,x)$ and thus $H(t,x) = h(t,x)$, that is, H extends h. Moreover, $H(0,\cdot) = h(0,\cdot) = f$. □

The homotopy extension property states, roughly speaking, that we can extend homotopies from closed subsets when we prescribe the "beginning" of the homotopy. The next result extends this observation inasmuch as we can also prescribe the "end" of the homotopy locally.

Theorem 4.44 (Both-Sided Homotopy Extension I). *Suppose that Y is an* ANE *(*CNE_Z*) for* $[0,1] \times X$. *Let $A \subseteq X$ be closed, and suppose that one of the following holds:*

(a) *X is T_4.*

(b) *X is T_{3a} and $A \subseteq X$ is compact.*

Let $h \in C([0,1] \times A, Y)$ (and be compact into Z), and let $f \in C(X,Y)$ (and be compact into Z) with $h(0,\cdot) = f|_A$. Let $X_0 \subseteq X$ be closed with $A \subseteq X_0$ and $g \in C(X_0, Y)$ (and compact into Z) with $h(1,\cdot) = g|_A$.

Then there is a continuous (and compact into Z) extension $H \colon [0,1] \times X \to Y$ of h with $H(0,\cdot) = f$ and an open neighborhood $V \subseteq X$ of A such that $H(1,\cdot)|_{\overline{V} \cap X_0} = g|_{\overline{V} \cap X_0}$.

Proof. The set $C := (\{0\} \times X) \cup ([0,1] \times A) \cup (\{1\} \times X_0)$ is closed in $P := [0,1] \times X$. We extend h continuously to $h \colon C \to Y$ by putting $h(0,\cdot) := f$ and $h(1,\cdot) := g$. Since Y is an ANE for P, there is a neighborhood $U \subseteq P$ of C such that h has a continuous (compact into Z) extension to some $h \colon U \to P$. By Corollary 2.113, there is an open neighborhood $V_0 \subseteq X$ of A with $[0,1] \times V_0 \subseteq U$.

Since X is T_4 (or T_3 and A is compact) there is an open neighborhood $V \subseteq X$ of A with $\overline{V} \subseteq V_0$. Applying Theorem 4.43, we find a continuous extension $H \colon [0,1] \times X \to Y$ of $h|_{[0,1] \times \overline{V}}$ with $h(0,\cdot) = f$. □

If either Y is an AE (not only an ANE) or if we are only interested in the *local* extension (on both sides), we do not have to require any separation properties:

Theorem 4.45 (Both-Sided Homotopy Extension II). *Suppose that Y is an* ANE *(*CNE_Z*) for* $[0,1] \times X$, *and that $A \subseteq X$ is closed.*

Let $h \in C([0,1] \times A, Y)$ (and be compact into Z). For $i = 0, 1$, let $X_i \subseteq X$ be closed with $A \subseteq X_i$, and let $f_i \in C(X_i, Y)$ (and be compact into Z) with $h(i, \cdot) = f_i|_A$.

Then there is a neighborhood $V \subset X$ of A and a continuous (and compact into Z) extension $H\colon [0,1] \times V \to Y$ of h with $H(i, \cdot)|_{X_0 \cap V} = f_i|_{X_0 \cap V}$ for $i = 0, 1$.

If Y is an AE *(CE_Z) for $[0,1] \times X$, it may be arranged that $V = X$.*

Proof. The set $C := (\{0\} \times X_0) \cup ([0,1] \times A) \cup (\{1\} \times X_1)$ is closed in $P := [0,1] \times X$. We extend h continuously to $h\colon C \to Y$ by putting $h(i, \cdot) := f_i$ ($i = 0, 1$). Since Y is an ANE (CNE_Z) for P, there is a neighborhood $U \subseteq P$ of C such that h has a continuous (compact into Z) extension $h\colon U \to P$. By Corollary 2.113, there is an open neighborhood $V \subseteq X$ of A with $[0,1] \times V \subseteq U$. Hence, $H := h|_{[0,1]\times V}$ has the required properties.

If Y is an AE (CE_Z) for $[0,1] \times X$, we can choose $U = P$ and thus $V_0 = X$, hence $V = X$. □

Recall that a subset M of a topological space X is called *contractible in X* if it is empty or if there is a homotopy $h\colon [0,1] \times M \to X$ with $h(0, \cdot) = \mathrm{id}_M$ and $h(1, \cdot) = x_0$ for some $x_0 \in M$.

Corollary 4.46. *Let $\mathcal{C}$ be a class of T_4 spaces. Let Y be an* ANE *for all spaces of the form $[0,1] \times X$ with X from $\mathcal{C}$. Consider the statements:*

(a) *Y is contractible in itself.*

(b) *Y is an* AE *for all spaces from $\mathcal{C}$.*

(c) *Y is an* AE *for the space $[0,1] \times Y$.*

Then (a)⇒(b) *and* (c)⇒(a). *Moreover,* (a)–(c) *are equivalent if $[0,1] \times Y$ belongs to $\mathcal{C}$ (and thus is T_4).*

Proof. To prove (a)⇒(b), suppose that Y is contractible in itself, that is, there is $H \in C([0,1] \times Y, Y)$ with $H(0, \cdot) = \mathrm{id}_Y$ and $H(1, \cdot) = y_0$ for some $y_0 \in Y$. Let X be a space from $\mathcal{C}$, $A \subseteq X$ be closed, and $f\colon A \to Y$ be continuous. We define $h\colon [0,1] \times A \to Y$ by $h(t,x) := H(1-t, f(x))$, and $f_0\colon X \to Y$ by $f_0(x) \equiv y_0$. Then $f_0|_A = h(0, \cdot)$, and so the homotopy extension property (Theorem 4.43) implies that h has an extension $h \in C([0,1] \times X, Y)$. Then $h(1, \cdot)|_A = f$, and so $h(1, \cdot) \in C(X, Y)$ is a required extension of f.

If Y belongs to $\mathcal{C}$, the implication (b)⇒(c) is trivial, and if (c) holds, we apply Theorem 4.45 with $A = \varnothing$, $f_0 = \mathrm{id}_Y$, and $f_1(y) \equiv y_0$ to find a homotopy $h\colon [0,1] \times Y \to Y$ satisfying $h(i, \cdot) = f_i$ ($i = 0, 1$), that is, Y is contractible in itself. □

Corollary 4.47. *An* ANR *is an* AR *if and only if it is contractible in itself.*

Proof. We apply Corollary 4.46 with the class $\mathcal{C}$ of metric spaces. □

4.4 UV$^\infty$ and R_δ Spaces and Homotopic Characterizations

Recall that two continuous maps $f_0, f_1: X \to Y$ are called *homotopic* if there is a homotopy $h: [0,1] \times X \to Y$ with $h(i, \cdot) = f_i$ for $i = 0, 1$. It is well-known and easy to see that being homotopic is an equivalence relation in $C(X, Y)$. The set of equivalence classes (homotopy classes) will be denoted by $[X, Y]$.

A crucial notion for us will be the following.

Definition 4.48. A compact Hausdorff space X is *weak* UV^∞ if for every completely regular space Y the following holds. If Y is an ANE for the class of paracompact Hausdorff spaces then any $f \in C(X, Y)$ is homotopic to a constant map.

This definition may appear a bit unusual, although it is sometimes simple to verify. The aim of this section is to give some other characterizations of that definition. One is the following which somewhat explains the terminology:

Definition 4.49. A compact space X is UV^∞ if it is homeomorphic to a closed subset X_0 of a space Y such that

(a) Y is a paracompact Hausdorff space and an ANE for the class of paracompact Hausdorff spaces.

(b) Every neighborhood $U \subseteq Y$ of X_0 contains a neighborhood $V \subseteq Y$ of X_0 such that V is contractible in U.

Remark 4.50. It is equivalent to require that U and/or V are open.

Theorem 4.51. *Every* UV^∞ *space is weak* UV^∞. *The converse holds if we assume* AC.

Proof. Let X be UV^∞ and $X_0 \subseteq Y$ be as in Definition 4.49. Let Z be a completely regular space which is an ANE for the class of paracompact Hausdorff spaces, and $f \in C(X, Z)$. We are to show that f is homotopic to a constant map. Let $g: X_0 \to X$ be the homeomorphism onto X which exists by hypothesis. Then $f \circ g: X_0 \to Z$ is continuous and thus has a continuous extension $F: U \to Z$ to a neighborhood $U \subseteq Y$ of X_0. By hypothesis, U contains a neighborhood $V \subseteq X$ such that there is a homotopy $h: [0,1] \times V \to U$ with $h(0,\cdot) = \mathrm{id}_V$ and $h(1,\cdot) = x_0 \in U$. Then $H: [0,1] \times X \to Z$, defined by $H(t,x) := F(h(t, g^{-1}(x)))$, is a homotopy satisfying $H(0,\cdot) = F \circ g^{-1} = f$ and $h(1,\cdot) = F(x_0)$, that is, $f: X \to Z$ is homotopic to a constant map.

For the converse implication, let X be weak UV^{∞}. By Proposition 2.70, there is a homeomorphism J of X onto a subset X_0 of $Y := [0,1]^I$. The space Y is Hausdorff, and if we assume AC, then Y is compact by Tychonoff's Theorem 2.71, and Y is an AE for the class of T_4 spaces by Corollary 2.67. In particular, Y is paracompact and normal by Propositions 2.26 and 2.45, and in view of Proposition 2.72, Y is an AE for the class of paracompact Hausdorff spaces. Now let $U \subseteq Y$ be an open neighborhood of X_0. We are to show that U contains a neighborhood $V \subseteq Y$ of X_0 which is contractible in U.

Proposition 4.5 shows that U is an ANE for the class of paracompact Hausdorff spaces, and since Y is completely regular by Urysohn's lemma, also $U \subseteq Y$ is completely regular by Theorem 2.42. Since X is weak UV^{∞} and $J: X \to X_0 \subseteq U$ is continuous, it follows that there is a homotopy $H: [0,1] \times X \to U$ with $H(0,\cdot) = J$ and $H(1,\cdot) = x_0 \in U$. Since $A := X_0 = J(X)$ is compact, it is closed in the Hausdorff space Y by Proposition 2.45. Define $h: [0,1] \times A \to U$ by $h(t,x) := h(t, J^{-1}(x))$. Since Y is T_4, Corollary 2.48 implies that there is an open neighborhood $V_0 \subseteq Y$ of $A \cup \{x_0\}$ with $Y_0 := \overline{V}_0 \subseteq U$. By Proposition 2.30, Y_0 is a paracompact Hausdorff space, and so also $[0,1] \times Y_0$ is a paracompact Hausdorff space by Corollary 2.76. Hence, U is an ANE for $[0,1] \times Y_0$. Define $f := \mathrm{id}_{Y_0}$ and $g: Y_0 \to Y_0$ by $g(x) \equiv x_0$. Then $h(0,\cdot) = J \circ J^{-1}|_A = f|_A$ and $h(1,\cdot) = g|_A$. Hence, the both-sided homotopy extension theorem (Theorem 4.44) implies that h has a continuous extension $h: [0,1] \times Y_0 \to U$ such that there is an open in Y_0 set $V_1 \subseteq Y_0$ with $A \subseteq V_1$, $h(0,\cdot) = \mathrm{id}_{V_1}$ and $h(1,\cdot)|_{V_1} = x_0$. In particular, $V := V_0 \cap V_1$ is open in Y with $X_0 \subseteq V \subseteq U$, and the restriction $h|_{[0,1]\times V}$ proves that V is contractible in U. □

Remark 4.52. Formally, our Definition 4.48 differs from the corresponding property stated in [93, Proposition 1.15(i)], since in [93] ANE spaces are required to be paracompact. This would be a problem, since it is not clear whether open subsets of $[0,1]^I$ are paracompact. However, according to personal communication, a slightly different definition for ANE spaces was meant than written in [93], and with this intended definition, our Definition 4.48/Theorem 4.51 corresponds exactly to [93, Proposition 1.15(i)].

We point out that the definitions of UV^{∞} spaces vary in literature in the classes of considered spaces. In particular, in many cases only metric spaces are considered, that is, it is required that X be metric, and the property (a) of Definition 4.49

is replaced by the requirement that Y is an ANR. Our Definition 4.49 corresponds to the definition of UV^∞ spaces from [93].

It is mathematical folklore that if one considers only metric spaces in the definition of UV^∞ spaces, one obtains exactly the so-called R_δ spaces:

Definition 4.53. A space X is called R_δ if it is the intersection of a decreasing sequence of compact metric spaces, each of it being contractible in itself.

For our slightly different definition of UV^∞ spaces, one can expect that folklore result at most if X is metrizable. However, even if X is metrizable, it is not immediately clear that the R_δ spaces are exactly the UV^∞ space in our sense, since we do not consider an ANR space Y in Definition 4.49: The class of spaces Y which we consider in Definition 4.49a is on the one hand broader, since Y might fail to be metrizable, and on the other hand smaller, since not every ANR is an ANE for the class of paracompact Hausdorff spaces (recall Theorem 4.24).

Thus, it might be somewhat surprising that the two definitions are actually equivalent for the case that X is metrizable. In fact, we will show this now simultaneously together with the mentioned folklore result.

We remark that the equivalence of UV^∞ spaces with R_δ spaces in the metric case was remarked without proof in [93], but maybe the change of the condition (a) in Definition 4.49 was ignored there.

Theorem 4.54. *For a compact metric space X the following statements are equivalent:*

(a) *X is a weak UV^∞ space.*

(b) *X is an UV^∞ space.*

(c) *X is homeomorphic to a subset X_0 of a compact* ANR *Y such that every neighborhood $U \subseteq Y$ of X_0 contains a neighborhood $V \subseteq Y$ of X_0 such that V is contractible in U.*

(d) *X is homeomorphic to a closed subset X_0 of a compact* AR *Y such that every neighborhood $U \subseteq Y$ of X_0 contains a neighborhood $V \subseteq Y$ of X_0 such that V is contractible in U.*

(e) *X is an R_δ space.*

(f) *X is the intersection of a decreasing sequence of compact* AR *spaces.*

(g) *For every ANR Y every $f \in C(X,Y)$ is homotopic to a constant map.*

(h) *For every compact* ANR *Y every $f \in C(X,Y)$ is homotopic to a constant map.*

We point out that in Theorem 4.54 the characterization (h) of R_δ spaces is new (to the author's knowledge). This characterization seems to be very convenient if one wants to prove that a given compact metric space is UV^{∞} (or, equivalently, R_δ).

Proof. The implication (g)⇒(f) is a deep result of D. M. Hyman [79]; we show only the remaining implications.

The implication (f)⇒(e) is a trivial consequence of the fact that every metric AR is contractible in itself by Corollary 4.47.

To see (e)⇒(g), let $X = \bigcap_{n=1}^{\infty} X_n$ with $X_1 \supseteq X_2 \supseteq \cdots$ being compact and contractible in itself metric spaces. Let Y be an ANR, and let $f \in C(X,Y)$. Since Y is an ANE for X_1, there is an open neighborhood $U \subseteq X_1$ of X and a continuous extension $F\colon U \to Y$ of f. Since $X_n \setminus U$ is a decreasing sequence of closed subsets of the compact space X_1 with an empty intersection, there is some n with $X_n \setminus U = \varnothing$ (Proposition 2.28), that is, $X_n \subseteq U$. Since X_n is contractible in itself, there is a homotopy $h\colon [0,1] \times X_n \to X_n \subseteq U$ with $h(0,\cdot) = \mathrm{id}_{X_n}$ and $h(1,\cdot) = x_0 \in X_n \subseteq U$. Then the homotopy $H := F \circ h|_{[0,1]\times X}\colon [0,1]\times X \to Y$ satisfies $H(0,\cdot) = f$ and $H(1,\cdot) = F(x_0)$, and so $f \in C(X,Y)$ is homotopic to a constant map.

To prove (f)⇒(d), we observe first that we have already obtained (g), and we know that X is a subset of a compact AR $Y = X_1$. It suffices to show that for every open neighborhood $U \subseteq X_1$ of X there is a neighborhood $V \subseteq U$ of X such that V is contractible in U. Since U is an ANR by Proposition 4.19, we obtain from (g) that the inclusion map $\mathrm{id}_X\colon X \to U$ is homotopic to a constant map, that is, there is a homotopy $h\colon [0,1] \times X \to U$ with $h(0,\cdot) = \mathrm{id}_X$ and $h(1,\cdot) = x_0 \in U$. Put $f := \mathrm{id}_U$, and let $g\colon U \to U$ be the constant map $g(x) := x_0$. Now we can apply Theorem 4.44 with $A := X$ and obtain that there is a continuous map $H\colon [0,1]\times U \to U$ with $H(0,\cdot) = f|_U$ and a neighborhood $V \subseteq U$ of X and $H(1,\cdot)|_V = g|_V$. Then the homotopy $H|_{[0,1]\times V}$ shows that V is contractible in U. Thus (d) holds with $Y = X_1$.

Since every AR is an ANR, the implication (d)⇒(c) is trivial.

For the implications (c)⇒(b) and (a)⇒(h), we observe that every compact ANR Y is completely regular and paracompact by Corollary 3.13 and Proposition 2.26. Moreover, Y is complete (Theorem 3.24) and thus an ANE for the class of paracompact Hausdorff spaces by Theorem 4.25 and Proposition 2.72. It follows that the space Y from Definition 4.48 or (c) can take the role of Y in (h) or Definition 4.49, respectively.

The implication (b)⇒(a) was already obtained in Theorem 4.51.

It remains to prove (h)⇒(g). Thus, suppose that (h) holds, and let Y be an ANR and $f \in C(X,Y)$ be continuous. Then $X_0 := f(X) \subseteq Y$ is compact.

Lemma 3.69 implies that there is a homeomorphism g of X_0 onto a compact and thus closed subset K of the Hilbert cube $[0,1]^{\mathbb{N}}$. Since Y is an ANR, there is a neighborhood $U \subseteq [0,1]^{\mathbb{N}}$ of K such that $g^{-1}: K \to Y$ has a continuous extension $G: U \to Y$. Theorem 4.31 implies that there is a compact ANR Y_0 with $K \subseteq Y_0 \subseteq U$. Applying the hypothesis (h), we find that there is a continuous map $h: [0,1] \times X \to Y_0$ with $h(0,\cdot) = g \circ f$ and $h(1,\cdot) = x_0 \in Y_0$. Since $Y_0 \subseteq U$, we obtain that $H := G \circ h: [0,1] \times X \to Y$ is continuous with $H(0,\cdot) = G \circ g \circ f = f$ and $H(1,\cdot) = G(x_0) \in Y$. Hence, H is the retraction which shows that (g) holds. □

We try to avoid usage of algebraic topology in this monograph. However, in order to give the reader familiar with homology theory an impression about the meaning of UV^∞ spaces, we sketch now briefly the relation with acyclic spaces.

Definition 4.55. For a topological space X, we denote by $\check{H}^n(X)$ the Čech cohomology group of X for dimension n with coefficients in the group $\mathbb{Z}$. The space X is called *acyclic* if $\check{H}^n(X) = \check{H}^n(*)$ for all n, where $*$ denotes the space consisting of only one point.

Proposition 4.56. *Every weak UV^∞ space is acyclic.*

Proof. Let X be weak UV^∞. In particular, X is Hausdorff and compact and thus paracompact and locally compact. Since $G = \mathbb{Z}$ is countable, it follows from [78] that there is a bijection of $\check{H}^n(X)$ into the set of homotopy classes $[X, K_n]$ where K_n denotes the n-th Eilenberg–MacLane space. Since X is weak UV^∞, every $f \in C(X, K_n)$ is homotopic to a constant map, that is, $[X, K_n]$ is a singleton. □

Remark 4.57. A more classical proof that every UV^∞ space is acyclic proceeds as follows: With the notation of Definition 4.49, since Y is Hausdorff and X_0 is compact, it follows that X_0 is the inverse limit of its contractible neighborhoods in Y. Since each U is acyclic and the Čech cohomology functor is continuous, see [49], also X_0 must be acyclic.

The converse of Proposition 4.56 does not hold:

Example 4.58 (Kahn). In [82] an example of an acyclic metric compact space X is given such that there is a map $\varphi \in C(X, S^3)$ which is not homotopic to a constant map. Hence, X cannot be UV^∞.

However, from the viewpoint of applications the difference between UV^∞ spaces and (compact Hausdorff) acyclic spaces is not very large: For practically all compact acyclic spaces which occur in analytical problems, one can also prove that they are UV^∞ spaces. So, "practically", it is convenient to think of UV^∞ spaces as the class of compact acyclic spaces.

Chapter 5

Advanced Topological Tools

5.1 Some Covering Space Theory

Let X, Y, and Z be topological spaces, and $p: X \to Y$.

Definition 5.1. A *lifting* of $f: Z \to Y$ (with respect to p) is a continuous map $g: Z \to Y$ satisfying $f = p \circ g$.

One can prove the existence and a certain uniqueness of liftings in many cases if one makes the following assumption about p:

Definition 5.2. Let $p: X \to Y$. An open set $U \subseteq Y$ is *evenly covered* if the set $p^{-1}(U)$ is the union of pairwise disjoint open subsets $X_i \subseteq X$ $(i \in I)$ such that $p|_{X_i}$ is a homeomorphism onto U for each $i \in I$. The map p is a *covering map* if each point of Y is contained in an evenly covered set.

It is admissible by this definition that p is not onto, since $I = \varnothing$ is not excluded.

If it can be arranged in Definition 5.2 that I has always the same (finite) number k of elements then p is called a *k-fold covering map*.

Proposition 5.3 (Uniqueness of Lifting). *Let $p: X \to Y$ be a covering map. If Z is connected and $f: Z \to Y$ is continuous then each two lifts of f are equal if they coincide in at least one point.*

Proof. Let g_1, g_2 be two lifts of f, and put

$$A_1 := \{z \in Z : g_1(z) = g_2(z)\}$$

and $Z := X \setminus A_2$. We are to show that A_k $(k = 1, 2)$ are both open. The assertion then follows from the connectedness of Z.

Thus, let $z \in A_k$ $(k = 1$ or $k = 2)$. Let $U \subseteq Y$ be an evenly covered open neighborhood of $f(z)$. Then $p^{-1}(U)$ is the union of pairwise disjoint open sets X_i $(i \in I)$ such that $p_i := p|_{X_i}$ is a homeomorphism onto U for all $i \in I$. Let i_j be such that $g_j(z) \in X_{i_j}$ $(j = 1, 2)$. By the continuity of g_j there is a neighborhood $V \subseteq Z$ of z with $g_j(V) \subseteq X_{i_j}$ $(j = 1, 2)$. In case $X_{i_1} = X_{i_2}$ (which happens for $k = 1$) we obtain from $p_{i_1} \circ g_1|_V = f|_V = p_{i_2} \circ g_2|_V$ that

$g_1|_V = g_2|_V$. It follows also for $k = 2$ that $X_{i_1} \neq X_{i_2}$ and thus $X_{i_1} \cap X_{i_2} = \varnothing$ implies $g_1(V) \cap g_2(V) = \varnothing$. □

Theorem 5.4 (Homotopy Lifting). *Let $p\colon X \to Y$ be a covering map. Then for each homotopy $h\colon [0,1] \times Z \to Y$ and each lift h_0 of $h(0,\cdot)$ there is a unique lift H of h with $H(0,\cdot) = h_0$.*

Proof. We will show that each $z_0 \in Z$ has a neighborhood $M \subseteq Z$ such that $h|_{[0,1]\times M}$ has a lift H_M with $H_M(0,\cdot) = h_0|_M$. Then for any $z \in M$ the map $H_M(\cdot,z)$ is a lift of $h(\cdot,z)$ with $H_M(0,z) = h_0(z)$ and thus uniquely determined by this property by Proposition 5.3, since $[0,1]$ is connected (Proposition 2.14). It follows that the only candidate for a lift is $H(\cdot,z) := H_M(\cdot,z)$, and that the latter is actually independent of the choice of M and thus H is well-defined. Since H_M is continuous, it follows that H is continuous and thus indeed a lift.

Let $\mathcal{O}$ denote the system of sets of the form $I \times Z_0$ where $I \subseteq [0,1]$ is an open (in $[0,1]$) interval, $Z_0 \subseteq Z$ is open and where $I \times Z_0 \subseteq h^{-1}(U)$ for some evenly covered open $U \subseteq Y$. Since the family of all sets $h^{-1}(U)$ with evenly covered open $U \subseteq Y$ is an open cover of $[0,1] \times Z$, the definition of the product topology implies that $\mathcal{O}$ is an open cover of $[0,1] \times Z$.

For any $z_0 \in Z$ the compact set $[0,1] \times \{z_0\}$ is covered by finitely many sets $I_1 \times Z_1, \dots, I_m \times Z_m \in \mathcal{O}$. Then $Z_0 := Z_1 \cap \dots \cap Z_m$ is an open neighborhood of z_0. We define a partition $0 = t_0 < t_1 < \dots < t_n = 1$ of $[0,1]$ by choosing for each finite nonempty intersection of the intervals $I_1, \dots, I_m$ some partition point in this intersection (and adding 0 and 1 to the partition). Then for each $k = 1, \dots, n$ there is some j with $[t_{k-1}, t_k] \in I_j$, and by construction there is some evenly covered open $U_k \subseteq Y$ with $h([t_{k-1}, t_k] \times Z_0) \subseteq U_k$.

Now we define inductively open neighborhoods $Z_0 \supseteq M_0 \supseteq M_1 \supseteq \dots \supseteq M_n$ of z_0 such that $h|_{[0,t_k]\times M_k}$ has a lift H_k with $H_k(0,\cdot) = h_0|_{M_k}$ $(k = 0, \dots, n)$. Then $M := M_n$ is the required neighborhood.

For $k = 0$, we can choose $M_0 := Z_0$. If M_{k-1} is already defined, we already have a partially defined lift $H_k|_{[0,t_{k-1}]\times M_{k-1}}$. Since U_k is evenly covered, $p^{-1}(U_k)$ is the union of pairwise disjoint open sets $X_i \subseteq X$ $(i \in I)$ such that $p_i := p|_{X_i}$ is a homeomorphism onto U_k for each $i \in I$. Let $i \in I$ be that index with $H_k(t_{k-1}, z_0) \in X_i$. The continuity of $H_k(t_{k-1},\cdot)$ implies that there is some open neighborhood $M_k \subseteq M_{k-1}$ of z_0 with $H_k(\{t_{k-1}\} \times M_k) \subseteq X_i$. Then $H_k(t,z) := p_i^{-1} \circ h(t,z)$ defines for $(t,z) \in [t_{k-1}, t_k] \times M_k$ a continuous map. By the glueing lemma (Lemma 2.93), H_k is even continuous on $[0,t_k] \times M_k$ and thus a required lift. □

As a special case, we obtain the so-called path-lifting property of covering maps:

Corollary 5.5 (Path-Lifting). *Let $p: X \to Y$ be a covering map. Then for each $f \in C([0,1], Y)$ and each $x \in p^{-1}(f(0))$ there is a unique lift F of f with $F(0) = x$.*

Proof. This is Theorem 5.4 with $Z := \{0\}$. □

Our aim now is to show that the space $[0,1]$ in Corollary 5.5 can actually be replaced by a much larger class of spaces.

Before we can do this, we use Corollary 5.5 to formulate a necessary and sufficient criterion for the existence of a lifting:

Theorem 5.6 (Lifting Criterion). *Let $p: X \to Y$ be a covering map, and Z be path-connected and locally path-connected. Let $f: Z \to Y$ be continuous, and $x_0 \in X$ and $z_0 \in Z$ be fixed with $p(x_0) = f(z_0)$.*

Then f has a lift F with $F(z_0) = x_0$ if and only if for each $\gamma \in C([0,1], Z)$ with $\gamma(0) = \gamma(1) = z_0$ the unique lift γ_0 of γ with $\gamma_0(0) = x_0$ satisfies $\gamma_0(1) = x_0$. In this case, the lift F is unique.

Proof. If the lift F exists, it must by Corollary 5.5 be defined at $z \in Z$ as follows: Let $\gamma: [0,1] \to Z$ be some path with $\gamma(0) = z_0$ and $\gamma(1) = z$, and let γ_0 the unique lift of γ with $\gamma_0(0) = x_0$. Then $F(z)$ must be $\gamma_0(1)$.

With the choice $z = z_0$, we obtain the necessity of the criterion. For the proof of the sufficiency, we note that the above defined map F is well-defined. Indeed, let δ be another path with a corresponding lift δ_0. The glueing lemma implies the continuity of

$$\hat{\gamma}(t) := \begin{cases} \gamma(2t) & \text{if } t \in [0, 1/2], \\ \delta(2-2t) & \text{if } t \in [1/2, 1], \end{cases}$$

which by hypothesis lifts to a path $\hat{\gamma}_0$ satisfying $\hat{\gamma}_0(0) = \hat{\gamma}_0(1) = x_0$. By the uniqueness of the lifting, it follows that $\gamma_0(1) = \hat{\gamma}_0(1/2) = \delta_0(1)$, and so $F(z)$ is well-defined.

To see that F is continuous at $z_1 \in Z$, let $U \subseteq Y$ be an evenly covered open neighborhood of $f(z_1)$. Let $V \subseteq f^{-1}(U)$ be a path-connected neighborhood of z_1. Since U is evenly covered, $p^{-1}(U)$ is the union of pairwise disjoint open sets $X_i \subseteq X$ ($i \in I$) such that $p_i := p|_{X_i}$ is a homeomorphism onto U for every $i \in I$. Let i be that index with $F(z_1) \in X_i$. For $z \in V$, there is $\gamma_0 \in C([0,1], V)$ with $\gamma_0(0) = z_1$ and $\gamma_0(1) = z$. Let $\gamma_1 \in C([0,1], Z)$ satisfy $\gamma_1(0) = z_0$ and

$\gamma_1(1) = z_1$. By the glueing lemma, we can define a continuous path from z_0 to z by

$$\gamma_2(t) := \begin{cases} \gamma_1(2t) & \text{if } t \in [0, 1/2], \\ \gamma_0(2t-1) & \text{if } t \in [1/2, 1]. \end{cases}$$

For $i = 0, 1, 2$, let $\widetilde{\gamma}_i$ be the unique lifting of γ_i with $\widetilde{\gamma}_1(0) = \widetilde{\gamma}_2(0) = x_0$ and $\widetilde{\gamma}_0(0) = \widetilde{\gamma}_1(1)$. By the uniqueness, we have $\widetilde{\gamma}_2(t) = \widetilde{\gamma}_0(2t-1)$ for $t \in [1/2, 1]$. Using this for $t = 1$, we obtain by the definition of F that $F(z_1) = \widetilde{\gamma}_1(0) = \widetilde{\gamma}_0(0)$ and $F(z) = \widetilde{\gamma}_2(0) = \widetilde{\gamma}_0(1)$. Note that $C := \gamma_0([0,1])$ is path-connected and by Proposition 2.14 thus connected. Since $A_1 := C \cap X_i$ and $A_2 := C \cap \bigcup\{X_j : X_j \neq X_i\}$ are open in C and disjoint with $C = A_1 \cup A_2$, it follows from $\gamma_0(z_1) \in A_1$ that $A_2 = \varnothing$ and thus $C \subseteq X_i$. Hence, $F(z) = \gamma_0(z) \in X_i$ and thus $F(z) = p_i^{-1}(f(z))$ for every $z \in V$. Since p_i is a homeomorphism onto U, it follows that $F|_V = p_i^{-1} \circ f|_V$ is continuous on V. In particular, F is continuous at z_1. □

Recall that a topological space Z is called *simply connected* at $z \in Z$ if for each path $\gamma\colon [0,1] \to Z$ with $\gamma(0) = \gamma(1) = z$ there is a homotopy $h\colon [0,1]^2 \to Z$ with $h(0,\cdot) = \gamma$ and $h(1,\cdot) = h(\cdot,0) = h(\cdot,1) = z$. Z is *simply connected* if it is simply connected at every $z \in Z$.

Theorem 5.7 (Unique Existence of Lifting). *Let $p\colon X \to Y$ be a covering map, and Z be path-connected, locally path-connected, and simply connected at $z \in Z$. Let $f \in C(Z,Y)$ and $x \in X$ be fixed with $p(x) = f(z)$. Then f has a unique lift F satisfying $F(z) = x$.*

Proof. By Theorem 5.6, it suffices to show that for each path $\gamma\colon [0,1] \to Z$ with $\gamma(0) = \gamma(1) = z$ the corresponding lift γ_0 with $\gamma_0(0) = x$ satisfies $\gamma_0(1) = x$.

By hypothesis, there is a homotopy $h\colon [0,1]^2 \to Z$ with $h(0,\cdot) = \gamma$ and $h(1,\cdot) = h(\cdot,0) = h(\cdot,1) = z$. By Theorem 5.4, h has a unique lift H with $H(0,\cdot) = \gamma_0$. By Corollary 5.5, the path $h(\cdot,0)$ has a unique lift f with $f(0) = x$. Since the constant function $f(\cdot) = x$ and $H(\cdot,0)$ are two such lifts, it follows that $H(\cdot,0) = f(\cdot) = x$, in particular $H(1,0) = x$. Considering the paths $h(1,\cdot)$, we obtain similarly $H(1,\cdot) = x$, in particular $H(1,1) = x$. Finally, considering the paths $h(\cdot,1)$, we obtain similarly $H(\cdot,1) = x$, in particular $\gamma_0(1) = H(0,1) = x$. □

The path-lifting property (Corollary 5.5) is the special case of Theorem 5.7 with $Z = [0,1]$. Note, however, that we needed the path-lifting property first in order to prove Theorem 5.7.

5.2 A Glimpse on Dimension Theory

In Section 5.3, we will formulate a homotopic Vietoris theorem which is our main tool for degree theory of multivalued maps. One of the hypotheses of that result concerns the notion of the large inductive dimension.

Essentially, all what we need to know about this notion for degree theory is that relatively compact subsets of finite-dimensional manifolds have finite large inductive dimension. We will develop dimension theory only so far that this can be proved conveniently.

Definition 5.8. The *large inductive dimension* $\operatorname{Ind} X$ of a topological space X is defined inductively as follows: $\operatorname{Ind} \varnothing = -1$, and for $X \neq \varnothing$, let $\operatorname{Ind} X$ be the smallest integer number $n \geq 0$ (or ∞ if no such number exists) such that for any closed set $A \subseteq X$ and any open neighborhood $U \subseteq X$ of A there is an open neighborhood $V \subseteq U$ of A with $\operatorname{Ind}(\partial V) \leq n - 1$.

A trivial induction by $\operatorname{Ind} X$ shows that if Y is homeomorphic to X then $\operatorname{Ind} Y = \operatorname{Ind} X$. In other words, the large inductive dimension is preserved under homeomorphisms and thus a topological invariant.

One might expect that $\operatorname{Ind} M \leq \operatorname{Ind} X$ if $M \subseteq X$. Unfortunately, this is not true without additional hypotheses: A counterexample due to Dowker can be found in e.g. [52, Example 2.2.12]. However, this so-called subspace theorem of dimension theory can be proved if X satisfies certain separation axioms or if M is closed. The latter is simple:

Proposition 5.9. *If $M \subseteq X$ is closed then* $\operatorname{Ind} M \leq \operatorname{Ind} X$.

Proof. We show by induction on n that if $\operatorname{Ind} X \leq n$ and $M \subseteq X$ is closed, then $\operatorname{Ind} M \leq n$. The case $n = -1$ is trivial, since we must have $X = \varnothing$. For the induction step, let $A \subseteq M$ be closed in M and $U_M \subseteq M$ be open in M with $A \subseteq U_M$. Proposition 2.10 implies that A is closed in X, and by definition of the subspace topology, there is an open in X set $U \subseteq X$ with $U_M = M \cap U$. Since $\operatorname{Ind} X \leq n$, there is an open neighborhood $V \subseteq U$ of A with $\operatorname{Ind}(\partial V) \leq n - 1$. Then $V_M := M \cap V$ is open in M, and the relative boundary $\partial_M V_M$ is a closed subset of $\partial_X V$. Hence, the induction hypothesis implies $\operatorname{Ind}(\partial_M V_M) \leq \operatorname{Ind}(\partial_X V) \leq n - 1$, and so $\operatorname{Ind} M \leq n$. $\square$

Corollary 5.10. *Let $X \neq \varnothing$ be T_4. Then the following conditions are equivalent:*

(a) $\operatorname{Ind} X \leq n$.

(b) *For each closed $A \subseteq X$ and each open neighborhood $U \subseteq X$ of A there is an open set $V \subseteq X$ with $A \subseteq V \subseteq \overline{V} \subseteq U$ and* $\operatorname{Ind}(\partial V) \leq n-1$.

(c) *For each disjoint closed sets $A, B \subseteq X$, X divides into three disjoint subsets U, V, C with U, V being open such that $A \subseteq U$, $B \subseteq V$, and* $\operatorname{Ind} C \leq n-1$.

Proof. Suppose that (a) holds. Corollary 2.48 implies that there is an open set $U_0 \subseteq X$ satisfying $A \subseteq U_0 \subseteq \overline{U}_0 \subseteq U$. Since $\operatorname{Ind} X \leq n$, there is an open set $U \subseteq U_0$ with $A \subseteq U$ and $\operatorname{Ind}(\partial U) \leq n-1$. Hence, (b) holds.

If (b) holds and $A, B \subseteq X$ are closed and disjoint, then there is an open set $U \subseteq X$ with $A \subseteq U \subseteq \overline{U} \subseteq X \setminus B$ such that $C := \partial U = \overline{U} \setminus U$ satisfies $\operatorname{Ind} C \leq n-1$. Then (c) holds with V being the complement of $\overline{U} = U \cup C$.

Suppose now that (c) holds. Let $A \subseteq X$ be closed, and be $U \subseteq X$ an open neighborhood of A. Then $B := X \setminus U$ is closed and disjoint from A. By hypothesis, X divides into pairwise disjoint open sets $V, W, C \subseteq X$ with V, W being open such that $A \subseteq V$, $B \subseteq W$, and $\operatorname{Ind} C \leq n-1$. Then $A \subseteq V \subseteq \overline{V} \subseteq X \setminus W \subseteq U$, and so $\partial V = \overline{V} \setminus V \subseteq C$. Since ∂V is closed in X, it is also closed in C (Proposition 2.10). Proposition 5.9 thus implies $\operatorname{Ind}(\partial V) \leq \operatorname{Ind} C \leq n-1$. Hence, (a) holds. □

For the proof of the mentioned subspace theorem (Proposition 5.9 if M fails to be closed), we need a technical result which is of independent interest and due to Dowker [42].

Theorem 5.11. *Let X be T_5. If there is a sequence of open sets $X_k \subseteq X$ satisfying $X = X_1 \supseteq X_2 \supseteq \cdots$, $\bigcap_{k=1}^{\infty} X_k = \varnothing$, and $\operatorname{Ind}(X_k \setminus X_{k+1}) \leq n$ for all k then* $\operatorname{Ind} X \leq n$.

Proof. The proof is by induction on n. For the induction start $n = -1$, we observe that we must have $X_k \setminus X_{k+1} = \varnothing$ for all k and thus $X = \varnothing$. Suppose now that the assertion holds for $n-1$ and that X_k are as in the theorem. We are to show that the property of Corollary 5.10(c) holds with n.

Thus, let $A, B \subseteq X$ be closed and disjoint. Since X is T_4, Corollary 2.49 implies that there are open sets $U_0, V_0 \subseteq X$ with $A \subseteq U_0$, $B \subseteq V_0$, and $\overline{U}_0 \cap \overline{V}_0 = \varnothing$. Putting $D_k := X_k \setminus X_{k+1}$, we define now by induction on $k = 1, 2, \ldots$ open sets $U_k, V_k \subseteq X$ and sets $C_k \subseteq X$ with $\operatorname{Ind} C_k \leq n-1$ satisfying the following properties:

$$\overline{U}_k \cap \overline{V}_k \cap X_{k+1} = \varnothing, \tag{5.1}$$

$$U_k \cap V_k = U_k \cap C_k = V_k \cap C_k = \varnothing \tag{5.2}$$

$$C_k \subseteq D_k \subseteq U_k \cup V_k \cup C_k \subseteq X_k \tag{5.3}$$

$$\overline{U}_{k-1} \cap X_k \subseteq U_k \quad \text{and} \quad \overline{V}_{k-1} \cap X_k \subseteq V_k. \tag{5.4}$$

This is possible. Indeed, assume that U_{k-1}, V_{k-1}, and in case $k \geq 2$ also C_{k-1} are already known, in particular (also for $k = 1$)

$$\overline{U}_{k-1} \cap \overline{V}_{k-1} \cap X_k = \varnothing. \tag{5.5}$$

The sets $A_k := \overline{U}_{k-1} \cap D_k$ and $B_k := \overline{V}_{k-1} \cap D_k$ are closed in $D_k \subseteq X_k$ and disjoint. Since $\operatorname{Ind} D_k \leq n$, Corollary 5.10(c) implies that D_k divides into three disjoint sets $\Omega_{0,k}, \Omega_{1,k}, C_k \subseteq D_k$ with $\Omega_{0,k}, \Omega_{1,k}$ being open in D_k such that $A_k \subseteq \Omega_{0,k}$, $B_k \subseteq \Omega_{1,k}$, and $\operatorname{Ind} C_k \leq n - 1$.

We consider now the space $Y_k := X_k \setminus C_k$. Note that $C_k = D_k \setminus (\Omega_{0,k} \cup \Omega_{1,k})$ is closed in D_k which is closed in X_k. Hence, Proposition 2.10 implies that C_k is closed in X_k and thus that Y_k is open in X_k and thus open in X. Moreover, since $C_k \subseteq D_k$ is disjoint from X_{k+1}, we have $X_{k+1} \subseteq Y_k$.

Note that $Z_k := D_k \setminus C_k$ divides into the disjoint open subsets $\Omega_{0,k}$ and $\Omega_{1,k}$ which thus are also closed in $Z_k = Y_k \setminus X_k$ which in turn is closed in Y_k. Using Proposition 2.10, we thus find that $E_k := (\overline{U}_{k-1} \cap Y_k) \cup \Omega_{0,k}$ and $F_k := (\overline{V}_{k-1} \cap Y_k) \cup \Omega_{1,k}$ are closed in Y_k. Moreover, (5.5) implies

$$\begin{aligned} &E_k \cap F_k \\ &\quad \subseteq (\overline{U}_{k-1} \cap \overline{V}_{k-1} \cap Y_k) \cup (\Omega_{0,k} \cap \Omega_{1,k}) \\ &\quad \cup (\Omega_{0,k} \cap \overline{V}_{k-1}) \cup (\Omega_{1,k} \cap \overline{U}_{k-1}) = \varnothing \cup \varnothing \\ &\quad \cup (\Omega_{0,k} \cap B_k) \cup (\Omega_{1,k} \cap A_k) = \varnothing. \end{aligned}$$

Using that Y_k is T_4, we obtain by Corollary 2.49 that there are disjoint open in Y_k and thus open in X sets $U_k, V_k \subseteq Y_k$ with $E_k \subseteq U_k$, $F_k \subseteq V_k$ such that the closures of U_k and V_k in Y_k is disjoint. By Proposition 2.10, the latter means $\overline{U}_k \cap \overline{V}_k \cap Y_k = \varnothing$. Since $X_{k+1} \subseteq Y_k$, this implies (5.1). From $Y_k \cap C_k = \varnothing$, we obtain (5.2). Since

$$U_k \cup V_k \cup C_k \supseteq E_k \cup F_k \cup C_k \supseteq \Omega_{0,k} \cup \Omega_{1,k} \cup C_k = D_k,$$

we find (5.3). Finally, since $C_k \subseteq D_k$ is disjoint from $\Omega_{0,k} \supseteq A_k = D_k \cap \overline{U}_{k-1}$, we obtain

$$\overline{U}_{k-1} \cap X_k = \overline{U}_{k-1} \cap Y_k \subseteq E_k \subseteq U_k,$$

and analogously also the second inclusion of (5.4) holds. Hence, the sets $U_k, V_k, C_k \subseteq X$ have all required properties.

Now we show that the sets

$$U := \bigcup_{k=0}^{\infty} U_k, \quad V := \bigcup_{k=0}^{\infty} V_k, \quad C := \bigcup_{k=1}^{\infty} C_k$$

have the property required for Corollary 5.10. Indeed, since U_k and V_k are open, it follows that U and V are open. Moreover, using (5.3), we find that $U \cup V \cup C$ contains all of the sets D_k $(k = 1, 2, \dots)$, and so $U \cup V \cup C = X$.

To see that U, V, and C are pairwise disjoint, we note that (5.4) implies that $U_\ell \cap X_k \subseteq U_k$ and $V_\ell \cap X_k \subseteq V_k$ for all $k \geq \ell$. Since $U_k, V_k, C_k \subseteq X_k$ and $C_k \subseteq D_k$ by (5.3), we thus obtain with (5.2) that $U_\ell \cap V_k \subseteq U_k \cap V_k = \varnothing$, $U_k \cap V_\ell \subseteq U_k \cap V_k = \varnothing$, $(U_\ell \cup V_\ell) \cap C_k \subseteq (U_k \cup V_k) \cap C_k = \varnothing$, and $(U_k \cup V_k) \cap C_\ell \subseteq X_k \cap D_\ell = \varnothing$ for all $k \geq \ell$. Hence, U, V, C are pairwise disjoint.

To see that $\operatorname{Ind} C \leq n - 1$, we apply the induction hypothesis in the space C which is T_5 by Theorem 2.42. Indeed, the open in C sets $\hat{X}_k := C \cap X_k$ satisfy $C = \hat{X}_1 \supseteq \hat{X}_2 \supseteq \cdots$ and $\bigcap_{k=1}^{\infty} \hat{X}_k = \varnothing$. Moreover, $\hat{D}_k := \hat{X}_k \setminus \hat{X}_{k-1} = C \cap D_k = C_k$. For the last equality, we used that $C_k \subseteq D_k$ by (5.3) and that the sets $D_\ell \supseteq C_\ell$ $(\ell = 1, 2, \dots)$ are pairwise disjoint. Since $\operatorname{Ind} C_k \leq n - 1$ by construction, we thus have $\operatorname{Ind} \hat{D}_k \leq n - 1$, and so by induction hypothesis $\operatorname{Ind} C \leq n - 1$. Hence, the sets U, V, C indeed satisfy all properties required in Corollary 5.10, and the induction step is complete. □

Corollary 5.12. *Let X be T_5. Let there be pairwise disjoint sets $A_1, A_2, \dots \subseteq X$ which constitute a countable cover of X and satisfy* $\operatorname{Ind} A_k \leq n$ *for all k and such that $A_1 \cup \cdots \cup A_k$ is closed for every k. Then* $\operatorname{Ind} X \leq n$.

Proof. Putting $X_k := X \setminus (A_1 \cup \cdots \cup A_k)$, we obtain the assertion from Theorem 5.11. □

If X satisfies a more restrictive separation axiom, the assertion of Corollary 5.12 holds even without the hypothesis that the sets A_k are pairwise disjoint. This assertion is called the sum theorem of dimension theory. It was Dowker who observed in [42] that the sum theorem and the special case of the announced subspace theorem for open sets are best proved together by an induction:

Theorem 5.13. *Let X be a T_6 space.*

(a) **(Sum Theorem of Dimension Theory).** *Let $A_1, A_2, \dots \subseteq X$ be closed sets which cover X and satisfy* $\operatorname{Ind} A_k \leq n$ *for all k. Then* $\operatorname{Ind} X \leq n$.

(b) *Let $M \subseteq X$ be open and* $\operatorname{Ind} X \leq n$. *Then* $\operatorname{Ind} M \leq n$.

Proof. The proof is by induction on n. For the induction start $n = -1$, we must have $A_k = \emptyset$ or $X = \emptyset$, respectively, and so the assertion is trivial. Assume now that the assertion is proved for $n - 1$.

We show first that (b) holds for n. Thus, let $M \subseteq X$ be open and $\operatorname{Ind} X \le n$. Since X is T_6, there is $f \in C(X, [0, 1])$ with $f^{-1}(0) = X \setminus M$ (and $f^{-1}(1) = \emptyset$). We put $P_0 := \emptyset$, and for $k = 1, 2, \dots$, we put $P_k := f^{-1}([1/k, 1])$ and $O_k := f^{-1}((1/k, 1])$. Then the sets P_k are closed, O_k are open, and

$$P_k \subseteq O_{k+1} \subseteq P_{k+1} \subseteq M \quad \text{for } k = 0, 1, 2, \dots. \tag{5.6}$$

Since for each $x \in M$ there is some $k \ge 2$ with $1/k \le f(x) \le 1/(k-1)$, we have

$$M = \bigcup_{k=2}^{\infty} (P_k \setminus O_{k-1}), \tag{5.7}$$

and Proposition 5.9 implies $\operatorname{Ind} P_k \le \operatorname{Ind} X \le n$ for all $k = 0, 1, \dots$

In order to prove that $\operatorname{Ind} M \le n$, we understand all topological notions relative to the space M unless we say something else. Thus, let $A \subseteq M$ be closed and $U \subseteq M$ be an open neighborhood of A. We define for $k = 2, 3, \dots$

$$A_k := A \cap (P_k \setminus O_{k-1}) \quad \text{and} \quad U_k := U \cap (O_{k+1} \setminus P_{k-2}).$$

Since (5.6) implies $P_k \setminus O_{k-1} \subseteq O_{k+1} \setminus P_{k-2} \subseteq P_{k+1}$, we have $A_k \subseteq U_k \subseteq P_{k+1}$. Moreover, U_k is open in X and thus open in P_{k+1}, and $A_k \subseteq M$ is closed and thus also closed in $P_{k+1} \subseteq M$. Since $\operatorname{Ind} P_{k+1} \le n$ we obtain from Corollary 5.10(b) that there is an open set V_k with $\operatorname{Ind}(\partial V_k) = n - 1$ such that $A_k \subseteq V_k \subseteq \overline{V}_k \subseteq U_k$.

Here, the topological notions concerning V_k must first be understood relative to the space P_{k+1}. However, since U_k is open in P_{k+1} and thus V_k is open in U_k which is open in M, it follows that V_k is open in M. Moreover, since P_{k+1} is closed, the closure of V_k in P_{k+1} and in X are the same. Since this closure is contained in $U_k \subseteq M$, we obtain from Proposition 2.10 that this is also the closure of V_k in M. Since V_k is open in M and P_{k+1} and its closures in the spaces M and P_{k+1} are the same, it follows that also its boundaries in the space M and P_{k+1} are the same so that the topological notions concerning V_k can also be understood relative to the space M.

Now the set $V := \bigcup_{k=2}^{\infty} V_k$ is open (in M). Since $A_k \subseteq V \subseteq U_k \subseteq U$, we have with (5.7) that $A \subseteq V \subseteq U$. Hence, $\operatorname{Ind} M \le n$ follows if we can show that $\operatorname{Ind}(\partial V) \le n - 1$. To see the latter, we observe that Theorem 2.42 implies that

$$X_0 := \bigcup_{k=2}^{\infty} (\partial V_k)$$

(all topological notions still refer to M) is T_6. Since $\operatorname{Ind}(\partial V_k) \leq n - 1$, we obtain from (a) for $n-1$ (induction hypothesis) that $\operatorname{Ind} X_0 \leq n-1$. In view of Proposition 5.9, it thus suffices to show that $\partial V \subseteq X_0$.

Thus, let $x \in \partial V$. We choose some index $\ell \geq 3$ with $x \in O_{\ell-2}$. Then $O_{\ell-2} \cap M$ is an open (in M) neighborhood of x which is disjoint from $V_k \subseteq U_k$ for all $k \geq \ell$, because $O_{\ell-2} \subseteq P_{k-2}$. Since $x \in \overline{V}$, it follows that x is already contained in

$$\overline{V_2 \cup \cdots \cup V_{\ell-1}} = \overline{V}_2 \cup \cdots \cup \overline{V}_{\ell-1}.$$

Since $x \notin V$, we find that x lies in some of the sets $\overline{V}_k \setminus V_k = \partial V_k$ for $k \leq \ell$. This shows $\partial V \subseteq X_0$. We thus have finished the proof that (b) holds for n.

It follows that also (a) holds for n. Indeed, if $A_1, A_2, \dots \subseteq X$ are closed sets which cover X and satisfy $\operatorname{Ind} A_k \leq n$, we put $B_k := A_k \setminus \bigcup_{\ell<k} A_\ell$. Then B_k is open in A_k, and so (a) implies $\operatorname{Ind} B_k \leq n$. Since the sets B_k are pairwise disjoint and $B_1 \cup \cdots \cup B_k = A_1 \cup \cdots \cup A_k$ is closed for every k, we obtain from Corollary 5.12 that $\operatorname{Ind} X \leq n$. □

In the above proof of Theorem 5.13, we partly followed the arguments of the proof of [52, Lemma 2.3.4].

Remark 5.14. (AC). In Theorem 5.13, the hypothesis that X be T_6 can be slightly relaxed, and the assertion of Theorem 5.13(a) holds also for certain classes of uncountable covers.

For details, we refer to Dowker's original paper [42] or further generalizations in [52, Section 2.3].

As a special case of Theorem 5.13(b), we obtain the announced subspace theorem of dimension theory. In fact, it was already observed in [42] that for T_5 spaces the assertion of Theorem 5.13(b) is equivalent to the subspace theorem:

Theorem 5.15 (Subspace Theorem of Dimension Theory). *Let X be a T_6 space. Then $M \subseteq X$ implies* $\operatorname{Ind} M \leq \operatorname{Ind} X$.

Proof. We show by induction on n that if $\operatorname{Ind} X \leq n$ and $M \subseteq X$, then $\operatorname{Ind} M \leq n$. The case $n = -1$ is trivial, since we must have $X = \varnothing$. For the induction step, we note that X is T_5 by Theorem 2.36, and so all subsets are T_4. We verify the property of Corollary 5.10(c) in the space $M \subseteq X$. Let $A, B \subseteq M$ be disjoint and closed in M. Then $M_0 := X \setminus (\overline{A} \cap \overline{B})$ is an open subset of X and so Theorem 5.13(b) implies $\operatorname{Ind} M_0 \leq \operatorname{Ind} X \leq n$. Note that by Proposition 2.10 the closure of A in M is $A = \overline{A} \cap M$, and similarly $B = \overline{B} \cap M$. Hence, $\overline{A} \cap \overline{B} \cap M = A \cap B = \varnothing$ which implies $M \subseteq M_0$.

The set $A_0 := M_0 \cap \overline{A}$ and $B_0 := M_0 \cap \overline{B}$ are disjoint closed subsets of M_0. Since $\operatorname{Ind} M_0 \leq n$, find by Corollary 5.10(c) that M_0 divides into disjoint subsets $U_0, V_0, C_0 \subseteq M_0$ with U_0, V_0 being open in M_0 such that $A_0 \subseteq U_0$, $B_0 \subseteq V_0$, and $\operatorname{Ind} C_0 \leq n - 1$. Then $M \subseteq M_0$ divides into the disjoint subsets $U := U_0 \cap M$, $V := V_0 \cap M$, and $C := C_0 \cap M$ where U and V are open in M and $\operatorname{Ind} C \leq n - 1$ by induction hypothesis, because $C \subseteq C_0$. □

As a historical note, we remark that for metric spaces the sum and subspace theorems were known much earlier than Dowker's results, see for instance [111, Theorem II.1]. However, although these proofs in metric spaces are rather different, they are not much easier.

It is well-known that $\operatorname{Ind} \mathbb{R}^n = n$, but we need actually only the upper estimate which follows from the sum theorem.

Proposition 5.16. $\operatorname{Ind} \mathbb{R}^n \leq n$.

Proof. We prove the assertion by induction on n. For $n = 0$, that is, for the space $X = \{0\}$, we can choose $V = \varnothing$ or $V = X$ in Definition 5.8. Suppose that the assertion is proved for $n - 1$ and that $A \subseteq U \subseteq \mathbb{R}^n$ with closed A and open U. Consider the intersection of A with the cube $K_k := [-k, k]^n$. By the Heine–Borel theorem (Proposition 3.59), this intersection is compact, and so it is covered by finitely many open cubes whose closures are contained in U. Joining successively cubes to this cover for larger k, we obtain that A has a countable cover by closed cubes which are contained in U. Let V denote the interior of the union of these cubes. Then ∂V has a countable cover by compact (hence closed) sets which are homeomorphic to compact (hence closed) subsets of $\mathbb{R}^{n-1}$. The induction hypothesis, Proposition 5.9, and the sum theorem thus implies $\operatorname{Ind}(\partial V) \leq n - 1$, and so $\operatorname{Ind} \mathbb{R}^n \leq n$. □

We will also formulate an alternative version of the homotopic Vietoris theorem which has slightly different hypotheses. This alternative version involves the covering dimension instead of the large inductive dimension. Since it seems that this alternative version plays a less important role in connection with degree theory, we do not develop the theory of covering dimension here. Nevertheless, we formulate the definition and main results:

Definition 5.17. The *covering dimension* $\dim X$ of a topological space X is the infimum of all integer numbers $n \geq -1$ (or ∞ if no such number exists) such that every finite open cover of X has a finite open refinement such that every point of X is contained in at most $n + 1$ elements of the refinement.

Clearly, also the covering dimension is a topological invariant.

Also for the covering dimension, a corresponding variant of the sum theorem and of the subspace theorem (for closed subsets) can be proved for T_4 spaces.

However, all what we need to know about the covering dimension concerning degree theory is that relatively compact subsets of finite-dimensional manifolds have finite covering dimension. This will follow from the subspace and sum theorem of Ind together with the following result:

Theorem 5.18. *If X is T_4 then* $\dim X \leq \operatorname{Ind} X$.

Proof. The proof requires the sum theorem for the covering dimension and several other deeper results concerning coverings and is therefore beyond the scope of this monograph. A very readable proof can be found in [52, Theorem 3.1.28]. For a slightly different approach, see e.g. [111, Section VII.2.A]. □

5.3 Vietoris Maps

In classical approaches to degree theory of multivalued maps a crucial role is played by so-called Vietoris maps:

Definition 5.19. Let X and Y be topological spaces. A perfect map $p\colon X \to Y$ is *Vietoris*, if for every $y \in Y$ the set $p^{-1}(y)$ is acyclic.

The reason for the importance of Vietoris maps is the famous Vietoris–Begle theorem:

Theorem 5.20 (Vietoris–Begle). *Let X and Y be paracompact and Hausdorff. If $p\colon X \to Y$ is a Vietoris map then the induced map $p^*\colon \check{H}^n(Y) \to \check{H}^n(X)$ is an isomorphism for all n.*

Proof. The proof of this result is beyond the scope of this monograph. The original proof (in a special case) is due to Vietoris [143], the general case goes back to Begle [12], [13]. □

Roughly speaking, this result states that – from the viewpoint of cohomology theory – the map p can be inverted, that is, p^{-1} can – from the cohomological point of view – be considered as a single-valued map.

Now if a multivalued map Φ can be written in the form $q \circ p^{-1}$ with a Vietoris map p (and such a representation does always exist), one can – on the homological level – consider it as a single-valued map and develop a corresponding degree theory as in the single-valued case. This is how classically the degree of multivalued maps is defined, see e.g. [23], [33], [55], [73], [83], [94], [120], [148]. The same idea was used in numerous results on topological fixed point theory for multivalued maps, e.g. [53], [70], [80] to name a few. Even the earliest paper on topological methods for fixed points of multivalued maps with nonconvex values [48] used this idea.

For our degree theory for function triples, it appears that the approach by homology theory is not sufficient: We must be able to apply some analogue on the homotopic (and not on the homological) level. Such an approach was initiated by W. Kryszewski who developed a corresponding degree in [91] (see also [89], [90]). We note that forerunners of this theory were also developed in a somewhat different setting by R. Bader, L. Gorniewicz, A. Granas, and others, see e.g. [11], [72], [74]. The corresponding variant of the Vietoris result can be found in [92] (see also [93]); we will go into details later.

Unfortunately, this approach requires the additional hypothesis that the covering dimension $\dim p^{-1}(y)$ is finite and even uniformly bounded with respect to y. This is often not the case for multivalued operators in infinite-dimensional spaces.

An alternative approach to a homotopical variant of the Vietoris theorem was also developed by W. Kryszewski in [34], [93] and is based on a trick of G. Kozlowski (in an unpublished result) about a construction with a so-called double mapping cylinder, see e.g. [45]. Roughly speaking, this approach allows us to replace the hypothesis that $p^{-1}(y)$ has finite dimension by the hypothesis that $p^{-1}(y)$ is a weak UV^∞-space.

In fact, as remarked earlier, it will not be sufficient for our approach to consider Vietoris maps: We either have to impose an additional restriction on the dimension of fibres, or we have to assume that the fibres are (weak) UV^∞. From the viewpoint of applications, this is not a severe additional restriction, but without any such restriction our approach will not work. Maps which satisfy this slightly more restrictive condition will be called Vietoris*:

Definition 5.21. Let X and Y be topological spaces. A perfect map $p: X \to Y$ is *Vietoris**, if one of the following holds:

(a) For every $y \in Y$ the set $p^{-1}(y)$ is weak UV^∞.

(b) For every $y \in Y$ the set $p^{-1}(y)$ is acyclic, and

$$\sup_{y \in Y} \dim p^{-1}(y) < \infty.$$

More precisely, in the first case, we call p a UV^∞-*Vietoris* map, and in the second case, we call p a *finite-dimensional Vietoris* map.

Proposition 5.22. *Every Vietoris* map is Vietoris.*

Proof. This follows from Proposition 4.56. □

We will also need the notion of a homotopy equivalence:

Definition 5.23. Two topological spaces X and Y are *homotopy equivalent* if there are continuous maps $f: X \to Y$ and $g: Y \to X$ such that the two maps $g \circ f, \mathrm{id}_X: X \to X$ are homotopic to each other and simultaneously the two maps $f \circ g, \mathrm{id}_Y: Y \to Y$ are homotopic to each other.

We will need this notion only in connection of the following example:

Example 5.24. In a normed space E, the set $X := E \setminus \{0\}$ is homotopy equivalent to the sphere $Y := S_1(0) \subseteq E$. Indeed, $f(x) := x/\|x\|$ and $g := \mathrm{id}_Y$ satisfy $f \circ g = \mathrm{id}_Y$, and the homotopy $h(t,x) := (t\|x\| + (1-t))f(x)$ proves that the map $g \circ f = f: X \to X$ is homotopic to $\mathrm{id}_X: X \to X$. This example proves in particular that the noncompact set $E \setminus \{0\}$ is homotopy equivalent to a compact ANR.

The homotopic variant of the Vietoris theorem which we are going to use is the following result from [92] and [93]:

Theorem 5.25 (Homotopic Vietoris). (AC). *Let X and Y be paracompact Hausdorff spaces, and $p: X \to Y$. Let Z be a paracompact Hausdorff space which is homotopy equivalent to an* ANR. *Assume that one of the following holds:*

(a) *p is UV^∞-Vietoris,* $\operatorname{Ind} Y < \infty$, *and Y is T_6.*

(b) *p is finite-dimensional Vietoris,* $\dim Y < \infty$, *and Y is compact or Z is homotopy equivalent to a compact* ANR.

Then p induces a bijection between the homotopy classes of $[Y, Z]$ and of $[X, Z]$, that is, for every continuous map $\varphi: X \to Z$ there is some continuous $\psi: Y \to Z$ such that φ is homotopic to $\psi \circ p$, and all maps ψ with this property are homotopic to each other.

Proof. The proof of this result is beyond the scope of this monograph. The second case (if p is finite-dimensional Vietoris) is in view of [93, Remark 2.16(ii)] a special case of [93, Theorem 2.17(i)] (alternatively, see [92]).

The first case (if p is UV^∞-Vietoris) follows from [46, Theorem 10.4.5] if X and Y are compact metric spaces. This special case is already sufficient for most of our applications, cf. also the subsequent Remark 5.26. The result for more general spaces (if p is UV^∞-Vietoris) was announced in [93, Theorem 2.19], and the details will probably appear in [34]. The author thanks W. Kryszewski for sending the details of that proof. However, the proof is not reproduced here, since the details are rather involved and require several results about homotopy and dimension theory which we have not discussed in this monograph (e.g. results about cofibrations, suspensions, mapping cylinder etc). We just remark that the proof is similar to that of [45] and is by induction on $\operatorname{Ind} X$, based on a construction with a so-called double mapping cylinder; the T_6 property is used in the induction step for the sum theorem of dimension theory (this or another hypothesis guaranteeing the validity of the sum theorem was mistakenly forgotten in the announcement in [93, Theorem 2.19]). □

Remark 5.26. With the exception of one remark (Remark 11.38), we will need Theorem 5.25 only for the case that Y (and thus also X) is compact. In this case, Theorem 5.25 can also be proved without AC.

Actually, in all our later applications to degree theory, we will have a compact metrizable space Y and $Z = E \setminus \{0\}$ with a finite-dimensional normed space E (recall Example 5.24).

Using classical ideas, we will now establish a correspondence between Vietoris maps and multivalued maps which are acyclic in the following sense.

Definition 5.27. Let Ω and Γ be topological spaces, and $\Phi\colon \Omega \multimap \Gamma$.

Then Φ is *acyclic* if it is upper semicontinuous and $\Phi(x)$ is nonempty, compact and acyclic for every $x \in \Omega$

We call $\Phi\colon \Omega \multimap \Gamma$ *acyclic** if it is upper semicontinuous and the following holds:

(a) For every $x \in \Omega$ the set $\Phi(x)$ is a nonempty weak UV^∞.

(b) For every $x \in \Omega$ the set $\Phi(x)$ is nonempty, compact, and acyclic, and

$$\sup_{x\in\Omega} \dim \Phi(x) < \infty.$$

More precisely, in the first case, we call Φ a UV^∞ map, and in the second case, we call Φ a *finite-dimensional acyclic* map.

Proposition 5.28. *Every acyclic* map is acyclic.*

Proof. This follows from Proposition 4.56. □

Proposition 5.29. *Let Φ be acyclic. Then all restrictions of Φ are acyclic. Moreover, if J_1, J_2 are homeomorphisms such that $\Psi := J_2 \circ \Phi \circ J_1$ is defined,* *then also* Ψ *is acyclic. Analogous statements hold when one replaces "acyclic" by "acyclic***".*

Proof. The first assertion follows from Proposition 2.90. It follows from the definition that the notions of UV^∞-sets, acyclic sets, and $\dim$ are topological, that is, invariant under homeomorphisms. The upper semicontinuity of Ψ follows from Proposition 2.94. □

Proposition 5.30. *If* $p\colon \Gamma \to \Omega$ *is Vietoris or Vietoris***, then* $\Phi := p^{-1}\colon \Omega \multimap \Gamma$ *is acyclic or acyclic***, respectively.*

Proof. Corollary 2.106 implies that Φ is upper semicontinuos. Hence, the claim follows from $\Phi(x) = p^{-1}(x)$ for all $x \in \Omega$. □

In Section 11.1 we will prove a certain converse of Proposition 5.30.

Part II

Coincidence Degree for Fredholm Maps

Chapter 6

Some Functional Analysis

6.1 Bounded Linear Operators and Projections

Throughout this section, let X and Y be normed spaces over $\mathbb{K} = \mathbb{R}$ or $\mathbb{K} = \mathbb{C}$; the norm will always be denoted by $\|\cdot\|$. We will consider only Banach spaces later on, but in this section, we also have to deal with incomplete spaces (for instance, for the proof of the Hahn–Banach extension theorem). The following test for completeness is rather convenient in normed spaces.

Proposition 6.1. *X is a Banach space if and only if for any sequence $x_n \in X$ with $\sum_{n=1}^{\infty} \|x_n\| < \infty$ the series $x = \sum_{n=1}^{\infty} x_n$ converges. If that series converges, we have*

$$\|x - \sum_{k=1}^{n-1} x_k\| \leq \sum_{k=n}^{\infty} \|x_k\| \quad \textit{for all } n = 1, 2, \ldots \tag{6.1}$$

We use throughout the convention that the empty sum is defined as zero. In particular, (6.1) becomes in case $n = 1$ the "triangle inequality for series"

$$\|\sum_{k=1}^{\infty} x_k\| \leq \sum_{k=1}^{\infty} \|x_k\|$$

(if the left-hand side exists.)

Proof. Putting $y_n := \sum_{k=1}^{n} x_k$ and $s_n := \sum_{k=1}^{n} \|x_k\|$, we obtain by the triangle inequality

$$\|y_m - y_{n-1}\| \leq \sum_{k=n}^{m} \|x_k\| = s_m - s_{n-1} \quad \text{for all } m \geq n \geq 1. \tag{6.2}$$

If $y_n \to x$, then letting $m \to \infty$ in (6.2) and using the continuity of the norm, we obtain (6.1). Moreover, if $\sum_{n=1}^{\infty} \|x_n\| < \infty$ then s_n is a Cauchy sequence in $\mathbb{R}$, and thus we obtain from (6.2) with Lemma 3.8(b) that y_n is a Cauchy sequence in X. Hence, if additionally X is a Banach space, then y_n converges.

Conversely, if X is incomplete, there is a Cauchy sequence $y_n \in X$ which does not converge. By Lemma 3.8(a), none of its subsequences is convergent. Hence, passing to a subsequence if necessary, we can assume by Lemma 3.8(b) that $\|y_{n+1} - y_n\| < (n+1)^{-2}$ for all n. Putting $y_0 := 0$ and $x_n := y_n - y_{n-1}$, we thus find that $\sum_{n=1}^{\infty} \|x_n\| < \infty$ while $\sum_{k=1}^{n} x_k = y_n$ fails to converge. □

Recall that the *product* $X \times Y$ of two normed spaces becomes a normed space in an obvious way with the *sum norm* $\|(x, y)\| := \|x\| + \|y\|$. It is easy to see that $X \times Y$ is a Banach space if X and Y are Banach spaces.

Recall also that if $U \subseteq X$ is a linear subspace then one can define an equivalence relation on X consisting of the equivalence classes $[x] := x + U$ $(x \in U)$. The corresponding *factor space* X/U is defined as the vector space of equivalence classes $x + U$ with the operations $[x] + [y] := [x + y]$ and $\lambda[x] := [\lambda x]$. It is easily seen that these operations are well-defined. Moreover, if U is closed, then X/U becomes a normed space with the norm

$$\|[x]\| := \inf_{[y]=[x]} \|y\| = \inf_{y \in x+U} \|y\| = \operatorname{dist}(x, U).$$

Proposition 6.2. *If X is a Banach space and U is closed then X/U is a Banach space.*

Proof. We apply Proposition 6.1. Thus, let $[x_n]$ be a sequence in X/U with $\sum_{n=1}^{\infty} \|[x_n]\| < \infty$. By definition of the norm, there are $y_n \in X$ with $[y_n] = [x_n]$ and $\|y_n\| \le \|[x_n]\| + 2^{-n}$. Then $\sum_{n=1}^{\infty} \|y_n\| < \infty$, and so Proposition 6.1 implies that $x = \sum_{n=1}^{\infty} y_n$ converges. Then $s_n := [x_1] + \cdots + [x_n]$ converges to $[x]$, because $z_n := y_1 + \cdots + y_n$ satisfies $s_n = [z_n]$, and so $\|[x] - s_n\| = \|[x - z_n]\| \le \|x - z_n\| \to 0$. □

For linear $A\colon X \to Y$, it is customary to omit the braces, that is, we use the notation $Ax := A(x)$. We denote by $\mathcal{L}(X, Y)$ the space of bounded linear operators $A\colon X \to Y$, that is, for which A is linear, and

$$\begin{aligned} \|A\| := \sup_{\|x\| \le 1} \|Ax\| &= \sup_{x \neq 0} \frac{\|Ax\|}{\|x\|} \\ &= \min\{L \in [0, \infty] : \|Ax\| \le L\|x\| \text{ for all } x \in X\} \end{aligned}$$

is finite. We also define the shortcut $\mathcal{L}(X) := \mathcal{L}(X, X)$.

It is straightforward to check that $\mathcal{L}(X, Y)$ is a linear space with $\|\cdot\|$ as a norm.

Proposition 6.3. *For linear $A\colon X \to Y$ the following statements are equivalent:*

(a) *A is continuous at* 0.

(b) $A \in \mathcal{L}(X, Y)$.

(c) *A is Lipschitz.*

In the latter case a Lipschitz constant is $\|A\|$.

Proof. In view of Corollary 3.6, this is a special case of Proposition 3.53. □

For linear $A: X \to Y$, we define the *null space* $\mathrm{N}(A) := A^{-1}(0)$ and the *range* $\mathrm{R}(A) := A(X)$.

Proposition 6.4. *For each $A \in \mathcal{L}(X, Y)$ the set $\mathrm{N}(A)$ is a closed linear subspace of X, and $\mathrm{R}(A)$ is a linear subspace of Y.*

Proof. The subspace properties are trivial, and $\mathrm{N}(A) = A^{+}(\{0\})$ is closed by Proposition 2.92. □

We recall that A is one-to-one if and only if $\mathrm{N}(A) = \{0\}$.

Lemma 6.5. *If $A \in \mathcal{L}(X, Y)$ and $U := \mathrm{N}(A)$ then there is a one-to-one $A_0 \in \mathcal{L}(X/U, Y)$ with $\mathrm{R}(A) = \mathrm{R}(A_0)$ defined by $A_0[x] := Ax$.*

Proof. A straightforward calculation shows that A_0 is well-defined and $\|A_0\| \le \|A\|$. Since $\mathrm{N}(A) = U$, it follows that $\mathrm{N}(A_0) = \{0\}$, and so A_0 is one-to-one. □

The *dual space* of X is defined as $X^* := \mathcal{L}(X, \mathbb{K})$.

Proposition 6.6. *X^* is a Banach space. If Y is a Banach space then so is $\mathcal{L}(X, Y)$.*

Proof. We prove the second assertion, since the first is a special case. Only the completeness requires a proof. Thus, let $A_n \in \mathcal{L}(X, Y)$ be a Cauchy sequence. Then for every $x \in X$ the sequence $A_n x \in Y$ is Cauchy and thus convergent to some $Ax \in Y$. Since A_n are linear, it follows that A is linear. For $\varepsilon > 0$ there is some N such that $\|A_n - A_m\| \le \varepsilon$ for all $n, m \ge N$. Then we have for all $x \in X$ and all $n, m \ge N$ that $\|A_n x - A_m x\| \le \varepsilon \|x\|$. Letting $m \to \infty$, we obtain $\|A_n x - Ax\| \le \varepsilon \|x\|$ for all $n \ge N$. In particular, $\|A_n - A\| \to 0$ which also implies $A = A_n - (A_n - A) \in \mathcal{L}(X, Y)$. □

Proposition 6.7. *If $U \subseteq X$ is a dense linear subspace and Y is a Banach space then every $A_0 \in \mathcal{L}(U, Y)$ has a unique extension to some $A \in \mathcal{L}(X, Y)$, and we have $\|A\| = \|A_0\|$.*

Proof. For $x \in X$ choose a sequence $u_n \in U$ with $u_n \to x$. Then u_n is a Cauchy sequence, and the estimate $\|A_0 u_n - A_0 u_m\| \le \|A_0\| \|u_n - u_m\|$ implies that also $A_0 u_n \in Y$ is a Cauchy sequence and thus convergent to some $y \in Y$.

This y is independent of the choice of the sequence, since if $v_n \in U$ is any sequence with $v_n \to x$, we have $\|A_0 v_n - A_0 u_n\| \le \|A_0\| \|v_n - u_n\| \to 0$, and so also $A_0 v_n \to y$. Hence, we can define $A: X \to Y$ by $Ax := y$, and by the continuity, it is clear that this is the only candidate for the required extension.

Since A_0 and forming limits are linear, it follows that A is linear. Moreover, the continuity of the norm implies in view of $\|A_0 u_n\| \le \|A_0\| \|u_n\|$ that $\|Ax\| \le \|A_0\| \|x\|$, and so $A \in \mathcal{L}(X, Y)$ and $\|A\| \le \|A_0\|$; the converse inequality $\|A\| \ge \|A_0\|$ follows from $A|_U = A_0$. □

We recall that for linear operators $A: X \to Y$ and $B: Y \to Z$, one defines a "multiplication" as the composition of the operators, that is $BA := B \circ A$. Then we have $\|BA\| \le \|B\| \|A\|$, and in case $Y = X$ in particular $\|A^n\| \le \|A\|^n$ $(n = 0, 1, \dots)$. We will need the continuity of the multiplication map:

Lemma 6.8. *The map $\mathcal{L}(X, Y) \times \mathcal{L}(Y, Z) \to \mathcal{L}(X, Z)$, $(A, B) \mapsto BA$, is continuous.*

Proof. The assertion follows from the estimates

$$\begin{aligned} \|BA - B_0 A_0\| &= \|B(A - A_0) + (B - B_0) A_0\| \\ &\le \|B\| \|A - A_0\| + \|B - B_0\| \|A_0\| \end{aligned}$$

and $\|B\| \le \|B_0\| + \|B - B_0\|$. □

A well-known important consequence of Proposition 6.6 is the Neumann series which we recall now.

Definition 6.9. We denote by $\mathrm{Iso}(X, Y)$ the set of all isomorphisms $J \in \mathcal{L}(X, Y)$ onto Y, that is, $J^{-1} \in \mathcal{L}(Y, X)$. In case $X = Y$, we put $\mathrm{Iso}(X) := \mathrm{Iso}(X, X)$.

Proposition 6.10 (Neumann Series). *Let X be a Banach space.*

(a) *If $A \in \mathcal{L}(X)$ satisfies $\sum_{n=0}^{\infty} \|A^n\| < \infty$ then* $\mathrm{id}_X - A \in \mathrm{Iso}(X)$. *Moreover,* $(\mathrm{id}_X - A)^{-1} = \sum_{n=0}^{\infty} A^n$, *and*

$$\|(\mathrm{id}_X - A)^{-1} - \sum_{k=0}^{n-1} A^k\| \le \sum_{k=n}^{\infty} \|A^k\| \quad \text{for all } n = 0, 1, \dots \tag{6.3}$$

(b) *The set* $\mathrm{Iso}(X, Y)$ *is open in* $\mathcal{L}(X, Y)$, *and $J \mapsto J^{-1}$ is continuous on* $\mathrm{Iso}(X, Y)$. *More precisely, if $J_0 \in \mathrm{Iso}(X, Y)$, then any $J \in \mathcal{L}(X, Y)$ with $\|J - J_0\| < \|J_0^{-1}\|^{-1}$ belongs to* $\mathrm{Iso}(X, Y)$, *and*

$$\|J^{-1} - J_0^{-1}\| \le \frac{\|J_0^{-1}(J - J_0)\| \|J_0^{-1}\|}{1 - \|J_0^{-1}(J - J_0)\|}. \tag{6.4}$$

Proof. For (a), we note that Proposition 6.1 implies in view of Proposition 6.6 that $S_n := \sum_{k=0}^{n} A^k$ converges in $\mathcal{L}(X)$ to some S, and

$$\|S - S_{n-1}\| = \|\sum_{k=n}^{\infty} A^k\| \le \sum_{k=n}^{\infty} \|A^k\|.$$

Since the hypothesis implies $\|A^n\| \to 0$ and thus $S_n(\mathrm{id}_X - A) = (\mathrm{id}_X - A)S_n = \mathrm{id}_X - A^{n+1} \to \mathrm{id}_X$, we find as $n \to \infty$ in view of Lemma 6.8 that $S(\mathrm{id}_X - A) = (\mathrm{id}_X - A)S = \mathrm{id}_X$, and so $S = (\mathrm{id}_X - A)^{-1}$.

Concerning (b), we apply (a) with $A := J_0^{-1}(J_0 - J)$, noting that we have $\|A^n\| \le q^n$ with $q := \|A\| < 1$. Hence, $J_1 := \mathrm{id}_X - A = J_0^{-1} J \in \mathrm{Iso}(X)$, and applying (6.3) with $n = 1$, we find $\|J_1^{-1} - \mathrm{id}_X\| \le \sum_{k=1}^{\infty} q^k = q/(1-q)$. It follows that $J = J_0 J_1 \in \mathrm{Iso}(X,Y)$, and since $J^{-1} - J_0^{-1} = (J_1^{-1} - \mathrm{id}_X)J_0^{-1}$ implies $\|J^{-1} - J_0^{-1}\| \le \|J_1^{-1} - \mathrm{id}_X\|\|J_0^{-1}\|$, we obtain that (6.4) holds which implies the continuity of $J \mapsto J^{-1}$ at J_0. □

We recall also the celebrated closed graph theorem.

Theorem 6.11 (Closed Graph). *Let X and Y be Banach spaces and $A\colon X \to Y$ be linear. Then $A \in \mathcal{L}(X,Y)$ if and only if $\mathrm{graph}(A)$ is closed in $X \times Y$.*

Proof. If $A \in \mathcal{L}(X,Y)$ then $\mathrm{graph}(A)$ is closed by Corollary 2.117 and Proposition 6.3. Conversely, let $\mathrm{graph}(A)$ be closed. We note that the sets $M_n := \{x \in X : \|Ax\| \le n\}$ satisfy $X \subseteq \bigcup_{n=1}^{\infty} M_n$. Applying Baire's category theorem (Theorem 2.3) with $N_n := \overline{M}_n$, we find that there is some n_0 such that $N_{n_0} = \overline{M}_{n_0}$ has an interior point $x_0 \in X$, that is, $K_r(x_0) \subseteq \overline{M}_{n_0}$ for some $r > 0$. Note that M_{n_0} is convex and symmetric (that is, $0 \in M_{n_0} = -M_{n_0}$), and so also $\overline{M}_{n_0}$ is convex (Proposition 3.54) and symmetric. Hence,

$$K_r(0) \subseteq \mathrm{conv}\big(K_r(x_0) \cup (-K_r(x_0))\big) \subseteq \overline{M}_{n_0} = \overline{A^{-1}\big(K_{n_0}(0)\big)},$$

and the linearity of A implies for each $x \in X$ that

$$0 = x - x \in x + \frac{\|x\|}{r} K_r(0) \subseteq x + \overline{A^{-1}\big(K_{\|x\|n_0/r}(0)\big)} = \overline{A^{-1}\big(K_{\|x\|n_0/r}(Ax)\big)}.$$

Thus, starting from any $x_0 \in X$, we obtain inductively that there are $x_n \in A^{-1}\big(K_{\|x_{n-1}\|n_0/r}(Ax_{n-1})\big)$ with $\|x_n\| \le \|x_{n-1}\|/2$. Then $\|Ax_n - Ax_{n-1}\| \le \|x_{n-1}\|n_0/r$ and $\|x_n\| \le 2^{-n}\|x_0\|$. Hence, $y_n := Ax_n$ satisfies for all $m > n$

$$\begin{aligned} \|y_m - y_n\| &\le \sum_{k=n+1}^{m} \|y_k - y_{k-1}\| \le \sum_{k=n+1}^{m} 2^{-(k-1)}\|x_0\|n_0/r \\ &= 2^{1-n} n_0 \|x_0\|/r. \end{aligned} \tag{6.5}$$

We obtain that y_n is a Cauchy sequence, and so $y_n \to y$ for some $y \in Y$. Since $(x_n, y_n) \in \operatorname{graph}(A)$, $(x_n, y_n) \to (0, y)$, the hypothesis implies $(0, y) \in \operatorname{graph}(A)$, that is, $y = A0 = 0$. Putting $n = 0$ in (6.5), we obtain as $m \to \infty$ that

$$\|Ax_0\| = \|-y_0\| \leq 2^{1-0} n_0 r^{-1} \|x_0\|.$$

Since $x_0 \in X$ was arbitrary, this shows that A is bounded with $\|A\| \leq 2n_0/r$. □

Theorem 6.12 (Bounded Inverse). *Let X and Y be Banach spaces. If $A \in \mathcal{L}(X, Y)$ is invertible then $A^{-1} \in \mathcal{L}(Y, X)$, that is $A \in \operatorname{Iso}(X, Y)$.*

Proof. Proposition 2.1 implies that $\operatorname{graph}(A^{-1})$ is closed if $\operatorname{graph}(A)$ is closed. Hence, the assertion follows from Theorem 6.11. □

The main assertion of Theorem 6.11 is not empty: Even if X and Y are Banach spaces, it is not true that every linear $A: X \to Y$ is automatically bounded, although it is impossible to prove this without AC [145].

However, using AC or dropping the hypothesis that X is complete, we can give a lot of examples of such operators. Using AC, we can even show that on *every* infinite-dimensional normed space X (and $Y \neq \{0\}$) there are unbounded linear $A: X \to Y$:

Example 6.13. (AC). Let X be infinite-dimensional. Corollary 2.6 implies that X has a Hamel basis $e_i \in X$ $(i \in I)$. We choose a sequence of pairwise different $i_n \in I$, and define

$$f(e_i) := \begin{cases} n\|e_i\| & \text{if } i = i_n, \\ 0 & \text{otherwise.} \end{cases}$$

Since f is defined on a Hamel basis, it extends to a linear functional $f: X \to \mathbb{K}$ by means of

$$f(\lambda_1 e_{j_1} + \cdots + \lambda_n e_{j_n}) := \lambda_1 f(e_{j_1}) + \cdots + \lambda_n f(e_{j_n})$$

which in view of $f(e_{i_n}) = n\|e_{i_n}\|$ is unbounded. For every normed space $Y \neq \{0\}$, $e \in Y \setminus \{0\}$, it follows that $A: X \to Y$, $Ax := f(x)e$ is linear and unbounded.

Note that AC was "only" used to find an infinite Hamel basis. In particular, replacing X by a subspace spanned by countably infinite many linearly independent vectors, we obtain A as above without referring to AC. However, the following Proposition 6.14 shows that such spaces are never Banach spaces.

Proposition 6.14. *If an infinite-dimensional normed space X has a countable Hamel basis $e_1, e_2, \ldots$ then X is not a Banach space.*

Proof. Assume by contradiction that X is a Banach space. The linear hull M_n of $\{e_1, \ldots, e_n\}$ is closed by Proposition 3.59. Since $\bigcup_{n=1}^{\infty} M_n = X$, Baire's category theorem implies that there is some n such that M_n has an interior point x_0. This is a contradiction since $x_0 + \varepsilon e_{n+1} \notin M_n$ for every $\varepsilon > 0$. □

Using the mentioned result of [145], we obtain from Example 6.13 that without AC even some opposite of Corollary 2.6 holds:

Proposition 6.15. *Without* AC*, it is impossible to prove (in* ZF+DC*) that there is an infinite-dimensional Banach space X with a Hamel basis.*

Proof. Otherwise, the construction of Example 6.13 would show that there is an unbounded linear $f: X \to \mathbb{K}$ which contradicts [145]. □

Example 6.13 shows in particular that also operators A with $\dim \mathrm{R}(A) < \infty$ can be unbounded. However, for such operators there is a boundedness criterion which is similar to the closed graph theorem but which does not require completeness of the spaces:

Proposition 6.16. *If $A: X \to Y$ is linear with $n = \dim \mathrm{R}(A) < \infty$ then $A \in \mathcal{L}(X, Y)$ if and only if $A^+(U)$ is closed for every subspace $U \subseteq \mathrm{R}(A)$ with $\dim U = n - 1$.*

In particular, a linear $f: X \to \mathbb{K}$ belongs to X^* if and only if $\mathrm{N}(f)$ is closed.

Proof. Note that U is closed by Proposition 3.59, and so necessity follows from Propositions 2.92 and 6.3.

For sufficiency, we consider first the case $A = f: X \to \mathbb{K}$. We can assume that there is $x_0 \in X$ with $c := f(x_0) \neq 0$. Since $\mathrm{N}(f)$ is closed by hypothesis, there is $r > 0$ such that $B_r(x_0)$ is disjoint from $\mathrm{N}(f)$. Since f is linear, it follows that $B_r(0)$ is disjoint from $f^{-1}(c)$. We claim that $f \in X^*$ with $\|f\| \leq c/r$. Indeed, assume by contradiction that there is $x \in X$ with $\|x\| \leq 1$ and $|f(x)| > c/r$. The latter implies that there is $\lambda \in \mathbb{K}$ with $|\lambda| < 1$ and $\lambda f(x) = c/r$. We obtain the contradiction $\lambda r x \in B_r(0) \cap f^{-1}(c)$.

For the general case, let $e_1, \ldots, e_n \in Y$ be a basis of $\mathrm{R}(A)$. Then there are unique $f_1, \ldots, f_n: X \to \mathbb{K}$ with $Ax = f_1(x)e_1 + \cdots + f_n(x)e_n$. Since A is linear, it follows that all f_k are linear. Moreover, $\mathrm{N}(f_k) = A^+(U_k)$ where $U_k \subseteq \mathrm{R}(A)$ is the linear hull of $\{e_1, \ldots, e_n\} \setminus \{e_k\}$. Since $\dim U_k = n-1$, the hypothesis implies that $\mathrm{N}(f_k)$ is closed, and so $f_k \in X^*$. In particular, $f_1, \ldots, f_n$ are continuous, and so also A is continuous and thus bounded by Proposition 6.3. □

In the following considerations the sum of subspaces will play an important role. Let us first give a simple sufficient criterion that such a sum is closed.

Proposition 6.17. *Let $U, V \subseteq X$ be linear subspaces with U being closed and $\dim V < \infty$. Then $U + V$ is closed in X.*

Proof. Let $A \in \mathcal{L}(X, X/U)$ be defined by $Ax := [x]$. Then $M := A(V)$ is finite-dimensional and thus closed in X/U by Proposition 3.59. Consequently, $A^{+}(M) = U + V$ is closed. □

Recall that if $U, V \subseteq X$ then the *direct sum* $U \oplus V$ is defined only if U and V are linear subspaces with $U \cap V = \{0\}$, and in this case $U \oplus V := U + V$. The elements of $U \oplus V$ are those $x \in X$ which have a unique representation $x = u + v$ with $u \in U$ and $v \in V$. A linear subspace $V \subseteq X$ is *complementary* to a linear subspace $U \subseteq X$ if $U \oplus V = X$.

Recall that a map $P \in \mathcal{L}(X)$ is called a *projection* if it is a retraction (that is, if $P^2 = P$).

Proposition 6.18. *For each projection $P \in \mathcal{L}(X)$ the spaces $U := \mathrm{R}(P)$ and $V := \mathrm{N}(P)$ are complementary and closed, and P is the unique projection with $U = \mathrm{R}(P)$ and $V = \mathrm{N}(P)$. Moreover, $Q := \mathrm{id}_X - P$ is the unique projection with $V = \mathrm{R}(Q)$ and $U = \mathrm{N}(Q)$.*

Conversely if $U, V \subseteq X$ are complementary and closed, and if X is a Banach space or if $\dim U < \infty$ then there is a projection $P \in \mathcal{L}(X)$ with $U = \mathrm{R}(P)$ and $V = \mathrm{N}(P)$.

Proof. If $U, V \subseteq X$ are complementary and closed then clearly each projection $P \in \mathcal{L}(X)$ with $U = \mathrm{R}(P)$ and $V = \mathrm{N}(P)$ must be defined by

$$P(u + v) := u \quad \text{for all } u \in U, v \in V. \tag{6.6}$$

This shows the uniqueness assertions (for P and Q).

For the existence assertion, we note that if we define $P: X \to X$ by (6.6) then P is linear, $P^2 = P$, $U = \mathrm{R}(P)$, and $V = \mathrm{N}(P)$. We have to show that $P \in \mathcal{L}(X)$.

In case $n = \dim U < \infty$, we have to show by Proposition 6.16 that for every subspace $U_0 \subseteq U$ with $\dim U_0 = n - 1$ the subspace $P^{-1}(U_0) = U_0 + V$ is closed. Since $\dim U_0 < \infty$, this follows immediately from Proposition 6.17.

In the case that X is a Banach space, we have to show by Theorem 6.11 that $\mathrm{graph}(P)$ is closed in $X \times X$. Thus, suppose that $(x_n, u_n) \in \mathrm{graph}(P)$ converge to $(x, u) \in X \times X$. Since $u_n \in \mathrm{R}(P) = U$ and U is closed, we have $u \in U$. Similarly, $v_n := x_n - u_n \in V$ implies $v := x - u \in V$. Hence, $P(x) = P(u + v) = u$, and so $(x, u) \in \mathrm{graph}(P)$.

We thus have shown the uniqueness resp. existence of P (and Q).

Conversely, if $P \in \mathcal{L}(X)$ is a projection then $V := \mathrm{N}(P)$ and $U := \mathrm{R}(P)$ are closed subspaces by Proposition 6.4 and Proposition 4.13, respectively. If $x \in U$ then $x = Px$ which implies that $x \in V$ holds only for $x = 0$. Each $x \in X$ has a representation $x = u + v$ with $u := Px \in U$ and $v := x - u \in V$, since $Pv = P(x - u) = u - Pu = 0$. To prove that $Q := \mathrm{id}_X - P$ is the required projection, we calculate for $u \in U$ and $v \in V$ that $Q(u + v) = (\mathrm{id}_X - P)u + (\mathrm{id}_X - P)v = 0 + \mathrm{id}_X v = v$. □

We call a subspace $U \subseteq X$ *complemented (in X)* if it is closed and if it has a *closed* complementary subspace.

Corollary 6.19. *For a linear subspace $U \subseteq X$, we consider the statements:*

(a) *U is complemented.*
(b) *U is the range of a projection $P \in \mathcal{L}(X)$.*
(c) *U is the null space of a projection $Q \in \mathcal{L}(X)$.*

Then (b)$\Leftrightarrow$(c)$\Rightarrow$(a).

If X is a Banach space or $\dim U < \infty$*, all three statements are equivalent.*

Proof. The assertion follows from Proposition 6.18. □

The following consequences will be used for Fredholm operators.

Corollary 6.20. *Let $U, V \subseteq X$ be linear subspaces. Suppose that X is a Banach space or that* $\dim U < \infty$.

If U is complemented in X and $W = U \cap V$ is complemented in U then $W \subseteq V$ is complemented in V.

Proof. In case $\dim U = \infty$, we note that U is closed and thus a Banach space by Lemma 3.8(d). Corollary 6.19 thus implies that there exists a projection $P_1 \in \mathcal{L}(X)$ onto U and a projection $P_2 \in \mathcal{L}(U)$ onto W. Then $P := P_2 P_1 \in \mathcal{L}(X)$ is a projection onto W, and so $P|_V \in \mathcal{L}(V, W) \subseteq \mathcal{L}(V)$ is a projection onto W. Hence, the assertion follows from Corollary 6.19. □

Proposition 6.21. *Let $X = U \oplus V$ with closed subspaces $U, V \subseteq X$. Assume that X is a Banach space or* $\dim U < \infty$ *or* $\dim V < \infty$. *Then for each $B \in \mathcal{L}(U, Y)$, $C \in \mathcal{L}(V, Y)$ there is exactly one additive map $A: X \to Y$ satisfying $A|_U = B$ and $A|_V = C$; we have $A \in \mathcal{L}(X, Y)$. If B and C are isomorphisms onto their respective ranges and $Y = \mathrm{R}(B) \oplus \mathrm{R}(C)$ then $A \in \mathrm{Iso}(X, Y)$.*

Proof. By Proposition 6.18 there are unique projections $P, Q \in \mathcal{L}(X)$ satisfying $\mathrm{R}(P) = \mathrm{N}(Q) = U$ and $\mathrm{N}(P) = \mathrm{R}(Q) = V$. Then $x = Px + Qx$ implies

$A(x) = A(Px) + A(Qx) = BPx + CQx$. Hence, A is uniquely defined, and since compositions of bounded linear operators are bounded, it follows that $A \in \mathcal{L}(X, Y)$.

For the second assertion, we apply the first assertion to find that there is a unique $A_0 \in \mathcal{L}(Y, X)$ satisfying $A_0|_{\mathrm{R}(B)} = B^{-1}$ and $A_0|_{\mathrm{R}(C)} = C^{-1}$. For $u \in U$ and $v \in V$ we have $A_0 A(u+v) = A_0 Au + A_0 Av = u + v$. Hence, $A_0 A = \mathrm{id}_X$ and similarly $AA_0 = \mathrm{id}_Y$. □

The Hahn–Banach extension theorem is usually formulated with sublinear functionals. For our purposes the following special case is more than sufficient.

Theorem 6.22 (Hahn–Banach Extension). (AC). *Let $U \subseteq X$ be a linear subspace and $f \in U^*$. Then f has an extension $F \in X^*$ with $\|F\| = \|f\|$.*

Proof. We show first the auxiliary assertion that in case $U \neq X$ the function f has an extension $F \in V^*$ with $\|F\| \leq \|f\|$ with $V \subseteq X$ being a strictly larger subspace than U. More precisely, we can let V be the linear hull of U and some given $e \in X \setminus U$.

We assume first that $\mathbb{K} = \mathbb{R}$. Then each $x \in V$ can be uniquely written as $x = \lambda e + u$ with $\lambda \in \mathbb{R}$ and $u \in U$, and so $F_c(\lambda e + u) := \lambda c + f(u)$ defines for fixed $c \in \mathbb{R}$ a linear $F_c \colon V \to \mathbb{R}$ which has the required property if and only if $|F_c(x)| \leq \|f\| \|x\|$ for all $x \in V$, that is, if and only if

$$|\lambda c + f(u)| \leq \|f\| \|\lambda e + u\|$$

holds for all $(\lambda, u) \in \mathbb{R} \times U$. Since this is satisfied in case $\lambda = 0$, we can assume $\lambda \neq 0$. Dividing the inequality by λ, we find that it is equivalent to require

$$-\|f\| \|e + \lambda^{-1} u\| \leq c + f(\lambda^{-1} u) \leq \|f\| \|e + \lambda^{-1} u\|.$$

Replacing $\lambda^{-1} u$ by v, we thus have to show that there is some $c \in \mathbb{R}$ satisfying

$$-\|f\| \|e + v\| - f(v) \leq c \leq \|f\| \|e + v\| - f(v) \tag{6.7}$$

for all $v \in U$. To see that such $c \in \mathbb{R}$ exists, we note that for all $u, v \in U$ we have

$$f(u) - f(v) = f\big((e + u) - (e + v)\big) \leq \|f\| \big(\|e + u\| + \|e + v\|\big),$$

and so

$$-\|f\| \|e + v\| - f(v) \leq \|f\| \|e + u\| - f(u)$$

holds for all $u, v \in U$. It follows that the supremum c of all $v \in U$ of the left-hand side is finite and satisfies (6.7). We thus have shown the auxiliary assertion in case $\mathbb{K} = \mathbb{R}$.

In case $\mathbb{K} = \mathbb{C}$, we understand U as a real normed space U_0 (by restricting the scalar multiplication) and define $g \in U_0^*$ by $g(u) := \operatorname{Re} f(u)$. Note that $\|g\| \le \|f\|$. Applying twice what we have shown above, we obtain an extension $G \in V_0^*$ of g to the (real) linear hull V_0 of U_0 and $e, ie \in X \setminus U$ with $\|G\| \le \|g\| \le \|f\|$. We can understand V_0 as the complex linear hull V of U and e. Then we can define $F\colon V \to \mathbb{C}$ by putting $F(v) := G(v) - iG(iv)$. Note that F is an extension of f, since for all $u \in U$ we have $\operatorname{Re} F(u) = G(u) = g(u) = \operatorname{Re} f(u)$ and

$$\operatorname{Im} F(u) = -G(iu) = -g(iu) = -\operatorname{Re} f(iu) = \operatorname{Re}\bigl(-if(u)\bigr) = \operatorname{Im} f(u).$$

The map F is linear, since for all $\alpha, \beta \in \mathbb{R}$ and $v \in V$ we have

$$\begin{aligned} F\bigl((\alpha + i\beta)v\bigr) &= G(\alpha v) - iG(i\alpha v) + G(i\beta v) - iG(i^2\beta v) \\ &= \alpha G(v) - i\alpha G(iv) - i^2\beta G(iv) + i\beta G(v) = \alpha F(v) + i\beta F(v). \end{aligned}$$

Moreover, for each $v \in V$ there is some $s \in \mathbb{C}$ with $|s| = 1$ such that $sF(v) \in \mathbb{R}$. It follows that $sF(v) = F(sv) = G(sv) - iG(isv)$ is real. Since $G(sv)$ and $G(isv)$ are real, we obtain that $G(isv) = 0$, and so

$$|F(v)| = |sF(v)| = |G(sv)| \le \|f\|\|sv\| = \|f\||s|\|v\| = \|f\|\|v\|.$$

Since $v \in V$ was arbitrary, we obtain $F \in V^*$ and $\|F\| \le \|f\|$, and so the auxiliary assertion is also shown in case $\mathbb{K} = \mathbb{C}$.

The general case is reduced to the auxiliary assertion by means of Hausdorff's maximality theorem: We let $\mathcal{F}$ denote the set of all couples (g, V) where $V \supseteq U$ is a linear subspace of X and $g \in V^*$ is an extension of f with $\|g\| \le \|f\|$. On $\mathcal{F}$, we introduce a partial order by letting $(g_0, V_0) \preceq (g_1, V_1)$ if and only if $V_0 \subseteq V_1$ and $g_1|_{V_0} = g_0$. By Hausdorff's maximality theorem, $\mathcal{F}$ contains a maximal chain $\mathcal{C}$. This chain has a maximal element (f_0, U_0). Indeed, we let U_0 denote the union of all linear subspaces $V \subseteq X$ with $(g, V) \in \mathcal{C}$ for some g, and for $x \in V$, we put $f_0(x) := g(x)$. We claim that we obtain a well-defined f_0 with $(f_0, U_0) \in \mathcal{F}$. To see this, we note that for all $x_0, x_1 \in U_0$, say $x_i \in V_i$ with $(g_i, V_i) \in \mathcal{C}$ $(i = 0, 1)$ we can assume without loss of generality that $(g_0, V_0) \preceq (g_1, V_1)$, because $\mathcal{C}$ is a chain. Hence, $g_0(x_0) = g_1(x_0)$ (which in case $x_0 \in U$ is $f(x_0)$), $g_1(x_0 + x_1) = g_1(x_0) + g_1(x_1)$, $g_1(\lambda x_0) = \lambda g_1(x_0)$, and $|g_1(x_0)| \le \|f\|\|x_0\|$. It follows that f_0 is well-defined on U_0, extends f, is linear, and moreover $f_0 \in U_0^*$ with $\|f_0\| \le \|f\|$. Thus, $(f_0, U_0) \in \mathcal{F}$, as required.

If $U_0 \ne X$, we find by the auxiliary assertion that there is some $(F, V) \in \mathcal{F}$ with $(f_0, U_0) \preceq (F, V)$ and U_0 being a proper subspace of V. Hence,

$\mathcal{C} \cup \{(F, V)\}$ would be a strictly larger chain containing $\mathcal{C}$ which is a contradiction. The contradiction shows that $U_0 = X$, and so $F = f_0$ is the required extension: Note that the converse inequality $\|f_0\| \geq \|f\|$ follows from $f_0|_U = f$. □

Remark 6.23. We will see in the subsequent Remark 6.26 that it is not possible to prove Theorem 6.22 without additional hypotheses without AC. However, if X is separable then AC is not needed for the proof (it suffices to use DC).

Indeed, let $\{x_1, x_2, \dots\}$ be dense in X. Using the auxiliary assertion of the proof of Theorem 6.22 repeatedly, we find successively $(F_n, U_n) \in \mathcal{F}$ with U_n being the linear hull of U and $\{x_1, \dots, x_n\}$. By DC, the elements (F_n, U_n) form a chain in $\mathcal{F}$, and as in the proof of Theorem 6.22, we find some $(f_0, U_0) \in \mathcal{F}$ with $U_n \subseteq U_0$ for all n. Since $\overline{U}_0 = X$, Proposition 6.7 implies that f_0 has a unique extension to some $F \in X^*$ satisfying $\|F\| = \|f_0\| = \|f\|$.

A similar argument can be used to show that also a more general form of the Hahn–Banach theorem holds without AC under some separability hypotheses, see [67, p. 183].

Definition 6.24. The *support* of X^* is the set of all $x \in X$ with the property that either $x = 0$ or that there is some $f \in X^*$ with $f(x) \neq 0$.

Corollary 6.25. (AC). *The support of X^* is X.*

Proof. If $e \in X \setminus \{0\}$, we let U denote the linear hull of e, and we define $f \in U^*$ by $f(\lambda e) := \lambda$ for $\lambda \in \mathbb{K}$. By Theorem 6.22, there is $F \in X^*$ satisfying $F(x) = f(x) = 1 \neq 0$. □

Remark 6.26. Even if we do not assume AC, Corollary 6.25 holds if X is separable (by Remark 6.23). Moreover, even if X fails to be separable, Corollary 6.25 can be shown for a large class of spaces without AC. For instance, Corollary 6.25 holds trivially if X is an inner product space. Another huge class of spaces for which Corollary 6.25 holds (without AC) is the class of preideal spaces [135, Corollary 3.4.8] (which includes ℓ_∞ and $L_\infty(S)$) if the underlying measure space S has the finite subset property (for instance, if S is σ-finite).

However, without AC, it cannot be excluded that there are nontrivial Banach spaces with $X^* = \{0\}$. For instance, for the spaces $X = L_\infty([0,1])/C([0,1])$ or $X = \ell_\infty/c_0$ the existence of a nonzero $f \in X^*$ cannot be proved in ZF+DC, see [101]. The subsequent Proposition 6.27 implies that in such spaces, we cannot even prove (in ZF+DC) that there is a nontrivial finite-dimensional complemented subspace.

On the other hand, the converse implication of Proposition 6.27 also implies that *if* X^* has full support X then all finite-dimensional subspaces are complemented.

Proposition 6.27. *A finite-dimensional subspace $U \subseteq X$ is complemented if and only if U is contained in the support of X^*.*

Proof. Let $U \subseteq X$ be a complemented finite-dimensional subspace and $e \in U \setminus \{0\}$. The linear hull V of e is a finite-dimensional subspace of U, and $U = V \oplus W$ with a finite-dimensional subspace W. By hypothesis, there is a closed subspace U_0 with $X = U \oplus U_0 = (V \oplus W) \oplus U_0 = V \oplus (W \oplus U_0)$. Since V and $W \oplus U_0$ are closed by Proposition 6.17, we obtain by Corollary 6.19 that there is a projection $P \in \mathcal{L}(X)$ onto V. Since V is spanned by e, we have $Px = f(x)e$ for some $f: X \to \mathbb{K}$. From $P \in \mathcal{L}(X)$, we obtain $f \in X^*$. Since $Pe = e \neq 0$, we have $f(e) = 1 \neq 0$, and so e is contained in the support of X^*.

Conversely, we prove by induction on $\dim U$ that U is complemented if it is contained in the support of X^*. For $U = \{0\}$, the space X is the required closed complement. Otherwise, U contains a subspace $U_0 \subseteq U$ with $\dim U_0 = \dim U - 1$. By induction hypothesis and Corollary 6.19 there is a projection $P_0 \in \mathcal{L}(X)$ onto U_0. Then $A := P_0|_U \in \mathcal{L}(U)$ is a linear map of the finite-dimensional space U which is not onto. By the dimension theorem of linear algebra, $\mathrm{N}(A) \subseteq U$ must contain some $e \neq 0$. Since U is contained in the support of X^*, there is $f \in X^*$ with $f(e) \neq 0$. We put $Px := P_0x + f(e)^{-1}(f(x) - f(P_0x))e$. Since $A|_{U_0} = \mathrm{id}_{U_0}$ and $Ae = 0 \neq e$, we have $e \in U \setminus U_0$. In view of $\dim U = \dim U_0 + 1$, each $x \in U$ can be written in the form $x = u + \lambda e$ with $u \in U_0$ and $\lambda \in \mathbb{K}$. Then $Px = Pu + \lambda Pe = (P_0u + 0) + \lambda(0 + e) = u + \lambda e = x$. Hence, $P|_U = \mathrm{id}_U$. Since $\mathrm{R}(P) \subseteq U$, $P \in \mathcal{L}(X)$ is a projection onto U, and Corollary 6.19 implies that U is complemented. □

Corollary 6.28. (AC). *Every finite-dimensional subspace $U \subseteq X$ is complemented.*

Proof. In view of Corollary 6.25, this follows from Proposition 6.27. □

The *codimension* of a linear subspace $U \subseteq X$ is defined as the dimension of X/U.

Proposition 6.29. *For $U \subseteq X$ and $n \in \mathbb{N}$ the following statements are equivalent:*

(a) *U has finite codimension n.*

(b) *There is a complementary subspace to U of finite dimension n.*

(c) *There is a complementary subspace to U, and all complementary subspaces have the same finite dimension n.*

Proof. If $[e_1], \dots, [e_n] \in X/U$ is a basis for X/U then $e_1, \dots, e_n$ are a basis for a linear subspace $V \subseteq X$ with $X = U \oplus V$, since for each $x \in X$ we have unique $\lambda_1, \dots, \lambda_n \in \mathbb{K}$. $[x] = \lambda_1[e_1] + \dots + \lambda_n[e_n]$ which means $x = u + \lambda_1 e_1 + \dots + \lambda_n e_n$ with $u \in U$.

Conversely, if $U \oplus V = X$ where V contains linear independent (or a basis) $e_1, \dots, e_n$, then $[e_1], \dots, [e_n]$ are linear independent (or a basis) in X/U. In fact $[x] = \lambda_1[e_1] + \dots + \lambda_n[e_n]$ means $x = u + \lambda_1 e_1 + \dots + \lambda_n e_n$ with $u \in U$, and the latter representation of x is unique (and exists for all $x \in X$, respectively). □

While finite-dimensional subspaces are automatically closed in view of Proposition 3.59, an analogous assertion for subspaces of finite codimension does not hold.

Example 6.30. Let $f: X \to \mathbb{K}$ be linear and unbounded (Example 6.13). We fix some $e \in X$ with $f(e) \neq 0$, and let V be the linear hull of e. Then $U := \mathrm{N}(f)$ satisfies $U \oplus V = X$, since $x = u + \lambda e$ with $u \in U$ and $\lambda \in \mathbb{K}$ holds if and only if $f(x) = \lambda f(e)$ (for sufficiency we note that $u := x - \lambda e$ belongs to U). Hence, U has (finite) codimension $\dim V = 1$, but Proposition 6.16 shows that U is not closed in X.

6.2 Linear Fredholm Operators

Throughout this section, let X and Y be Banach spaces over $\mathbb{K} = \mathbb{R}$ or $\mathbb{K} = \mathbb{C}$; the norm will always be denoted by $\|\cdot\|$.

Definition 6.31. $A \in \mathcal{L}(X, Y)$ is a *linear Fredholm operator* if the following holds:

(a) $\mathrm{N}(A)$ is a complemented subspace of finite dimension n.

(b) $\mathrm{R}(A)$ is a subspace of finite codimension m.

In this case, $k := n - m$ is called the *Fredholm index* of A. The set of all linear Fredholm operators of index k is denoted by $\mathcal{L}_k(X, Y)$. In case $X = Y$, we put $\mathcal{L}_k(X) := \mathcal{L}_k(X, Y)$.

Note that if one assumes AC then $\mathrm{N}(A)$ is automatically complemented if it has finite dimension (Corollary 6.28). However, the above definition will not require us to invoke AC in the proofs.

Example 6.32. The dimension theorem of linear algebra implies that in case $\dim X, \dim Y < \infty$ we have $\mathcal{L}(X,Y) = \mathcal{L}_k(X,Y)$ with $k := \dim X - \dim Y$.

The name Fredholm operator comes from the famous Fredholm alternative. We say that a linear $A\colon X \to Y$ satisfies the Fredholm alternative if one of following two alternatives holds.

(a) The equation $Ax = y$ is well-posed, that is, uniquely solvable for every $y \in Y$, and moreover, the solution x depends continuously on y.

(b) The equation $Ax = y$ is not solvable for every $y \in Y$, and for those $y \in Y$ for which it is solvable, the solution is not unique.

Proposition 6.33. *For an operator $A \in \mathcal{L}_0(X,Y)$ the following statements are equivalent:*

(a) $\mathrm{N}(A) = \{0\}$, *that is, A is one-to-one.*

(b) $\mathrm{R}(A) = Y$, *that is, A is onto Y.*

(c) $A \in \mathrm{Iso}(X,Y)$.

In particular, A satisfies the Fredholm alternative.

Proof. $\mathrm{N}(A) = \{0\}$ if and only if $n = 0$, and A is onto if and only if $m = 0$. If A is one-to-one and onto then A is even an isomorphism onto Y by Theorem 6.12. □

In many text books, it is required in the definition of Fredholm operators that $\mathrm{R}(A)$ be closed in Y. Since we assume that A is bounded, this requirement holds automatically:

Proposition 6.34. *If $A \in \mathcal{L}(X,Y)$ and $\mathrm{R}(A)$ has finite codimension then $\mathrm{R}(A)$ is closed in Y. In particular, every linear Fredholm operator has a closed range.*

Proof. We put $U := \mathrm{R}(A)$. By Proposition 6.29 there is a finite-dimensional subspace $V \subseteq Y$ with $Y = U \oplus V$. Actually, we will only use that V is closed and thus a Banach space.

Without loss of generality, we can assume that A is one-to-one, since, otherwise we can replace A by the operator A_0 from Lemma 6.5. Then $B\colon X \times V \to Y$, $B(x,v) := Ax + v$ is onto and one-to-one, since $Y = \mathrm{R}(A) \oplus V$ (in particular, $B(x,v) = 0$ implies $Ax = v = 0$). By the theorem on the bounded inverse (Theorem 6.12), it follows that $C := B^{-1} \in \mathcal{L}(Y, X \times V)$. In particular, C is continuous (Theorem 6.11). Since $M := X \times \{0\}$ is closed in $X \times V$, Proposition 2.92 implies that $\mathrm{R}(A) = B(M) = C^{-1}(M)$ is closed in Y. □

Note that Example 6.30 implies that Proposition 6.34 cannot follow alone from the fact that the codimension of $\mathrm{R}(A)$ is finite. In fact, the proof uses the boundedness of A essentially.

The most important theorem about linear Fredholm maps is the relation with locally compact operators.

By $\mathcal{K}(X,Y)$ (or $\mathcal{K}(X)$ in case $X = Y$), we denote the locally compact linear operators $A: X \to Y$. We collect first the most important properties of $\mathcal{K}(X,Y)$. We recall that we consider only Banach spaces in this section.

Proposition 6.35. (a) *A linear operator $A: X \to Y$ belongs to $\mathcal{K}(X,Y)$ if and only if $\gamma\big(A(M)\big) = 0$ for all bounded $M \subseteq X$ and all $\gamma \in \{\chi_Y, \alpha, \beta\}$.*

(b) *A linear $A: X \to Y$ belongs to $\mathcal{K}(X,Y)$ if and only if $\gamma\big(A\big(S_r(0)\big)\big) = 0$ for some $r > 0$ and some $\gamma \in \{\chi_Y, \alpha, \beta\}$.*

(c) *$\mathcal{K}(X,Y)$ is a closed linear subspace of $\mathcal{L}(X,Y)$.*

(d) *If Z is a Banach space over $\mathbb{K}$, $A: X \to Y$ and $B: Y \to Z$ are linear, one being bounded and the other locally compact then $BA \in \mathcal{K}(X,Y)$.*

(e) *If $A \in \mathcal{L}(X,Y)$ satisfies $\dim \mathrm{R}(A) < \infty$ then $A \in \mathcal{K}(X,Y)$.*

Proof. Corollary 3.52 implies the first two assertions and also that each $A \in \mathcal{K}(X,Y)$ is continuous at 0. Hence, $\mathcal{K}(X,Y) \subseteq \mathcal{L}(X,Y)$ by Proposition 6.3. By (b) and Proposition 3.30, we obtain that $\mathcal{K}(X,Y)$ is a linear subspace of $\mathcal{L}(X,Y)$. To see that $\mathcal{K}(X,Y)$ is closed, suppose that $A_n \in \mathcal{K}(X,Y)$ satisfy $\|A_n - A\| \to 0$ for some $A \in \mathcal{L}(X,Y)$. Note that for $M := S_1(0)$ and $\varepsilon > 0$ the set $A_n(M)$ is an $(\varepsilon + \|A_n - A\|)$-net for $A(M)$. Hence, (3.3) implies $\chi_X\big(A(M)\big) \le \chi_X\big(A_n(M)\big) + \|A_n - A\| = \|A_n - A\|$, and so $A \in \mathcal{K}(X,Y)$.

If $B \in \mathcal{L}(Y,Z)$ is continuous and $A \in \mathcal{K}(X,Y)$ then $BA \in \mathcal{K}(X,Y)$ follows from Proposition 2.100, and if $B \in \mathcal{K}(Y,Z)$ and $A \in \mathcal{L}(X,Y)$ then for any bounded $M \subseteq X$ the set $N := A(M)$ is bounded and thus $\chi_Z\big(B(N)\big) = 0$. The last assertion follows from Proposition 3.59. □

The most important result about Fredholm operators of index 0 is the following.

Theorem 6.36. *For $A \in \mathcal{L}(X,Y)$ the following statements are equivalent:*

(a) *$A \in \mathcal{L}_0(X,Y)$.*

(b) *A can be written in the form $A = J - K$ with $J \in \mathrm{Iso}(X,Y)$ and $K \in \mathcal{K}(X,Y)$.*

(c) *A can be written in the form* $A = J - K$ *with* $J \in \mathrm{Iso}(X,Y)$ *and* $K \in \mathcal{L}(X,Y)$ *with* $\dim \mathrm{R}(K) < \infty$.

In the latter case, we have $K \in \mathcal{K}(X,Y)$ *and call* K *a* corrector *of* A.

We prepare the proof of Theorem 6.36 by some results of independent interest.

Lemma 6.37. *Let* $A = J - K$ *where* $J \in \mathrm{Iso}(X,Y)$ *and* $K \in \mathcal{L}(X,Y)$*, Suppose that there are* $N \in \mathbb{N}$ *and* $\gamma \in \{\chi_X, \alpha, \beta\}$ *such that* $(J^{-1}K)^N|_{S_1(0)}$ *is* $(1, {}^{\gamma^c}_{\gamma^c})$*-condensing. Then* $\dim \mathrm{N}(A) < \infty$*, and* $\mathrm{R}(A)$ *is closed.*

Proof. Let $U := \mathrm{N}(A)$. For $x \in U$, we have $Jx = Kx$ and thus $x = J^{-1}Kx$ which implies $(J^{-1}K)^N|_{S_1(0)\cap U} = \mathrm{id}_{S_1(0)\cap U}$ after a trivial induction by N. If $\dim U < \infty$ fails then we find by Proposition 3.32 a countable set $C \subseteq S_1(0) \cap U$ with $\beta(C) > 0$. Then $\gamma\big((J^{-1}K)^N(C)\big) = \gamma(C) > 0$ by (3.1), contradicting the hypothesis that $(J^{-1}K)^N|_{S_1(0)}$ is $(1, {}^{\gamma^c}_{\gamma^c})$-condensing.

We define the one-to-one and onto map $A_0 \in \mathcal{L}(X/U, \mathrm{R}(A))$ as in Lemma 6.5 and show that there is a constant $c > 0$ with $\|A_0[x]\| \geq c\|[x]\|$. Then Proposition 3.56 implies that A_0 is an isomorphism and that $\mathrm{R}(A)$ is a Banach space and thus closed. Thus, assume by contradiction that there is a sequence $[x_n] \in X/U$ with $\|[x_n]\| = 1$ and $y_n := A_0[x_n] \to 0$. By definition of the norm in X/U, we can assume that $1 \leq \|x_n\| \leq 1 + n^{-1}$, and by definition of A_0, we have $y_n = Ax_n = Jx_n - Kx_n$. Putting $e_n := x_n/\|x_n\|$, we obtain $J^{-1}Ke_n = e_n - \|x_n\|^{-1}J^{-1}y_n$ which implies (induction by N)

$$(J^{-1}K)^N e_n = e_n - s_{N,n} \quad \text{with} \quad s_{N,n} := \|x_n\|^{-1}\sum_{k=1}^{N}(J^{-1}K)^{k-1}J^{-1}y_n.$$

Note that $\|x_n\| \geq 1$ and $y_n \to 0$ imply that $s_{N,n} \to 0$ as $n \to \infty$. Hence, Lemma 3.33 implies for $C := \{e_1, e_2, \dots\}$ that $\gamma\big((J^{-1}K)^N(C)\big) = \gamma(C)$. Since $C \subseteq S_1(0)$ is $(1, {}^{\gamma^c}_{\gamma^c})$-condensing, we obtain $\gamma(C) = 0$. Hence, C is relatively compact, and so, passing to a subsequence if necessary, we can assume that $e_n \to e$ for some $e \in X$. Since $\|x_n\| \to 1$, we find $x_n \to e$ and thus $\|[x_n] - [e]\| = \|[x_n - e]\| \leq \|x_n - e\| \to 0$. Hence, $[x_n] \to [e]$. Since $\|[x_n]\| = 1$, we thus have $\|[e]\| = 1$. Moreover, the continuity of A_0 implies $y_n = A_0[x_n] \to A_0[e]$. Since $y_n \to 0$, we have $0 \neq [e] \in \mathrm{N}(A_0)$, contradicting the fact that A_0 is one-to-one. □

Lemma 6.38 (Riesz–Schauder Theory). *Let* $A = \mathrm{id}_X - K$ *where* $K \in \mathcal{K}(X)$*. Then* $R_n := \mathrm{R}(A^n)$ *is closed,* $N_n := \mathrm{N}(A^n)$ *has finite dimension* $(n = 1, 2, \dots)$*, and moreover:*

(a) $A(R_n) = R_{n+1} \subseteq R_n$ *and* $N_n \subseteq N_{n+1} = A^-(N_n)$ *for all* n*, and the inclusions are strict at most for finitely many* n.

(b) *If* n *is sufficiently large, then* $X = R_n \oplus N_n$, $A(N_n) \subseteq N_n$, *and* $A|_{R_n} \in \operatorname{Iso}(R_n)$.

(c) *If* n *is sufficiently large, then there is a unique* $J \in \operatorname{Iso}(X)$ *with* $J|_{R_n} = A|_{R_n}$ *and* $J|_{N_n} = \operatorname{id}_{N_n}$.

Proof. Proposition 6.35 implies after a trivial induction that $K_n := \operatorname{id}_X - A^n \in \mathcal{K}(X)$ for $n = 1, 2, \dots$. Hence, the first assertion follows from Lemma 6.37. The inclusion $N_n \subseteq N_{n+1} = A^-(N_n)$ is immediate from the definition, and the proof of $A(R_n) = R_{n+1} \subseteq R_n$ is just a trivial induction.

Assume by contradiction that the inclusion $R_{n_k+1} \subseteq R_{n_k}$ is strict for all k ($n_1 < n_2 < \cdots$). Since R_{n_k+1} is closed, Lemma 3.31 implies that there are $x_k \in R_{n_k}$ with $x_k \in S := S_1(0)$ and $\operatorname{dist}(x_k, R_{n_k+1}) \geq 1/2$. Since $Ax_j \in R_{n_j+1} \subseteq R_{n_k+1}$ ($j \geq k$) and $x_j \in R_{n_j} \subseteq R_{n_k+1}$ ($j > k$), we have $y_{j,k} := x_j - Ax_j + Ax_k \in R_{n_k+1}$ ($j > k$), and so $\|Kx_k - Kx_j\| = \|x_k - y_{j,k}\| \geq 1/2$ ($j > k$). Hence, $\beta\big(K(S)\big) \geq 1/2$, contradicting Proposition 6.35.

Similarly, if the inclusion $N_{n_k} \subseteq N_{n_k+1}$ is strict for all k ($n_1 < n_2 < \cdots$), Lemma 3.31 would imply that there are $x_k \in N_{n_k+1} = A^-(N_{n_k})$ with $x_k \in S := S_1(0)$ and $\operatorname{dist}(x_k, N_{n_k}) \geq 1$. Since $Ax_k \in N_{n_k} \subseteq N_{n_j}$ ($k \leq j$) and $x_k \in N_{n_j}$ ($k < j$), we have $y_{k,j} := x_k - Ax_k + Ax_j \in N_{n_j}$ ($k < j$), and so $\|Kx_j - Kx_k\| = \|x_j - y_{k,j}\| \geq 1$ ($k < j$). Hence, $\beta\big(K(S)\big) \geq 1$, contradicting Proposition 6.35. So we have established (a).

Concerning (b) and (c), we fix n such that $A(R_n) = R_{n+1} = R_n$ and $N_n = N_{n+1} = A^-(N_n)$ which holds for all sufficiently large n by (a). Then $A(N_n) \subseteq N_n$, $B := A|_{R_n} \colon R_n \to R_n$ is onto. Moreover, we have $\mathrm{N}(B) = \{0\}$. Indeed, $y \in \mathrm{N}(B) \subseteq R_n$ implies $By = 0$ and $y = A^n x$ for some $x \in X$. Hence, $A^{n+1}x = 0$, and so $x \in N_{n+1} = N_n$ which implies $y = A^n x = 0$.

Since R_n is closed and thus a Banach space, Theorem 6.12 implies that $B \in \operatorname{Iso}(R_n)$. In view of $B^n \in \operatorname{Iso}(R_n)$, we find for each $x \in X$ in view of $y := A^n x \in R_n$ exactly one $z \in R_n$ satisfying $y = B^n z$. The latter is equivalent to $A^n x = A^n z$ and thus equivalent to $x - z \in N_n$. Hence, $X = R_n \oplus N_n$. Since $B \in \operatorname{Iso}(R_n)$ and $\operatorname{id}_{N_n} \in \operatorname{Iso}(N_n)$, we find by Proposition 6.21 a unique $J \in \operatorname{Iso}(X)$ with $J|_{R_n} = B$ and $J|_{N_n} = \operatorname{id}_{N_n}$. □

Proof of Theorem 6.36. The last assertion follows from Proposition 6.35, and with this the implication (c)⇒(b) is trivial. If (b) holds then $K_0 := J^{-1}K \in \mathcal{K}(X)$ by Proposition 6.35, and so we can apply Lemma 6.38 with $A_0 := \operatorname{id}_X - K_0$. We find that there are closed subspaces $X_i \subseteq X$ ($i = 1, 2$) with $X = X_1 \oplus X_2$, $A_0(X_i) \subseteq X_i$ ($i = 1, 2$), $\mathrm{N}(A_0) \subseteq X_2$, $\dim X_2 < \infty$, and $J_0 \in \operatorname{Iso}(X)$ such that $B := J_0|_{X_1} = A_0|_{X_1} \in \mathrm{Iso}(X_1)$ and $J_0|_{X_2} = \mathrm{id}_{X_2}$. Applying the dimension theorem of linear algebra with $C := A_0|_{X_2}: X_2 \to X_2$ in the finite-dimensional space X_2, we obtain linear subspaces $U_i, V_i \subseteq X_2$ $(i = 1, 2)$ with $X_2 = U_i \oplus V_i$ $(i = 1, 2)$, $U_1 = \mathrm{N}(C)$, $V_2 = \mathrm{R}(C)$, and with $\dim U_1 = \dim U_2$. Since $\mathrm{N}(A_0) \subseteq X_2$ and $A = JA_0$, we have $\mathrm{N}(A) = \mathrm{N}(A_0) = \mathrm{N}(C) = U_1$. Together with $\mathrm{R}(A_0) = \mathrm{R}(B) \oplus \mathrm{R}(C) = X_1 \oplus V_2$, we obtain from $X = X_1 \oplus X_2$ that $X = \mathrm{N}(A) \oplus (V_1 \oplus X_1)$ and $X = \mathrm{R}(A_0) \oplus U_2$. In view of $A = JA_0$, the latter implies $Y = \mathrm{R}(A) \oplus J(U_2)$. It follows in view of Propositions 6.17 that $\mathrm{N}(A) = U_1$ is complemented in X and that its dimension is $\dim U_1 = \dim U_2 = \dim J(U_2)$ which by Proposition 6.29 is the codimension of $\mathrm{R}(A)$. Hence, $A \in \mathcal{L}_0(X, Y)$.

Conversely, let $A \in \mathcal{L}_0(X, Y)$. Using Proposition 6.29 and 6.34, we find closed subspaces $U_i, V_i \subseteq X$ $(i = 1, 2)$ with $X = U_i \oplus V_i$ $(i = 1, 2)$, $U_1 = \mathrm{N}(A)$, $V_2 = \mathrm{R}(A)$, and $\dim U_1 = \dim U_2 < \infty$. Note that $B := A|_{V_1} \in \mathcal{L}(V_1, V_2)$ is one-to-one and onto an thus $B \in \mathrm{Iso}(V_1, V_2)$ by Theorem 6.12. By Corollary 3.58, there is $J_1 \in \mathrm{Iso}(U_1, U_2)$. By Proposition 6.21, there is a unique $J \in \mathrm{Iso}(U_1 \oplus V_1, U_2 \oplus V_2) = \mathrm{Iso}(X)$ satisfying $J|_{U_1} = J_1$ and $J|_{V_1} = B$. Then $K := J - A$ satisfies $\mathrm{R}(K) = K(U_1 \oplus V_1) = K(U_1)$. Since $\dim U_1 < \infty$, we obtain $\dim \mathrm{R}(K) < \infty$. Hence, K is a corrector for A. □

Proposition 6.39. *Let $A_0 \in \mathcal{L}_0(X, Y)$. Then for each corrector K_0 of A_0 there is $\varepsilon > 0$ such that every $A \in \mathcal{L}(X, Y)$ with $\|A - A_0\| \le \varepsilon$ belongs to $\mathcal{L}_0(X, Y)$ and, moreover, every $K \in \mathcal{L}(X, Y)$ satisfying $\dim \mathrm{R}(K) < \infty$ and $\|K - K_0\| \le \varepsilon$ is a corrector for A.*

Proof. By hypothesis, $J_0 := A_0 + K_0 \in \mathrm{Iso}(X, Y)$. If $0 < 2\varepsilon < \|J_0^{-1}\|^{-1}$ and $A, K \in \mathcal{L}(X, Y)$ satisfy $\|A - A_0\| \le \varepsilon$ and $\|K - K_0\| \le \varepsilon$ then $J := A + K$ satisfies $\|J - J_0\| < \|J_0^{-1}\|^{-1}$, and so Proposition 6.10 implies that $J \in \mathrm{Iso}(X, Y)$. □

We point out that perturbation results like Proposition 6.39 immediately carry over to Fredholm maps of nonzero index:

Theorem 6.40. *Let $A_0 \in \mathcal{L}_k(X, Y)$. Then there is some $\varepsilon > 0$ such that all $A \in \mathcal{L}(X, Y)$ belong to $\mathcal{L}_k(X, Y)$ if one of the following holds:*

(a) $\|A - A_0\| \le \varepsilon$ *or*

(b) $A - A_0 \in \mathcal{K}(X, Y)$.

Proof. The case $k = 0$ follows immediately from Proposition 6.39 or from Theorem 6.36, respectively. To see the latter in case $A - A_0 \in \mathcal{K}(X, Y)$, we note that $A_0 = J - K_0$ with some corrector K_0, and so $K := K_0 - (A - A_0)$ belongs to $\mathcal{K}(X, Y)$ and satisfies $A = J - K$.

The case $k > 0$ can be reduced to the case $k = 0$ as follows: We associate to A and A_0 the operators $B, B_0 \in \mathcal{L}(X, Y \times \mathbb{K}^k)$, defined by $Bx := (Ax, 0)$ and $B_0 x := (A_0 x, 0)$. Then $A \in \mathcal{L}_k(X, Y)$ implies $B \in \mathcal{L}_0(X, Y \times \mathbb{K}^k)$. We have already shown that the latter implies $B_0 \in \mathcal{L}_0(X, Y \times \mathbb{K}^k)$ which then implies $A_0 \in \mathcal{L}_k(X, Y)$.

Similarly, we can reduce the case $k < 0$ to the case $k = 0$ by associating to A and A_0 the operators $B, B_0 \in \mathcal{L}(X \times \mathbb{K}^{-k}, Y)$, defined by $B(x, y) := Ax$ and $B_0(x, y) := A_0 x$. □

Using an analogous reduction to the case of index zero, it is possible to give a very simple proof of the famous index formula:

Theorem 6.41. *If $A \in \mathcal{L}_k(X, Y)$ and $B \in \mathcal{L}_\ell(Y, Z)$ then $BA \in \mathcal{L}_{k+\ell}(X, Z)$.*

Proof. Assume first $k = \ell = 0$. By Theorem 6.36, there are correctors K_A and K_B of A and B, respectively. Then $J_A := A + K_A \in \operatorname{Iso}(X, Y)$ and $J_B := B + K_B \in \operatorname{Iso}(Y, Z)$. It follows that

$$J := BA + BK_A + K_B A + K_B K_A = J_B J_A \in \operatorname{Iso}(X, Z).$$

Since Proposition 6.35 implies $K := J - BA = BK_A + K_B A + K_B K_A \in \mathcal{K}(X, Z)$, we obtain from Theorem 6.36 that $BA \in \mathcal{L}_0(X, Z)$.

The general case reduces to the special case similarly as in the proof of Theorem 6.40. To avoid case distinctions, we choose $n \in \mathbb{N}$ with $n + k, n + k + \ell \geq 0$ and define $A_0 \in \mathcal{L}(X \times \mathbb{K}^n, Y \times \mathbb{K}^{n+k})$ and $B_0 \in \mathcal{L}(Y \times \mathbb{K}^{n+k}, Z \times \mathbb{K}^{n+k+\ell})$ by $A_0(x, u) := (Ax, 0)$ and $B_0(y, v) := (By, 0)$ for $x \in X$, $y \in Y$, $u \in \mathbb{K}^n$, $v \in \mathbb{K}^{n+k}$. Then A_0 and B_0 are Fredholm of index 0, and so by what we just proved $B_0 A_0 \in \mathcal{L}_0(X \times \mathbb{K}^n, Z \times \mathbb{K}^{n+k+\ell})$. Since $B_0 A_0(x, u) = (BAx, 0)$, this implies $BA \in \mathcal{L}_{k+\ell}(X, Z)$. □

For the degree on manifolds, an important role is played by transversality. Since we will restrict our attention to Fredholm operators, the following definition of transversality for linear Fredholm operators is appropriate for our purposes.

Definition 6.42. Consider a linear $A \in \mathcal{L}_k(X, Y)$. We call a closed subspace $Y_0 \subseteq Y$ *transversal* to A if $\mathrm{R}(A) + Y_0 = Y$ and if at least one of the following holds:

(a) Y_0 is complemented in Y.

(b) $A^{-1}(Y_0)$ is complemented in X.

The following result implies in particular that then actually both of the properties of Definition 6.42 hold.

Proposition 6.43. *Let $A \in \mathcal{L}_k(X,Y)$. Let $Y_0 \subseteq Y$ be a closed subspace with* $\mathrm{R}(A) + Y_0 = Y$, *and put* $X_0 := A^{-1}(Y_0)$. *Then* $A_0 := A|_{X_0} \in \mathcal{L}_k(X_0, Y_0)$, *and in case* $\dim Y_0 < \infty$ *also* $\dim X_0 = \dim Y_0 + k$.

Moreover, Y_0 is transversal to A if and only if there are closed subspaces $Y_1 \subseteq Y$ and $X_1 \subseteq X$ with $Y = Y_0 \oplus Y_1$ and $X = X_0 \oplus X_1$ such that $A_1 := A|_{X_1} \in$ $\mathrm{Iso}(X_1, Y_1)$.

Proof. We put $R := \mathrm{R}(A)$ and $R_0 := R \cap Y_0$. Since $m := \dim(Y_0/R) \leq \dim(Y/R) < \infty$, Proposition 6.29 implies that there is a space $V \subseteq Y_0$ of dimension m with $Y_0 = R_0 \oplus V$. Since $R \cap V \subseteq R_0$, we have $R \cap V = \{0\}$, and $R + V = (R + R_0) + V = R + Y_0 = Y$. Hence, $Y = R \oplus V$. In particular, $m = \dim V$ is the codimension of R. From $A \in \mathcal{L}_k(X,Y)$, we obtain that $N := \mathrm{N}(A)$ has dimension $k + m$, and that there is a closed subspace $U \subseteq X$ with $X = N \oplus U$. The operator $J := A|_U \in \mathcal{L}(U, R)$ is one-to-one and onto R. Since Proposition 6.34 implies that R is closed, U and R are Banach spaces. Hence, $J \in \mathrm{Iso}(U, R)$ by Theorem 6.12. We put $U_0 := J^{-1}(R_0) \subseteq U$. Then $X_0 = A^{-1}(R_0) = N \oplus U_0$. Since $Y_0 = R_0 \oplus V$, it follows that the codimension of $\mathrm{R}(A_0) = R_0$ in Y_0 is $\dim V = m$. Moreover, $\mathrm{N}(A_0) = N$ is complemented in X_0 and has dimension $k+m$. Thus, $A_0 \in \mathcal{L}_k(X_0, Y_0)$. In case $\ell := \dim Y_0 < \infty$, we have $\ell = \dim R_0 + \dim V = \dim\big(J^{-1}(R_0)\big) + m = \dim U_0 + m$ which implies $\dim X_0 = \dim N + \dim U_0 = (k+m) + \ell - m = k + \dim Y_0$.

Note that $X_0 = N \oplus U_0$ implies that U_0 is complemented in X_0. Note also that $U_0 \subseteq U$ and $X = N \oplus U$ imply that $U_0 = (N \oplus U_0) \cap U = X_0 \cap U$.

Hence, if X_0 is complemented in X then Corollary 6.20 implies that U_0 is complemented in U, that is, there is a closed (in U) $X_1 \subseteq U$ with $U = U_0 \oplus X_1$. Since J is an isomorphism, we obtain that $Y_1 := J(X_1)$ is closed in R.

Similarly, if Y_0 is complemented in Y then, since R_0 is complemented in Y_0, Corollary 6.20 implies that R_0 is complemented in R, that is, there is a closed subspace $Y_1 \subseteq R$ with $R = R_0 \oplus Y_1$. Then $X_1 := J^{-1}(Y_1)$ is closed in U.

In both cases, we thus have found a closed subspace $Y_1 \subseteq R$ and a closed subspace $X_1 \subseteq U$ with $Y_1 = J(X_1)$ such that $U = U_0 \oplus X_1$ and $R = R_0 \oplus Y_1$. Note that Proposition 2.10 implies that X_1 is closed in X and Y_1 is closed in Y. We have $X = N \oplus U_0 \oplus X_1 = X_0 \oplus X_1$, $Y = R_0 \oplus V \oplus Y_1 = Y_0 \oplus Y_1$, and $A_1 = J|_{X_1} \in \mathrm{Iso}(X_1, Y_1)$. □

Proposition 6.44. *Let $Y_0 \subseteq Y$ be transversal to $A_0 \in \mathcal{L}_k(X,Y)$. Then there is some $\varepsilon > 0$ such that for all $A \in \mathcal{L}(X,Y)$ with $\|A - A_0\| \leq \varepsilon$ we have $A \in \mathcal{L}_k(X,Y)$, and Y_0 is transversal to A.*

Proof. We put $X_0 := A_0^{-1}(Y_0)$. By Proposition 6.43, there are closed subspaces $X_1 \subseteq X$ and $Y_1 \subseteq Y$ with $X = X_0 \oplus X_1$, $Y = Y_0 \oplus Y_1$, and such that

$J_0 := A_0|_{X_1} \in \mathrm{Iso}(X_1, Y_1)$. By Proposition 6.10 there is $\delta > 0$ such that every map $J \in \mathcal{L}(X_1, Y_1)$ with $\|J - J_0\| < \delta$ is an isomorphism onto Y_1. By Proposition 6.18, there is a projection $P \in \mathcal{L}(Y)$ with $\mathrm{R}(P) = Y_1$ and such that $Q := \mathrm{id}_Y - P$ is a projection with $\mathrm{R}(Q) = Y_0$. Let $\varepsilon \in (0, \delta/\|P\|)$. Decreasing $\varepsilon > 0$ if necessary, we can in view of Theorem 6.40 assume in addition that for all $A \in \mathcal{L}(X, Y)$ with $\|A - A_0\| \le \varepsilon$ we have $A \in \mathcal{L}_k(X, Y)$.

We show that Y_0 is transversal to each such A. Indeed, putting $J := PA|_{X_1} \in \mathcal{L}(X_1, Y_1)$, we have $\|J - J_0\| = \|PA|_{X_1} - PJ_0\| \le \|P\|\varepsilon < \delta$, and so $J \in \mathrm{Iso}(X_1, Y_1)$. Since $Y = Y_0 \oplus Y_1$, we find that Y_0 is complemented in Y and, moreover, each $y \in Y$ can be written in the form $y = y_0 + y_1$ with $y_i \in Y_i$ $(i = 0, 1)$. Put $x_1 := J^{-1} y_1 \in X_1 \subseteq X$. Then $z := Ax_1 \in \mathrm{R}(A)$, $w := y_0 - Qz \in Y_0$, and $z + w = (Pz + Qz) + (y_0 - Qz) = y_0 + Pz = y_0 + Jx_1 = y_0 + y_1 = y$. Hence, $Y = \mathrm{R}(A) + Y_0$. □

Chapter 7

Orientation of Families of Linear Fredholm Operators

The aim of this chapter is to define a notion of orientation of families of linear Fredholm operators of index 0. This will later be used to define the orientation of nonlinear Fredholm maps and, in finite dimensions, also for continuous maps. This notion will be one of the key tools in the definition of the degree of Fredholm maps and the Brouwer degree on manifolds, respectively.

We ask the reader to be patient in this chapter, since these two important motivations for the definitions cannot immediately be explained, and since we collect a lot of minor results in this chapter for usage in later chapters. In particular, we will apply the collected results straightforwardly in Chapters 8 and 9 to obtain results about the orientation of nonlinear Fredholm operators in Banach manifolds or continuous maps in finite-dimensional manifolds, respectively.

The approach we take for the definition of orientation is in principle that from [17], [18] but with a crucial refinement: It is much more convenient for us to treat general Banach bundles than to consider only the special case of tangent bundles of manifolds. This generalization seems to be completely new.

This novelty has not only the advantage that we can use our approach without any difficulties also for continuous maps in finite-dimensional manifolds. It also has the advantage that the presentation of the topic is less technical, since we do not have to use implicitly the involved definition of tangent bundles of manifolds as was required in the presentation in [17], [18]. In fact, by our approach it will not be necessary at all to speak about manifolds in this chapter.

Throughout this chapter, we are interested in (families of) linear maps between Banach spaces over the field $\mathbb{K} = \mathbb{R}$ or $\mathbb{K} = \mathbb{C}$. In our later applications, the case $\mathbb{K} = \mathbb{R}$ is the only case of interest for us. However, it makes no difference in the presentation to allow also $\mathbb{K} = \mathbb{C}$, so we will not exclude this, although it is not clear whether the orientation is very useful for this case.

7.1 Orientation of a Linear Fredholm Operator

In this section, we are only interested in a single Fredholm operator $A \in \mathcal{L}_0(X, Y)$ between Banach spaces X and Y over the field $\mathbb{K}$. In the case $\mathbb{K} = \mathbb{R}$ which interests us mainly, the notion of orientation of A will appear rather simple: It just consists in picking one of (at most) two equivalence classes in the set of correctors.

Let us first define the corresponding equivalence relation and prove that there are exactly two equivalence classes (in case $\mathbb{K} = \mathbb{R}$) except in degenerate situations.

Let $A \in \mathcal{L}(X, Y)$, and let for a moment $\mathcal{C}(A)$ denote the set of all correctors of A, that is, the set of all $K \in \mathcal{L}(X, Y)$ with $\dim \mathrm{R}(K) < \infty$ such that $A + K \in \mathrm{Iso}(X, Y)$. Recall that according to Theorem 6.36 we have $\mathcal{C}(A) \neq \varnothing$ if and only if $A \in \mathcal{L}_0(X, Y)$.

Now if $K_1, K_2 \in \mathcal{C}(A)$, we consider the auxiliary operator

$$\begin{aligned} K_{1,2} &:= \mathrm{id}_X - (A + K_1)^{-1}(A + K_2) \\ &= (A + K_1)^{-1}(A + K_1) - (A + K_1)^{-1}(A + K_2) \\ &= (A + K_1)^{-1}(K_1 - K_2). \end{aligned} \tag{7.1}$$

Since $\dim \mathrm{R}(K_1 - K_2) < \infty$, we have $\dim \mathrm{R}(K_{1,2}) < \infty$. Hence, whenever $X_0 \subseteq X$ is a finite-dimensional subspace containing $\dim \mathrm{R}(K_{1,2})$ then $J_{X_0} := \mathrm{id}_{X_0} - K_{1,2}|_{X_0} \in \mathcal{L}(X_0)$. Since $J_{X_0} = (A + K_1)^{-1}(A + K_2)|_{X_0}$ is one-to-one and $\dim X_0 < \infty$, we have automatically $J_{X_0} \in \mathrm{Iso}(X_0)$, and

$$\det\bigl((A + K_1)^{-1}(A + K_2)\bigr) := \det\bigl((A + K_1)^{-1}(A + K_2)|_{X_0}\bigr) \neq 0.$$

(This idea of defining a determinant in certain infinite-dimensional situations goes probably back to Kato [84].) We call K_1 and K_2 equivalent if

$$\det\bigl((A + K_1)^{-1}(A + K_2)\bigr) > 0. \tag{7.2}$$

Proposition 7.1. *The above definition is independent of the particular choice of X_0. The above definition is indeed an equivalence relation on $\mathcal{C}(A)$ which in case $\mathbb{K} = \mathbb{R}$ consists of at most two equivalence classes. For $A \in \mathcal{F}_0(X, Y)$ and $X^* \neq \{0\}$ there are at least two equivalence classes.*

Recall that the Hahn–Banach theorem (Corollary 6.25) implies that $X^* \neq \{0\} \iff X \neq \{0\}$. However, the above formulation avoids AC.

Proof. To see the independence of X_0, we write $X_0 = U \oplus V$ with $V := \mathrm{R}(K_{1,2})$. Since $J_{X_0} = \mathrm{id}_{X_0} - K_{1,2}|_{X_0} \in \mathcal{L}(X_0)$ satisfies for $u \in U$ and $v \in V$ that $J_{X_0}(u + v) = J_{X_0}u + J_{X_0}v = (u - K_{1,2}u) + (v - K_{1,2}v) = u + (J_V v - K_{1,2}u)$, we can write J_{X_0} in matrix form

$$J_{X_0}\begin{pmatrix} u \\ v \end{pmatrix} = \begin{pmatrix} \mathrm{id}_U & 0 \\ -K_{1,2}|_U & J_V \end{pmatrix}\begin{pmatrix} u \\ v \end{pmatrix},$$

and so $\det J_{X_0} = \det J_V$, the latter being independent of X_0.

Let us write for the moment $K_1 \sim K_2$ if (7.2) holds, that is, if $\det J_{X_0} > 0$. In this case $\det J_{X_0}^{-1} > 0$. Since $J_{X_0}^{-1} = (A + K_2)^{-1}(A + K_1)|_{X_0}$, this means $K_2 \sim K_1$, and so $\sim$ is symmetric. In case $K_1 = K_2$, we have $J_{X_0} = \mathrm{id}_{X_0}$ and thus $\det J_{X_0} = 1$. Hence, $\sim$ is reflexive.

Now let $K_1, K_2, K_3 \in \mathcal{C}(A)$. We put $K_{i,j} := \mathrm{id}_X -(A + K_i)^{-1}(A + K_j)$ $(i, j = 1, 2, 3)$. As we have seen above, $\dim \mathrm{R}(K_{i,j}) < \infty$, and so there is a finite-dimensional linear subspace $X_0 \subseteq X$ which contains $\mathrm{R}(K_{i,j})$ $(i, j = 1, 2, 3)$. Putting $J_{i,j} = \mathrm{id}_{X_0} - K_{i,j}|_{X_0} \in \mathcal{L}(X_0)$, we have $\det J_{i,j} \neq 0$ and $K_i \sim K_j$ if and only if $\det J_{i,j} > 0$. Note now that $J_{1,3} = J_{1,2}J_{2,3}$, and so $\det J_{1,3} = \det J_{1,2} \det J_{2,3}$. Hence, $K_1 \sim K_3$ holds if either simultaneously both $K_1 \sim K_2$ and $K_2 \sim K_3$ hold or if simultaneously $K_1 \not\sim K_2$ and $K_2 \not\sim K_3$ and $\mathbb{K} = \mathbb{R}$. Hence, $\sim$ is transitive and thus an equivalence relation and, moreover, in case $\mathbb{K} = \mathbb{R}$ there are at most two equivalence classes.

Now if $A \in \mathcal{L}_0(X, Y)$ there is some $K_1 \in \mathcal{C}(A)$ by Theorem 6.36, in particular $J := A + K_1 \in \mathrm{Iso}(X, Y)$. We fix some $f \in X^* \setminus \{0\}$ and some $x_0 \in X$ with $f(x_0) = 2$. Now we can define $C \in \mathcal{L}(X, Y)$ by $Cx := f(x)Jx_0$. Then $\dim \mathrm{R}(C) = 1$. We put $K_2 := K_1 - C$.

If $x \in \mathrm{N}(J - C)$ then $Jx = Cx$ implies $x = J^{-1}Cx = f(x)x_0$. Putting $\lambda := f(x)$, we thus have $x = \lambda x_0$, and so $\lambda = f(x) = \lambda f(x_0) = 2\lambda$ which implies $x = \lambda x_0 = 0$. Hence, $\mathrm{N}(J - C) = \{0\}$. Since $J - C \in \mathcal{L}_0(X, Y)$ by Theorem 6.36, Proposition 6.33 thus implies that $A + K_2 = J - C \in \mathrm{Iso}(X, Y)$. Since $\dim \mathrm{R}(K_2) \leq \dim \mathrm{R}(K_1) + \dim \mathrm{R}(C) < \infty$, we thus have shown that $K_2 \in \mathcal{C}(A)$.

With the previous notations, we have $K_{1,2} = (A + K_1)^{-1}(K_1 - K_2) = J^{-1}C = f(\cdot)x_0$. Hence, we can choose X_0 as the linear hull of $\{x_0\}$, and obtain $\det(\mathrm{id}_{X_0} - K_{1,2}) = 1 - 2 < 0$. Consequently, $K_2 \not\sim K_1$, and so $\mathcal{C}(A)$ contains at least two equivalence classes. □

Definition 7.2. An *orientation* of $A \in \mathcal{L}_0(X, Y)$ is an equivalence class of correctors of A according to the equivalence relation (7.2). In case $\mathbb{K} = \mathbb{R}$ and $X^* \neq \{0\}$, we denote the complement as the *opposite orientation*.

An *oriented linear Fredholm operator* is a couple (A, σ) consisting of an $A \in \mathcal{L}_0(X, Y)$ and an orientation σ for A. In a misuse of notation, we will usually denote this couple by A and refer to σ as "the orientation" of A.

Definition 7.3. For $i = 1, 2$ let X_i and Y_i be Banach spaces and $A_i \in \mathcal{L}_0(X_i, Y_i)$ be oriented with orientations σ_i. Then the *product orientation* of $A_1 \otimes A_2 \in \mathcal{L}_0(X_1 \times X_2, Y_1 \times Y_2)$ is that orientation which contains $K_1 \otimes K_2$ for some $K_i \in \sigma_i$ $(i = 1, 2)$.

Proposition 7.4. *The product orientation σ is well-defined and, in the above notation, the following holds for $K_i \in \mathcal{L}(X_i, Y_i)$ $(i = 1, 2)$.*

(a) *If one of the two relations $K_1 \in \sigma_1$ and $K_2 \in \sigma_2$ holds then the other relation holds if and only if $K_1 \otimes K_2 \in \sigma$.*

(b) *If $\mathbb{K} = \mathbb{R}$ and K_i are correctors of A_i with $K_i \notin \sigma_i$ for $i = 1, 2$ then $K_1 \otimes K_2 \in \sigma$.*

Proof. Let K_i and $\hat{K}_i$ be correctors of A_i $(i = 1, 2)$. Then clearly $K_1 \otimes K_2$ and $\hat{K}_1 \otimes \hat{K}_2$ are correctors of $A_1 \otimes A_2$. We are to show that if K_i and $\hat{K}_i$ belong to the same (or in case $\mathbb{K} = \mathbb{R}$ alternatively opposite) equivalence class for $i = 1$ or $i = 2$ then this holds for $i = 1$ *and* $i = 2$ if and only if $K_1 \otimes K_2$ and $\hat{K}_1 \otimes \hat{K}_2$ belong to the same equivalence class. To see this, we put for $i = 1, 2$,

$$K_{1,2,i} := \mathrm{id}_{X_i} - (A_i + K_i)^{-1}(A_i + \hat{K}_i),$$

and let $X_{0,i} \subseteq X_i$ be finite-dimensional subspaces containing $\mathrm{R}(K_{1,2,i})$. Then $X_0 := X_{0,1} \times X_{0,2} \subseteq X_1 \times X_2$ is a finite-dimensional subspace containing the range of

$$K_{1,2} := \mathrm{id}_{X_1} \otimes \mathrm{id}_{X_2} - (A_1 \otimes A_2 + K_1 \otimes K_2)^{-1}(A_1 \otimes A_2 + \hat{K}_1 \otimes \hat{K}_2),$$

and in the splitting $X_0 = X_{0,1} \times X_{0,2}$, we can write

$$\begin{aligned} \det(\mathrm{id}_{X_0} - K_{1,2}) &= \det \begin{pmatrix} \mathrm{id}_{X_{0,1}} - K_{1,2,1} & 0 \\ 0 & \mathrm{id}_{X_{0,2}} - K_{1,2,2} \end{pmatrix} \\ &= \det(\mathrm{id}_{X_{0,1}} - K_{1,2,1}) \det(\mathrm{id}_{X_{0,2}} - K_{1,2,2}). \end{aligned}$$

Hence, if one of the determinants on the right-hand side is positive (or in case $\mathbb{K} = \mathbb{R}$ negative) then the determinant on the left-hand side is positive if and only if also the other determinant on the right-hand side is positive (negative). This is the assertion which we wanted show. □

Definition 7.5. Let $A \in \mathcal{L}_0(X, Y)$ and $B \in \mathcal{L}_0(Y, Z)$ be oriented with orientation σ_A and σ_B. Then the *composite orientation* of $BA \in \mathcal{L}(X, Z)$ is defined by

$$\sigma_{BA} := \{BK_A + K_B A + K_B K_A : K_A \in \sigma_A, K_B \in \sigma_B.\}$$

Proposition 7.6. *In the above situation $BA \in \mathcal{L}_0(X, Z)$, and σ_{BA} is an orientation. For every $K_A \in \sigma_A$, we have*

$$\sigma_{BA} = \{BK_A + K_B A + K_B K_A : K_B \in \sigma_B\},$$

and for every $K_B \in \sigma_B$*, we have*

$$\sigma_{BA} = \{BK_A + K_B A + K_B K_A : K_A \in \sigma_A\}.$$

Moreover, if either σ_B *or* σ_A *are given, then for any orientation* σ_{BA} *of* BA *there is exactly one corresponding orientation* σ_A *or* σ_B *for* A *or* B*, respectively, such that* σ_{BA} *is the composite orientation. For every* $K_{BA} \in \sigma_{BA}$ *this orientation is given by*

$$\begin{aligned}\sigma_A &:= \{(B + K_B)^{-1}(K_{BA} - K_B A) : K_B \in \sigma_B\},\\ \sigma_B &:= \{(K_{BA} - BK_A)(A + K_A)^{-1} : K_A \in \sigma_A\},\end{aligned}$$

respectively. Equivalently, for every $K_B \in \sigma_B$ *or* $K_A \in \sigma_A$ *respectively, this orientation is given by*

$$\begin{aligned}\sigma_A &:= \{(B + K_B)^{-1}(K_{BA} - K_B A) : K_{BA} \in \sigma_{BA}\},\\ \sigma_B &:= \{(K_{BA} - BK_A)(A + K_A)^{-1} : K_{BA} \in \sigma_{BA}\},\end{aligned}$$

respectively.

In particular: If two of the three maps (A, B, BA) are oriented then the remaining map can be oriented uniquely such that BA carries the corresponding composite orientation.

Proof. $BA \in \mathcal{L}_0(X, Z)$ follows from Theorem 6.41. Let $K_A \in \sigma_A$. If $K_B \in \mathcal{C}(B)$ then

$$K_{BA} := BK_A + K_B A + K_B K_A \in \mathcal{C}(BA), \tag{7.3}$$

because $BA + K_{BA} = (B + K_B)(A + K_A)$, and conversely, if $K_{BA} \in \mathcal{C}(BA)$ then

$$K_B := (K_{BA} - BK_A)(A + K_A)^{-1}$$

is the unique map with (7.3), and this map belongs to $\mathcal{C}(B)$ since $B + K_B = (BA + K_{BA})(A + K_A)^{-1}$. If $K_0 \in \mathcal{C}(B)$, we put

$$K_{1,2} := \mathrm{id}_Y - (B + K_0)^{-1}(B + K_B),$$

$K_{0A} := BK_A + K_0 A + K_0 K_A \in \mathcal{C}(BA)$, and

$$\begin{aligned}\hat{K}_{1,2} &:= \mathrm{id}_X - (BA + K_{0A})^{-1}(BA + K_{BA})\\ &= \mathrm{id}_X - \big((B + K_0)(A + K_A)\big)^{-1}(B + K_B)(A + K_A)\\ &= (A + K_A)^{-1} K_{1,2}(A + K_A).\end{aligned}$$

Putting $Y_0 := \mathrm{R}(K_{1,2})$, we obtain that the finite-dimensional space $X_0 := (A + K_A)^{-1}(Y_0)$ contains $\mathrm{R}(\hat{K}_{1,2})$. Hence, $J_A := (A + K_A)|_{X_0} \in \mathcal{L}(X_0)$. Since J_A is one-to-one and $\dim X_0 < \infty$, it follows that $J_A \in \mathrm{Iso}(X_0)$, and

$$\det(\mathrm{id}_{X_0} - \hat{K}_{1,2}|_{X_0}) = \det\big(J_A^{-1}(\mathrm{id}_{Y_0} - K_{1,2})J_A\big) = \det(\mathrm{id}_{Y_0} - K_{1,2}).$$

Hence, $K_{BA}, K_{0A} \in \mathcal{C}(BA)$ are in the same equivalence class if and only if $K_B, K_0 \in \mathcal{C}(B)$ are in the same equivalence class. It follows that (7.3) runs through an orientation of BA if and only if K_B runs through an orientation of B.

Conversely, let $K_B \in \sigma_B$. If $K_A \in \mathcal{C}(A)$ then it follows as above that (7.3) holds, and conversely, if $K_{BA} \in \mathcal{C}(BA)$ then

$$K_A := (B + K_B)^{-1}(K_{BA} - K_B A)$$

is the unique map with (7.3), and this map belongs to $\mathcal{C}(A)$ since $A + K_A = (B + K_B)^{-1}(BA + K_{BA})$. If $K_0 \in \mathcal{C}(A)$, we put

$$K_{1,2} := \mathrm{id}_X - (A + K_0)^{-1}(A + K_A),$$

$K_{B0} := BK_0 + K_B A + K_B K_0 \in \mathcal{C}(BA)$, and note that

$$\begin{aligned}\hat{K}_{1,2} &:= \mathrm{id}_X - (BA + K_{B0})^{-1}(BA + K_{BA}) \\ &= \mathrm{id}_X - \big((B + K_B)(A + K_0)\big)^{-1}(B + K_B)(A + K_A) = K_{1,2}.\end{aligned}$$

For $X_0 := \mathrm{R}(K_{1,2}) = \mathrm{R}(\hat{K}_{1,2})$, we obtain

$$\det(\mathrm{id}_{X_0} - \hat{K}_{1,2}|_{X_0}) = \det(\mathrm{id}_{X_0} - K_{1,2}|_{X_0}).$$

Hence, $K_{BA}, K_{B0} \in \mathcal{C}(BA)$ are in the same equivalence class if and only if $K_A, K_0 \in \mathcal{C}(A)$ are in the same equivalence class. It follows that (7.3) runs through an orientation of BA if and only if K_A runs through an orientation of A. □

If an oriented Fredholm operator is an isomorphism, then 0 is a natural corrector. For this reason, we define:

Definition 7.7. If $J \in \mathrm{Iso}(X, Y)$ then the *natural orientation* σ of $J \in \mathcal{L}_0(X, Y)$ is that with $0 \in \sigma$.

Corollary 7.8. *Let $J_X \in \mathrm{Iso}(X_0, X)$ and $J_Y \in \mathrm{Iso}(Y, Y_0)$. For $A \in \mathcal{L}(X, Y)$, we put $A_0 := J_Y A J_X \in \mathcal{L}(X_0, Y_0)$. Then $A \in \mathcal{L}_0(X, Y)$ if and only if $A_0 \in \mathcal{L}_0(X_0, Y_0)$, and in this case the orientations of A_0 are exactly those of the form $\sigma_0 = J_Y \circ \sigma \circ J_X$ where σ is an orientation of A. In this case, σ_0 is the composite orientation of σ with the natural orientations of J_X and J_Y.*

Proof. The equivalence of $A \in \mathcal{L}_0(X,Y)$ and $A_0 \in \mathcal{L}_0(X_0,Y_0)$ is trivial (or can also be obtained from Theorem 6.41, using that J_X, J_Y, J_X^{-1}, and J_Y^{-1} are Fredholm operators of index 0.) Choosing the zero corrector for J_X and J_Y, we see that the formulas of Proposition 7.6 for the composite orientation reduce to $\sigma_0 = J_Y \circ \sigma \circ J_X$ and also that σ is uniquely determined by σ_0. □

In case $\mathbb{K} = \mathbb{R}$, we would like to consider the natural orientation as "positive", the opposite orientation as "negative". Hence, we define:

Definition 7.9. If (A,σ) is an oriented Fredholm operator of index 0 then

$$\operatorname{sgn} A := \begin{cases} 1 & \text{if } 0 \in \sigma, \\ -1 & \text{if } 0 \notin \sigma \text{ and } A \in \operatorname{Iso}(X,Y), \\ 0 & \text{if } A \notin \operatorname{Iso}(X,Y). \end{cases}$$

Here, $0 \in \sigma$ means that σ is the equivalence class of the corrector $K = 0$. The notation "sgn A" requires some care, since the orientation σ (which is crucial for the definition) is suppressed in the notation: A more precise notation would be $\operatorname{sgn}(A,\sigma)$ but this would turn out notationally rather cumbersome in our later applications.

Proposition 7.10. *For the product orientation, we have in case* $\mathbb{K} = \mathbb{R}$

$$\operatorname{sgn}(A_1 \otimes A_2) = \operatorname{sgn}(A_1)\operatorname{sgn}(A_2).$$

Proof. Let $A_i \in \mathcal{L}_0(X_i,Y_i)$ $(i = 1,2)$ with orientation σ_i, $X := X_1 \times X_2$, $Y := Y_1 \times Y_2$, $A := A_1 \otimes A_2$, and let σ denote the product orientation.

Clearly, $A \in \operatorname{Iso}(X,Y)$ if and only if $A_i \in \operatorname{Iso}(X_i,Y_i)$ for $i = 1,2$. Hence, $\operatorname{sgn} A \neq 0$ if and only if $\operatorname{sgn} A_1 \operatorname{sgn} A_2 \neq 0$. Thus, it suffices to consider the case $A_i \in \operatorname{Iso}(X_i,Y_i)$ for $i = 1,2$.

Suppose first that $\operatorname{sgn}(A_i) = 1$ for $i = 1$ or $i = 2$, that is $0 \in \sigma_i$. Proposition 7.4(a) implies in this case that $(0,0) \in \sigma$ if and only if $0 \in \sigma_1$ and $0 \in \sigma_2$ holds, that is $\operatorname{sgn} A = 1$ if and only if $\operatorname{sgn}(A_1) = \operatorname{sgn}(A_2) = 1$.

Conversely, suppose that $\operatorname{sgn}(A_1) = \operatorname{sgn}(A_2) = -1$. Then $0 \notin \sigma_i$ is a corrector of A_i for $i = 1,2$ and so Proposition 7.4(b) implies that $(0,0) \in \sigma$, hence, $\operatorname{sgn} A = 1$. □

In the finite-dimensional case there is a strict relation between oriented Fredholm operators and orientations of the space. Recall that on a finite-dimensional vector space X a basis $(e_1,\dots,e_n)$ of X is equivalent to another basis

$(x_1, \dots, x_n)$ if the unique isomorphism sending e_k to x_k $(k = 1, \dots, n)$ has positive determinant. An *orientation* of X is the choice of one of these equivalence classes.

Let now X and Y be oriented (hence finite-dimensional). Recall that by Example 6.32, a map $A \in \mathcal{L}(X, Y)$ belongs to $\mathcal{L}_0(X, Y)$ if and only if $\dim X = \dim Y$.

Definition 7.11. Let X and Y be oriented. Then $A \in \operatorname{Iso}(X, Y)$ is *orientation preserving* if for each basis $(e_1, \dots, e_n)$ of the orientation of X the image $(Ae_1, \dots, Ae_n)$ is an orientation of Y.

If $A \in \mathcal{L}_0(X, Y)$ then the *induced orientation* on A is the family of all $K \in \mathcal{L}(X, Y)$ such that $A + K \in \operatorname{Iso}(X, Y)$ is orientation preserving.

Proposition 7.12. *In case* $X = Y$ *(with the same orientation),* $J \in \operatorname{Iso}(X)$ *is orientation preserving if and only* $\det J > 0$.

Proof. J is the transformation of the corresponding basis. □

Proposition 7.13. *The induced orientation on A is an orientation. Conversely, if* $A \in \mathcal{L}_0(X, Y)$ *is oriented and one of the spaces X or Y is oriented then there is exactly one orientation on the other space so that the induced orientation on A is the given orientation of A.*

In other words: If $A \in \mathcal{L}_0(X, Y)$ and two items from (A, X, Y) are oriented then there is a unique *induced orientation* for the third item.

Proof. Let K_1 belong to the induced orientation, and let K_2 be a corrector of A. Then K_1 and K_2 are equivalent if and only if (7.2) holds. Applying Proposition 7.12 with $J := (A + K_1)^{-1}(A + K_2)$, we see that this is the case if and only if for each $(e_1, \dots, e_n)$ of the orientation of X also $(x_1, \dots, x_n) = (Je_1, \dots, Je_n)$ belongs to the orientation of X. Note that $y_k := (A + K_2)e_k = (A + K_1)x_k$. Hence, the basis $(y_1, \dots, y_n)$ belongs to the orientation of Y if and only if $(x_1, \dots, x_n)$ belongs to the orientation of X. This shows the first assertion. Conversely, if K_1 runs through all elements of an orientation of A, it is clear that if X is oriented then $(y_1, \dots, y_n)$ runs through an orientation of Y and conversely, if Y is oriented then $(x_1, \dots, x_n)$ runs through an orientation of Y, since by the above considerations, we find for any orientation preserving $J \in \operatorname{Iso}(X)$ some K_2 in the same equivalence class as K_1 with $J = (A + K_1)^{-1}(A + K_2)$. □

Corollary 7.14. *If X and Y are oriented and* $A \in \operatorname{Iso}(X, Y)$ *is equipped with the induced orientation then*

$$\operatorname{sgn} A = \begin{cases} 1 & \text{if } A \text{ is orientation preserving,} \\ -1 & \text{otherwise.} \end{cases}$$

Proof. This follows from the definition of induced orientation. □

In the infinite-dimensional situation, one cannot equip X and Y with an orientation but only $A \in \mathcal{L}_0(X, Y)$. However, to define a degree, we will need some sort of orientations also on X and Y. The approach of P. Benevieri and M. Furi is to use the following trick: X and Y are replaced by certain finite-dimensional subspaces $X_0 \subseteq X$ and $Y_0 \subseteq Y$ with $A|_{X_0} \in \mathcal{L}_0(X_0, Y_0)$: One can then work nicely with orientations if the orientation of A induces an orientation of $A|_{X_0}$ in a reasonable way. The latter is the purpose of the following definition:

Definition 7.15. Consider a linear oriented $A \in \mathcal{L}_0(X, Y)$. Let $Y_0 \subseteq Y$ be transversal to A, and put $X_0 := A^{-1}(Y_0)$ and $A_0 := A|_{X_0} \in \mathcal{L}(X_0, Y_0)$. Then the *inherited orientation* of A_0 (inherited from the orientation σ of A), is given by

$$\{K|_{X_0} \in \mathcal{L}(X_0, Y_0) : K \in \sigma \text{ and } \mathrm{R}(K) \subseteq Y_0\}.$$

Proposition 7.16. *In the above setting, we have $A_0 \in \mathcal{L}_0(X_0, Y_0)$, and the correctors of A_0 are exactly those from*

$$\{K|_{X_0} \in \mathcal{L}(X_0, Y_0) : K \text{ is a corrector of } A \text{ with } \mathrm{R}(K) \subseteq Y_0\},$$

and conversely, there is a projection $P \in \mathcal{L}(X)$ with $\mathrm{R}(P) = X_0$ such that for any corrector K_0 of A_0 the operator $K := K_0 P$ is a corrector of A with $\mathrm{R}(K) \subseteq Y_0$.

The inherited orientation is an orientation of A_0. Conversely, if K is a corrector of A which does not belong to the orientation of A and $\mathrm{R}(K) \subseteq Y_0$ then $K|_{X_0}$ does not belong to the orientation of A.

Proof. By Proposition 6.43, we have $A_0 \in \mathcal{L}_0(X_0, Y_0)$, and there are closed subspaces $X_1 \subseteq X$ and $Y_1 \subseteq Y$ with $X = X_0 \oplus X_1$, $Y = Y_0 \oplus Y_1$, such that $A_1 := A|_{X_1} \in \mathrm{Iso}(X_1, Y_1)$. By Proposition 6.18, there is a projection $P \in \mathcal{L}(X)$ with $\mathrm{R}(P) = X_0$ and $\mathrm{N}(P) = X_1$.

Now if $K_0 \in \mathcal{L}(X_0, Y_0)$ is a corrector of A_0, then $K := K_0 P$ is a corrector of A. Indeed, $(A + K)|_{X_0} = A_0 + K_0 \in \mathrm{Iso}(X_0, Y_0)$, and $(A + K)|_{X_1} = A_1 \in \mathrm{Iso}(X_1, Y_1)$, and so Proposition 6.21 implies that $A + K$ is an isomorphism of $X = X_0 \oplus X_1$ onto $Y = Y_0 \oplus Y_1$. Hence, K is a corrector of A which satisfies $\mathrm{R}(K) \subseteq Y_0$ and $K_0 = K|_{X_0}$.

Conversely, if K is a corrector of A with $\mathrm{R}(K) \subseteq Y_0$ then $K_0 := K|_{X_0} \in \mathcal{L}(X_0, Y_0)$ is a corrector of A_0. Indeed, $J := A + K \in \mathrm{Iso}(X, Y)$. Since $\mathrm{R}(K) \subseteq Y_0$, we have $J^{-1}(Y_0) \subseteq A^{-1}(Y_0) = X_0$, and since $J \in \mathrm{Iso}(X, Y)$, we must have $J(X_0) = Y_0$. Hence, $J|_{X_0} = A_0 + K_0 \in \mathrm{Iso}(X_0, Y_0)$.

Now if K_i $(i = 1, 2)$ are two correctors of A with $\mathrm{R}(K_i) \subseteq Y_0$, we put $K_{i,0} := K_i|_{X_0} \in \mathcal{L}(X_0, Y_0)$. Since we have just calculated that $(A + K_1)^{-1}(Y_0) = X_0$ and that $A_0 + K_{1,0} \in \mathrm{Iso}(X_0, Y_0)$, we obtain that (7.1) assumes its range in a finite-dimensional subspace $X_0' \subseteq X_0$, and

$$\begin{aligned} K_{1,2,0} &:= \mathrm{id}_{X_0} - (A_0 + K_{1,0})^{-1}(A_0 + K_{2,0}) \\ &= (A_0 + K_{1,0})^{-1}(K_{2,0} - K_{1,0}) = K_{1,2}|_{X_0}. \end{aligned}$$

In particular, X_0' contains the range of $K_{1,2,0}$, and since $K_{1,2,0}|_{X_0'} = K_{1,2}|_{X_0'}$, we find that

$$\det(\mathrm{id}_{X_0'} - K_{1,2,0}|_{X_0'}) = \det(\mathrm{id}_{X_0'} - K_{1,2}|_{X_0'})$$

and thus

$$\det\bigl((A_0 + K_{1,0})^{-1}(A_0 + K_{2,0})\bigr) = \det\bigl((A + K_1)^{-1}(A + K_2)\bigr).$$

It follows that K_i $(i = 1, 2)$ belong to the same equivalence class of correctors of A if and only if $K_{i,0}$ $(i = 1, 2)$ belong to the same equivalence class of correctors of A_0. Hence, if we consider all correctors $K \in \mathcal{C}(A)$ of the same equivalence class with $\mathrm{R}(K) \subseteq Y_0$, the corresponding operators $K_0 := K|_{X_0} \in \mathcal{C}(A_0)$ constitute exactly an equivalence class of correctors of A_0. □

Corollary 7.17. *Consider an oriented $A \in \mathcal{L}_0(X, Y)$. Let $Y_0 \subseteq Y$ be transversal to A, and put $X_0 := A^{-1}(Y_0)$ and $A_0 := A|_{X_0} \in \mathcal{L}_0(X_0, Y_0)$ with the inherited orientation. Then*

$$\operatorname{sgn} A = \operatorname{sgn} A_0.$$

In particular, $A \in \mathrm{Iso}(X, Y)$ if and only if $A_0 \in \mathrm{Iso}(X_0, Y_0)$.

Proof. Proposition 7.16 implies that if 0 is a corrector of A then $0 = 0|_{X_0}$ is a corrector of A_0, and conversely if 0 is a corrector of A_0 then $0 = 0P$ is a corrector of A. This means $A \in \mathrm{Iso}(X, Y)$ if and only if $A_0 \in \mathrm{Iso}(X_0, Y_0)$. In this case, Proposition 7.16 implies further that 0 belongs to the orientation of A if and only if $0 = 0|_{X_0}$ belongs to the orientation of A_0. □

7.2 Orientation of a Continuous Family

In later sections, our aim will be to consider nonlinear operators. To this end, we have to replace A by a family of operators (which will later e.g. be a family of derivatives of the considered nonlinear operators), and we have to define a notion of orientation for such a family.

We understand $\mathcal{L}_0(X, Y)$ equipped with the topology inherited from $\mathcal{L}(X, Y)$.

Definition 7.18. Let I be a topological space, and $A\colon I \to \mathcal{L}_0(X,Y)$ be continuous. An *orientation* of A is a lower semicontinuous map $\sigma\colon I \multimap \mathcal{L}(X,Y)$ such that $\sigma(t)$ is an orientation for $A(t)$ for each $t \in I$.

We call A *orientable* if such an orientation exists.

Formally, "lower semicontinuous" appears to be a much weaker requirement than the continuity requirement imposed in [17]. However, we will show now that actually our definition is the same as that from [17], and we obtain also rather convenient equivalent characterizations (some are mentioned in [17]). The key to this equivalence is the following lemma which follows from the continuity (6.4) of the inversion map.

Lemma 7.19. *Let $\mathbb{K} = \mathbb{R}$. Let $\hat{K}_1, \hat{K}_2$ be correctors of $\hat{A} \in \mathcal{L}_0(X,Y)$ of the same (or opposite) equivalence class. Then there is some $\varepsilon > 0$ such that for all $A, K_1, K_2 \in \mathcal{L}(X,Y)$ with $\|A - \hat{A}\| \le \varepsilon$, $\|K_k - \hat{K}_k\| \le \varepsilon$ and $\dim \mathrm{R}(K_k) < \infty$ $(k = 1,2)$, we have $A \in \mathcal{L}_0(X,Y)$, and K_1, K_2 are both correctors of A of the same (or opposite, respectively) equivalence class.*

Proof. Put $\hat{J}_k := \hat{A} + \hat{K}_k$ and $J_0 := \hat{J}_1^{-1}\hat{J}_2$. By Proposition 6.10, there is $\varepsilon > 0$ such that for all $J_k \in \mathcal{L}(X,Y)$ with $\|J_k - \hat{J}_k\| \le 2\varepsilon$ we have $J_k \in \mathrm{Iso}(X,Y)$ $(k = 1,2)$ and

$$2\|\hat{J}_1^{-1}\|\varepsilon + \|J_1^{-1} - \hat{J}_1^{-1}\|(\|\hat{J}_2\| + 2\varepsilon) < \|J_0^{-1}\|^{-1}. \tag{7.4}$$

Now let $A, K_1, K_2 \in \mathcal{L}(X,Y)$ satisfy $\|A - \hat{A}\| \le \varepsilon$, $\|K_k - \hat{K}_k\| \le \varepsilon$, and $\dim \mathrm{R}(K_k) < \infty$ $(k = 1,2)$. We put $J_k := A + K_k$, and note that $\|J_k - \hat{J}_k\| \le 2\varepsilon$ implies that $J_k \in \mathrm{Iso}(X,Y)$ satisfy (7.4). In particular, K_k are correctors of A, and so $A \in \mathcal{L}_0(X,Y)$ by Theorem 6.36. We define $K_{1,2}$ by (7.1) and analogously $\hat{K}_{1,2}$ by

$$\hat{K}_{1,2} = \mathrm{id}_X - \hat{J}_1^{-1}\hat{J}_2 = \hat{J}_1^{-1}(\hat{K}_1 - \hat{K}_2).$$

Note that in particular $\mathrm{R}(K_{1,2})$ and $\mathrm{R}(\hat{K}_{1,2})$ are finite-dimensional and thus contained in a finite-dimensional subspace $X_0 \subseteq X$. Moreover,

$$K_{1,2} - \hat{K}_{1,2} = \hat{J}_1^{-1}\hat{J}_2 - J_1^{-1}J_2 = \hat{J}_1^{-1}(\hat{J}_2 - J_2) + (\hat{J}_1^{-1} - J_1^{-1})J_2$$

implies by (7.4) and Proposition 6.10 that

$$J(t) := \mathrm{id}_X - tK_{1,2} - (1-t)\hat{K}_{1,2} = J_0 - t(K_{1,2} - \hat{K}_{1,2})$$

belongs to $\mathrm{Iso}(X)$ for every $t \in [0,1]$. In particular, $J(t)|_{X_0}$ is one-to-one. Since $K_{1,2}$ and $\hat{K}_{1,2}$ assume their range in X_0, it follows that $A_t := J(t)|_{X_0} \in \mathrm{Iso}(X_0)$

for every $t \in [0,1]$. In particular, $\det(A_t) \neq 0$ for all $t \in [0,1]$. The continuity of the determinant in the space X_0 implies that $\det(A_t)$ does not change its sign for $t \in [0,1]$. Since

$$\det(A_0) = \det(\mathrm{id}_{X_0} - \hat{K}_{1,2}) = \det\big((\hat{A}+\hat{K}_1)^{-1}(\hat{A}+\hat{K}_2)\big)$$

is positive (negative) by hypothesis, it follows that also $\det(A_1)$ is positive (negative). The latter determinant is just (7.2). □

Now we can prove the announced equivalence:

Theorem 7.20. *Let $\mathbb{K} = \mathbb{R}$, and $A\colon I \to \mathcal{L}_0(X,Y)$ be continuous and $\sigma\colon I \multimap \mathcal{L}(X,Y)$ such that $\sigma(t)$ is an orientation for $A(t)$ for every $t \in I$. Then for $t_0 \in I$ the following statements are equivalent:*

(a) *σ is lower semicontinuous at t_0.*

(b) *There is some $K \in \sigma(t_0)$ such that for each neighborhood of $U \subseteq \mathcal{L}(X,Y)$ of K the set $\sigma^+(U)$ is a neighborhood of t_0.*

(c) *There is a neighborhood $V \subseteq I$ of t_0 with $\bigcap_{t\in V}\sigma(t) \neq \varnothing$.*

(d) *For each $K \in \sigma(t_0)$ there is a neighborhood $V \subseteq I$ of t_0 with $K \in \bigcap_{t\in V}\sigma(t)$.*

(e) *The (pointwise) opposite orientation is lower semicontinuous at t_0.*

In particular, σ is an orientation for A if and only if some (equivalently all) of the above properties hold for every $t_0 \in I$.

For (e), *we require of course that the opposite orientation exists pointwise, e.g. $X^* \neq \{0\}$.*

The property (c) is the one used in [17] as the definition of orientation.

Proof. The implications (d)⇒(a) and (c)⇒(b) are trivial. Since $\sigma(t_0) \neq \varnothing$, we have also trivially (d)⇒(c) and (a)⇒(b). We show now (b)⇒(d).

Thus, let $K \in \sigma(t_0)$ be as in (b), and let $K_0 \in \sigma(t_0)$. By Lemma 7.19 there is some $\varepsilon > 0$ and a neighborhood $V_0 \subseteq I$ of t_0 such that for all $t \in V_0$ and all $K_1, K_2 \in \mathcal{L}(X,Y)$ with $\|K_1 - K\| < \varepsilon$, $\|K_2 - K_0\| < \varepsilon$, the operators K_1, K_2 are correctors of $A(t)$ of the same equivalence class. Applying (b) with $U := B_\varepsilon(K)$, we find that there is a neighborhood $V \subseteq V_0$ of t_0 such that for all $t \in V$ there is $K_1 \in \sigma(t)$ with $\|K_1 - K\| < \varepsilon$. Since $t \in V_0$, we obtain that $K_2 := K_0$ and $K_1 \in \sigma(t)$ are both correctors of $A(t)$ of the same equivalence class, and so $K_0 = K_2 \in \sigma(t)$. Hence, $K_0 \in \bigcap_{t\in V}\sigma(t)$.

Concerning (e), let $\sigma_-(t)$ denote the opposite orientation of $\sigma(t)$ for all $t \in I$, and for $t_0 \in I$ choose $K_1 \in \sigma_-(t_0)$ and $K_2 \in \sigma(t_0)$. Then K_1, K_2 are correctors of $A(t_0)$ of different equivalence classes. Lemma 7.19 and the continuity of A implies that there is a neighborhood $V \subseteq I$ of t_0 such that for each $t \in V$ the operators K_1, K_2 are correctors of $A(t)$ of different equivalence classes. Hence, σ has the property (c) if and only if σ_- has the property (c). □

Corollary 7.21. *Let $\mathbb{K} = \mathbb{R}$ and $A: I \to \mathcal{L}_0(X, Y)$ be constant, $A(t) \equiv A_0$. Then any orientation of A is constant on the components of I.*

Proof. Fix a component C of I and some $t_0 \in C$. The set

$$A_1 := \{t \in I : \sigma(t) = \sigma(t_0)\} = \{t \in I : \sigma(t) \cap \sigma(t_0) \neq \emptyset\}$$

and its complement

$$A_2 := \{t \in I : \sigma(t) \neq \sigma(t_0)\} = \{t \in I : \sigma(t) \setminus \sigma(t_0) \neq \emptyset\}$$

are both open in I by Theorem 7.20(d). Hence, Corollary 2.18 implies $C \subseteq A_1$. □

One implication of the continuity of orientations is that we can easily compute the change of signs in finite dimensions:

Proposition 7.22. *Let $A: I \to \mathcal{L}_0(X, Y)$ be continuous with an orientation σ: $I \multimap \mathcal{L}(X, Y)$. Suppose $\dim X = \dim Y < \infty$, $K = \mathbb{R}$, and let t_1, t_2 belong to the same component of I with $A(t_1) \in \operatorname{Iso}(X, Y)$. Then*

$$\operatorname{sgn} A(t_1) = \operatorname{sgn} A(t_2) \iff \operatorname{sgn} \det\bigl(A(t_1)^{-1} A(t_2)\bigr) > 0.$$

Proof. Since $\dim X = \dim Y < \infty$ there is $J \in \operatorname{Iso}(Y, X)$. For $B_1 \in \operatorname{Iso}(X, Y)$ and $B_2 \in \mathcal{L}(X, Y)$, we calculate by the composition formula for determinants that

$$\det(B_1^{-1} B_2) = \det\bigl(B_1^{-1} J^{-1} J B_2\bigr) = \det\bigl((J B_1)^{-1} J B_2\bigr) = \frac{\det(J B_2)}{\det(J B_1)}. \quad (7.5)$$

If $K_1, K_2 \in \mathcal{L}(X, Y)$ are correctors of $A(t)$, it follows with $B_i := A(t) + K_i$ that K_1 and K_2 are equivalent if and only if $\det(J(A(t) + K_i))$ have the same sign for $i = 1, 2$. Note that $g(t) := \det(J(A(t) + \sigma(t)))$ is the composition of lower semicontinuous functions and thus lower semicontinuous. We have just shown that either $g(t) \subseteq (0, \infty)$ or $g(t) \subseteq (-\infty, 0)$, and so the lower semicontinuity of g implies that the single-valued function $f(t) := \operatorname{sgn} g(t)$ is actually continuous, that is, locally constant. Proposition 2.19 thus implies $f(t_1) = f(t_2)$. Note that

$$\operatorname{sgn} A(t_i) > 0 \iff 0 \in \sigma(t_i) \iff f(t) = \operatorname{sgn} \det\bigl(J A(t_i)\bigr)$$

for $i = 1, 2$. Applying (7.5) with $B_i := A(t_i)$, we find

$$\operatorname{sgn}\det\bigl(A(t_1)^{-1}A(t_2)\bigr) = \frac{\operatorname{sgn}\det\bigl(JA(t_2)\bigr)}{\operatorname{sgn}\det\bigl(JA(t_1)\bigr)} = \frac{f(t_2)\operatorname{sgn} A(t_2)}{f(t_1)\operatorname{sgn} A(t_1)}.$$

Hence, the assertion follows from $f(t_1) = f(t_2)$. □

There are several results about the existence and uniqueness of orientations for continuous $A\colon I \to \mathcal{L}_0(X, Y)$. However, we will not formulate them here but prove these results immediately in a more general context in the following section.

7.3 Orientation of a Family in Banach Bundles

For our later applications, it is not enough to consider maps $A\colon I \to \mathcal{L}_0(X, Y)$ where X and Y are constant. We will instead have to consider families of maps $A(t) \in \mathcal{L}_0(X_t, Y_t)$ $(t \in I)$ where X_t and Y_t depend on $t \in I$ and are isomorphic to a fixed space E_X and E_Y, respectively, and where the dependence of X_t and Y_t on t is "continuous" in a certain sense.

The easiest of these cases is if X_t and Y_t are mapped to fixed spaces by a *norm-continuous* family of isomorphisms. This case is reduced to the setting of the previous section straightforwardly by the following simple observation.

We equip $\operatorname{Iso}(X, Y)$ with the norm inherited from $\mathcal{L}(X, Y)$.

Lemma 7.23. *Let X, Y, X_0, and Y_0 be Banach spaces, I be a topological space, and let $J_X\colon I \to \operatorname{Iso}(X_0, X)$ and $J_Y\colon I \to \operatorname{Iso}(Y, Y_0)$ be continuous. Then $\Phi\colon I \multimap \mathcal{L}(X, Y)$ is lower semicontinuous at $t_0 \in I$ if and only if $\Phi_0\colon I \multimap \mathcal{L}(X_0, Y_0)$, $\Phi_0(t) := J_Y(t) \circ \Phi(t) \circ J_X(t)$, is lower semicontinuous at t_0.*

Proof. The multiplication mapping $M\colon \operatorname{Iso}(Y, Y_0) \times \mathcal{L}(X, Y) \times \mathcal{L}(X_0, X) \to \mathcal{L}(X_0, Y_0)$, $M(C, B, A) := CBA$ is continuous by Lemma 6.8. If Φ is lower semicontinuous at t_0, then also $\Psi(t) := \{J_Y(t)\} \times \Phi(t) \times \{J_X(t)\}$ is lower semicontinuous at t_0 by Proposition 2.95, and so the composition $\Phi_0 = M \circ \Psi$ is lower semicontinuous at t_0 by Proposition 2.94.

Conversely, if Φ_0 is lower semicontinuous at t_0 then $\Phi(t) = J_Y(t)^{-1} \circ \Phi_0(t) \circ J_X(t)^{-1}$ is lower semicontinuous at t_0 by what we just proved, since $t \mapsto J_i(t)^{-1}$ $(i = 1, 2)$ are continuous by Proposition 6.10. □

Proposition 7.24. *Let X, Y, X_0, and Y_0 be Banach spaces, I be a topological space, and let $J_X\colon I \to \operatorname{Iso}(X_0, X)$ and $J_Y\colon I \to \operatorname{Iso}(Y_0, Y)$ be continuous. Then $A\colon I \to \mathcal{L}_0(X, Y)$ is continuous if and only if $A_0\colon I \to \mathcal{L}_0(X_0, Y_0)$, $A_0(t) := J_Y(t) \circ A(t) \circ J_X(t)$ is continuous, and in this case the orientations of A_0 are* *exactly those of the form* $\sigma_0(t) := J_Y(t) \circ \sigma(t) \circ J_X(t)$ *where* σ *is an orientation of* A.

Proof. Lemma 7.23 implies immediately the first assertion and that σ is lower semicontinuous if and only if σ_0 is lower semicontinuous. Hence, the assertion follows from Corollary 7.8. □

The difficulty in a more general setting which we will meet soon is that X_t and Y_t are not equipped with a norm but are only isomorphic to fixed Banach spaces E_X and E_Y as topological vector spaces: For this reason, it is not clear how to define a concept of norm-continuity for families $A(t) \in \mathcal{L}(X_t, Y_t)$ $(t \in I)$ as above.

To define such a concept, we have to speak about Banach bundles. The details of the formulation of the definition of the latter vary slightly in literature. We use the following terminology.

Definition 7.25. Let X be a topological space, E be a Banach space over $\mathbb{K}$, and X_x $(x \in X)$ be a family of pairwise disjoint vector spaces. Let $T = \bigcup_{x \in X} X_x$ be equipped with a fixed topology.

A *trivialization* of T at $x_0 \in X$ is a homeomorphism h of $U \times E$ onto $\bigcup_{x \in U} X_x$ where $U \subseteq X$ is an open neighborhood of x_0 and $h(x, \cdot)$ acts from E into X_x and is linear for every $x \in U$.

The *bundle projection* is the map which associates to each $p \in T$ the unique $x \in X$ with $p \in X_x$.

Proposition 7.26. *In the above situation, if* $x \in U$ *then* X_x *is a topological vector space (with the inherited topology), and* $h_x := h(x, \cdot)$ *is an isomorphism of* X *onto* X_x *with*

$$h_x^{-1}(p) = h^{-1}(x, p) \quad \textit{for all } x \in U,\ p \in X_x. \tag{7.6}$$

The bundle projection π *satisfies*

$$\pi^{-1}(M) = \bigcup_{x \in M} X_x \quad \textit{for each } M \subseteq X, \tag{7.7}$$

and if at each $x_0 \in X$ *there is a trivialization, the following statements are equivalent:*

(a) π *is continuous.*

(b) *At each* $x_0 \in X$ *there is a trivialization whose image is open in* X.

(c) *Every trivialization has an open image in* X.

Proof. Since h maps onto $\bigcup_{x \in U} X_x$, and h_y maps into X_y which is disjoint from X_x for $y \neq x$, it follows that h_x maps onto X_x and satisfies (7.6). Since $h_x\colon E \to X_x$ is a linear homeomorphism, it is straightforward to verify that the vector operations on X_x are continuous and thus that X_x is a topological vector space and that h_x is an isomorphism. The formula (7.7) is immediate from the definition of π. Since the range of trivializations has the form (7.7) with open $M = U \subseteq X$, it is open T if π is continuous. Conversely, if $p \in T$, $x_0 := \pi(p)$, and if there is a trivialization $h\colon U \times E \to \bigcup_{x \in U} X_x$ at x_0 with $h(U \times E)$ being open in T then π is continuous at x. Indeed, if $O \subseteq X$ is an open neighborhood of x_0 then $M := h((U \cap O) \times E)$ is open in $h(U \times E)$ and thus open in T by Proposition 2.10. Hence, M is a neighborhood of x with $\pi(M) \subseteq O$. □

If $\dim E < \infty$, a definition of vector bundles found in literature in the above situation is the requirement that at each $x_0 \in X$ there is a trivialization and that π is continuous. Unfortunately, in case $\dim X = \infty$ this is not enough for our purpose: We have to restrict our attention to a certain family of trivializations which are compatible in the sense that a "change of trivializations" is not only continuous but even continuous with respect to the operator norm topology. Therefore, our definition is more cumbersome:

Definition 7.27. A *Banach bundle* over a Banach space E is a space $T = \bigcup_{x \in X} X_x$ as in Definition 7.25 together with a family $\mathcal{H}$ of trivializations such that for each $x_0 \in X$ there is some trivialization from $\mathcal{H}$ at x_0 with open image in T and such that for each $h, H \in \mathcal{H}$ the corresponding maps $h_x := h(x, \cdot)$ and $H_x := H(x, \cdot)$ (defined for x on open subsets $U, V \subseteq X$) are such that $x \mapsto H_x^{-1} h_x$ is continuous as a map from $U \cap V$ into $\mathcal{L}(E)$. In case $\dim E < \infty$, we call T a *vector bundle*.

Thus, formally, a Banach bundle is a couple $(T, \mathcal{H})$. However, we usually do not mention $\mathcal{H}$ explicitly but just speak about $\mathcal{H}$ as the family of associated trivializations.

In the case $\dim E < \infty$ the continuity of $x \mapsto H_x^{-1} h_x$ follows automatically from (7.6), and therefore, one could let $\mathcal{H}$ be the system of all trivializations: The particular choice of $\mathcal{H}$ plays actually no crucial role in the definition of the Banach bundle. However, in case $\dim E = \infty$, the set $\mathcal{H}$ must be part of the definition.

Let us point out once more the reason for this technical requirement about $\mathcal{H}$: Since X_x are not normed, we have no natural concept of norm-continuity of the maps $x \mapsto h_x$. However, E is normed, and so we have a notion of norm-continuity of the maps $x \mapsto H_x^{-1} h_x$.

The most important examples of Banach bundles for us are the following:

Example 7.28. The constant bundle $T := X \times E$. This is a Banach bundle with $X_x := \{x\} \times E_x$. The map id_T is a trivialization at every $x_0 \in X$, and the bundle projection is given by $\pi(x, p) = x$.

This is the example which corresponds to the case that X_x is actually normed and independent of x.

Example 7.29. Let X be a Banach manifold over E, and X_x $(x \in X)$ be the corresponding tangent spaces. (We will give precise definitions of these notions in Section 8.3.) Then $T = \bigcup_{x \in X} X_x$ is a Banach bundle when we equip it with a natural topology and trivialization which we will describe in Definition 8.20. This will be the bundle which we use in the definition of nonlinear Fredholm maps. In this example, the spaces X_x are not equipped with a norm.

Remark 7.30. The spaces $X_x = \pi^{-1}(x)$ are the *fibres* of π and thus in literature often not mentioned explicitly in the definition of a vector bundle, since they are determined by π: Formally, it is simpler to define a Banach bundle just as the map π together with the family $\mathcal{H}$ of trivializations. In fact, without mentioning π or explicitly repeating the definition of trivialization, the notation $T = \bigcup_{x \in X} X_x$ is ambiguous.

However, the abstract definition using only π is a bit strange in the context of the above examples which are the main case of interest to us, since π itself is not the crucial object: The important objects are the spaces $X_x = \pi^{-1}(x)$ and how they are "coupled" with the topology of X (and in case $\dim E = \infty$ by the trivializations) which is what is described by our definition of Banach bundles.

Now we define what we mean by "norm-continuous" families of linear mappings in Banach bundles.

Definition 7.31. Let T and S be Banach bundles over E_X and E_Y with fibres X_x $(x \in X)$ and Y_y $(y \in Y)$ and associated trivializations $\mathcal{H}_T$ and $\mathcal{H}_S$, respectively. Let I be a topological space.

By $\mathcal{L}(I, T, S)$, we denote the families of all maps A which associate to each $t \in I$ a map $A(t) \in \mathcal{L}(X_{x(t)}, Y_{y(t)})$ such that the following holds:

(a) $x\colon I \to X$ and $y\colon I \to Y$ are continuous.

(b) For each $t_0 \in I$ there are trivializations $h_T \in \mathcal{H}_T$ and $h_S \in \mathcal{H}_S$ at $x(t_0)$ and $y(t_0)$, respectively such that the map

$$A_{h_S,h_T}(t) := h_S\big(y(t), \cdot\big)^{-1} \circ A(t) \circ h_T\big(x(t), \cdot\big) \in \mathcal{L}(E_X, E_Y) \quad (7.8)$$

is continuous at t_0 as a map from a neighborhood of $t_0 \in I$ into $\mathcal{L}(E_X, E_Y)$.

The sets $\mathcal{L}_k(I, T, S)$ and $\mathrm{Iso}(I, T, S)$ are the subsets of all $A \in \mathcal{L}(I, T, S)$ which additionally satisfy $A(t) \in \mathcal{L}_k(X_{x(t)}, Y_{y(t)})$ or $A(t) \in \mathrm{Iso}(X_{x(t)}, Y_{y(t)})$ for all $t \in I$, respectively.

For the following simple observation it is actually not necessary that $h_T \in \mathcal{H}_T$ and $h_S \in \mathcal{H}_S$.

Proposition 7.32. *A map $A \in \mathcal{L}(I, T, S)$ belongs to $\mathcal{L}_k(I, T, S)$ or $\mathrm{Iso}(I, T, S)$ if and only if for each $t_0 \in I$ there are trivializations h_S and h_T at $x(t_0)$ or $y(t_0)$, respectively such that the above maps $A_{h_S,h_T}(t_0)$ belongs to $\mathcal{L}_k(E_X, E_Y)$ or $\mathrm{Iso}(E_X, E_Y)$, respectively.*

More general, if h_T and h_S are trivializations defined on $U \times E_X$ or $V \times E_Y$, respectively, then the set of all $t \in I$ for which (7.8) is defined is the open set $x^{-1}(U) \cap y^{-1}(V) \subseteq I$, and for every t from this set, we have

$$A(t) \in \mathcal{L}_k(X_{x(t)}, Y_{y(t)}) \iff A_{h_S,h_T}(t) \in \mathcal{L}_k(E_X, E_Y),$$
$$A(t) \in \mathrm{Iso}(X_{x(t)}, Y_{y(t)}) \iff A_{h_S,h_T}(t) \in \mathrm{Iso}(E_X, E_Y).$$

Proof. The formula (7.8) shows that $A_{h_S,h_T}(t)$ and $A(t)$ differ only by the composition with isomorphisms. Such compositions preserve the property of being Fredholm of index k or being an isomorphism. □

In the same manner as in Definition 7.31, we can define orientations for $A \in \mathcal{L}_0(I, T, S)$:

Definition 7.33. Let T and S be Banach bundles over E_X and E_Y with fibres X_x ($x \in X$) and Y_y ($y \in Y$) and associated families of trivializations $\mathcal{H}_T$ and $\mathcal{H}_S$, respectively. Let I be a topological space.

An *orientation* of $A \in \mathcal{L}_0(I, T, S)$ is a map σ which associates to each $t \in I$ an orientation $\sigma(t) \subseteq \mathcal{L}(X_t, Y_t)$ of $A(t) \in \mathcal{L}(X_{x(t)}, Y_{y(t)})$ such that for each $t_0 \in I$ there are trivializations $h_T \in \mathcal{H}_T$ and $h_S \in \mathcal{H}_S$, respectively such that the multivalued map

$$\sigma_{h_S,h_T}(t) := h_S\big(y(t), \cdot\big)^{-1} \circ \sigma(t) \circ h_T\big(x(t), \cdot\big) \subseteq \mathcal{L}(E_X, E_Y) \tag{7.9}$$

is lower semicontinuous at t_0 as a multivalued map from a neighborhood of $t_0 \in I$ into $\mathcal{L}(E_X, E_Y)$.

We call $A \in \mathcal{L}_0(I, T, S)$ *orientable* if such an orientation exists.

In the above definitions, we have required the continuity resp. lower semicontinuity only for *one* choice of the trivializations, and only at t_0. The important point of $\mathcal{H}$ in Definition 7.27 is that it implies as a consequence the continuity *everywhere* for *each* trivialization associated from $\mathcal{H}$.

Proposition 7.34. *Let T and S be Banach bundles over E_X and E_Y with fibres X_x $(x \in X)$ and Y_y $(y \in Y)$ and associated trivializations $\mathcal{H}_T$ and $\mathcal{H}_S$, respectively. Let I be a topological space.*

(a) *If $A \in \mathcal{L}(I, T, S)$ then for each $h_S \in \mathcal{H}_S$ and $h_T \in \mathcal{H}_T$ the map* (7.8) *is continuous on the open set where it is defined.*

(b) *If σ is an orientation of $A \in \mathcal{L}_0(I, T, S)$ then for each $h \in \mathcal{H}$ the map* (7.9) *is an orientation of* (7.8) *(on the open set where this map is defined).*

Proof. Let h_T and h_S be defined on $U \times E_X$ and $V \times E_Y$, respectively. Proposition 7.32 implies that (7.8) is defined on the open sets $U_0 := x^{-1}(U) \cap y^{-1}(V) \subseteq I$, and the same holds for (7.9) by an analogous argument. We are to show that these map are continuous (resp. lower semicontinuous) at every $t_0 \in U_0$. By hypothesis there are trivializations $H_T \in \mathcal{H}_T$ and $H_S \in \mathcal{H}_S$ at $x(t_0)$ and $y(t_0)$, respectively such that the corresponding maps A_{H_S,H_T} and σ_{H_S,H_T} are lower semicontinuous at t_0.

There is some open neighborhood $U_1 \subseteq U_0$ of t_0 such that H_T and H_S are defined on $x(U_1) \times E_X$ and $y(U_1) \times E_Y$, respectively. For $t \in U_1$, we put

$$J_X(t) := h_T\big(x(t), \cdot\big)^{-1} H_T\big(x(t), \cdot\big)$$

and

$$J_Y(t) := H_S\big(y(t), \cdot\big)^{-1} h_S\big(y(t), \cdot\big).$$

The choice of $\mathcal{H}_S$ and $\mathcal{H}_T$ in Banach bundles implies that $J_X \colon U_0 \to \mathcal{L}(E_X)$ and $J_Y \colon U_0 \to \mathcal{L}(E_Y)$ are continuous, in particular continuous at $t_0 \in T$. Hence, Lemma 7.23 implies that $A_{h_S,h_T}(t) = J_Y(t) \circ A_{H_S,H_T}(t) \circ J_X(t)$ and $\sigma_{h_S,h_T}(t) = J_Y(t) \circ \sigma_{H_S,H_T}(t) \circ J_X(t)$ are continuous (resp. lower semicontinuous) at t_0. □

An analogous assertion to Theorem 7.20(e) holds also in case of Banach bundles:

Proposition 7.35. *Let $\mathbb{K} = \mathbb{R}$, $X^* \neq \{0\}$, and $A \in \mathcal{L}_0(I, T, S)$. Then σ is an orientation of A if and only if the (pointwise) opposite orientation is an orientation of A.*

Proof. Let X_x $(x \in X)$ and Y_y $(y \in Y)$ denote the fibres, that is $A(t) \in \mathcal{L}_0(X_{x(t)}, Y_{y(t)})$. Let σ be an orientation for A. If $h_T \in \mathcal{H}_T$, $h_S \in \mathcal{H}_S$, then Proposition 7.34 implies that (7.9) is an orientation for (7.8). Theorem 7.20(e) implies that the pointwise opposite orientation is also an orientation for A_{h_T,h_S}, and by Corollary 7.8, this corresponds to the pointwise opposite orientation of $\sigma(t)$. Hence, the pointwise opposite orientation is an orientation of A. The converse holds by symmetry reasons. □

Proposition 7.36. *Let* $\mathbb{K} = \mathbb{R}$, $X^* \neq \{0\}$, *and* $A \in \operatorname{Iso}(I, T, S)$ *be oriented. Then* $\operatorname{sgn} A(\cdot)$ *is constant on the components of* I.

Proof. We let σ denote the orientation, and σ_- the opposite orientation according to Proposition 7.35, and put

$$B_1 := \{t \in I : \operatorname{sgn} A(t) > 0\} = \sigma^+(\{0\}).$$

Then the complement of B_1 in I is

$$B_2 := \{t \in I : \operatorname{sgn} A(t) < 0\} = \sigma_-^+(\{0\}).$$

Assume that $A(t) \in \mathcal{L}(X_{x(t),y(t)})$ for $t \in I$. If $t_0 \in B_k$ ($k = 1$ or $k = 2$), we find by Proposition 7.34 that there are trivializations h_T and h_S at $x(t_0)$ or $y(t_0)$ respectively such that the map (7.9), is an orientation for (7.8). The pointwise opposite orientation is also an orientation by Theorem 7.20(c) which then implies that B_k contains a neighborhood of t_0. Thus, B_1 and B_2 are both open in I, and so Corollary 2.18 implies that each component of I is contained in either B_1 or in B_2. □

There always do exist orientations as in Proposition 7.36. To see this, we define:

Definition 7.37. The *natural orientation* of $A \in \operatorname{Iso}(I, T, S)$ is that which associates to each $A(t)$ the natural orientation.

Proposition 7.38. *The natural orientation of* $A \in \operatorname{Iso}(I, T, S)$ *is an orientation in the sense of Definition* 7.33.

Proof. Using the notation of Definition 7.33, we are to show that (7.9) defines an orientation for (7.8). Since $0 \in \sigma_{h_S,h_T}(t)$ for every t, this follows from Theorem 7.20(c). □

It is useful to know that we can consider the orientation of $A \in \mathcal{L}_0(I, T, S)$ as a lifting. To simplify considerations, we work with a space depending on A.

Let T and S be a Banach bundles over E_X and E_Y with fibres X_x ($x \in X$) and Y_y ($y \in Y$) and associated trivializations $\mathcal{H}_T$ and $\mathcal{H}_S$.

Let $A \in \mathcal{L}_0(I, T, S)$ in the sense of Definition 7.31, in particular $A(t) \in \mathcal{L}_0(X_{x(t)}, Y_{y(t)})$. We consider the space

$$I_A := \{(t, \sigma) : \sigma \text{ is an orientation of } A(t)\}.$$

Inspired by [18], we consider for $h_T \in \mathcal{H}_T$, $h_S \in \mathcal{H}_S$ and open $O \subseteq I$ such that h_T and h_S are defined at least on $x(O) \times E_X$ and $y(O) \times E_Y$, respectively, and for any $K \in \mathcal{L}(E_X, E_Y)$ the (possibly empty) subset

$$U_{h_T,h_S,O,K} := \{(t,\sigma) \in I_A : t \in O \text{ and } K \in h_S\big(y(t),\cdot\big)^{-1} \circ \sigma \circ h_T\big(x(t),\cdot\big)\}.$$

By Proposition 2.57, there is a unique topology on I_A for which the family of all sets $U_{h_T,h_S,O,K}$ produced in the above manner constitute a subbasis. We define the projections $p_A\colon I_A \to I$ by $p_A(t,\sigma) := t$ and $q_A\colon I_A \multimap \bigcup_{t\in I} \mathcal{L}(X_{x(t)}, Y_{y(t)})$ by $q_A(t,\sigma) := \sigma$.

Lemma 7.39. *With the above notation, the orientations of A are exactly the maps $q_A \circ S$ where S is a lifting of id_I with respect to p_A, and the orientations σ of A and liftings S are in a one-to-one correspondence given by $S(t) := (t, \sigma(t))$. In particular, A is orientable if and only if id_I has a lifting.*

Proof. If σ is an orientation of A then $S(t) := (t,\sigma(t))$ defines a lifting of id_I. To see that S is continuous at every $t_0 \in I$, we let $U \subseteq I_A$ be a neighborhood of $S(t_0)$ with $U \neq I_A$. By definition of the topology, U contains a finite intersection of sets $U_{h_{T,k},h_{S,k},O_k,K_k}$ $(k = 1,\dots,n)$ containing $S(t_0)$. The latter implies $K_k \in \sigma_k(t_0)$ where

$$\sigma_k(t) := h_{S,k}\big(x(t),\cdot\big)^{-1} \circ \sigma(t) \circ h_{T,k}\big(x(t),\cdot\big).$$

Since Proposition 7.34 shows that σ_k is an orientation for

$$A_k(t) := h_{S,k}\big(x(t),\cdot\big)^{-1} \circ A(t) \circ h_{T,k}\big(x(t(,\cdot\big),$$

we obtain from Theorem 7.20(d) that $K_k \in \sigma_k(t)$ for all t in a neighborhood $V_k \subseteq I$ of t_0. It follows that $S(t) \in U_{h_{T_k},h_{S_k},O_k,K_k}$ for all $t \in V_k$ and thus $S(t) \in U$ for all $t \in V_1 \cap \dots \cap V_n$.

Conversely, if S is a lifting of id_I then $\sigma(t) := q_A(S(t))$ is an orientation. Indeed, given $t_0 \in I$ and trivializations $h_T \in \mathcal{H}_T$ and $h_S \in \mathcal{H}_S$ at $x(t_0)$ and $y(t_0)$, defined on $U \times E_X$ and $V \times E_Y$, respectively, we have to show that (7.9) is lower semicontinuous at $t_0 \in O := x^{-1}(U) \cap y^{-1}(V)$. Thus, let $K \in \sigma_{h_S,h_T}(t_0)$. Then $U_{h_T,h_S,O,K}$ is a neighborhood of $S(t_0)$, and so the continuity of S implies that there is a neighborhood $O_0 \subseteq O$ of t_0 with $S(O_0) \subseteq U_{h_T,h_S,O,K}$, in particular $K \in \sigma_{h_S,h_T}(t)$ for every $t \in O_0$. □

Lemma 7.40. *If $\mathbb{K} = \mathbb{R}$ and $X^* \neq \{0\}$ then the above considered map p_A is a 2-fold covering map.*

Proof. For $t_0 \in I$, let $h_T \in \mathcal{H}_T$ and $h_S \in \mathcal{H}_S$ be trivializations at $x(t_0)$ or $y(t_0)$, defined on $U \times E_X$ and $V \times E_Y$, respectively. We let $B(t)$ denote the operator (7.8), that is, B is defined on the open neighborhood $I_0 := x^{-1}(U) \cap y^{-1}(V)$ of t_0. Proposition 7.1 implies that the operator $B(t_0)$ has correctors K_1, K_2 of the two opposite equivalence classes. Lemma 7.19 and the continuity of B at t_0 imply that there is an open neighborhood $O \subseteq I_0$ of t_0 such that K_1 and K_2 are correctors for each $B(t)$ $(t \in O)$ of opposite equivalence classes. We claim that this set O is evenly covered by p_A: Indeed, $p_A^{-1}(O)$ consists of all (t, σ) with $t \in O$, where the two possible choices of σ are represented by $h_S(y(t), \cdot) \circ K_k \circ h_T(x(t), \cdot)^{-1}$ $(k = 1, 2)$ in view of Corollary 7.8. Hence, $p^{-1}(O)$ is the disjoint union of the two sets $X_k := U_{h_S, h_T, O, K_k}$ $(k = 1, 2)$, and $p_A \colon X_k \to O$ $(k = 1, 2)$ is one-to-one and onto. By definition of the topology, X_k are open in I_A, and the open in X_k subsets of X_k are exactly those of the form $\{(t, \sigma) \in X_k : t \in M\} = p|_{X_k}^{-1}(M)$ where $M \subseteq O$ is open. Hence, $p|_{X_k}$ is a homeomorphism onto O. □

Theorem 7.41. *Let $\mathbb{K} = \mathbb{R}$ and $X^* \neq \{0\}$. Let T and S be Banach bundles, and $A \in \mathcal{L}_0(I, T, S)$.*

(a) *If $C \subseteq I$ is connected and σ is an orientation of $A|_C \in \mathcal{L}_0(C, T, S)$ then σ is uniquely determined if its value is known in one point of C.*

(b) *If I is path-connected, locally path-connected, and simply connected then A is orientable.*

(c) *If $I = [0, 1] \times I_0$ then A is orientable if and only if $A(t_0, \cdot)$ is orientable for some $t_0 \in [0, 1]$. In this case, the orientation of A is uniquely determined by the orientation of $A(t_0, \cdot)$.*

Proof. In view of Lemma 7.39, we can reformulate the assertions in terms of liftings of id_I (or of the homotopy $\mathrm{id}_{[0,1] \times I_0}$) for the covering map p_A (or p_H) considered above in an obvious manner. Lemma 7.40 implies that these reformulations are special cases of Proposition 5.3, Theorem 5.7, and Theorem 5.4, respectively. □

If T_i are Banach bundles over E_i $(i = 1, 2)$ then $T_1 \times T_2$ becomes a Banach bundle over $E_1 \times E_2$, called the *product bundle*, in an obvious manner with trivializations given by

$$h\big((t_1, t_2), (x_1, x_2)\big) = \big(h_1(t_1, x_1), h_2(t_2, x_2)\big),$$

where h_i are trivializations of T_i $(i = 1, 2)$. In situations as the above, we write $h = h_1 \otimes h_2$ although, formally, this holds only under a permutation of the arguments.

Definition 7.42. For $i = 1, 2$, let I_i be topological spaces, T_i and S_i be Banach bundles, and $A_i \in \mathcal{L}(I_i, T_i, S_i)$. Then $A_1 \otimes A_2 \in \mathcal{L}(I_1 \times I_2, T_1 \times T_2, S_1 \times S_2)$ is defined in the obvious pointwise manner. If $A_i \in \mathcal{L}_0(I_i, T_i, S_i)$ are oriented with orientations σ_i $(i = 1, 2)$, then the *product orientation* $\sigma(t_1, t_2)$ of $A_1 \otimes A_2 \in \mathcal{L}_0(I_1 \times I_2, T_1 \times T_2, S_1 \times S_2)$ is for $t_i \in I_i$ defined as the product orientation of $\sigma_1(t_1)$ and $\sigma_2(t_2)$.

Proposition 7.43. *The product orientation is an orientation in the sense of Definition* 7.33.

Proof. For $t_i \in I_i$, let $A_i(t_i) \in \mathcal{L}(X_{i,x_i(t_i)}, Y_{i,y_i(t_i)})$ $(i = 1, 2)$, and let h_{T_i} and h_{S_i} be trivializations of T_i and S_i at $x_i(t_i)$ or $y_i(t_i)$, respectively. Let h_T and h_S be the corresponding trivializations of $T := T_1 \times T_2$ and $S := S_1 \times S_2$, and put $x := x_1 \otimes x_2$, $y := y_1 \otimes y_2$, and $A(t_1, t_2) := A_1(t_1) \otimes A_2(t_2)$. We have to show that (7.9) is lower semicontinuous at a given (t_1, t_2). Since

$$\sigma_{i,0}(t) := h_{S_i}\big(y_i(t), \cdot\big)^{-1} \circ \sigma_i(t) \circ h_{T_i}\big(x_i(t), \cdot\big)$$

are orientations for

$$A_{i,0}(t) := h_{S_i}\big(y_i(t), \cdot\big)^{-1} \circ A_i(t) \circ h_{T_i}\big(x_i(t), \cdot\big),$$

we obtain from Theorem 7.20(c) that there are neighborhoods V_i of t_i such that $\bigcap_{s_i \in V_i} \sigma_{i,t_0}(s_i)$ contains some K_i. Then $V_1 \times V_2$ is a neighborhood of (t_1, t_2) with

$$K_1 \otimes K_2 \in \bigcap_{(s_1,s_2) \in V_1 \times V_2} \sigma_{h_T, h_S}(s_1, s_2).$$

Since (7.9) are pointwise orientations for (7.8), we thus obtain from Theorem 7.20(c) that (7.9) is indeed lower semicontinuous at (t_1, t_2). □

Definition 7.44. Let T, S, and R be Banach bundles with fibres X_x $(x \in X)$, Y_y $(y \in Y)$, and Z_z $(z \in Z)$, respectively. Let I and J be a topological spaces. Let $A \in \mathcal{L}(I, T, S)$ and $B \in \mathcal{L}(J, S, R)$ such that $A(t) \in \mathcal{L}(X_{x(t)}, Y_{y_A(t)})$ and $B(s) \in \mathcal{L}(Y_{y_B(s)}, Z_{z(s)})$. If there is a continuous $\tau\colon I \to J$ with $y_B(\tau(t)) = y_A(t)$ then $B \circ_\tau A \in \mathcal{L}(I, T, R)$ is defined by

$$(B \circ_\tau A)(t) := B(\tau(t))A(t) \in \mathcal{L}(X_{x(t)}, Z_{z(\tau(t))}).$$

If additionally $A \in \mathcal{L}_0(I, T, S)$ and $B \in \mathcal{L}_0(I, S, R)$ have orientations σ_A and σ_B, respectively, then the *composite orientation* σ_{BA} for $B \circ_\tau A$ is defined such that $\sigma_{BA}(t)$ is the composite orientation of $B(\tau(t))$ and $A(t)$.

Proposition 7.45. *The composite orientation is an orientation for $B \circ_\tau A$ in the sense of Definition 7.33. Conversely, if σ_{BA} is any orientation of $B \circ_\tau A$ then:*

(a) *If B is oriented then there is a unique orientation of A such that σ_{BA} is the composite orientation.*

(b) *If A is oriented and τ is a homeomorphism onto its range $J_0 \subseteq J$ then there is a unique orientation of $B|_{J_0}$ such that σ_{BA} is the composite orientation.*

In particular, if τ is a homeomorphism onto J and if two of the three maps $(A, B, B \circ_\tau A)$ are oriented then the remaining map can be uniquely oriented such that $B \circ_\tau A$ carries the composite orientation.

Proof. Proposition 7.6 implies that the corresponding orientations are uniquely defined pointwise. We have to show that these pointwise orientation are actually orientations in the sense of Definition 7.33. Thus, assume that T, S, and R are Banach bundles over E_X, E_Y, and E_Z, and let h_T, h_S, and h_R be trivializations of T, S, and R at $x(t_0)$, $y(\tau(t_0))$, or $z(\tau(t_0))$, respectively. We are to show for the three assertions that the corresponding multivalued map

$$\begin{aligned} s_{BA}(t) &:= h_R\big(z(\tau(t)), \cdot\big)^{-1} \circ \sigma_{BA}(t) \circ h_T\big(x(t), \cdot\big) \subseteq \mathcal{L}(E_X, E_Z), \\ s_A(t) &:= h_S\big(y(\tau(t)), \cdot\big)^{-1} \circ \sigma_A(t) \circ h_T\big(x(t), \cdot\big) \subseteq \mathcal{L}(E_X, E_Y), \\ s_B(s) &:= h_R\big(y(s), \cdot\big)^{-1} \circ \sigma_B(s) \circ h_S\big(x(s), \cdot\big) \subseteq \mathcal{L}(E_Y, E_Z) \end{aligned}$$

is lower semicontinuous at t_0, t_0, or $\tau(t_0)$, respectively. By Proposition 7.34, we know that the other two maps are orientations of the respective Fredholm maps

$$\begin{aligned} C_{BA}(t) &:= h_R\big(z(\tau(t)), \cdot\big)^{-1} \circ B(\tau(t))A(t) \circ h_T\big(x(t), \cdot\big) \in \mathcal{L}_0(E_X, E_Z), \\ C_A(t) &:= h_S\big(y(\tau(t)), \cdot\big)^{-1} \circ A(t) \circ h_T\big(x(t), \cdot\big) \in \mathcal{L}_0(E_X, E_Y), \\ C_B(s) &:= h_R\big(y(s), \cdot\big)^{-1} \circ B(s) \circ h_S\big(x(s), \cdot\big) \in \mathcal{L}_0(E_Y, E_Z). \end{aligned}$$

Since the orientation $\sigma_{BA}(t)$ is the composite orientation of the oriented maps $(B(\tau(t)), \sigma_B(\tau(t)))$ and $(A(t), \sigma_A(t))$, it follows straightforwardly from Definition 7.5 that the orientation $s_{BA}(t)$ is the composite orientation of the oriented maps $(C_B(\tau(t)), s_B(\tau(t)))$ and $(C_A(t), s_A(t))$.

In particular, each $K_{BA} \in s_{BA}(t_0)$ satisfies $K_{BA} = K_1(t_0)$ with

$$K_1(t) := C_B(\tau(t))K_A + K_B C_A(t) + K_B K_A$$

for some $K_A \in s_A(t_0)$, $K_B \in s_B(\tau(t_0))$. If σ_A and σ_B are orientations then s_A and s_B are orientations for C_A and C_B, and so Theorem 7.20(d) and the continuity of τ implies that there is a neighborhood $U \subseteq I$ of t_0 with $K_A \in s_A(t)$ and

$K_B \in s_B(t)$ for all $t \in U$. Hence, $K_1(t) \in s_{BA}(t)$ for all $t \in U$. Since K_1 is continuous at t_0, it follows that s_{BA} is lower semicontinuous at t_0, and so σ_{BA} is indeed an orientation.

The proof for the other two assertions is similar: By Proposition 7.6, each $K_A \in s_A(t_0)$ or each $K_B \in s_B(\tau(t_0))$ satisfies $K_A = K_2(t_0)$ or $K_B = K_3(\tau(t_0))$ with

$$K_2(t) := \big(B(\tau(t)) + K_B\big)^{-1}(K_{BA} - K_B A(t))$$
$$K_3(s) := (K_{BA} - B(s)K_A)\big(A(\tau^{-1}(s)) + K_A\big)$$

for some $K_{BA} \in s_{BA}(t_0)$ and some $K_B \in s_B(\tau(t_0))$ or $K_A \in s_A(t_0)$, respectively. If σ_{BA} and σ_B or σ_A, respectively, are orientations then s_{BA} and s_B or s_A are orientations for C_{BA} and C_B or C_A, respectively. Theorem 7.20(d) and the continuity of τ implies that there is a neighborhood $U \subseteq I$ of t_0 with $K_{BA} \in s_{BA}(t)$ and $K_B \in s_B(\tau(t))$ or $K_A \in s_A(t)$ for all $t \in U$, respectively. The continuity of K_2 and K_3 at t_0 or $\tau(t_0)$, respectively, implies as above that s_A and $s_B|_{J_0}$ are lower semicontinuous at t_0 or $\tau(t_0)$, respectively. Hence, σ_A or $\sigma_B|_{J_0}$ are orientations for A or $B|_{J_0}$, respectively. □

In finite dimensions, we can also speak about orientations of Banach bundles:

Definition 7.46. Let $\dim E < \infty$, and T be a vector bundle over E with fibres T_x $(x \in X)$. An *orientation* of the vector bundle T on $M \subseteq X$ is an orientation of every T_x $(x \in M)$ with the following property.

For every $x_0 \in M$ there is a trivialization h for x_0, defined on $U \times E$, such that the isomorphism $h(x, \cdot)$ induces by the orientation of T_x an orientation $\sigma(x)$ on E (for every $x \in U \cap M$) with $\sigma|_{U \cap M}$ being lower semicontinuous at x_0.

Proposition 7.47. *Let $\mathbb{K} = \mathbb{R}$ and T be a vector bundle with fibres T_x $(x \in X)$ which is oriented on $M \subseteq X$. Then for every trivialization h defined on $U \times E$ the orientation $\sigma(x)$ on E induced by $h(x, \cdot)$ has the property that $\sigma|_{M \cap U}$ is locally constant and constant on the components of $M \cap U$.*

Proof. We show first that $\sigma|_{M \cap U}$ is lower semicontinuous at every $x_0 \in M \cap U$. By hypothesis there is a trivialization h_0 at x_0 defined on some $U_0 \times E$, such that the orientation $\sigma_0(x)$ on E induced by $h_0(x, \cdot)$ has the property that $\sigma_0|_{U_0 \cap M}$ is lower semicontinuous at X_0. Define $J\colon U_0 \cap U \cap M \to \operatorname{Iso}(E)$ by $J(x) := h(x, \cdot)^{-1} \circ h_0(x, \cdot)$. Since $\dim E < \infty$, the map J is continuous. Moreover, $\sigma(x) = J(\sigma_0(x))$. Proposition 2.94 implies that $\sigma|_{U_0 \cap U \cap M}$ is lower semicontinuous at x_0. Hence, also $\sigma|_{U \cap M}$ is lower semicontinuous at x_0.

Now if $(e_1, \dots, e_n)$ represents the orientation of $\sigma|_{M \cap U}$ there is some $\varepsilon > 0$ such that for any basis $(x_1, \dots, x_n)$ with $\|e_k - x_k\| < \varepsilon$ $(k = 1, \dots, n)$ the unique $J \in \operatorname{Iso}(E)$ satisfying $J(e_k) = x_k$ satisfies $\|\mathrm{id}_E - J\| < 1$. Proposition 6.10 implies that $G(t) = \mathrm{id}_E - t(\mathrm{id}_E - J) \in \operatorname{Iso}(E)$ for every $t \in [0, 1]$. Hence, $\det G(t) \neq 0$ for all $t \in [0, 1]$. By the continuity of G and det, we obtain that $\det J = \det G(1)$ has the same sign as $\det G(0) = \det \mathrm{id}_E = 1$. Thus, $(x_1, \dots, x_n)$ represents the same orientation as $(e_1, \dots, e_n)$. Since $\sigma|_{M \cap U}$ is lower semicontinuous at x_0, it follows that $\sigma|_{M \cap U}$ is constant in a neighborhood of X_0. The last assertion follows from Proposition 2.19. □

We see now that in the finite-dimensional case there is a canonical correspondence between the orientation of maps from $A \in \mathcal{L}_0(I, T, S)$ and the orientations of T and S:

Proposition 7.48. *Let $\mathbb{K} = \mathbb{R}$. Let T and S be vector bundles with finite-dimensional fibres X_x $(x \in X)$ and Y_y $(y \in Y)$, and $A \in \mathcal{L}_0(I, T, S)$, $A(t) \in \mathcal{L}_0(X_{x(t)}, Y_{y(t)})$. Then for $M \subseteq I$ the following holds:*

(a) *If T and S are oriented on $x(M)$ and $y(M)$, respectively, then the pointwise induced orientation on $A(t)$ $(t \in M)$ is an orientation of A on M.*

(b) *If A is oriented on M, S is oriented on $y(M)$, and $x|_M$ is a homeomorphism onto $x(M) \subseteq T$ then the pointwise induced orientation of T on $x(M)$ is an orientation on $x(M)$.*

(c) *If A is oriented on M, T is oriented on $x(M)$, and $y|_M$ is a homeomorphism onto $y(M) \subseteq S$ then the pointwise induced orientation of S on $y(M)$ is an orientation on $y(M)$.*

Proof. We denote the given (or pointwise induced) orientations of A, T, and S by σ_A, σ_T, and σ_S, respectively. In all cases, we have to show the lower semicontinuity of the pointwise induced orientation at $t_0 \in M$, $x(t_0)$, or $y(t_0)$, respectively. We choose trivializations h_T and h_S of T and S at $x(t_0)$ or $y(t_0)$, respectively, and put $J_T(t) := h_T(x(t), \cdot)^{-1} \in \operatorname{Iso}(T_{x(t)}, E)$ and $J_S(t) := h_S(y(t), \cdot)^{-1} \in \operatorname{Iso}(T_{y(t)}, E)$. Then J_T and J_S are defined in a neighborhood $U \subseteq I$ of t_0.

Proposition 7.47 implies that if orientations σ_T or σ_S are given then $s_T(t) := J_T(t) \circ \sigma_T(x(t))$ or $s_S(t) := J_S(t) \circ \sigma_S(y(t))$ are locally constant on $M \cap U$. Conversely, if σ_T or σ_S is only pointwise defined on $x(M \cap U)$ or $y(M \cap U)$, and if $x|_{M \cap U}$ or $y|_{M \cap U}$ is a homeomorphism and s_T or s_S is lower semicontinuous on $x(M \cap U)$ or $y(M \cap U)$ then $s_T \circ x|_{M \cap U}^{-1}$ or $s_S \circ y|_{M \cap U}^{-1}$ is lower semicontinuous on $x(M \cap U)$ or $y(M \cap U)$, respectively, and so by definition σ_T or σ_S are orientations on $x(M \cap U)$ or $y(M \cap U)$.

Similarly, Proposition 7.34 implies that the pointwise orientation σ_A is an orientation on $M \cap U$ if and only if the pointwise orientation $\sigma(t) = J_S(t) \circ \sigma_A(t) \circ J_T^{-1}(t)$ is lower semicontinuous on $M \cap U$.

Since $\sigma(t)$ corresponds to the induced orientations of s_S or s_T (or vice versa), the assertions follow by combining the above observations. □

Chapter 8

Some Nonlinear Analysis

Throughout this chapter, let E, Z, E_X, and E_Y be Banach spaces over $\mathbb{K} = \mathbb{R}$ or $\mathbb{K} = \mathbb{C}$.

8.1 The Pointwise Inverse and Implicit Function Theorems

Recall that if $U \subseteq E$ then a map $F: U \to Z$ is called *Gateaux differentiable* at $x \in \overset{\circ}{U}$ with derivative $dF(x) = A$ if

$$Ah := \lim_{t \to 0+} \frac{F(x + th) - F(x)}{t}$$

exists for all $h \in E$ and defines some $A \in \mathcal{L}(U, Z)$.

For functions $F: U \to \mathbb{R}^m$ with $U \subseteq \mathbb{R}^n$, the following result becomes the well-known assertion that if the Gateaux derivative $dF(x)$ exists then it is given by the Jacobi-matrix which consists of the k-th partial derivative (at x) as the k-th row:

Proposition 8.1. *Let $x = (x_1, \ldots, x_n)$ be an interior point of $U \subseteq \mathbb{R}^n$, and let $F: U \to Z$ be Gateaux differentiable at x. Then all partial derivatives*

$$d_k F(x) := d(F(x_1, \ldots, x_{k-1}, \cdot, x_{k+1}, \ldots, x_n))(x_k)$$

exist for $k = 1, \ldots, n$, and

$$dF(x)(h_1, \ldots, h_n) = \sum_{k=1}^{n} d_k F(x) h_k. \tag{8.1}$$

Proof. Considering $h = \pm e_k$ with $e_k = (0, \ldots, 0, 1, 0, \ldots, 0) \in \mathbb{R}^n$ (with 1 at the k-th position) in the definition of the Gateaux derivative, we find that $d_k F(x)$ exists and is equal to $dF(x)e_k$. Hence, $d_k F(x) h_k = dF(x) h_k e_k$. Summing up over all k, we obtain (8.1) by the linearity of $dF(x)$. □

Gateaux differentiable functions satisfy the following form of the mean value theorem:

Theorem 8.2 (Mean Value). *Let $M \subseteq U \subseteq E$ where M is convex. If $F: U \to Z$ is Gateaux differentiable at each $x \in M$ with $\|dF(x)\| \leq L$ then F is Lipschitz on M with constant L.*

Proof. Let $x_0, y_0 \in M$ and $L_0 > L$. Define $\varphi: [0,1] \to M$ by $\varphi(t) := x_0 + t(y_0 - x_0)$. For each $t \in [0,1]$ the map F is Gateaux differentiable on $x = \varphi(t)$ with $\|dF(x)\| \leq L$. Then $\varphi(s) = x + (s-t)(y_0 - x_0)$ for all $s \in [0,1]$, and so there is some $\varepsilon > 0$ such that

$$\|F(\varphi(s)) - F(\varphi(t))\| \leq L_0 \|y_0 - x_0\| |s - t|$$

for $s \in K_{2\varepsilon}(t)$ (understood in the metric space $X := [0,1]$). Let $r(t)$ denote the supremum of all these $\varepsilon \in (0,1]$. Since X is compact, it is covered by finitely many of the balls $B_{r(t)}(t)$ with $t \in [0,1]$, that is, there is a partition $0 = t_0 \leq t_1 \leq \cdots \leq t_n = 1$ such that $t_k - t_{k-1} < r(t_{k-1}) + r(t_k) \leq \max\{2r(t_{k-1}), 2r(t_k)\}$ for $k = 1, \ldots, n$. By definition of $r(t)$, this implies

$$\|F(y_0) - F(x_0)\| = \left\| \sum_{k=1}^{n} (F(\varphi(t_k)) - F(\varphi(t_{k-1}))) \right\| \leq$$

$$\sum_{k=1}^{n} L_0 \|y_0 - x_0\| (t_k - t_{k-1}) = L_0 \|y_0 - x_0\|.$$

Since $L_0 > L$ was arbitrary, we obtain $\|F(y_0) - F(x_0)\| \leq L \|y_0 - x_0\|$. □

Recall that if $U \subseteq E$ and $F: U \to Z$, then F is called *Fréchet differentiable* (or just *differentiable*) at $x \in \mathring{U}$ if there is some $A \in \mathcal{L}(E, Z)$ with

$$\lim_{h \to 0} \frac{\|F(x+h) - F(x) - Ah\|}{\|h\|} = 0. \tag{8.2}$$

Replacing h by th in (8.2), we see that each Fréchet differentiable operator is necessarily Gateaux differentiable, and that its derivative A is uniquely defined and must be the Gateaux derivative.

Proposition 8.3. *If $F = z + A|_U$ where $z \in Z$, $A \in \mathcal{L}(E, Z)$, and $U \subseteq E$ is a neighborhood of $x \in E$ then F is differentiable at x with $dF(x) = A$.*

Proof. We have $F(x+h) - F(x) - Ah = 0$, hence (8.2) holds. □

Proposition 8.4. *If $F_i: U \to Z_i$ are differentiable at x and $A_i := dF_i(x)$ for $i = 1, 2$ then also $F_1 \times F_2: U \to Z_1 \times Z_2$ is differentiable at x with $d(F_1 \times F_2)(x) = A_1 \times A_2$*

Proof. We have

$$\begin{aligned}&\|(F_1 \times F_2)(x+h) - (F_1 \times F_2)(x) - (A_1 \times A_2)h\| \\ &\quad = \|F_1(x+h) - F_1(x) - A_1 h\| + \|F_2(x+h) - F_2(x) - A_2 h\|,\end{aligned}$$

and so (8.2) holds. □

We recall the continuity and the chain rule for differentiable functions:

Proposition 8.5. *If F is differentiable at x then F is continuous at x. If additionally G is differentiable at $y = F(x)$ then $H := G \circ F$ is differentiable at x with derivative $C := dG(y)dF(x)$.*

Proof. Put $k_h := F(x+h) - F(x)$ and $r_h := k_h - dF(x)h$. Then (8.2) implies $r_h/\|h\| \to 0$ as $h \to 0$. In particular, there is some finite number K and some $\delta > 0$ with $\|r_h\|/\|h\| \le K$ whenever $0 < \|h\| < \delta$. Putting $K_0 := K + \|dF(x)\|$, we thus have $\|k_h\| \le K_0\|h\|$ for $\|h\| < \delta$. In particular, $k_h \to 0$ as $h \to 0$. Hence, F is continuous at x, and

$$\begin{aligned}&\frac{H(x+h) - H(x) - Ch}{\|h\|} \\ &\quad = \frac{\|k_h\|}{\|h\|}\frac{G(y+k_h) - G(y) - dG(y)k_h}{\|k_h\|} + dG(y)\frac{r_h}{\|h\|} \to 0\end{aligned}$$

as $h \to 0$ by definition of $dG(y)$, since $\|k_h\|/\|h\| \le K_0$, $r_h/\|h\| \to 0$ and $k_h \to 0$ as $h \to 0$. □

To formulate the chain rule in terms of partial derivatives, let X, Y and I be Banach spaces. For $H\colon W \to Y$ with $W \subseteq I \times X$, we denote the *partial derivatives* by

$$\begin{aligned}d_I H(t,x) &:= d(H(\cdot\,,x))(t) \in \mathcal{L}(I,Y)\\ d_X H(t,x) &:= d(H(t,\,\cdot))(x) \in \mathcal{L}(X,Y).\end{aligned}$$

Proposition 8.6. *Let $H\colon W \to Y$ as above be differentiable at $(t,x) \in W$. Then $d_I H(t,x)$ and $d_X H(t,x)$ exist as Fréchet derivatives and satisfy*

$$d_X H(t,x)(h_I, h_X) = d_I H(t,x)h_I + d_X H(t,x)h_X \quad \textit{for all } (h_I,h_X) \in I \times X.$$

For any functions $F\colon U \to I$, $G\colon U \to X$ with $U \subseteq Z$, $(F \times G)(U) \subseteq W$ which are differentiable at $z \in U$ and satisfy $F(z) = t$, $G(z) = x$, the composition $c(u) := H(F(u), G(u))$ is differentiable at z with

$$dc(z) = d_I H(t,x)dF(z) + d_X H(t,x)dG(z).$$

Proof. Putting $A := dH(t,x)$, we define $A_I \in \mathcal{L}(I,Z)$ by $A_I h_I := A(h_I, 0)$ and $A_X \in \mathcal{L}(X,Z)$ by $A_X h_X := A(0, h_X)$. Then

$$\|H(t+h_I, x) - H(t,x) - A_I h_I\| = \|H((t,x)+(h_I,0)) - H(t,x) - A(h_I,0)\|$$

implies that $d_I H$ exists at (t,x) with $A_I = d_I H(t,x)$. Similarly one obtains $A_X = d_X H(t,x)$. Since A is linear, we have $A(h_I, h_X) = A_I h_I + A_X h_X$. The second assertion follows by applying the chain rule and Proposition 8.4 to $c = H \circ (F \times G)$. □

The following form of a "pointwise inverse function theorem" without any continuity requirement on dF is perhaps not so well-known.

Theorem 8.7 (Pointwise Inverse Function). *Let $U_0 \subseteq E$ and $F: U_0 \to Z$ be differentiable at x. Let $V \subseteq Z$ be a neighborhood of $y = F(x)$, and suppose that $F: U \to V$ is invertible for some $U \subseteq U_0$. Then $G := F^{-1}: V \to U$ is differentiable at $y = F(x)$ if and only if G is continuous at y, and $A := dF(x) \in$* Iso(E,Z). *In this case $B := dG(y) \in$* Iso(Z,E) *satisfies $B = A^{-1}$.*

For the subsequent application it is important that we do not require in Theorem 8.7 that U is a neighborhood of x; only U_0 must be a neighborhood of x by definition of the derivative.

Proof. For necessity, we observe that Proposition 8.5 implies that G is continuous. Moreover, Proposition 8.5 applied to $\mathrm{id}_U = G \circ F$ and $\mathrm{id}_V = F \circ G$ implies $\mathrm{id}_E = BA$ and $\mathrm{id}_Z = AB$. It follows that $B = A^{-1} \in \mathrm{Iso}(Z,E)$.

For sufficiency, we put $B := A^{-1}$. Putting $r(h) := G(y+h) - x - Bh$, we are to show that $r(h)/\|h\| \to 0$ as $h \to 0$. Let $\rho > 0$ be such that $K_\rho(y) \subseteq V$. For $h \in K_\rho(0)$, put $k_h := G(y+h) - x$. The continuity of G implies $k_h \to 0$ as $h \to 0$, and we have $x + k_h = G(y+h) \in U$ and $r(h) = k_h - Bh$. Since F is differentiable at x, we thus find for $s(h) := F(x+k_h) - y - Ak_h$ that $s(h)/\|k_h\| \to 0$ as $h \to 0$. Shrinking $\rho > 0$ if necessary, we thus can assume that $\|Bs(h)\| \le \|k_h\|/2$ for $h \in K_\rho(0)$.

Note that $F(x+k_h) = F(G(y+h)) = y+h$ and thus $s(h) = h - Ak_h$, hence $k_h = Bh - Bs(h)$. This implies $r(h) = -Bs(h)$ and $\|k_h\| \le \|Bh\| + \|k_h\|/2$ for $h \in K_\rho(0)$. Hence, $\|k_h\| \le 2\|Bh\| \le 2\|B\|\|h\|$ for $h \in K_\rho(0)$. We obtain

$$\frac{\|r(h)\|}{\|h\|} \le \frac{\|Bs(h)\|}{\|k_h\|/(2\|B\|)} \le 2\|B\|^2 \frac{\|s(h)\|}{\|k_h\|} \to 0$$

as $h \to 0$, as required. □

There is a corresponding "implicit function theorem". In this result, let X, Y and I be Banach spaces.

Theorem 8.8 (Pointwise Implicit Function). *Let $W \subseteq I \times X$, and $H\colon W \to Y$ be differentiable at $(t_0, x_0) \in W$. Suppose that there are neighborhoods $I_0 \subseteq I$ of t_0 and $Y_0 \subseteq Y$ of $y_0 := H(t_0, x_0)$ such that for each $(t, y) \in I_0 \times Y_0$ there is exactly one $x = G(t, y)$ with $(t, x) \in H^{-1}(y)$. Then $G\colon I_0 \times X_0 \to W$ is differentiable at (t_0, y_0) if and only if G is continuous at (t_0, y_0) and $d_X H(t_0, x_0) \in \operatorname{Iso}(X, I)$. In this case, $d_I G(t_0, y_0)$ and $d_Y G(t_0, y_0)$ exist and satisfy*

$$\begin{aligned} d_I G(t_0, y_0) &= -d_X H(t_0, x_0)^{-1} d_I H(t_0, x_0) \in \mathcal{L}(I, X) \\ d_Y G(t_0, y_0) &= d_X H(t_0, x_0)^{-1} \in \operatorname{Iso}(Y, X). \end{aligned} \tag{8.3}$$

Proof. The assertion follows from Theorem 8.7 with $E := I \times X$, $Z := I \times Y$, $V := I_0 \times Y_0$, $U_0 := W$, $U := I_0 \times G(V)$, and $F(t, x) := (t, H(t, x))$. Indeed, as already pointed out, we need not require that U is a neighborhood of (t_0, x_0). Note that in operator matrix representation we have by Proposition 8.6 (applied with $H = F$) that

$$A := dF(t_0, x_0) = \begin{pmatrix} \mathrm{id}_I & 0 \\ d_I H(t_0, x_0) & d_X H(t_0, x_0) \end{pmatrix}$$

and so $dF(t_0, x_0) \in \operatorname{Iso}(E, Z)$ if and only if $d_X H(t_0, x_0) \in \operatorname{Iso}(X, I)$. The form of F implies for $J := F|_{U_0}^{-1}$ that $J(t, y) = (t, G(t, y))$. Hence, J is continuous/differentiable at (t_0, y_0) if and only if G is continuous/differentiable at (t_0, y_0). Hence, the assertion about the differentiability follows from Theorem 8.7. Moreover, if G is differentiable at (t_0, y_0) then Proposition 8.6 (applied with $H = J$) shows that in operator matrix representation

$$B := dJ(t_0, y_0) = \begin{pmatrix} \mathrm{id}_I & 0 \\ d_I G(t_0, y_0) & d_Y G(t_0, y_0) \end{pmatrix}.$$

We note that Theorem 8.7 implies $AB = \mathrm{id}_E$. Considering the lower line in this operator matrix equality, we obtain (8.3). □

Recall that if $U \subseteq E$ is open, then $F\colon U \to Z$ is called *continuously differentiable* on U (and we write $F \in C^1(U, Z)$ or $F \in C^1$) if F is differentiable on U and the derivative $x \mapsto dF(x)$ is continuous as a map from U into the normed space $\mathcal{L}(E, Z)$.

The classes C^n, $n \geq 2$, are defined analogously by induction: We write $F \in C^n(U, Z)$ if $x \mapsto dF(x)$ belongs to $C^{n-1}(U, \mathcal{L}(E, Z))$. As customary, we define C^n in case $n = 0$ as the class C of continuous functions, and in case $n = \infty$ as those functions which are of class C^n for every $n \in \mathbb{N}$.

Proposition 8.9. *If $0 \le m \le n \le \infty$ then $C^n \subseteq C^m$, and the composition of two functions of class C^n is of class C^n.*

Proof. The assertion follows from Proposition 8.5 and obvious inductions. □

We consider $I_0 := \mathrm{Iso}(E, Z)$ as a subset of the Banach space $I := \mathcal{L}(E, Z)$. Recall that Proposition 6.10 implies that I_0 is open. Applying Theorem 8.8 with $H: I_0 \times E \to Z$, $H(A, x) := Ax$, we see that the implicit function $G(A, y) = A^{-1}y$ is differentiable with respect to A (the continuity of G follows from Proposition 6.10), and the derivative is in view of Proposition 8.3 given by (with $X := E$)

$$d_I G(A, y)U = -d_X H(A, A^{-1}y)^{-1} d_I H(A, A^{-1}y)U = -A^{-1}UA^{-1}y.$$

One can verify that this derivative holds actually even in operator norm. One way to see this is to apply the above calculation for $H: I_0 \times \mathcal{L}(E) \to \mathcal{L}(Z, E)$, $H(A, U) := A^{-1}U$, but it is perhaps simpler to verify this directly:

Proposition 8.10. *The map $J: \mathrm{Iso}(E, Z) \to \mathrm{Iso}(Z, E)$, defined by $J(A) := A^{-1}$, is of class C^∞, and*

$$dJ(A)U = -J(A)UJ(A). \tag{8.4}$$

Proof. For $A \in \mathrm{Iso}(E, Z)$, we put $r(U) := J(A + U) - J(A) + J(A)UJ(A)$. Then

$$(A+U)r(U) = \mathrm{id}_E - (\mathrm{id}_E + UA^{-1}) + (UA^{-1} + UJ(A)UJ(A)) = UJ(A)UJ(A)$$

implies that for all $U \in \mathcal{L}(E, Z) \setminus \{0\}$

$$\frac{\|r(U)\|}{\|U\|} = \frac{\|(A + U)^{-1}UJ(A)UJ(A)\|}{\|U\|} \le \|J(A + U)\| \|J(A)\|^2 \|U\|.$$

Since Proposition 6.10 implies that $J(A + U)$ remains bounded as $\|U\| \to 0$, we obtain $r(U)/\|U\| \to 0$. Hence, dJ is differentiable at A, and the formula (8.4) holds. Now an induction by $n = 0, 1, \ldots$ implies that J is of class C^n. Indeed, the case $n = 0$ follows from Proposition 8.5 (or alternatively from Proposition 6.10), and if J is of class C^n then (8.4) shows that dJ is of class C^n, and so J is of class C^{n+1}. □

8.2 Oriented Nonlinear Fredholm Maps

Definition 8.11. If $\Omega \subseteq E$ is open then $F\colon \Omega \to Z$ is *Fredholm* of *index* k (in symbols: $F \in \mathcal{F}_k(\Omega, Z)$), if $F \in C^1(\Omega, Z)$ and $dF\colon \Omega \to \mathcal{L}_k(E, Z)$. In case $k = 0$ and if $M \subseteq \Omega$, an *orientation* for F on M is an orientation of $dF|_M$ according to Definition 7.18.

F is called *orientable* (on M) if $F \in \mathcal{F}_0(\Omega, Z)$ and an orientation σ exists on Ω (or M). The couple (F, σ) is called an *oriented Fredholm operator*. Notationally, we just write F and refer to σ as "the orientation" of F.

Note that if $X = E$ and $F = A$ is linear then $dF(x) = A$ for every $x \in X$. Since E is connected, all orientations of F are actually constant by Corollary 7.21 so that the choice of an orientation comes down to the choice of an orientation of the linear $A \in \mathcal{F}_0(E, Z)$. Hence, although formally the orientations for F as a C^1-map and for $F = A$ as a linear map are different items (one is a constant multivalued map, the other only the image of this map in each point), the actual orientation in each point means the same so that there is not really an ambiguity with Definition 7.2 when we talk about orientations in this case.

We will also need (oriented) homotopies for the (oriented) Fredholm maps:

Definition 8.12. Let I be a topological space, $W \subseteq I \times E$ be open, and $W_t := \{u : (t, u) \in E\}$. A *generalized Fredholm homotopy* (of index k) is a continuous map $H\colon W \to Z$ with open $W \subseteq E$ and the property that $H(t, \cdot) \in \mathcal{F}_k(W_t, Z)$ for every $t \in I$ such that the corresponding partial derivative $d_X H(t, x) := d(H(t, \cdot))(x)$ is continuous as a map from W into $\mathcal{L}(E, Z)$. In case $k = 0$ and if $M \subseteq W$, an *orientation* for H on M is an orientation for $d_X H|_M$ according to Definition 7.18.

H is called *orientable* (on M) if such an orientation σ exists on W (or M). We call the couple (H, σ) an *oriented generalized Fredholm homotopy*. Notationally, we just write H and refer to σ as "the orientation" of H.

For the case $W = [0, 1] \times \Omega$ with some open $\Omega \subseteq E$, we call H an (oriented/orientable) *Fredholm homotopy*.

For easier reference, we collect some properties of the previous definitions:

Theorem 8.13. *Let $\mathbb{K} = \mathbb{R}$, $E^* \neq \{0\}$, and $\Omega \subseteq E$, I be a topological space, and $W \subseteq I \times E$ be open. Let $F \in \mathcal{F}_0(\Omega, Z)$ and $H\colon W \to Z$ be a generalized Fredholm homotopy of index* 0.

(a) *σ is an orientation for F or H if and only if the (pointwise) opposite orientation is an orientation for F or H, respectively.*

(b) *If F (or H) are oriented on $C \subseteq \Omega$ (or $C \subseteq W$) is connected with $dF(C) \subseteq \mathrm{Iso}(E, Z)$ (or $d_X H(C) \subseteq \mathrm{Iso}(E, Z)$) then $\operatorname{sgn} dF(x)$ (or $\operatorname{sgn} d_X H(t, \cdot)(x)$) are the same for all $x \in C$ (or $(t, x) \in C$).*

(c) *If $C \subseteq \Omega$ (or $C \subseteq W$) is connected and σ is an orientation for F (or H) on C then σ is uniquely determined if its value is known in one point of C.*

(d) *Let I be locally path-connected. If Ω or W is simply connected and if in some point of each component of Ω or W an orientation for F or H is given then F and H have a unique corresponding orientation on Ω or W.*

(e) *If H is a Fredholm homotopy then H is orientable if and only if $H(t_0, \cdot)$ is orientable for some $t_0 \in [0, 1]$. In this case, the orientation of H is uniquely determined by the orientation of $H(t_0, \cdot)$.*

Proof. The assertions follow straightforwardly from Propositions 7.35, Proposition 7.36, and Theorem 7.41, respectively. Concerning (d) note that E and $I \times E$ are locally path-connected by Corollary 2.66. Proposition 2.23 implies that the components of Ω and of $I \times \Omega$ are open. Hence, the orientation on the components can be fixed independent of each other. For the uniqueness, recall that components are connected by Proposition 2.17. □

Proposition 8.14. *Let $\mathbb{K} = \mathbb{R}$ and $\Omega \subseteq E$, I be a topological space, and $W \subseteq I \times E$ be open. Let $F \in \mathcal{F}_0(\Omega, Z)$, and $H \colon W \to Z$ be a generalized Fredholm homotopy of index 0. If $C \subseteq \Omega$ or $C \subseteq W$ and dF or $d_X H$ are constant on C then the orientations of F or H on C are exactly the constant orientations.*

Proof. The nontrivial implication follows from Corollary 7.21. □

8.3 Oriented Fredholm Maps in Banach Manifolds

In most applications of degree theory, one will have maps defined on an open subset Ω of a Banach space E with values in a Banach space Z. In such a situation the notions of Section 8.2 are sufficient to apply degree theory.

Unfortunately, they are not sufficient for us in this monograph, since we must first define the degree before we can apply it. And in order to define the degree, we need also degree theory on finite-dimensional manifolds.

We point this out once more: Even in order to develop only degree theory for Fredholm maps on Banach *spaces*, we need as a tool degree theory on finite-dimensional *manifolds*. So we cannot avoid to speak about manifolds. (This is not the case if one only wants to *apply* the theory, but it is necessary if one wants to develop it.)

Since it makes no difference in the formal definitions, we can actually consider even Banach manifolds. The definitions of manifolds differ slightly in literature, so let us make precise what we mean by manifolds.

Definition 8.15. A Hausdorff space X is a *Banach manifold* of class C^n over E, $n \in \{0, 1, \ldots, \infty\}$, if X has an open cover $\mathscr{U} = \{U_i : i \in I\}$ with a corresponding family $\{c_i : i \in I\}$ (the *atlas*) such that the following holds:

(a) Each c_i is a homeomorphisms (the *chart*) of U_i onto an open subset of E.

(b) For each $i, j \in I$ the map $c_j \circ c_i^{-1} : c_i(U_i \cap U_j) \to E$ is C^n.

A *chart for a point* x is a chart $c_i : U_i \to E$ with $x \in U_i$.

We point out that we require in Definition 8.15 that X is Hausdorff, but in contrast to many text books, we do neither assume that X is paracompact nor that X is second countable. (The latter would be a restrictive requirement anyway in the context of Banach manifolds.)

Of course, when we speak about a Banach manifold X, we actually mean the couple (X, A) where A is an atlas for X. However, we do not mentioned the atlas explicitly, but only implicitly take corresponding charts if required. Moreover, we do *not* assume that the atlas is maximal, that is, when we speak of a chart we really mean a member of the atlas (or in some cases the restriction of such a map).

Every open subset X of a Banach space E becomes a C^∞-Banach manifold over E with $\{\mathrm{id}_X\}$ as its atlas. We tacitly understand this atlas (and not any larger atlas!) when we interpret a Banach space as a Banach manifold.

Proposition 8.16. *Every Banach manifold X is locally path-connected. All subsets of X are first countable and in particular compactly generated.*

Proof. Since E is locally path-connected, every open subset of E and thus also any space homeomorphic to such a set is locally path-connected. Hence, every point in X has a locally path-connected neighborhood. Since E is first countable, also X is first countable. Hence, also all subsets of X are first countable. The last assertion follows from Proposition 2.108. □

With the above convention of interpreting open $X \subseteq E$ as a Banach manifold, the following definition of C^n-maps and of Fredholm maps extends our definitions of Section 8.2.

Definition 8.17. Let X and Y be C^n Banach manifolds over E_X and E_Y, respectively, and $\Omega \subseteq X$ be open. Then a map $F : \Omega \to Y$ is of class C^n (and we write $F \in C^n(\Omega, Y)$ or $F \in C^n$) if for each $x \in \Omega$ there are charts c_X and c_Y

for x and $F(x)$, respectively, such that $c_Y \circ F \circ c_X^{-1}$ is of class C^n in a neighborhood of $c_X(x)$. In case $n \geq 1$ and if each of the derivatives of $c_Y \circ F \circ c_X^{-1}$ in every point belongs to $\mathcal{L}_k(E_X, E_Y)$, we call F *Fredholm* of *index k* and write $F \in \mathcal{F}_k(\Omega, Y)$.

If I is a topological space and $W \subseteq I \times X$ is open, then a continuous map $H: W \to Y$ is a *generalized partial C^r homotopy* ($1 \leq r \leq \infty$) or *generalized Fredholm homotopy* of index k, respectively, if the following holds. For each $(t_0, x_0) \in W$ there are charts c_X and c_Y for x_0 and $H(t_0, x_0)$, respectively, such that there are open neighborhoods $I_0 \subseteq I$ of t_0 and $U \subseteq E_X$ of $c_X(x_0)$ such that

$$G_t := c_Y \circ H(t, \cdot) \circ c_X^{-1} \in C^r(U, E_Y) \quad \text{for all } t \in I_0,$$

and $(t, x) \mapsto dG_t(x)$ is continuous as a map from $(t, x) \in W$ into $\mathcal{L}(E_X, E_Y)$ or $\mathcal{L}_k(E_X, E_Y)$, respectively. In case $r = 0$, a *generalized partial C^0 homotopy* is just a continuous map $H: W \to Y$.

In case $W = [0, 1] \times \Omega$, we speak about a *partial C^r homotopy* or *Fredholm homotopy* of index k, respectively.

Note that we require in this definition that X and Y are of class C^n. It is easy to see by this requirement that then actually the same condition holds for *any* choice of the charts.

Unfortunately, the definition of orientation for Fredholm maps/homotopies in Banach manifolds is much more involved and requires to introduce the tangent bundle of a Banach manifold. To this end, we recall some of many equivalent definitions of the tangent space.

Definition 8.18. Let X be a C^1-Banach manifold over E, and $x \in X$. Consider on the set of all couples (c, u) where c is a chart for x and $u \in E$ the equivalence relation

$$(c, u) \sim (c_0, u_0) \iff d(c_0 \circ c^{-1})(c(x))u = u_0. \tag{8.5}$$

Let $T_{x,0}$ denote the set of corresponding equivalence classes $[(c, u)]_x$. Then $T_{x,0}$ becomes a vector space with the operations $[(c, u_1)]_x + [(c, u_2)]_x := [(c, u_1 + u_2)]_x$ and $\lambda[(c, u)]_x := [(c, \lambda u)]_x$. The *tangent space* T_x (or, more verbosely, $T_x X$) is defined as the vector space $\{x\} \times T_{x,0}$ with the obvious inherited operations.

The consideration of $\{x\} \times T_{x,0}$ instead of $T_{x,0}$ is just a formal trick to make sure that $T_x \cap T_y = \varnothing$ for $x \neq y$ so that T_x are candidates for fibres of a Banach bundle. Actually, also the definition of $T_{x,0}$ is only a formal trick which we explain in a moment.

Proposition 8.19. *The relation* (8.5) *is indeed an equivalence relation, and the above vector space operations are well-defined. Moreover, for each chart c for x the map $L_{c,x}: E \to T_{x,0}$, defined by $L_{c,x}u := [(c,u)]_x$, is linear, one-to-one, and onto, and if c_0 is another chart for x then $L_{c_0,x}^{-1} L_{c,x} = d(c_0 \circ c^{-1})(c(x))$.*

Proof. If $(c,u) = (c_0,u_0)$, we have $d(c_0 \circ c^{-1})(c(x))u = d(\mathrm{id})(c(x))u = \mathrm{id}_E\, u = u_0$, hence (8.5) is reflexive. Note that if c and c_0 are two charts for x then Theorem 8.7 implies that

$$d(c_0 \circ c^{-1})(c(x)) \in \mathcal{L}(E) \text{ is invertible with inverse } d(c \circ c_0^{-1})(c_0(x)). \quad (8.6)$$

In particular, $(c,u) \sim (c_0,u_0)$ implies

$$d(c \circ c_0^{-1})(c_0(x))u_0 = (d(c_0 \circ c^{-1})(c(x)))^{-1}u_0 = u,$$

and so $(c_0,u_0) \sim (c,u)$. If $(c,u) \sim (c_0,u_0)$ and $(c_0,u_0) \sim (c_1,u_1)$ then Proposition 8.5 implies

$$\begin{aligned} d(c_1 \circ c^{-1})(c(x))u &= d((c_1 \circ c_0^{-1}) \circ (c_0 \circ c^{-1}))(c(x))u \\ &= d(c_1 \circ c_0^{-1})(c_0(x))d(c_0 \circ c^{-1})(c(x))u = d(c_1 \circ c_0^{-1})(c_0(x))u_0 = u_1, \end{aligned}$$

and so $(c,u) \sim (c_1,u_1)$. Hence, (8.5) is an equivalence relation. If c, c_0 are charts for x and $[(c,u_i)]_x = [(c_0,v_i)]_x$ $(i = 0,1,2)$ then $d(c_0 \circ c^{-1})(c(x))(u_1 + u_2) = v_1 + v_2$ and $d(c_0 \circ c^{-1})(c(x))(\lambda u_0) = \lambda v_0$ which implies $[(c,u_1 + u_2)]_x = [(c_0, v_1 + v_2)]_x$ and $[(c,\lambda u_0)]_x = [(c_0,\lambda v_0)]_x$. Hence, the vector space operations are well-defined. It is clear from the definitions that $L_{c,x}$ is linear. $L_{c,x}$ is one-to-one since $[(c,u_1)]_x = [(c,u_2)]_x$ implies $d(\mathrm{id})(c(x))u_1 = u_2$ and thus $u_1 = \mathrm{id}_E\, u_1 = u_2$. To see that $L_{c,x}$ is onto, let $[(c_0,u_0)]_x \in T_{x,0}$. By (8.6), we can define $u := (d(c_0 \circ c^{-1})(c(x)))^{-1}u_0$, and then obtain $d(c_0 \circ c^{-1})(c(x))u = u_0$. Hence, $(c,u) \sim (c_0,u_0)$ which implies $L_{c,x}u = [(c_0,u_0)]_x$. The latter shows also that $L_{c,x}u = L_{c_0,x}u_0$ and thus $u_0 = L_{c_0,x}^{-1}L_{c,x}u$. Since $u_0 = d(c_0 \circ c^{-1})(c(x))u$, we obtain the formula for $L_{c_0,x}^{-1}L_{c,x}$. □

Proposition 8.19 implies that if we equip T_x with the topology induced by $L_{c,x}$ (we show in the moment that this topology is independent of the choice of c), then T_x is isomorphic to E. Hence, we could actually have defined $T_{x,0} := E$ (considering E as a topological vector space, not as a Banach space).

However, in this way it would be harder to describe the relations of $T_{x,0}$ when x varies which is what we will do now: The idea is that, although we do *not* want to require that X be embedded as a "surface" in some other space, $T_{x,0}$ (and T_x) should take the role of the "tangent plane" to this surface in the point x (with origin of $T_{x,0}$ in x).

It is a crucial technical difficulty in this connection that the given data is only sufficient to describe the topology of T_x (which makes it isomorphic to E) but that there is no "canonical" norm with which we could equip T_x, since each natural candidate for a norm depends on the choice of the chart c. Fortunately, for Banach bundles we only need to know the topology. So let us now give the family T_x ($x \in X$) the full structure of a Banach bundle:

Definition 8.20. $TX := \bigcup_{x \in X} T_x$ denotes the *tangent bundle* of the C^1-manifold X. If c is a chart defined on $U \subseteq X$, we define $Tc: \bigcup_{x \in U} T_x \to c(U) \times E$ by $Tc(x, [(c,u)]_x) := (c(x), u)$. We equip TX with the topology whose basis consists of the sets $(Tc)^{-1}(O_1 \times O_2)$ where c is a chart and $O_1, O_2 \subseteq E$ are open. The Banach bundle structure of Definition 7.27 comes with $I := X$ and $h_c(x,u) := (x, [(c,u)]_x)$ as trivializations of the bundle for $x \in U$.

For a subset $M \subseteq X$, it will be convenient to define TM as the corresponding restriction of the tangent bundle (with $I = M$).

Even if the reader prefers to take another definition of T_x (for instance, in [77] the equivalence relation is defined immediately for triples (x, c, u) which avoids the distinction between $T_{x,0}$ and T_x of Definition 8.18), all natural definitions equip TX with the same topology and the corresponding natural trivializations h_c coming from charts of X. Later, we will only use h_c to describe TX and avoid the usage of the details of TX. However, first we have to prove that we obtain indeed a Banach bundle.

Proposition 8.21. *Tc is well-defined, one-to-one and onto. The topology on TX is well-defined. If X is a C^n-Banach manifold over E ($n \geq 1$) then TX is a Banach bundle over E and simultaneously a C^{n-1}-Banach manifold over $E \times E$ with Tc as charts. We have $h_c(x, \cdot) \in \operatorname{Iso}(E, T_x)$.*

Proof. With $L_{c,x}$ from Proposition 8.19, we have $Tc(x,z) = (c(x), L_{c,x}^{-1} z)$, and so Tc is well-defined, one-to-one, and onto $c(U) \times L_c^{-1}(T_{x,0}) = c(U) \times E$ with $(Tc)^{-1}(y,u) = (x, L_{c,x} u)$ for $y = c(x)$ and $u \in E$.

If c_0 is another chart defined on $U_0 \subseteq X$ and $x \in U \cap U_0$ then c and c_0 are charts for x, and for $y = c(x)$ and $u \in E$, we have by Proposition 8.19 that

$$(Tc_0) \circ (Tc)^{-1}(y,u) = (c_0(x), L_{c_0,x}^{-1} L_{c,x} u) = (c_0(x), d(c_0 \circ c^{-1})(y) u).$$

Hence, $(Tc_0) \circ (Tc)^{-1}: c(U \cap U_0) \times E \to c_0(U \cap U_0) \times E$ is one-to-one and onto and given by

$$(Tc_0) \circ (Tc)^{-1}(y,u) = ((c_0 \circ c^{-1})(y), d(c_0 \circ c^{-1})(y) u). \tag{8.7}$$

This map is continuous, and so also the inverse $(Tc) \circ (Tc_0)^{-1}$ is continuous, since it is the same map with just c and c_0 exchanged. Thus, $(Tc_0) \circ (Tc)^{-1}$ is a homeomorphism. It follows that the topology defined in Definition 8.20 is locally independent of the particular choice of c, and that it is actually the unique topology such that $(Tc)^{-1}$ is a homeomorphism onto an open subset of TX. In particular, Tc are homeomorphisms, and $\bigcup_{x\in U} Tx$ is open.

Since (8.7) shows also that $(Tc_0) \circ (Tc)^{-1}$ is of class C^{n-1}, we find that TX is a C^{n-1}-Banach manifold with Tc as charts if we can show that TX is Hausdorff. Thus, let $z_1, z_2 \in TX$, $z_1 \neq z_2$ with $z_i = (x_i, [c_i, u_i]_{x_i})$ $(i = 1, 2)$. If $x_1 = x_2$, let c be a chart for $x_1 = x_2$. Since $Tc(z_1)$ and $Tc(z_2)$ have disjoint neighborhoods, and Tc is a homeomorphism, it follows that z_1 and z_2 have disjoint neighborhoods. In case $x_1 \neq x_2$, we find disjoint neighborhood $U_i \subseteq X$ of x_i $(i = 1, 2)$, and then $\bigcup_{x\in U_i} T_x$ are disjoint neighborhoods of z_i $(i = 1, 2)$.

We have $h_c = (Tc)^{-1} \circ g_c$ with $g_c \colon U \times E \to c(U) \times E$ defined by $g_c(x, u) := (c(x), u)$. Since g_c and Tc are homeomorphisms, it follows that h_c is a homeomorphism. Moreover, $h_c(x, \cdot) = (x, L_{c,x}(\cdot))$ is linear from E into T_x. Hence, h_c are trivializations with open images, and we can apply Proposition 7.26 to obtain the continuity of the bundle projection.

To see that TX is a Banach bundle, it remains to show that the trivializations h_c satisfy the compatibility required in Definition 7.27. Hence, we have to show that if c and c_0 are two charts on U and U_0, respectively, then $f(x) := h_{c_0}(x, \cdot)^{-1} \circ h_c(x, \cdot)$ is continuous as a map from $U \cap U_0$ into $\mathcal{L}(E)$. Using Proposition 8.19, we calculate

$$f(x)u = (x, L_{c_0,x}^{-1} \circ L_{c,x}u) = (x, d(c_0 \circ c^{-1})(c(x))u).$$

Hence, we have to show that $x \mapsto d(c_0 \circ c^{-1})(c(x)) \in \mathcal{L}(E)$ is continuous. This holds since c is continuous and $c_0 \circ c^{-1} \in C^1(c(U \cap U_0), E)$. □

For the rest of this section, let X and Y be Banach manifolds over the Banach space E_X and E_Y over the field $\mathbb{K} = \mathbb{R}$ or $\mathbb{K} = \mathbb{C}$.

In order to avoid confusion, we use the notation T_xX and T_yY for the tangent spaces at $x \in X$ or $y \in Y$ in the manifolds X and Y, respectively.

Definition 8.22. Let $\Omega \subseteq X$ be open $F \in C^1(\Omega, Y)$, we define $dF(x) \in \mathcal{L}(T_xX, T_{F(x)}Y)$ by

$$dF(x) := h_{c_Y}(F(x), \cdot) \circ d(c_Y \circ F \circ c_X^{-1})(c_X(x)) \circ h_{c_X}(x, \cdot)^{-1}, \tag{8.8}$$

where c_X and c_Y are charts at x and $F(x)$, and where h_{c_X} and h_{c_Y} denote the corresponding trivializations of the tangent bundles according to Definition 8.20 (for X and Y, respectively).

Similarly, if I is a topological space, $W \subseteq I \times X$ is open and $H\colon W \to Y$ is a generalized C^1 homotopy, we define the *partial derivative* $d_X H(t,x) \in \mathcal{L}(T_xX, T_{H(t,x)}Y)$ as the value of the derivative of $H(t,\cdot)$ at x.

For the following proof, we have of course to resort to the definition of h_{c_X} and h_{c_Y}.

Proposition 8.23. *The above definition is independent of the particular choice of charts. Moreover,* $dF \in \mathcal{L}(\Omega, TX, TY)$ *and* $d_X H \in \mathcal{L}(W, TX, TY)$.

We have $dF(x) \in \mathcal{L}_k(T_xX, T_{F(x)}Y)$ *for all* $x \in \Omega$ *if and only if* $F \in \mathcal{F}_k(X,Y)$. *Similarly,* $d_X H(t,x) \in \mathcal{L}_k(T_xX, T_{H(t,x)}Y)$ *for all* $(t,x) \in W$ *if and only if* H *is a generalized Fredholm homotopy of index* k.

Proof. By definition of h_{c_X} and h_{c_Y}, we have

$$dF(x)(x,[(c_X,u)]_x) = (F(x),[(c_Y, d(c_Y \circ F \circ c_X^{-1})(c_X(x))u)]_{F(x)}).$$

By the definition of the equivalence class $[\cdot]_{F(x)}$, the right-hand side is independent of the choice of c_Y, and by definition of $[(c_X,u)]_x$ it is also independent of the choice of representatives (c_X,u) from that class.

For the proof of $d_X H \in \mathcal{L}(W, TX, TY)$, we choose the trivializations $h_T := h_{c_X}$ and $h_S := h_{c_y}$ in Definition 7.31. We have to show that

$$\begin{aligned} h(t,x) &:= h_{c_y}(H(t,x),\cdot)^{-1} d_X H(t,x) \circ h_{c_x}(x,\cdot) \\ &= d(c_Y \circ H(t,\cdot) \circ c_X^{-1})(c_X(x)) \in \mathcal{L}(E_X, E_Y) \end{aligned}$$

is continuous. Since $x \mapsto c_X(x)$ is continuous, this is clear by the definition of a generalized partial C^1 homotopy.

For the special choice $H(t,x) := F(x)$ it follows that $dF \in \mathcal{L}(\Omega, TX, TY)$.

The last assertion follows from Proposition 7.32. □

Proposition 8.24. *The chain rule holds also on manifolds, that is, if* G *and* F *are* C^1 *maps between open subsets of manifolds then*

$$d(G \circ F)(x) = dG(F(x))dF(x),$$

in every point x *for which* $(G \circ F)(x)$ *is defined.*

Proof. Let c_X, c_Y, and c_Z be charts for x, $y := F(x)$, and $z := G(y)$ in the respective manifolds. Then

$$dF(x)(x,[(c_X,u)]_x) = (y,[(c_Y,d(c_Y \circ F \circ c_X^{-1})(c_X(x))u)]_y),$$
$$dG(y)(y,[(c_Y,v)]_y) = (z,[(c_Z,d(c_Z \circ G \circ c_Y^{-1})(c_Y(y))v)]_z),$$

and Proposition 8.5 implies

$$\begin{aligned} d(G \circ F)(x)(x,[(c_X,u)]_x) &= (z,[(c_Z,d(c_Z \circ G \circ F \circ c_X^{-1})(c_X(x))u)]_z) \\ &= (z,[(c_Z,d(c_Z \circ G \circ c_Y^{-1} \circ c_Y \circ F \circ c_X^{-1})(c_X(x))u)]_z) \\ &= (z,[(c_Z,d(c_Z \circ G \circ c_Y^{-1})(c_Y(y))d(c_Y \circ F \circ c_X^{-1})(c_X(x))u)]_z) \end{aligned}$$

A comparison of these formulas shows the assertion. □

Due to Proposition 8.23, we are now finally in a position to define the orientation of Fredholm operators in Banach manifolds.

Definition 8.25. If $\Omega \subseteq X$ is open and $F \in \mathcal{F}_0(\Omega, Y)$, then an *orientation* for F on $M \subseteq \Omega$ is an orientation for $dF|_M \in \mathcal{L}_0(M, TX, TY)$ according to Definition 7.33.

F is called *orientable* on M if an orientation exists on M. The couple (F, σ) is called *oriented Fredholm* on M. Notationally, we just write F and refer to σ as "the orientation" of F on M.

Definition 8.26. If $H: W \to Y$ is a generalized Fredholm homotopy of index 0 then an *orientation* for H on $M \subseteq W$ is an orientation for $d_X H|_M \in \mathcal{L}_0(M, TX, TY)$ according to Definition 7.33.

H is called an *orientable* on M if such an orientation σ exists, and we call the couple (H, σ) an *oriented Fredholm homotopy* on M. Notationally, we just write H and refer to σ as "the orientation" of H on M.

All results of Theorem 8.13 hold also on Banach manifolds:

Theorem 8.27. *Let $\mathbb{K} = \mathbb{R}$, $E^* \neq \{0\}$, $\Omega \subseteq X$ be open, I be a topological space, and $W \subseteq I \times X$ be open. Let $F \in \mathcal{F}_0(\Omega, Y)$ and $H: W \to Y$ be a generalized Fredholm homotopy of index* 0.

(a) *σ is an orientation for F or H if and only if the (pointwise) opposite orientation is an orientation for F or H, respectively.*

(b) *Let F (or H) be oriented and $C \subseteq \Omega$ (or $C \subseteq W$) be connected. Suppose that $dF(x) \subseteq \mathrm{Iso}(T_X X, T_{F(x)} Y)$ for all $x \in C$ (or $d_X H(t,x) \subseteq \mathrm{Iso}(T_X X, T_{H(t,x)} Y)$ for all $(t,x) \in C$). Then $\operatorname{sgn} F(x)$ (or $\operatorname{sgn} H(t,\cdot)(x)$) are the same for all $x \in C$ (or $(t,x) \in C$).*

(c) *If $C \subseteq \Omega$ (or $C \subseteq W$) is connected and σ is an orientation for F (or H) on C then σ is uniquely determined if its value is known in one point of C.*

(d) *Let I be locally path-connected. If Ω is simply connected and if in some point of each component of Ω or W an orientation for F or H is given then F and H have a unique corresponding orientation on Ω or W.*

(e) *If $W = [0,1] \times \Omega$ then H is orientable if and only if $H(t_0, \cdot)$ is orientable for some $t_0 \in [0,1]$. In this case, the orientation of H is uniquely determined by the orientation of $H(t_0, \cdot)$.*

Proof. In view of Proposition 8.16, the proof of Theorem 8.13 carries over word by word. □

The general analogue of Proposition 8.14 is a bit clumsy to formulate in case of Banach manifolds since "constantness" of the derivative can only be defined with the aid of charts and, more severe, also depends on the choice of the chart. We note only a special case for homotopies which is easy to formulate and sufficient for most practical purposes.

Proposition 8.28. *Let $\mathbb{K} = \mathbb{R}$, I be a topological space, $W \subseteq I \times X$ be open, and let $M \subseteq W$ be of the particular form $M = C \times \{x\}$ with some connected set $C \subseteq I$.*

Let $H: W \to Y$ be a generalized Fredholm homotopy of index 0 which is oriented on M. If $d_X H(t,x)$ is independent of $t \in C$ then also the orientation of $d_X H(t,x)$ is independent of $t \in C$.

Proof. Let c_X and c_Y be charts for x and $H(t,x)$, respectively. Let h_{c_X} and h_{c_Y} be as in Definition 8.20. Put $A(t) := c_Y \circ d(H(t, c_X^{-1}(\cdot)))(c_X(x))$, $J_1 := h_{c_X}(x, \cdot) \in \operatorname{Iso}(T_x X, E_X)$ and $J_2 := h_{c_Y}(H(t,x), \cdot) \in \operatorname{Iso}(T_{H(t,x)} Y, E_Y)$. Note that J_2 is independent of $t \in C$. Then $d_X H(t,x) = J_2 \circ A(t) \circ J_1^{-1}$ for all $t \in C$. Since the left-hand side and J_2 are independent of $t \in C$, also $A(t)$ is independent of $t \in C$. If σ denotes the orientation of $d_X H$ on M then by definition $s(t) := J_2 \circ \sigma(t,x) \circ J_1$ must be an orientation for A. Corollary 7.21 implies that s is constant on C, and so σ is constant on M. □

Let X, Y, and Z be Banach manifolds.

Definition 8.29. Let $\Omega \subseteq X$ and $U \subseteq Y$ be open, and $F \in \mathcal{F}_0(\Omega, U)$, $G \in \mathcal{F}_0(U, Z)$ be oriented on $M \subseteq \Omega$ and $F(M)$, respectively. Then the *composite orientation* of $H := G \circ F$ on M is defined at $x \in M$ for $dH(x) = dG(F(x))dF(x) \in \mathcal{L}(T_x X, T_{H(x)} Z)$ as the composite orientation of the maps $dF(x) \in \mathcal{L}(T_x X, T_{F(x)} Y)$ and $dG(F(x)) \in \mathcal{L}(T_{F(x)} Y, T_{H(x)} Z)$.

Proposition 8.30. *Let X, Y, and Z be Banach manifolds. Then the composite orientation is an orientation on M. Moreover:*

(a) *If $G \circ F$ and G are oriented on M and $F(M)$, respectively then there is a unique orientation of F on M such that the orientation on $G \circ F$ is the composite orientation.*

(b) *If $G \circ F$ and F are oriented on M and if $F: M \to F(M)$ is a homeomorphism then there is a unique orientation of G on $F(M)$ such that the orientation of $G \circ F$ is the composite orientation.*

Proof. The assertion follows from Proposition 7.45 with $\tau = F$. □

We also have a corresponding result for generalized Fredholm homotopies.

Let X, Y, and Z be Banach manifolds, and I be a topological space. Let $V \subseteq I \times X$ and $W \subseteq [0,1] \times Y$ be open and $H_1: V \to Y$ and $H_2: W \to Z$ be Fredholm homotopies of index 0. Let $M \subseteq V$ be such that with $\hat{H}_1(t,x) := (t, H_1(t,x))$ we have $\hat{H}_1(M) \subseteq V$. Then $H := H_2 \circ \hat{H}_1$ is a generalized Fredholm homotopy defined in an open neighborhood of M. We define the composite orientation of $H(t, \cdot)$ as the composite orientation of $H_2(t, \cdot) \circ H_1(t, \cdot)$.

Proposition 8.31. *Consider the above situation. If H_1 and H_2 are oriented on M and M_1, respectively, then the composite orientation of $H = H_2 \circ \hat{H}_1$ is an orientation of H on M. Moreover:*

(a) *If H and H_2 are oriented on M and $\hat{H}_1(M)$, respectively, then there is a unique orientation of H_1 on M such that $H_2 \circ \hat{H}_1$ carries the composite orientation.*

(b) *If H and H_1 are oriented on M and if $\hat{H}_1: M \to \hat{H}_1(M)$ is a homeomorphism then there is a unique orientation of H_2 on $\hat{H}_1(M)$ such that $H_2 \circ \hat{H}_1$ carries the composite orientation.*

Proof. The assertion follows from Proposition 7.45 with $\tau = \hat{H}_1$. □

If X_i are manifolds over Banach spaces E_{X_i} $(i = 1, 2)$, we define the *product manifold* $X := X_1 \times X_2$ over $E_X := E_{X_1} \times E_{X_2}$ such that the charts are given by $\varphi_1 \otimes \varphi_2$ where φ_i are charts of X_i for $i = 1, 2$. By a canonical identification, it follows that $T_x X = T_{x_1} X_1 \times T_{x_2} X_2$ and that $TX = TX_1 \times TX_2$ as a product bundle.

Definition 8.32. If X_i, Y_i $(i = 1, 2)$ are Banach manifolds, $\Omega_i \subseteq X_i$ are open and $F_i \in \mathcal{F}_0(\Omega_i, Y_i)$ have orientations σ_i on $M_i \subseteq \Omega_i$, then the *product orientation* $\sigma(x_1, x_2)$ of $F_1 \otimes F_2 \in \mathcal{F}_0(\Omega_1 \times \Omega_2, Y_1 \times Y_2)$ is for $(x_1, x_2) \in M_1 \times M_2$ defined as the product orientation of $\sigma_1(x_1)$ and $\sigma_2(x_2)$.

Proposition 8.33. *The above defined product orientation is an orientation of* $F_1 \otimes F_2$ *on* $M_1 \times M_2$.

Proof. The assertion follows from Proposition 7.43. □

8.4 A Partial Implicit Function Theorem in Banach Manifolds

Throughout this section, let X and Y be C^1 Banach manifolds over Banach spaces E_X and E_Y over the field $\mathbb{K} = \mathbb{R}$ or $\mathbb{K} = \mathbb{C}$.

Definition 8.34. Let $U \subseteq X$ be open. A map $F \in C^1(U, Y)$ is a diffeomorphism onto an open subset $V \subseteq Y$ if $F \in C^1(U, V)$ is invertible and $F^{-1} \in C^1(V, U)$. If X and Y are of class C^r and if also F and F^{-1} are of class C^r, we call F a C^r-diffeomorphism $(1 \le r \le \infty)$.

Proposition 8.35. *Let F be a diffeomorphism from an open subset $U \subseteq X$ onto an open subset of Y. Then the following holds:*

(a) $dF(x) \in \mathrm{Iso}(T_xX, T_{F(x)}Y)$ *for every* $x \in U$.

(b) *If X and Y are of class C^r $(1 \le r \le \infty)$ then F is a C^r-diffeomorphism if and only if F is of class C^r.*

Proof. Let c_X and c_Y be charts defined on open neighborhoods $U_0 \subseteq X$ and $V_0 \subseteq Y$ of x and $F(x)$, respectively, without loss of generality $F(U_0) \subseteq V_0$. Then $G := c_Y \circ F \circ c_X^{-1}$ is defined on $U_X := c_X(U_0)$. The continuity of F^{-1} implies that $U_Y := G(U_X)$ is open in V_0 and thus open in E_Y. By hypothesis, $H := c_X \circ F^{-1} \circ c_Y^{-1}$ is differentiable in a neighborhood of $G(c_X(x))$. Clearly, $G: U_X \to U_Y$ is invertible with inverse H. Theorem 8.7 implies that

$$d(c_Y \circ F \circ c_X^{-1})(c_X(x)) = dG(c_X(x)) \in \mathrm{Iso}(E_X, E_Y).$$

Hence, the first assertion follows from (8.8). For the second assertion, we have to show that if G is of class C^r then also $H = G^{-1}$ is of class C^r. Theorem 8.7 implies $dH(v) = dG(H(v))^{-1}$. Defining J as in Proposition 8.10, we thus proved $dH = J \circ dG \circ H$. For all $n < r$ the functions J and dG are of class C^n (by Proposition 8.10 or hypothesis, respectively). It follows that if also H is of class C^n then dH is of class C^n, and so H is of class C^{n+1}. Hence, an induction by n implies that H is of class C^n for every finite $n \leq r$. □

This has an important consequence on orientations:

Definition 8.36. Let $U \subseteq X$ be open and $J \in C^1(U, V)$ be a diffeomorphism onto an open subset $V \subseteq Y$. The *natural orientation* of J is defined for $x \in U$ pointwise such that $dJ(x) \in \mathrm{Iso}(T_x X, T_{J(x)} Y)$ carries the natural orientation in the sense of Definition 7.7.

Proposition 8.37. *The natural orientations of J is an orientation in the sense of Definition* 8.25.

Proof. The assertion follows from Proposition 7.38. □

Let X, Y, X_0 and Y_0 be Banach manifolds, J_X a diffeomorphism of an open subset $\Omega_0 \subseteq X$ onto $\Omega \subseteq X$, and J_Y a diffeomorphism of an open subset $V \subseteq Y$ onto a subset of Y_0. For $F \in \mathcal{F}_0(\Omega, V)$ and $F_0 := J_Y \circ F \circ J_X$, we have by Proposition 8.24 that

$$dF_0(x) = dJ_Y(F(J_X(x)))dF(J_X(x))dJ_X(x).$$

Proposition 8.38. *In the above setting, there is a one-to-one correspondence between the orientations σ of F and the orientations σ_0 of F_0 which is given by*

$$\sigma_0(x) = dJ_Y(F(J_X(x))) \circ \sigma(J_X(x)) \circ dJ_X(x).$$

In this case, σ_0 is the composite orientation of σ with the natural orientations of J_X and J_Y.

Similarly, if I is a topological space and $W_0 \subseteq I \times \Omega_0$ is open then there is a one-to-one correspondence between the orientations σ of generalized Fredholm maps $H\colon W \to Y$ with $W := \{(t, J_X(x)) : (t, x) \in W_0\}$ and the orientations σ_0 of $H_0\colon W_0 \to Y_0$,

$$H_0(t, x) := J_Y(H(t, J_X(x))),$$

which is given by

$$\sigma_0(t, x) = dJ_Y(H(t, J_X(x))) \circ \sigma(t, J_X(x)) \circ dJ_X(x).$$

Proof. Corollary 7.8 implies the pointwise one-to-one correspondence, and that σ_0 is pointwise the composite orientation. Since the natural orientation is an orientation by Proposition 8.37, the assertion follows from the respective Propositions 8.30 and 8.31. □

The converse of Proposition 8.35 is usually called the inverse function theorem. In Banach spaces the inverse function theorem reads as follows:

Theorem 8.39 (Inverse Function). *Let X and Y be of class C^r $(1 \le r \le \infty)$, Let $U \subseteq X$ be open, and $F \in C^r(U,Y)$. Then $x \in U$ has an open neighborhood $U_0 \subseteq X$ such that $F|_{U_0}$ is a C^r-diffeomorphism onto an open subset $V_0 \subseteq Y$ if and only if $dF(x) \in \operatorname{Iso}(T_xX, T_{F(x)}Y)$. In this case*

$$dF^{-1}(y) = dF(F^{-1}(y))^{-1} \quad \text{for all } y \in V_0. \tag{8.9}$$

Proof. This is the special case $H(t,x) = F(x)$ of the equivalence (a)⇔(b) of the subsequent Theorem 8.40; the formula (8.9) follows from (8.10). □

Theorem 8.40 (Partial Implicit Function). *Let X and Y be of class C^r $(1 \le r \le \infty)$, Let I be a topological space, $W \subseteq I \times X$ be open, and $H\colon W \to Y$ be a generalized partial C^r homotopy. Then for $(t_0, x_0) \in W$ and $y_0 := H(t_0, x_0)$ the following statements are equivalent.*

(a) *$H(t_0, \cdot)$ is a C^r-diffeomorphism of an open neighborhood of x_0 onto an open neighborhood of y_0.*

(b) *$d_X H(t_0, x_0) \in \operatorname{Iso}(T_{x_0}X, T_{y_0}Y)$.*

(c) *There are open neighborhoods $I_0 \subseteq I$ of t_0, $U \subseteq X$ of x_0, and $V \subseteq Y$ of y_0, such that $I_0 \times U \subseteq W$ and for each $(t,y) \in I_0 \times V$ there is exactly one $x = G(t,y) \in U$ with $(t,x) \in H^{-1}(y)$, and G is a generalized partial C^r homotopy with*

$$d_Y G(t,y) = (d_X H(t, G(t,y)))^{-1} \quad \text{for all } (t,y) \in I_0 \times V. \tag{8.10}$$

The map $J(t,y) := (t, G(t,y))$ is a homeomorphism of $I_0 \times V$ onto the open set $H^{-1}(V) \cap (I_0 \times U)$ with inverse $H_0(t,x) := (t, H(t,x))$, and $G(t,\cdot)$ is a C^r-diffeomorphism onto the open set $U \cap H(t,\cdot)^{-1}(V)$ with inverse $H(t,\cdot)$ for each $t \in I_0$.

Proof. The implication (a)⇒(b) follows from Proposition 8.35. The implication (c)⇒(a) is trivial. Hence, we have to prove (b)⇒(c). Note that the last part of (c) follows automatically from the first part, since $H(t,\cdot) \circ G(t,\cdot) = \mathrm{id}_V$ and $H_0 \circ J = \mathrm{id}_{I_0 \times V}$ by the definition of G, H_0, and J.

We show now that it suffices to prove the first part of (c) for the special case that $W = I \times \Omega$ with some open neighborhood $\Omega \subseteq X$ of x_0, $X = E_X$, $Y = E_Y$, and $d_X H(t,x) \in \mathrm{Iso}(X,Y)$ for all $(t,x) \in W$.

To see this, note that there are open neighborhoods $I_1 \subseteq I$ of t_0 and $\Omega \subseteq X$ of x_0 with $I_1 \times \Omega \subseteq W$ and such that there is a chart c_X of X defined on Ω and a chart c_Y of Y defined on $H(I_1 \times \Omega)$. Since H is a generalized partial C^r homotopy, we obtain that the restriction of $H_0(t,u) := c_Y(H(t, c_X^{-1}(x)))$ to $W_0 := I_1 \times c_X(\Omega)$ is a generalized partial C^r homotopy on a neighborhood of $(t_0, c_X(x_0))$. Moreover, from (8.8), we obtain for $u_0 := c_X(x_0)$ that $d_X H_0(t_0,u_0) \in \mathrm{Iso}(E_X, E_Y)$. Proposition 6.10 implies that we can assume, shrinking I_1 and Ω if necessary that H is a generalized partial C^n homotopy on W_0, and $d_X H_0(W_0) \subseteq \mathrm{Iso}(E_X, E_Y)$. Using the special case $X = E_X$ and $Y = E_Y$ with H_0 (with I replaced by I_1 and W by W_0) we obtain open neighborhoods $I_0 \subseteq I_1$ of t_0, $U_0 \subseteq c_X(\Omega)$ of u_0, and $V_0 \subseteq Y$ of $v_0 := c_Y(y_0)$ such that for each $(t,v) \in I_0 \times V_0$ there is exactly one $x = G_0(t,v) \in U_0$ with $(t,x) \in H_0^{-1}(v)$, and G_0 is a generalized partial C^r homotopy satisfying

$$d_Y G_0(t,y) = (d_X H_0(t, G_0(t,y)))^{-1}.$$

We must have $V_0 \subseteq H(I_1 \times \Omega)$, and so $V := c_Y^{-1}(V_0)$ and $U := c_X^{-1}(U_0)$ have the required properties with $G(t,y) = c_X^{-1}(G_0(t, c_Y(y)))$

Now we prove the mentioned special case $W = I \times \Omega$, $X = E_X$, $Y = E_Y$ and $d_X H(W) \subseteq \mathrm{Iso}(E_X, E_Y)$. Let $L \in (0,1)$ be fixed. Let $I_0 \subseteq I$ be a neighborhood of t_0 and $\rho > 0$ such that $K_\rho(x_0) \subseteq \Omega$ and

$$\|d_X H(t,x) - d_X H(t_0,x_0)\| \le L / \|d_X H(t_0,x_0)^{-1}\| \text{ for all } (t,x) \in I_0 \times K_\rho(x_0).$$

Shrinking I_0 if necessary, we find some $\varepsilon > 0$ such that, for all $(t,y) \in I_0 \times K_\varepsilon(y_0)$,

$$\rho_{t,y} := (1-L)^{-1} \|d_X H(t_0,x_0)^{-1}\| (\|y - y_0\| + \|y_0 - H(t,x_0)\|) < \rho.$$

We show that $U := B_\rho(x_0)$ and $V := B_\varepsilon(y_0)$ have the required properties.

To see this, we define for $(t,y) \in I_0 \times V$ the auxiliary function

$$F_{t,y}(x) := x + d_X H(t_0,x_0)^{-1}(y - H(t,x))$$

on $K_\rho(x_0)$, noting that the points $x = G(t,y)$ satisfying $(t,x) \in H^{-1}(y)$ are exactly the fixed points of $F_{t,y}$. We have

$$\begin{aligned} dF_{t,y}(x) &= \mathrm{id}_X - d_X H(t_0,x_0)^{-1} d_X H(t,x) \\ &= d_X H(t_0,x_0)^{-1}(d_X H(t_0,x_0) - d_X H(t,z)), \end{aligned}$$

and so Theorem 8.2 implies that $F_{t,y}\colon K_r(x) \to X$ is Lipschitz with constant L. Moreover,

$$(1 - L)^{-1} \|F_{t,y}(x_0) - x_0\| \leq \rho_{t,y} < r.$$

Theorem 3.10 implies that $F_{t,y}$ has exactly one fixed point in $K_r(x_0)$ which, moreover, belongs to $K_{\rho_{t,y}}(x_0) \subseteq B_r(x_0)$. Thus, $G\colon I_0 \times V \to U$ is well-defined as in the assertion (c), and $G(t, y) \in K_{\rho_{t,y}}(x_0)$. Since $\rho_{t,y} \to 0$ as $(t, y) \to (t_0, y_0)$, we obtain that G is continuous at (t_0, y_0). Putting $U_0 := U \cap H(t_0, \cdot)^{-1}(V)$, we have $G(t_0, \cdot) = H(t_0, \cdot)|_{U_0}^{-1}$, and Theorem 8.7 implies that $G(t_0, \cdot)$ is differentiable at y_0 and that (8.10) holds at least with $(t, y) = (t_0, y_0)$.

Since $d_X H(I_0 \times V) \subseteq \operatorname{Iso}(E_X, E_Y)$, we can apply the partial assertion which we proved so far when we replace (t_0, y_0) by any other $(t, y) \in I_0 \times V$. We obtain for every $(t, y) \in I_0 \times V$ that G is continuous at (t, y), $G(t, \cdot)$ is differentiable at y, and the derivative satisfies (8.10). Since $J \mapsto J^{-1}$ is continuous by (6.4), and since $d_X H$ is continuous and also (as we have just shown) G is continuous on $I_0 \times V$, we obtain from (8.10) that $d_Y G$ is continuous, and so G is a generalized partial C^1 homotopy. Proposition 8.35 implies that, since $H(t, \cdot)$ is of class C^r, also the inverse $G(t, \cdot)$ is of class C^r. □

We point out that Theorem 8.40 is more general than the one which is usually found in text books where it is usually assumed that H is even C^r with respect to both variables. In the latter case, one obtains of course a stronger conclusion in view of our pointwise implicit function theorem (Theorem 8.8). For completeness, we sketch how to derive this result from our previous ones:

Theorem 8.41 (Implicit Function). *Let in Theorem* 8.40 *also I be a Banach manifold of class C^r and $H \in C^r(W, Y)$. Then the function G in Theorem* 8.40(c) *is C^r, and also $d_I G(t, y)$ exists and satisfies*

$$d_I G(t, y) = -d_X H(t, G(t, y))^{-1} d_I H(t, y). \tag{8.11}$$

Proof. To see that G is of class C^r if Theorem 8.40(b) holds, we apply Theorem 8.39 with the auxiliary map $F(t, x) := (t, H(t, x))$ whose inverse is given by $F^{-1}(t, y) := (t, G(t, y))$: An analogous calculation as in the proof of Theorem 8.8 shows by the operator matrix representation that

$$dF(t, x) \in \operatorname{Iso}(T_t I \times T_x X, T_t I \times T_{H(t,x)} Y) = \operatorname{Iso}(T_{(t,x)}(I \times X), T_{F(t,x)}(I \times Y))$$

if $d_X H(t, x) \in \operatorname{Iso}(T_x X, T_{H(t,x)} Y)$. Hence, Theorem 8.39 implies that F^{-1} is of class C^r which implies that G is of class C^r. The formula (8.11) follows from (8.9) and from the operator matrix representation of $dF(t, x)$ analogously to the proof of Theorem 8.8. □

We will now prepare corresponding results for Banach manifolds with boundaries. Let us first introduce the latter.

Definition 8.42. Any $f \in E^*$ defines a *closed halfspace* $E_+ \subseteq E$ by $E_+ := \{x \in E : \operatorname{Re} f(x) \geq 0\}$.

It is formally easier for us to allow also $f = 0$ in the above definition, although then the notion "halfspace" is slightly misleading.

Note that $\partial E_+ = \{x \in E : \operatorname{Re} f(x) = 0\}$ if $f \neq 0$. We need to define derivatives of maps on ∂E_+:

Proposition 8.43. *Let $U \subseteq E$ and $F \colon E_+ \cap U \to Y$. Suppose that $x_0 \in \partial E_+ \cap \mathring{U}$. If F has an extension $F \colon U \to Y$ which is Gateaux differentiable at x_0 then $dF(x_0)$ is independent of this extension.*

Proof. From the definition of the Gateaux derivative, we find that $dF(x_0)h$ is uniquely determined for all $h \in E_+$. Moreover, for $h \in E \setminus E_+$, we have $-h \in E_+$, and thus also $dF(x_0)h = -dF(x_0)(-h)$ is uniquely determined. □

Proposition 8.43 implies that if an extension of F exists, we can speak about $dF(x_0)$ uniquely.

Definition 8.44. A *manifold with boundary* is defined analogously to a manifold with the difference that the charts map not onto open subsets of E but onto open subsets of closed halfspaces of E. The C^n smoothness of $c_j \circ c_i^{-1}$ is understood in the sense that this map is required to map the boundary parts of the corresponding halfspaces onto each other and has an extension to an invertible C^n-map of an open subset of E with a C^n inverse. The boundary ∂X of X consists by definition of all points which are mapped by a chart onto a boundary point of a halfspace.

Remark 8.45. In case $n \geq 1$, the inverse function theorem implies that interior points of halfspaces have open neighborhoods which are mapped by $c_j \circ c_i^{-1}$ onto open subsets of E so that the assumption that the boundary points of halfspaces are mapped onto the corresponding boundary points of halfspaces is satisfied automatically. In case of finite-dimensional manifolds, the same holds also for $n = 0$, but we will prove this only much later (Theorem 9.94).

Remark 8.46. In case $\mathbb{K} = \mathbb{R}$, the boundary ∂X is a Banach manifold over $\mathrm{N}(f)$ for $f \in E^* \setminus \{0\}$.

Definition 8.47. In Definition 8.44, we understand $d(c_j \circ c_i^{-1})$ also for points from $c_i(\partial X)$ in the sense of Proposition 8.43. With this obvious extension, we define the tangent bundle TX analogously as before. In particular, T_x is defined even if $x \in \partial X$.

The main reason why we need manifolds with boundaries is the following example:

Example 8.48. If X is a C^n-manifold (without boundary) then $[0, 1] \times X$ becomes in a natural way a C^n-manifold with boundary $(\{0\} \times X) \cup (\{1\} \times X)$ (in case $\mathbb{K} = \mathbb{R}$).

We recall that X is assumed in this section to be a C^1 manifold over a Banach space E_X. We now also allow that X is a C^1 manifold over E_X with boundary.

Definition 8.49. If X is of class C^n and $\Omega \subseteq X$ is open then a map $F \colon \Omega \to Y$ is of class C^n (and we write $F \in C^n(\Omega, Y)$ or $F \in C^n$) if for any chart c_X and c_Y of X and Y the composition $c_Y \circ F \circ c_X^{-1}$ is C^n (where defined), where the latter means on the boundary of a halfspace that this map has a C^n-extension defined on an open subset of E. In the sense of Proposition 8.43 we define then $dF(x) \in \mathcal{L}(T_x, T_y)$ even if $x \in \partial X$.

In case $n \geq 1$ and if each of the derivatives of these maps in every point of the halfspace (including the boundary) belongs to $\mathcal{L}_k(E, Y)$, we call F *Fredholm* of *index* k and write $F \in \mathcal{F}_k(\Omega, Y)$.

Remark 8.50. In view of Theorem 6.40, we can assume for $F \in \mathcal{F}_k(\Omega, Y)$ that the extension of $c_Y \circ F \circ c_X^{-1}$ to the open subset of E is Fredholm of index k.

Theorem 8.51 (Implicit Function with Boundary). *The assertion of Theorem* 8.41 *remains valid if* I *is a manifold with boundary.*

Proof. Note that we already know by Theorem 8.40 that G is a homeomorphism, that is, we really only have to prove that G is of class C^r and satisfies (8.11). However, the latter follows as in Theorem 8.41 by just extending the considered functions. □

8.5 Transversal Submanifolds

Throughout this section, we assume that X and Y are C^1 Banach manifolds over Banach spaces E_X and E_Y over the field $\mathbb{K} = \mathbb{R}$ or $\mathbb{K} = \mathbb{C}$. The Banach manifold X might have a boundary, but Y is assumed to be without boundary.

Definition 8.52. A subset $X_0 \subseteq X$ is called a *submanifold of dimension m* of class C^r $(0 \leq r \leq \infty)$ if for each $x \in X_0$ there is an open neighborhood $U \subseteq X$ of x, a C^r-diffeomorphism c of U onto an open subset of E_X, and a complemented subspace $E_0 \subseteq E_X$ of dimension m such that $X_0 \cap U = c^{-1}(E_0)$. We call such a couple (c, E_0) a *submanifold chart* for x.

A subset $X_0 \subseteq X$ is a *submanifold with boundary* of dimension m, if for each $x \in X_0$ there is an open neighborhood $U \subseteq X$ of x, a diffeomorphism c of U onto an open subset of E, a complemented subspace $E_0 \subseteq E$ of dimension M and a halfspace $E_{0,+} \subseteq E_0$ of E_0 such that $X_0 \cap U = c^{-1}(E_{0,+})$.

Note that we do *not* define infinite-dimensional submanifolds, and for this reason we do not have to care whether E_0 is dependent on x or not: Every submanifold of dimension m can be considered as a manifold over $\mathbb{K}^m$ in the following manner.

Given a submanifold X_0 of X of dimension m and if (c, E_0) is a submanifold chart, we compose the maps $c|_{X_0 \cap U}$ with an isomorphism $J_c \in \mathrm{Iso}(E_0, \mathbb{K}^m)$. Then X_0 becomes a manifold over $\mathbb{K}^m$ of class C^r whose atlas consists of all these compositions $c_0 := J_c \circ c|_{X_0 \cap U}$. Concerning the tangent bundle, we note that the definition of TX changes only formally when we add the map c to the atlas of X for a moment.

We do this only to define h_c correspondingly according to Definition 8.20; once h_c is known, we can return to our previous atlas of X. Now it is clear that

$$I_{X_0}(x, \cdot) := h_c(x, \cdot) \circ (h_{c_0}(x, \cdot) \circ J_c)^{-1} \in \mathrm{Iso}(T_x X_0, T_x X),$$

where h_{c_0} denotes the trivialization of TX_0 according to Definition 8.20.

Proposition 8.53. *I_{X_0} is independent of the particular choice of c and J_c, hence it defines a map $I_{X_0}\colon TX_0 \to TX$.*

Proof. We fix $x \in U$. Putting $g := h_{c_0}(x, \cdot) \circ J_c$, we calculate $g(x,u) = (x, [(c_0, J_c u)]_{x,X_0})$, where the notation should indicate that the equivalence relation has to be understood in X_0. Hence,

$$I_{X_0}(x, [(J_c \circ c|_{X_0 \cap U}, J_c u)]_{x,X_0}) = (x, [(c,u)]_{x,X}).$$

Now if $\hat{c}$ and $\hat{J}$ are possibly different choices, we have with $\hat{u} := d(\hat{c} \circ c^{-1})(c(x))u$ that $[(\hat{c}, \hat{u})]_{x,X} = [(c,u)]_{x,X}$, and so we are to show that

$$[(\hat{J} \circ \hat{c}|_{X_0 \cap U}, \hat{J}\hat{u})]_{x,X_0}) = [(J_c \circ c|_{X_0 \cap U}, J_c u)]_{x,X_0}.$$

This means, we are to show that

$$d(\hat{J} \circ \hat{c}|_{X_0 \cap U} \circ c|_{X_0 \cap U}^{-1} \circ J_c^{-1})(J_c c(x)) J_c u = \hat{J} \hat{u}.$$

By the chain rule, the left-hand side becomes

$$\hat{J} d(\hat{c}|_{X_0 \cap U} \circ c|_{X_0 \cap U}^{-1})(c(x))u = \hat{J} d(\hat{c} \circ c^{-1})(c(x))u,$$

and this is $\hat{J}\hat{u}$, as required. □

Since I_{X_0} is a homeomorphism onto its range and $I_{X_0}(x, \cdot)$ is an isomorphism, we can identify TX_0 with its image under I_{X_0}. We make this identification from now on, hence understanding TX_0 as a subset of TX.

With this identification, $T_x X_0$ is a subspace of $T_x X$. Note that the trivializations of the tangent bundle TX_0 are by definition of I_{X_0} given by

$$h_{c_0}(x, \cdot) = I_{X_0}(x, \cdot)^{-1} \circ h_c(x, \cdot) \circ J_c^{-1},$$

where we still use the above notation. In the sense of the identification, we can thus say that the trivializations are given by

$$\hat{h}_c(x, \cdot) = h_c(x, \cdot) \circ J_c^{-1},$$

where (c, E_0) is a submanifold chart and $J_c \in \mathrm{Iso}(E_0, \mathbb{K}^m)$. In particular, in the sense of the identification, we have

$$T_x X_0 = h_c(\{x\} \times E_0).$$

Definition 8.54. Let $\Omega \subseteq X$ be open, $F \in \mathcal{F}_k(\Omega, Y)$, and $Y_0 \subseteq Y$ be a submanifold. Then Y_0 is called *transversal* to F on $M \subseteq X$ if for each $x \in M \cap F^{-1}(Y_0)$ the subspace $T_{F(x)} Y_0 \subseteq T_{F(x)} Y$ is transversal to $dF(x) \in \mathcal{L}_k(T_x X, T_{F(x)} Y)$.

In finite-dimensional spaces, it is mathematical folklore that the following result is essentially a reformulation of the inverse function theorem, see e.g. [77, Chapter 1, Section 3] or [1, Theorem 1.7 and §17]. We have chosen the definitions such that essentially the same results hold in infinite-dimensional spaces. Note that [1, §17] treats also infinite-dimensional spaces, but e.g. the definition of transversality is slightly different, and it seems that it is much harder to verify in case of Fredholm operators than our definition. Moreover, the case of manifolds with boundary is not treated in [1].

Theorem 8.55 (Transversality). *Let $F \in \mathcal{F}_k(X,Y) \cap C^r(X,Y)$ where X and Y are C^r manifolds ($1 \le r \le \infty$), Y without boundary. Let $Y_0 \subseteq Y$ be a submanifold of class C^r of dimension m without boundary which is transversal to F on an open subset $M \subseteq X$. Let $X_0 := M \cap F^{-1}(Y_0) \neq \varnothing$.*

If X_0 does not contain a point from ∂X then $k+m \ge 0$, and X_0 is a submanifold of X of dimension $k+m$ of class C^r without boundary.

If X_0 contains a point from ∂X, assume in addition that $\mathbb{K} = \mathbb{R}$ and that Y_0 is transversal to $F|_{\partial X} \in C^r(\partial X, Y)$ on ∂X. Then $k+m \ge 1$, and X_0 is a submanifold of dimension $k+m$ of class r with boundary $\partial X_0 = X_0 \cap \partial X$.

In both cases, $F_0 := F|_{X_0} \in \mathcal{F}_k(X_0, Y_0)$, $T_x X_0 = dF(x)^{-1}(T_{F(x)} Y_0)$ in the sense of the above described identification of the tangent spaces, and $dF_0(x) = dF(x)|_{T_x X_0}$.

Proof. Let $x_0 \in X_0$, $V_0 \subseteq Y$ be open, and (c_Y, Z_0) be a submanifold chart for $y := F(x_0)$, $c_Y\colon V_0 \to E_Y$, of the submanifold $Y_0 \subseteq Y$. Without loss of generality, we assume $c_Y(y) = 0$. Let h_{c_Y} denote the trivialization of Y described above. By hypothesis, the subspace $T_y Y_0 = h_{c_Y}(\{y\} \times Z_0)$ is transversal to $dF(x_0)$.

Let $c\colon U_0 \to E_X$ be a chart for x_0, without loss of generality $c(x_0) = 0$. Replacing U_0 by $U_0 \cap M \cap F^{-1}(V_0)$ if necessary, we can assume that $U_0 \subseteq M$ and $F\colon U_0 \to V_0$. Then $G := c_Y \circ F \circ c^{-1}$ is by hypothesis a C^r Fredholm map from the open neighborhood $U := c(U_0)$ of $0 = c(x_0)$ into E_Y with $G(0) = 0$ (in case of $x_0 \in \partial X$ we consider an extension of G). In particular, $A := dG(0) \in \mathcal{L}_k(E_X, E_Y)$. Since $T_y Y_0$ is transversal to $dF(x_0)$, we obtain from (8.8) that $Z_0 = h_{c_Y}(x_0, \cdot)^{-1}(T_y Y_0)$ is transversal to A. Applying Proposition 6.43, we find that $E_0 := A^{-1}(Z_0)$ has dimension $m + k \ge 0$, $A_0 := A|_{E_0} \in \mathcal{L}_k(E_0, Z_0)$, and that there are closed subspaces $E_1 \subseteq E_X$ and $Z_1 \subseteq E_Y$ with $E_X = E_0 \oplus E_1$, $E_Y = Z_0 \oplus Z_1$ and such that $A_1 := A|_{E_1} \in \operatorname{Iso}(E_1, Z_1)$.

By Proposition 6.18, there is a projection $P \in \mathcal{L}(E_X)$ with $\mathrm{R}(P) = E_0$ and $\mathrm{N}(P) = E_1$, and $Q := \mathrm{id}_{E_X} - P$ is the unique projection with $\mathrm{R}(Q) = E_1$ and $\mathrm{N}(Q) = E_0$. There is also a projection $S \in \mathcal{L}(E_Y)$ with $\mathrm{R}(S) = Z_1$ and $\mathrm{N}(S) = Z_0$. Then $Q_1 := A_1^{-1} S A \in \mathcal{L}(E_X)$ is a projection with $\mathrm{R}(Q_1) = E_1$ and $E_0 \subseteq \mathrm{N}(Q_1)$. Since $E_X = E_0 \oplus E_1$, it follows that $E_0 = \mathrm{N}(Q_1)$ and thus $Q_1 = Q$, that is, $A_1^{-1} S A = Q$.

We define $J \in C^r(U, E_X)$ by $J(u) := Pu + A_1^{-1} S G(u)$. Then $J(0) = 0$ and $dJ(0) = P + A_1^{-1} S A = P + Q = \mathrm{id}_{E_X} \in \operatorname{Iso}(E_X)$. Shrinking U_0 and thus $U = c(U_0)$ if necessary, we find by the inverse function theorem for Banach spaces that J is a diffeomorphism onto an open neighborhood $V_1 \subseteq c_Y(V_0)$ of $0 = c_Y(y)$. For $u = c(x) \in U$, we have

$$x \in X_0 \iff c_Y(F(x)) \in Z_0 \iff G(u) \in Z_0 \iff J(u) \in E_0.$$

Hence, if $x_0 \notin \partial X$, a required submanifold chart for x_0 is given on U_0 by $(\hat{c}, E_0)$ with $\hat{c} = J \circ c^{-1}$.

Concerning the last assertion, we note that with h_c as in Definition 8.20, we have

$$\begin{aligned} dF(x_0) &= h_{c_Y}(F(x_0), \cdot) \circ dG(0) \circ h_c(x_0, \cdot)^{-1} \\ &= h_{c_Y}(F(x_0), \cdot) \circ A \circ h_c(x_0, \cdot)^{-1} \in \mathcal{L}(T_{x_0}X, T_yY). \end{aligned} \tag{8.12}$$

Hence,

$$dF(x_0)^{-1}(T_yY_0) = h_c(\{x_0\} \times A^{-1}(Z_0)) = h_c(\{x_0\} \times E_0) = T_{x_0}X_0.$$

We put now $G_0 := c_Y \circ F_0 \circ (\hat{c}|_{X_0})^{-1} = G|_{E_0}$. Then $G_0 \in C^r(E_0, Z_0)$, and so $F_0 \in C^r(X_0, Y_0)$. Moreover, $dG_0(0) = dG(0) \circ \mathrm{id}_{E_0} = A_0 \in \mathcal{L}_k(E_0, Z_0)$, and so $dF_0(x_0) \in \mathcal{L}_k(X_0, Y_0)$. Since Proposition 8.21 implies similarly as above that

$$\begin{aligned} dF_0(x_0) &= h_{c_Y}(y, \cdot) \circ dG_0(v_0) \circ h_{\hat{c}}(x_0, \cdot)^{-1} \\ &= h_{c_Y}(y, \cdot) \circ A_0 \circ h_{\hat{c}}(x_0, \cdot)^{-1} \in \mathcal{L}(T_xX_0, T_yY_0), \end{aligned}$$

we obtain by comparison with (8.12) that $dF_0(x_0) = dF(x_0)|_{T_{x_0}X_0}$. Since $x_0 \in X_0$ was arbitrary, we thus have shown the last assertion.

If $x_0 \in \partial X$ (hence $\mathbb{K} = \mathbb{R}$), we work throughout with extensions of the considered maps, thus assuming that U is open in E_X. Applying first what we proved so far with the manifold ∂X, we find that $U_0 \cap X_0 \cap \partial X$ is a submanifold of ∂X of dimension $m + k - 1 \geq 0$. This is a submanifold of X. Now using the above construction in X, we find that this submanifold is mapped by the diffeomorphism $\hat{c}$ onto a submanifold M_0 of E_0 of the same dimension $m + k - 1$.

Using a chart of the submanifold M_0 and shrinking U if necessary, we find that there is a C^r-diffeomorphism J_0 of $E_0 \cap U$ onto an open subset of E_0 such that $J_0(M_0 \cap U)$ is an open subset of a linear subspace $E_{0,0} \subseteq E_0$ of dimension $m + k - 1 = \dim E_0 - 1$, and $J_0(0) = 0$.

Then $J_1 := Q + (J_0 \circ P)$ is C^r and satisfies $dJ_1(0) = Q + dJ_0(0)P$, in particular $dJ_1(0)|_{E_0} = dJ_0(0) \in \mathrm{Iso}(E_0)$ and $dJ_1(0)|_{E_1} \in \mathrm{Iso}(E_1)$. Proposition 6.21 implies that $dJ_1(0) \in \mathrm{Iso}(E_X)$, and by the implicit function theorem, we can thus assume (shrinking U if necessary) that $J_1: U \to E_X$ is a C^r-diffeomorphism onto an open subset of E_X, $J_1|_{E_0 \cap U} = J_0|_{E_0 \cap U}$, and $J_1^{-1}(E_0) = E_0 \cap U$.

We thus have constructed a C^r-diffeomorphism $\varphi := J_1 \circ \hat{c}$ with $\varphi^{-1}(E_0) = X_0 \cap U_0$ and $\varphi^{-1}(E_{0,0}) = X_0 \cap U_0 \cap \partial X$. There is $f \in E_0^* \setminus \{0\}$ with $E_{0,0} = \mathrm{N}(f)$. Shrinking U_0 if necessary, we can assume that $I := U_0 \setminus \partial X$ is connected, and so also $(f \circ \varphi)(I) \subseteq \mathbb{R} \setminus \{0\}$ is connected. Replacing f by $-f$ if necessary, we thus can assume without loss of generality that $(f \circ \varphi)(I) \subseteq (0, \infty)$. Putting $E_{0,+} := \{x \in E_0 : f(x) \geq 0\}$, we have then $U_0 = \varphi^{-1}(E_{0,+})$ and $U_0 \cap \partial X = \varphi^{-1}(\partial E_{0,+})$. This shows that φ is the required submanifold chart, and $U_0 \cap X_0 \cap \partial X = U_0 \cap \partial X_0$. □

The special case $Y_0 = \{y\}$ of Theorem 8.55 deserves a separate formulation. In [77], Theorem 8.55 is even reduced to this special case (in the finite-dimensional setting), but we have chosen to prove Theorem 8.55 independently, since it seems that in the infinite-dimensional setting this reduction would be more complicated than our above proof of Theorem 8.55.

Definition 8.56. A point $x \in X$ is called a *regular point* of $F \in C^1(X, Y)$ if $dF(x) \in \mathcal{L}(T_x X, T_{F(x)} Y)$ is onto $T_{F(x)} Y$; otherwise, x is a *critical point.* A point $y \in Y$ is called a *regular value* of $F \in C^1(X, Y)$ if each point of $F^{-1}(y)$ is regular, otherwise *critical value.*

In other words: y is regular value of F if and only if $\{y\}$ is transversal to F on X.

Corollary 8.57. *Let $F \in \mathcal{F}_k(X, Y)$. Let $y \in Y$ be a regular value of F, and $X_0 := F^{-1}(y) \neq \varnothing$. If X_0 does not contain a point from ∂X then $k \geq 0$, and X_0 is a submanifold of X of dimension k without boundary, and $T_x X_0 = \mathrm{N}(dF(x))$ $(x \in X_0)$.*

If X_0 contains a point from ∂X, we assume in addition that $\mathbb{K} = \mathbb{R}$ and that y is a regular value for $F|_{\partial X} \in C^1(\partial X, Y)$. Then $k \geq 1$, and X_0 is a submanifold of dimension k with boundary $\partial X_0 = X_0 \cap \partial X$, and $T_x X_0 = \mathrm{N}(dF(x))$ $(x \in X_0)$.

Proof. This is the special case $Y_0 = \{y\}$ of Theorem 8.55 □

Remark 8.58. In case $k = 0$ the statement that $F^{-1}(y)$ is a submanifold of dimension 0 means that $F^{-1}(y)$ is discrete, that is, each point $x \in F^{-1}(y)$ has an open neighborhood which is disjoint from $F^{-1}(y) \setminus \{x\}$.

Of course, this special case could have been obtained much more straightforwardly by the inverse function theorem.

8.6 Parameter-Dependent Transversality and Partial Submanifolds

We define the notion of transversality also for generalized Fredholm homotopies in the obvious manner:

Definition 8.59. Let I be a topological space, X and Y be a Banach manifolds, $W \subseteq I \times X$ be open, and $H: W \to Y$ be a generalized Fredholm homotopy. Then a submanifold $Y_0 \subseteq Y$ is *transversal* to H on $M \subseteq W$ if Y_0 is transversal to $H(t, \cdot)$ on $M_t := \{x : (t, x) \in M\}$ for every $t \in I$.

The remainder of this section serves mainly as a preparation of tools needed to prove the homotopy invariance of the Benevieri-Furi coincidence degree. The reader not interested in this proof (or confined with the proof for C^1 homotopies) may skip this section (or read only Remark 8.64, respectively).

The main observation is that there is also a parameter-dependent version of Theorem 8.55 for Fredholm homotopies. We formulate it only for manifolds X without boundaries.

Theorem 8.60. *Let X and Y be C^r manifolds without boundary with $1 \le r \le \infty$. Let I be a topological space, $W \subseteq I \times X$ be open, and $H: W \to Y$ be a generalized Fredholm homotopy of index k. Let $Y_0 \subseteq Y$ be a submanifold of class C^r of dimension m without boundary which is transversal to H on W. Let $X_t := H(t, \cdot)^{-1}(Y_0)$ and $x_0 \in X_{t_0}$. Then $k + m \ge 0$, and there are open neighborhoods $I_0 \subseteq I$ of t_0, $U \subseteq X$ of x_0, an open neighborhood $V \subseteq E_X$ of 0, a complemented subspace $E_0 \subseteq E_X$ of dimension $k + m$, and a generalized partial C^r homotopy $H_0: I_0 \times V \to X$ with the following properties.*

$H_0(t_0, 0) = x_0$ and $H_0(\{t\} \times (E_0 \cap V)) \cap U = X_t \cap U$ for all $t \in I_0$. $H_0(t, \cdot)$ is a C^r-diffeomorphism onto an open subset of X, and $H_1(t, x) := (t, H_0(t, x))$ is a homeomorphism onto an open subset of W. The inverse of H_1 is also a generalized partial C^r homotopy. If I is a manifold (with or without boundary) of class C^q $(1 \le q \le r)$ and if H is of class C^q then H_1 is even a C^q-diffeomorphism.

Proof. Let (c_Y, Z_0) be a submanifold chart for $y := H(t_0, x_0)$ where $c_Y: V_0 \to E_Y$ with some open $V_0 \subseteq E_Y$. Without loss of generality, let $c_Y(y) = 0$. Let h_{c_Y} denote the trivialization of Y described after Proposition 8.53. Since Y_0 is transversal of H on W, the subspace $T_y Y_0 = h_{c_Y}(\{y\} \times Z_0)$ is transversal to $d_X H(t_0, x_0)$.

Let $U_0 \subseteq X$ be an open neighborhood of x_0 and $c: U_0 \to E_X$ be a chart of X, without loss of generality $c(x_0) = 0$. Shrinking U_0 if necessary, we can assume that there is an open neighborhood $I_0 \subseteq I$ of t_0 such that $I_0 \times U_0 \subseteq W$ and $H(I_0 \times U_0) \subseteq V_0$. Then $G(t,u) := c_Y(H(t,c^{-1}(u)))$ is by hypothesis a partial Fredholm homotopy of index k from the open neighborhood $I_0 \times c(U_0)$ of $(t_0, 0)$ into E_Y with $G(t_0, 0) = 0$. In particular, $A := d_X G(t_0, 0) \in \mathcal{L}_k(E_X, E_Y)$. By (8.8), we have that $Z_0 = h_{c_Y}(x_0, \cdot)^{-1}(T_y Y_0)$ is transversal to A. Applying Proposition 6.43, we find with $E_0 := A^{-1}(Z_0)$ that $A_0 := A|_{E_0} \in \mathcal{L}_k(E_0, Z_0)$, and that there are closed subspaces $E_1 \subseteq E_X$ and $Z_1 \subseteq E_Y$ with $E_X = E_0 \oplus E_1$, $E_Y = Z_0 \oplus Z_1$ and such that $A_1 := A|_{E_1} \in \operatorname{Iso}(E_1, Z_1)$.

By Proposition 6.18, there is a projection $P \in \mathcal{L}(E_X)$ with $\mathrm{R}(P) = E_0$ and $\mathrm{N}(P) = E_1$, and $Q := \mathrm{id}_{E_X} - P$ is the unique projection with $\mathrm{R}(Q) = E_1$ and $\mathrm{N}(Q) = E_0$. There is also a projection $S \in \mathcal{L}(E_Y)$ with $\mathrm{R}(S) = Z_1$ and $\mathrm{N}(S) = Z_0$. Then $Q_1 := A_1^{-1} S A \in \mathcal{L}(E_X)$ is a projection with $\mathrm{R}(Q_1) = E_1$ and $E_0 \subseteq \mathrm{N}(Q_1)$. Since $E = E_0 \oplus E_1$, it follows that $E_0 = \mathrm{N}(Q_1)$ and thus $Q_1 = Q$, that is, $A_1^{-1} S A = Q$.

We define a partial C^r homotopy $G_0 \colon I_0 \times c(U_0) \to E_X$ by $G_0(t,u) := Pu + A_1^{-1} S G(t,u)$. Then $d_X G_0(t_0, 0) = P + d_X G(t_0, 0) = P + A_1^{-1} S A = P + Q = \mathrm{id}_{E_X} \in \operatorname{Iso}(E_X)$. Shrinking I_0 and U_0 if necessary, we find by the partial implicit function theorem (Theorem 8.40) that there is a neighborhood $V \subseteq E_X$ of 0 and a uniquely defined partial C^1 homotopy $J \colon I_0 \times V \to c(U_0)$ satisfying $G_0(t, J(t,v)) = v$ such that $\hat{H}_1(t,x) := (t, J(t,v))$ is a homeomorphism onto an open subset of $I \times E_X$ and $J(t, \cdot)$ is a diffeomorphism onto an open subset of E_X. Theorem 8.51 implies that $\hat{H}_1$ is even a C^q-diffeomorphism if I is a manifold of class C^q and H and thus G and G_0 are of class C^q.

Shrinking I_0 if necessary, we can assume that there is some open neighborhood $U_1 \subseteq c(U_0)$ of 0 with $G_0(I_0 \times U_1) \subseteq V$. For $(t,u) = (t, c(x)) \in I_0 \times U_1$, we calculate

$$\begin{aligned} x \in X_t &\iff c_Y(H(t,x)) \in Z_0 \iff G(t,u) \in Z_0 \\ &\iff G_0(t,u) \in E_0 \iff u \in J(\{t\} \times (E_0 \cap V)). \end{aligned}$$

Hence, $c^{-1}(J(\{t\} \times (E_0 \cap V)) \cap U_1) = X_t \cap c^{-1}(U_1)$. The assertion thus follows with $U := c^{-1}(U_1)$ and $H_0 := c^{-1} \circ J$. □

Theorem 8.60 motivates the following definition which we will extend in a moment to submanifolds:

Definition 8.61. Let I be a topological space, and $Z \subseteq I \times X$. For $t \in I$, let $X_t \subseteq X$ be manifolds of class C^r $(0 \le r \le \infty)$, $Z_t := \{t\} \times X_t$, and $Z = \bigcup_{t \in [0,1]} Z_t$.

We call Z a *partial C^r manifold* over a Banach space E if it is equipped with a family of maps (the *partial charts*) such that for each $(t_0, x_0) \in Z$ there is a partial chart $c: U \to I \times E$ defined on an open neigbhorhood $U \subseteq Z$ of (t_0, x_0) such that c has the form $c(t,x) = (t, c_0(t,x))$ where $c_0(t,\cdot): U \cap X_t \to E$ is the restriction of a chart of X_t. Moreover, we require that for each two charts c_1, c_2 the composition $c_2 \circ c_1^{-1}$ is a generalized partial C^r homotopy.

In case $r \geq 1$, the *partial tangent bundle* $T_X Z$ over E of such a partial manifold is defined as the union $\bigcup_{(t,x)\in Z} T_x Z_t$ where the topology is chosen such for each chart c the trivializations are given by $H_c((t,x),u) := (t, h_{c(t,\cdot)}(u))$ where $h_{c(t,\cdot)}$ denotes the trivialization of TX_t.

We call a continuous map $H: Z \to Y$ a *generalized partial C^r homotopy* if for each partial chart c of Z the map $H \circ c^{-1}$ is a generalized partial C^r homotopy.

Concerning submanifolds, the definition is as follows:

Definition 8.62. In the above situation, we call Z a *partial C^r submanifold* of $I \times X$ of dimension n if for each $(t_0, x_0) \in Z$ there is *partial submanifold chart* (c, E_0) for (t_0, x_0), that is: $E_0 \subseteq E_X$ is a complemented subspace of dimension n, there is an open neighborhood $U \subseteq Z$ of (t_0, x_0), $c: U \to I \times E_X$ is a homoemorphism onto an open subset of $I \times E_X$ of the form $c(t,x) = (t, c_0(t,x))$ where $(c_0(t,\cdot), E_0)$ is a submanifold chart for $U \cap X_t$ of class C^r and c_0 is a partial C^r homotopy.

Analogously to the case of ordinary submanifolds, we obtain that each partial submanifold is a partial manifold over $\mathbb{K}^n$ by defining the partial charts as $\hat{c} = (\mathrm{id}_I \otimes J) \circ c$ where (c, E_0) is a submanifold chart and $J \in \mathrm{Iso}(E_0, \mathbb{K}^n)$. Note that the finite-dimensionality of $\mathbb{K}^n$ implies automatically that for each two such partial charts $\hat{c}_i$ $(i = 1, 2)$ the map $\hat{c}_2 \circ \hat{c}_1^{-1}$ is a generalized partial C^r homotopy.

Corollary 8.63. *Let X and Y be C^r manifolds without boundary $(1 \leq r \leq \infty)$, let I be a topological space, and let $W \subseteq I \times X$ be open, and $H: W \to Y$ be a generalized Fredholm homotopy of index k, partial C^r. Let $Y_0 \subseteq Y$ be a submanifold of dimension m without boundary which is transversal to H on W. Then $Z := H^{-1}(Y_0)$ is empty or a partial C^r submanifold of $I \times X$ of dimension $k + n$.*

If in addition I is a C^q manifold of dimension n $(0 \leq q \leq r)$, possibly with boundary, and if in addition $H \in C^q(W, Y)$ then Z is also a submanifold of $I \times X$ of class C^q of dimension $k + m + n$ with boundary $\partial Z = Z \cap ((\partial I) \times X)$.

Proof. The inverse of the map H_1 of Theorem 8.60 serves as a submanifold chart at $(t_0, x_0) \in Z$. □

Remark 8.64. For the case $q = r$, Corollary 8.63 is contained in Theorem 8.55 since Y_0 is transversal to $J(t,x) := (t, H(t,x))$, and moreover, by hypothesis also transversal to the restriction of J to $\partial W = W \cap (\partial I \times X)$. However, it seems that the assertion for $q < r$ (in particular, $q = 0 < 1 \leq r$ is needed for the proof of the homotopy invariance of the Benevieri–Furi coincidence degree) cannot be obtained from Theorem 8.55 directly.

8.7 Orientation on Submanifolds and on Partial Submanifolds

Throughout this section, we assume that X and Y are Banach manifolds without boundary over Banach spaces E_X and E_Y, respectively, and that I is a topological space.

In case $k = 0$ in Theorem 8.55, we have $F_0 := F|_{X_0} \in \mathcal{F}_0(X_0, Y_0)$ and that dF_0 is the restriction of $A := dF(x)$ to $T_x X_0$. Note that $T_{F(x)} Y_0$ is transversal to A. Hence, if F is an oriented Fredholm operator, then we obtain an inherited orientation for $dF_0(x)$ for every $x \in X_0$ by Definition 7.15.

Definition 8.65. If $F \in \mathcal{F}_0(X, Y)$ in Theorem 8.55 is an oriented Fredholm operator then for $x \in X_0$ the *inherited orientation* $\sigma(x)$ of $F_0 := F|_{X_0} \in \mathcal{F}_0(X_0, Y_0)$ is defined as the inherited orientation of $dF_0(x)$ as above.

We will show in a moment that this is indeed an orientation.

However, we do this in the more general context of Fredholm homotopies, since we will need the latter for the proof of the homotopy invariance of the Benevieri-Furi coincidence degree.

Note that if $W \subseteq I \times X$ is open, and if Y_0 is transversal to a generalized Fredholm homotopy $H: W \to Y$ of index 0, then we can apply Theorem 8.55 with $F = H(t, \cdot)$. In particular, $X_t := H(t, \cdot)^{-1}(Y_0)$ is a submanifold.

Definition 8.66. If H is an oriented generalized Fredholm homotopy of index 0 then the *inherited orientation* $\sigma(t,x)$ for $H(t, \cdot)|_{X_t} \in \mathcal{F}_0(X_t, Y_0)$ is defined as in Definition 8.65 with $F = H(t, \cdot)$.

Proposition 8.67. *The inherited orientation of a Fredholm map is an orientation in the sense of Definition* 8.25. *The inherited orientation of a generalized Fredholm homotopy has the property that it is an orientation of $d_X H$, considered as a map of the partial tangent bundle $T_X H^{-1}(Y_0)$ into the tangent bundle TY.*

Proof. We show the assertion for generalized homotopies, since the assertion for maps is the special case $H(t,\cdot) = F$. For $(t_0, x_0) \in Z := H^{-1}(Y_0)$ let H_1, V, and E_0 be as in Theorem 8.60. Putting $c = H_1^{-1}$, we have that (c, E_0) is a partial submanifold chart of $T_X Z$. Moreover, c is a partial chart for $T_X W$. Then $d_X H_1(t,x) \in \mathcal{L}(E_X, T_{(t,x)}W)$ can be considered as a trivialization of $T_X Z$ (up to a composition with isomorphisms). Moreover, for $H_0 := H_1|_{V\cap E_0} \in C(V, Z)$ the map $h_1(t,x) := d_X H_0(t,x) \in \mathcal{L}(E_0, T_{(t,x)}Z)$ is a trivialization of $T_X Z$. It is crucial for us that h_1 is just a restriction of $d_X H_1$.

Similarly, let (c_Y, Z_0) denote a submanifold chart of Y_0 for $h(t_0, x_0)$, and let h_{c_Y} denote the corresponding trivialization of TY. Then a trivialization h_2 of TY_0 is just given by a restriction of h_{c_Y} similarly as above.

Let σ and σ_0 denote the original and inherited orientation, respectively. By Proposition 7.34, we know that

$$s(t,x) := h_{c_Y}(H(t,x),\cdot)^{-1} \circ \sigma(t,x) \circ d_X H_1(t,x) \subseteq \mathcal{L}(E_X, E_Y)$$

is an orientation for

$$A(t,x) := h_{c_Y}(H(t,x),\cdot)^{-1} \circ d_X H(t,x) \circ d_X H_1(t,x) \in \mathcal{L}(E_X, E_Y)$$

in a neighborhood of (t_0, x_0), and we have to prove that

$$s_0(t,x) := h_2(H(t,x),\cdot)^{-1} \circ \sigma_0(t,x) \circ dH_1(t,x) \subseteq \mathcal{L}(E_0, Z_0)$$

is lower semicontinuous at x_0. Thus, let $K_0 \in s(t_0, x_0)$, that is

$$\hat{K}_0(t,x) := h_2(H(t,x),\cdot) \circ K_0 \circ d_X H_1(t,x)^{-1}$$

satisfies $\hat{K}_0(t_0, x_0) \in \sigma_0(x)$. By Definition 7.15, there is some $K \in \sigma(t_0, x_0)$ with $\mathrm{R}(K) \subseteq T_y Y_0$ and $\hat{K}_0(t_0, x_0) = K|_{T_{x_0} X_{t_0}}$. Then

$$K_1 := h_{c_Y}(H(t_0, x_0),\cdot)^{-1} \circ K \circ d_X H_1(t_0, x_0)$$

belongs to $s(t_0, x_0)$. By Theorem 7.20, there is an open neighborhood U_0 of (t_0, x_0) with $K_1 \in s(t,x)$ for all $(t,x) \in U_0$. It follows that

$$\hat{K}(t,x) := h_{c_Y}(H(t,x),\cdot) \circ K_1 \circ h_c(x,\cdot)^{-1} \in \sigma(t,x)$$

for all $(t,x) \in U_0$. Since $\mathrm{R}(K_1) \subseteq Y_0$, we have $\mathrm{R}(\hat{K}(t,x)) \subseteq T_{H(t,x)} Y_0$. Hence, $\hat{K}(t,x)|_{T_x X_0} \in \sigma_0(t,x)$ for all $(t,x) \in U_0$ which means $K_0 \in s_0(t,x)$ for all $(t,x) \in U_0$. In particular, s_0 is lower semicontinuous at (t_0, x_0), as required. □

For regular values, the orientations obtain the natural meaning:

Proposition 8.68. *Let $\Omega \subseteq X$ be open, $F \in \mathcal{F}_0(\Omega, Y)$, and $Y_0 \subseteq Y$ a finite-dimensional submanifold which is transversal to F on Ω. Then $X_0 := F^{-1}(Y_0)$ is empty or a submanifold of the same dimension as Y_0, and $F_0 := F|_{X_0} \in \mathcal{F}_0(X_0, Y_0)$. Moreover:*

(a) *A point $x \in X_0$ is a regular point of F_0 if and only if it is a regular point of F.*

(b) *A value $y \in Y_0$ is a regular value of F_0 if and only if it is a regular value of F.*

(c) *If F is oriented on $x \in X_0$ and F_0 carries the inherited orientation then*

$$\operatorname{sgn} dF(x) = \operatorname{sgn} dF_0(x) \tag{8.13}$$

Proof. The assertions about X_0 and F_0 follow from Theorem 8.55. The Fredholm alternative (Proposition 6.33) implies that $x \in X_0$ is a regular point of F_0 or F if and only if $dF_0(x) \in \operatorname{Iso}(T_x X_0, T_{F_0(x)} Y_0)$ or $dF(x) \in \operatorname{Iso}(T_x X, T_{F(x)} Y_0)$, respectively. The latter two are equivalent by Corollary 7.17, because Theorem 8.55 implies $dF_0(x) = dF(x)|_{T_x X_0}$, and because $T_{F_0(x)} Y_0 = T_{F(x)} Y_0$ is transversal to $dF(x)$. Corollary 7.17 implies also (8.13) if $dF(x)$ is oriented and $dF_0(x)$ carries the inherited orientation. The assertion about regular values follows from the assertion about regular points and the fact that $F^{-1}(y) = F_0^{-1}(y)$ for every $y \in Y_0$. □

8.8 Existence of Transversal Submanifolds

Throughout this section, we assume that X and Y are Banach manifolds without boundary over Banach spaces E_X and E_Y, respectively, and that I is a topological space.

Note that by a closed submanifold of Y we just mean a submanifold which is closed in Y in the sense of the topology.

Proposition 8.69 (Stability of Transversality). *Let $W \subseteq I \times X$ be open, and $Y_0 \subseteq Y$ be a closed submanifold of finite dimension which is transversal to a generalized Fredholm homotopy $H: W \to Y$ on $M \subseteq W$. Then there is an open neighborhood $U \subseteq W$ of M such that Y_0 is transversal to H on U.*

Proof. Let U be the union of all open sets $O \subseteq W$ such that Y_0 is transversal to H on O. Then Y_0 is transversal to F on U, and we have to show that $M \subseteq U$. Thus, let $(t_0, x_0) \in M$.

In the case $(t_0, x_0) \notin H^{-1}(Y_0)$, the continuity of H implies that $O := H^{-1}(Y \setminus Y_0)$ is an open neighborhood of (t_0, x_0). Trivially, Y_0 is transversal to H on O, and so $O \subseteq U$.

In the case $(t_0, x_0) \in H^{-1}(Y_0)$, let $c: U \to E_X$ be a chart for $x_0 \in X$, and let (c_Y, E_0), $c_Y: V \to E_Y$, be a submanifold chart for $y = H(t_0, x_0) \in Y$. Let h_{c_Y} be the corresponding trivializations of TY at y which by our identification (after composition with an isomorphism) can be understood also as a trivialization of TY_0 at y. We put

$$A(t, x) := d(c_Y \circ H(t, \cdot) \circ c^{-1})(c(x)) \in \mathcal{L}(E_X, E_Y).$$

Since $T_y Y_0$ is transversal to $d_X H(t_0, x_0)$, it follows in view of (8.8) that $Z_0 := h_Y(y, \cdot)^{-1}(T_y Y_0)$ is transversal to $A(t_0, x_0)$. Proposition 6.44 and the continuity of A imply that there is a neighborhood $O \subseteq W$ of (t_0, x_0) such that for every $(t, x) \in O$ the operator $A(t, x)$ is defined with Z_0 being transversal to $A(t, x)$. It follows by (8.8) that for every $(t, x) \in O$ with $H(t, x) \in Y_0$ the space $h_Y(\{H(t, x)\} \times Z_0) = T_{H(t,x)} Y_0$ is transversal to $d_X H(t, x)$. Hence, also in this case $O \subseteq U$. □

Concerning the existence of transversal manifolds, it suffices to consider the case of generalized Fredholm homotopies H, since the case of Fredholm maps F follows for constant homotopies $H(t, \cdot) = F$.

We consider first the case that Y is a Banach space.

Proposition 8.70 (Existence of Transversal Subspaces). (AC). *Let $W \subseteq I \times X$ be open, and $H: W \to Y = E_Y$ be a generalized Fredholm homotopy. Then for each compact $K \subseteq W$ there is an open neighborhood $U \subseteq W$ of K and a finite-dimensional subspace $Y_0 \subseteq Y$ which is transversal to $d_X H(t, x)$ for every $(t, x) \in U$. In particular for each finite-dimensional subspace $Y_1 \subseteq Y$ containing Y_1 and each $y \in Y$, the set $y + Y_1$ is transversal to H on U. If $Y \neq \{0\}$, it may be arranged that $Y_0 \neq \{0\}$.*

Proof. Let $\mathcal{O}$ denote the system of all open subsets $O \subseteq W$ with the property that there is a finite-dimensional subspace $V \subseteq Y$ such that V is transversal to $d_X H(t, x)$ for every $(t, x) \in O$. Proposition 6.29 implies for every $(t_0, x_0) \in K$ that there is some finite-dimensional subspace $V \subseteq Y$ with $Y = \mathrm{R}(d_X H(t_0, x_0)) \oplus V$. This V is transversal to $d_X H(t_0, x_0)$, and by Proposition 8.69 there is an open neighborhood $O \subseteq W$ such that V is transversal to $d_X H(t, x)$ for every $(t, x) \in O$. Hence, $\mathcal{O}$ is an open cover of K.

Since K is compact, it is thus covered by finitely many open sets $U_1, \dots, U_n \in \mathcal{O}$. There are finite-dimensional subspaces $V_1, \dots, V_n \subseteq Y$ such that V_k is transversal to $d_X H(t,x)$ for every $(t,x) \in U_k$ if $k = 1,\ldots,n$. Then $Y_0 := V_1 + \cdots + V_n$ is finite-dimensional and satisfies $\mathrm{R}(d_X H(t,x)) + Y_0 = Y$ for every $(t,x) \in U := U_1 \cup \cdots \cup U_n$. Proposition 6.43 implies in view of Corollary 6.28 that Y_0 is transversal to $d_X H(t,x)$ for every $(t,x) \in U$. For the last assertion, we note that in case $Y \neq \{0\}$ but $V_k = \{0\}$ for all $k = 1,\ldots,n$, we can replace Y_0 by any linear subspace of Y of positive finite dimension in the above argument. □

Remark 8.71. Note that AC is used only in the last step in our proof of Proposition 8.70. Using Proposition 6.27 instead of Corollary 6.28, we see that AC is not needed if one of the following holds:

(a) E_Y^* has full support E_Y.

(b) E_X^* has full support E_X.

Under this hypothesis, AC is also unnecessary for the following consequence.

Corollary 8.72 (Existence of Transversal Submanifolds). (AC). *Let I be a topological space, $W \subseteq I \times X$ be open, and $H\colon W \to Y$ be a generalized Fredholm homotopy. Let $K \subseteq W$ be compact, and suppose that there is a diffeomorphism of an open neighborhood of $H(K)$ onto an open subset of E_Y. Then there is an open neighborhood $U \subseteq W$ of K and a finite-dimensional submanifold $Y_0 \subseteq Y$ with the property that for every $(t,x) \in U$ the subspace $T_{H(t,x)}Y_0 \subseteq T_{H(t,x)}Y$ is transversal to $d_X H(t,x)$. In particular, Y_0 and every submanifold of Y containing Y_0 is transversal to H on U. In case $E_Y \neq \{0\}$, it may be arranged that Y_0 has positive dimension.*

Proof. Let J be the diffeomorphism of an open neighborhood $V \subseteq Y$ of $H(K)$ onto an open subset of E_Y. We apply Proposition 8.70 with W replaced by $W_0 := H^{-1}(V)$ and $H_0 := J \circ H_{W_0}$. Note that $K \subseteq W_0$. Hence, there exists a finite-dimensional subspace $E_0 \subseteq E_Y$ ($E_0 \neq \{0\}$ in case $E_Y \neq \{0\}$) and an open subset $U \subseteq W_0$ containing K such that E_0 is transversal to $d_X H_0(t,x)$ for every $(t,x) \in U$. Theorem 8.55 implies that $Y_0 := J^{-1}(E_0)$ is a finite-dimensional submanifold of Y with $T_y Y_0 = dJ(y)^{-1}(E_0)$ for every $y \in V$. For all $(t,x) \in U$ we have with $y := H(t,x) \in V$ by the chain rule on manifolds (Proposition 8.24) that $d_X H_0(t,x) = dJ(y) d_X H(t,x)$. It thus follows that $T_y Y_0$ is transversal to $d_X H(t,x)$. □

8.9 Properness of Fredholm Maps

For the Leray–Schauder triple degree, we will need that Fredholm maps are locally proper. In this section we obtain a corresponding result even for Fredholm homotopies from the partial implicit function theorem and some properties of Fredholm maps.

Let E_X and E_Y be Banach spaces over $\mathbb{K} = \mathbb{R}$ or $\mathbb{K} = \mathbb{C}$, and let X and Y be Banach manifolds without boundary over E_X and E_Y, respectively.

Theorem 8.73 (Fredholm Homotopies are Locally Proper). *Let I be a locally compact Hausdorff space, $W \subseteq I \times X$ be open and $H: W \to Y$ be a generalized Fredholm homotopy. Then H is locally proper in the sense that for each compact $K \subseteq W$ there is a neighborhood $N \subseteq W$ of K such that $H|_N$ is proper.*

Proof. Assume first that $K = \{(t_0, x_0)\}$. Let c and c_0 be charts for x_0 and $y_0 := H(t_0, x_0)$, respectively. Shrinking W if necessary, we can assume that $W = I_0 \times \Omega$ where $I_0 \subseteq I$ is an open neighborhood of t_0 and $\Omega \subseteq X$ is an open neighborhood of x_0 and such that $H_0(t, u) := c_0(H(t, c^{-1}(u)))$ is defined on $I_0 \times c(\Omega)$. Without loss of generality, we assume that $c(x_0) = 0$ and $c_0(y_0) = 0$.

By hypothesis, H_0 is a generalized partial Fredholm homotopy, in particular $A := d_X H_0(t_0, x_0) \in \mathcal{L}_k(E_X, E_Y)$ for some $k \in \mathbb{Z}$.

Let $n := |k|$, $E_0 := E_X \times \mathbb{K}^n$, $E_1 := E_Y \times \mathbb{K}^{n+k}$, and define $H_1: I_0 \times \Omega \times \mathbb{K}^n \to E_1$ by $H_1(t, u, z) := (H_0(t, u), 0)$. Then the partial derivative $A_1 := d(H(t_0, \cdot))(x_0)$ satisfies $A_1(u, z) = (Au, 0)$ and thus is a generalized partial Fredholm homotopy of index 0. By Theorem 6.36, there is a corrector $C \in \mathcal{L}(E_0, E_1)$ of A_1, that is, $A_1 + C \in \operatorname{Iso}(E_0, E_1)$, and $\mathrm{R}(C)$ has finite dimension.

We define $G: I_0 \times c(\Omega) \times \mathbb{K}^n \to E_1$ by $G(t, u, z) := H_1(t, u, z) + C(u, z)$ and apply Theorem 8.40 in $(t_0, 0, 0)$. Noting that $d(G(t_0, \cdot))(0) = A_1 + C$, we find open neighborhoods $I_1 \subseteq I$ of t_0, $U_0 \subseteq c(\Omega)$ of 0, and $U_1 \subseteq \mathbb{K}^n$ of 0 such that $G(t, u, z) := (t, G(t, u, z))$ is a homeomorphism of $U := I_1 \times U_0 \times U_1$ onto an open neighborhood $V_0 \subseteq I \times E_1$ of $(t_0, 0, 0)$. In particular, if $M_0 \subseteq V_0$ is compact then $U \cap G_0^{-1}(M_0)$ is compact. Since I is locally compact, we find by Corollary 2.50 some compact neighborhood $I_2 \subseteq I$ of t_0 with $I_2 \subseteq I_1$. There is a neigbhorhood $V_1 \subseteq E_1$ of $(0, 0)$ with $I_2 \times V_1 \subseteq V_0$. Put $N := I_2 \times U_0 \times U_1$. If $M_1 \subseteq V_1$ is compact then $M_0 := I_2 \times M_1 \subseteq V_0$ is compact by Theorem 2.63, and so $U \cap G_0^{-1}(M_0) = N \cap G^{-1}(M_1)$ is compact.

Let $D_0 \subseteq E_X$ and $D_1 \subseteq \mathbb{K}^n$ be closed bounded neighborhoods of 0 with $D_i \subseteq U_i$ $(i = 0, 1)$. Then $C_0 := \overline{C(D_0 \times D_1)}$ is compact. Shrinking D_0 and D_1 if necessary, we can assume that $K_r(C_0) \subseteq V_1$ for some $r > 0$. Shrinking

$r > 0$ if necessary, we may assume that $K_r(0) \subseteq E_Y$ is contained in the range of the map c_0. Moreover, shrinking I_2 and D_0 if necessary, we may assume for $N_0 := I_2 \times D_0$ that $H_0(N_0) \subseteq K_r(0)$.

Now if $M_1 \subseteq K_r(0)$ is compact then $M_0 := M_1 - C_0 \subseteq V_1$ is compact, and so $N \cap G^{-1}(M_0)$ is compact. The definition of G implies that for each $(t, u, z) \in N_1 := I_2 \times D_1 \times D_2$ with $H_1(t, z, u) \in M_1$, we have $(t, u, z) \in N \cap G^{-1}(M_0)$. Thus,

$$N_1 \cap H_1^{-1}(M_1) \subseteq N \cap G^{-1}(M_0).$$

Since $N_1 \cap H_1^{-1}(M_1)$ is a closed subset of the compact set $N \cap G^{-1}(M_0)$, we obtain that $N_1 \cap H_1^{-1}(M_1)$ is compact. By the definition of H_1, we conclude that for any compact $M \subseteq K_r(0) \subseteq E_Y$ the set $N \cap H_0^{-1}(M) = (I_2 \times D_0) \cap H_0^{-1}(M)$ is compact. Now if $M \subseteq E_Y$ is an arbitrary compact set, we obtain in view of $H_0(N_0) \subseteq K_r(0)$ that $N_0 \cap H_0^{-1}(M) = N_0 \cap H_0^{-1}(M \cap K_r(0))$ is compact. Hence, $H_0|_{N_0}$ is proper. We obtain with $N_2 := \{(t, c(u)) : (t, u) \in N_0\}$ that $H|_{N_2}$ is proper. Hence, the assertion is proved in the case $K = \{(t_0, x_0)\}$.

In general, let $\mathscr{U}$ denote the system of all open sets $U \subseteq W$ which are contained in some $N \subseteq W$ such that $H|_N$ is proper. By what we just proved, $\mathscr{U}$ is an open cover of K. By the compactness, K is covered by finitely many $U_1, \dots, U_n \in \mathscr{U}$. There are $N_1, \dots, N_n \subseteq W$ with $U_k \subseteq N_k$ and such that $H|_{N_k}$ is proper for $k = 1, \dots, n$. Then $N := N_1 \cup \dots \cup N_k$ is the required neighborhood of K, since for any compact set $M \subseteq Y$ the set $N \cap F^{-1}(M)$ is the union of the compact sets $N_k \cap F^{-1}(M)$ and thus compact. □

Corollary 8.74 (Fredholm Maps are Locally Proper). *Let $\Omega \subseteq X$ be open and $F \in \mathscr{F}_k(\Omega, Y)$. Then F is locally proper in the sense that for each compact $K \subseteq \Omega$ there is a neighborhood $N \subseteq \Omega$ of K such that $F|_N$ is proper.*

Proof. This is a special case of Theorem 8.73. □

Corollary 8.74 was probably first observed in [131].

Chapter 9

The Brouwer Degree

It is somewhat surprising to the author that apparently no thorough exposition of the Brouwer degree on manifolds (initiated by Brouwer in [29]) can be found in literature: Most text books about the topic which the author found, including e.g. [30], [37], [44], [50], [57], [77], [99], [107], [112], [122], [124], [147], treat only special cases (like $\mathbb{R}^n$ or C^∞ maps or by considering only connected, paracompact, or even closed manifolds), and the uniqueness of the degree is rarely proved, usually only under much stronger assumptions like by requiring a normalization property for diffeomorphisms which in case of a Banach space is not the natural notion of a normalization. The aim of this chapter is to close this gap.

The classical Brouwer degree on manifolds requires to consider oriented manifolds. We will also take the opportunity to define the Brouwer degree in the more general and more natural situation of oriented continuous maps instead of oriented manifolds. Since we restrict our attention to C^1 manifolds, we will be able to give a definition of orientation of continuous maps which is rather simple and which for C^1 maps can be directly related to our previous notion of orientation of Fredholm operators.

9.1 Finite-Dimensional Manifolds

In this section, we recall some well-known results for finite-dimensional manifolds.

We will frequently use the trick that instead of the whole manifold we consider only the submanifold which consists in a certain open neighborhood of a given compact set. It is due to this trick that we do not have to require that the considered manifolds be paracompact or second countable. In the following result, we collect all important topological properties which we can obtain by this trick.

Proposition 9.1. *Let $0 \leq r \leq \infty$. Each C^r manifold X (with or without boundary) over a finite-dimensional space E is locally compact and regular. Moreover, second countable subsets of X are paracompact and metrizable. All relatively compact subsets $M \subseteq X$ are second countable and satisfy $\dim M \leq \operatorname{Ind} M \leq n$, where n denotes the dimension of E over $\mathbb{R}$ in the sense of linear algebra.*

If $K \subseteq X$ is compact, and $U \subseteq X$ is an open neighborhood of K then the following holds.

(a) *K is contained in only finitely many components of the space U.*

(b) *There is an open neighborhood $\Omega \subseteq X$ of K with compact $\overline{\Omega} \subseteq U$ and such that $\Omega \cap U_m$ and $\overline{\Omega} \cap U_m$ are path-connected for every component U_m of the space U.*

If K consists of only one point not from the boundary, it may be arranged for any given chart that this chart maps Ω and $\overline{\Omega}$ onto some ball $B_\rho(x)$ or $K_\rho(x)$ in E, respectively.

Proof. Proposition 3.59 implies that X is locally compact. To see that X is regular, recall that X is Hausdorff, by hypothesis. Let $A \subseteq X$ be closed and $x \in X \setminus A$. There is an open set $U \subseteq X \setminus A$ such that there is a homeomorphism c of U onto an open subset $U_0 \subseteq E$. In view of $\dim E < \infty$, we can apply Corollary 3.58 and Proposition 3.59 to find that there is a neighborhood $V_0 \subseteq E$ of $c(x)$ such that $\overline{V}_0 \subseteq U_0$ is compact. Then $c^{-1}(\overline{V}_0)$ is compact by Proposition 2.100 and thus closed by Proposition 2.45, since X is Hausdorff. Hence, $V := X \setminus c^{-1}(\overline{V}_0)$ is an open neighborhood of A which is disjoint from $c^{-1}(V_0) \subseteq U$. Since Proposition 2.10 implies that $c^{-1}(V_0)$ is open in X and thus a neighborhood of x, we have shown that X is regular.

Theorem 2.42 implies that also all subsets of X are regular. In particular, second countable subsets of X, since they are Lindelöf by Proposition 2.61, are paracompact and metrizable by Theorems 3.76 and 3.79.

Recall now that X is locally path-connected by Proposition 8.16. Hence, also U is locally path-connected, and each component of U is path-connected and open in U by Proposition 2.17. In particular, the components of U are open. Thus, if $\mathcal{M}$ denotes the family of all components of U then it is an open cover of K and thus has a finite subcover which shows that K is contained in only finitely many components $U_1, \ldots, U_k$.

Let $\mathcal{O}$ denote the family of all open subsets $\Omega_0 \subseteq X$ such that Ω_0 and $\overline{\Omega}_0$ are C^r-diffeomorphic to an open ball in E or its intersection with a closed halfspace, and its closure, respectively, and $\overline{\Omega}_0 \subseteq U$. Then $\mathcal{O}$ is an open cover of U. Indeed, for every $x \in U$ there is an open subset $V \subseteq U$ and a chart c which maps V (C^r-diffeomorphic) onto an open subset of a closed halfspace E_+ (we can replace E_+ by E if $x \notin \partial X$). There is some $r > 0$ with $C := K_r(c(x)) \cap E_+ \subseteq E$. Put $\Omega := c^{-1}(B_r(c(x)) \cap E_+)$. Note that $c^{-1}(C)$ is compact and contains Ω. Since X is Hausdorff, Proposition 2.45 implies that $c^{-1}(C)$ is closed in X and thus contains $\overline{\Omega}$. The continuity of c now implies $\overline{\Omega} = c^{-1}(C)$. It follows that $x \in \Omega \in \mathcal{O}$.

Since K is compact, it is covered by finitely many $\Omega_1, \dots, \Omega_n \in \mathcal{O}$. If K consists of only one point, we can assume $n = 1$.

For the assertions about relatively compact subsets $M \subseteq X$, we can assume that $M \subseteq K$. Note that Propositions 5.9 and 5.16 imply that $\operatorname{Ind} \overline{\Omega}_k \le \operatorname{Ind} E \le n$. The sum theorem (Theorem 5.13) thus implies that $\operatorname{Ind}(\overline{\Omega}_1 \cup \dots \cup \overline{\Omega}_n) \le n$, and so Proposition 5.9 shows that $\operatorname{Ind} K \le n$. Hence, Theorem 5.18 and the subspace theorem (Theorem 5.15) imply the estimates $\dim M \le \operatorname{Ind} M \le \operatorname{Ind} K \le n$. Similarly, since all $\overline{\Omega}_k$ are second countable, it follows that also their finite union is second countable, and thus also the subset $M \subseteq K$ is second countable by Corollary 3.17.

Recall that we have shown that each component U_m of U is path-connected. Hence, we can connect those Ω_k belonging to the same component of U pairwise in U by (compact) paths in U. Every of the compact paths is covered by finitely many sets from $\mathcal{O}$. Joining finitely many such sets to the considered finite family $\Omega_1, \dots, \Omega_n$, if necessary, it may thus be arranged, that for each $k, j \in \{1, \dots, n\}$ for which Ω_k and Ω_j belong to the same U_m there is a chain $k_0 = k, \dots, k_\ell = j \in \{1, \dots, n\}$ such that $\Omega_{k_i} \cap \Omega_{k_{i-1}} \neq \varnothing$ for $i = 1, \dots, \ell$. It follows now easily that $\Omega := \Omega_1 \cup \dots \cup \Omega_n$ has the property that $\Omega \cap U_m$ is path-connected, and then also that $\overline{\Omega} = \overline{\Omega}_1 \cup \dots \cup \overline{\Omega}_n$ has the property that $\overline{\Omega} \cap U_m$ is path-connected. Since each $\overline{\Omega}_k$ is compact, we obtain also that $\overline{\Omega}$ is compact. □

Remark 9.2. Since every (relatively) compact subset $M \subseteq X$ is metrizable, we actually have $\dim M = \operatorname{Ind} M$ in Proposition 9.1, see e.g. [52, Theorem 4.1.3] (since M is second countable and thus separable, even the simpler result [52, Theorem 1.7.7] can be used to see this). Moreover, if in addition M has an interior point then $\dim M = \operatorname{Ind} M = n$. This follows, for instance, from the subset Theorem 5.15, since a subset of M is homeomorphic to $E \cong \mathbb{R}^n$, because one can show that actually $\operatorname{Ind} \mathbb{R}^n = n$. However, we will not need these facts.

Corollary 9.3. *Let X be a C^0 manifold over a finite-dimensional space E, $\Omega \subseteq X$ be open, I be a topological space, and $K \subseteq I \times X$ be compact. Then there is an open $\Omega_0 \subseteq X$ with compact $\overline{\Omega}_0 \subseteq \Omega$ such that $K \subseteq I \times \Omega_0$. If K is connected, it may be arranged that Ω_0 and $\overline{\Omega}_0$ are path-connected.*

Proof. By Corollary 2.101 there is a compact (and connected if K is connected) $K_0 \subseteq \Omega$ with $K \subseteq I \times K_0$. Proposition 9.1 implies that there is some open neighborhood $\Omega_0 \subseteq \Omega$ of K_0 with compact $\overline{\Omega}_0 \subseteq \Omega$. Moreover, if K_0 is connected then it is contained in one component of Ω and so, Proposition 9.1 implies also that it may be arranged that Ω_0 and $\overline{\Omega}_0$ are path-connected. □

It follows that $F \in C(X, Y)$ with compact $K = F^{-1}(y)$ are "locally proper" in neighborhoods of y and K in a sense: The idea is that if $F^{-1}(y)$ is contained in a compact set and does not intersect the boundary of that set then it must be compact. For later usage, we formulate this even for the case that F is replaced by a homotopy and show that the same holds even for all homotopies which are only sufficiently close.

Proposition 9.4. *Let Ω be a topological space, Y_0 a metric space, I be a compact space, $H \in C(I \times \Omega, Y)$, $y \in Y_0$, $N \subseteq \Omega$ be compact, and $M \subseteq N$ have the property that $H^{-1}(y) \subseteq I \times \mathring{M}$.*

Then there is $\varepsilon > 0$ such that for every X_0 satisfying $\overline{M} \subseteq X_0 \subseteq N$ and every $H_0 \in C(I \times X_0, Y_0)$ satisfying $d(H(t,x), H_0(t,x)) \leq \varepsilon$ for all $(t,x) \in I \times (X_0 \setminus \mathring{M})$, and every closed $C \subseteq K_\varepsilon(y)$ the set $H_0^{-1}(C)$ is contained in $I \times \mathring{M}$ and compact.

Proof. Since $B := N \setminus \mathring{M}$ is closed in N, it is compact by Proposition 2.29. Proposition 2.100 implies that $A := H(I \times B)$ is compact. Since $H^{-1}(y) \subseteq I \times \mathring{M}$, we have $y \notin A$, and so Corollary 3.14 implies $\delta := \operatorname{dist}(y, A) > 0$. Let $0 < \varepsilon < \delta/2$.

Now if X_0 and H_0 are as in the assertion, we have

$$d(H_0(t,x), y) \geq d(H(t,x), y) - d(H_0(t,x), H(t,x)) \geq \delta - \varepsilon > \varepsilon$$

for all $(t,x) \in I \times (X_0 \setminus \mathring{M})$. Hence, if $C \subseteq K_\varepsilon(y)$ is closed, we have for all $(t,x) \in I \times (X_0 \setminus \mathring{M})$ that $H(t,x) \cap C = \emptyset$, that is $H^{-1}(C) \subseteq I \times \mathring{M}$. In particular, $H_0^{-1}(C) = H_0|_{I \times \overline{M}}^{-1}(C)$ is compact, since the latter is a closed subset of the compact space $I \times \overline{M}$. □

We recall a simple observation of elementary calculus:

Lemma 9.5. *There is a function $g: \mathbb{R} \to [0,1]$ of class C^∞ which satisfies $g^{-1}(0) = (-\infty, 0]$.*

Proof. A function with the required properties is given by

$$g(t) := \begin{cases} e^{-1/t} & \text{if } t > 0, \\ 0 & \text{if } t \leq 0. \end{cases}$$

Indeed, an induction by $n = 0, 1, \ldots$ shows that $g^{(n)}|_{(0,\infty)}$ is a linear combination of the functions $h_k(t) := t^k e^{-1/t}$ $(t > 0)$ with $k \in \mathbb{Z}$. Since $h_k(t)/t \to 0$ as $t \to 0^+$, the definition of the derivative implies with an induction by $n = 0, 1, \ldots$ that $g^{(n+1)}(0)$ exists and is zero. □

Lemma 9.5 implies that in finite-dimensional spaces the Lemma of Urysohn has a smooth variant. For the moment, the following special case is sufficient for us.

Lemma 9.6. *If E is a real finite-dimensional space, $A \subseteq E$ is compact, and $B \subseteq E$ is closed and disjoint from A then there is a bounded function $f \in C^\infty(E, \mathbb{R})$ with $f(E) \subseteq [0, \infty)$, $f(A) \subseteq (0, \infty)$ and $f(B) = \{0\}$.*

Proof. Let $e_1, \dots, e_n$ be a basis of E. Recall that Corollary 3.58 implies that the norm on E is equivalent to the "Euclidean" norm

$$\|\sum_{k=1}^{n} \xi_n e_n\|^2 := \xi_1^2 + \dots + \xi_n^2,$$

and so without loss of generality we can assume that E is equipped with that norm. Let g be as in Lemma 9.5. For $x_0 \in E$ and $r > 0$, it follows that $f_{x_0,r}\colon E \to \mathbb{R}$, $f_{x_0,r}(x) := g(r^2 - \|x - x_0\|^2)$, is of class C^∞. Moreover, $f_{x_0,r}(x) \neq 0$ if and only if $x \in B_r(x_0)$. The compact set A is covered by finitely many balls $B_{r_k}(x_k)$ $(k = 1, \dots, m)$ which are disjoint from B. Then $f := f_{x_1,r_1} + \dots + f_{x_m,r_m}$ is the required function. □

The reason why second countable (hence paracompact) manifolds play such an important role is the following:

Lemma 9.7. *Let X be a second countable C^r manifold over a real finite-dimensional space E with $0 \leq r \leq \infty$. Then every open cover of X has a subordinate countable partition of unity consisting of functions of class C^r.*

Proof. Recall that X is regular and Lindelöf by Propositions 9.1 and 2.61. Let $\mathcal{U}$ be an open cover of X. Let $\mathcal{O}$ be the family of all open sets $O \subseteq X$ with the property that $O \subseteq U$ for some $U \in \mathcal{U}$ and that there is a chart c which is at least defined on $\overline{O}$ and such that $\overline{O}$ is compact. Proposition 9.1 implies that $\mathcal{O}$ is an open cover of X. By definition, $\mathcal{O}$ is a refinement of $\mathcal{U}$.

Theorem 3.80 implies that there are open sets $U_n, O_n \subseteq X$ with $X = \bigcup_{n=1}^{\infty} U_n$, $\overline{U}_n \subseteq O_n$, and such that $\overline{O}_n$ $(n \in \mathbb{N})$ is a refinement of $\mathcal{O}$. Hence, for each n there is some $O \in \mathcal{O}$ with $\overline{O}_n \subseteq O$. Let c_n be a chart defined on $\overline{O}$. Then $A_n := c_n(\overline{U}_n)$ is compact, $B_n := E \setminus c_n(O_n)$ is closed, and so Lemma 9.6 implies that there is a bounded $f_n \in C^\infty(E, \mathbb{R})$ with $f_n(E) \subseteq [0, \infty)$, $f_n(A_n) \subseteq (0, \infty)$, and $f_n(B_n) = \{0\}$. Since c_n is a C^r-diffeomorphism, it follows that

$$g_n(x) := \begin{cases} f_n(c_n(x)) & \text{if } x \in \overline{O}, \\ 0 & \text{if } x \notin O, \end{cases}$$

defines a function of class C^r such that $S_n := \{x \in X : g_n(x) \neq 0\}$ is contained in O_n, and $U_n \subseteq S_n$. Since $\overline{O}_n$ ($n \in \mathbb{N}$) is locally finite, it follows that $\overline{S}_n = \operatorname{supp} g_n$ ($n \in \mathbb{N}$) is locally finite. It follows that $g(x) = \sum_{n=1}^{\infty} g_n(x)$ is C^∞ near every $x \in X$, since x has a neighborhood which intersects only finitely many S_n. Since U_n ($n \in \mathbb{N}$) is a cover of X, we have $g(x) > 0$ for every $x \in X$. Hence, $\lambda_n(x) := g_n(x)/g(x)$ is a required partition of unity which is subordinate to $\mathscr{O}$ and thus subordinate to $\mathscr{U}$. □

As a first application of Lemma 9.7, we obtain now a full smooth generalization of Urysohn's lemma and even of the Tietze–Urysohn theorem, even in finite-dimensional manifolds.

Theorem 9.8 (Smooth Tietze–Urysohn). *Let X be a second countable C^r manifold over a real finite-dimensional space E with $0 \leq r \leq \infty$. Let $A, B \subseteq X$ be closed and disjoint with open neighborhoods $U, V \subseteq X$, respectively, and let Y be a real Banach space. Then for every $f \in C^r(U, Y)$, $g \in C^r(V, Y)$, $y \in Y$, there is $F \in C^r(X, Y)$ satisfying $F|_A = f|_A$, $F|_B = g|_B$, and*

$$F(X) \subseteq \operatorname{conv}(f(U) \cup \{y\}) \cup \operatorname{conv}(g(V) \cup \{y\}).$$

The smooth variant of Tietze's lemma is obtained with $y \in f(U)$ and $B = V = \emptyset$ (and g being the empty function), and Urysohn's lemma is the special case $Y = \mathbb{R}$, $f \equiv 0$, $g \equiv 1$, and $y = 0$.

Proof. Note that Proposition 9.1 implies that X is metrizable and thus T_4. Hence, without loss of generality, we can assume that U and V are disjoint. Lemma 9.7 implies that X has a countable partition of unity λ_n of class C^r which is subordinate to the open cover $\{U, V, X \setminus (A \cup B)\}$. Let N_U, N_V be the (disjoint) set of all indices n with $\operatorname{supp} \lambda_n \subseteq U$ or $\operatorname{supp} \lambda_n \subseteq V$, respectively, and N_0 the set of the remaining indices. Then

$$F(x) := \sum_{n \in N_U} \lambda_n(x) f(x) + \sum_{n \in N_V} \lambda_n(x) g(x) + \sum_{n \in N_0} \lambda_n(x) y$$

defines a function with the required properties. □

Using Lemma 9.7, one can also show results like the following.

Theorem 9.9. *Let E_X and E_Y be real finite-dimensional normed spaces. Let X be a second countable C^r manifold over E_X ($0 \leq r \leq \infty$), $K \subseteq X$ be compact, and $f: K \to E_Y$ be continuous. Then for each $\varepsilon > 0$ there is $g \in C^r(X, E_Y)$ with $\|f(x) - g(x)\| < \varepsilon$ for all $x \in K$.*

Proof. The family $O_x := f^{-1}(B_\varepsilon(f(x))) \cup (X \setminus K)$ $(x \in K)$ is an open cover of K. Let $O_{x_1}, \dots, O_{x_m}$ be a finite subcover of K which by definition of O_x is then even a cover of X. By Lemma 9.7 there is a subordinate partition of unity $\lambda_n \in C^r(X, \mathbb{R})$ $(n \in \mathbb{N})$. For each n there is a minimal index $k = k_n$ such that $\operatorname{supp} \lambda_n$ is contained in O_{x_k}. We define $y_n := f(x_{k_n})$. Then

$$g(x) := \sum_{n=1}^{\infty} \lambda_n(x) y_n$$

has the required properties. Indeed, if $x \in K$ and $n \in \mathbb{N}$ are such that $\lambda_n(x) \neq 0$ then $x \in \operatorname{supp} \lambda_n \subseteq O_{x_{k_n}}$. Hence,

$$f(x) \in f(O_{k_n} \cap K) \subseteq B_\varepsilon(f(x_{k_n})) = B_\varepsilon(y_n).$$

It follows that

$$\begin{aligned}\|g(x) - f(x)\| &= \Big\|\sum_{n=1}^{\infty} \lambda_n(x)(y_n - f(x))\Big\| \\ &\leq \sum_{n=1}^{\infty} \lambda_n(x)\|f(x) - y_n\| < \sum_{n=1}^{\infty} \lambda_n(x)\varepsilon = \varepsilon\end{aligned}$$

for all $x \in K$. □

Remark 9.10. The only reason why we do not obtain higher smoothness than r in Lemma 9.7 and Theorem 9.9 is that higher smoothness is not well-defined for manifolds of class C^r: The maps constructed in the proof are as smooth as the available charts. In particular, if X is a partial C^r manifold, we obtain that the partition of unity in Lemma 9.7 and the map g in Theorem 9.9 are partial C^r homotopies.

The proof of that statement is completely identical to our above proofs: One only has to consider partial charts instead of charts.

The approach to the Brouwer degree from differential topology which we take actually requires manifolds of class C^2 (or smoother), as we will see. Therefore, we will need a tool which allows us to pass from manifolds of class C^1 to manifolds of class C^2 (or smoother). Such a tool is the following result. For simplicity, we formulate it only for manifolds without boundary, since this is the only case which we need.

Theorem 9.11. (AC). *Every C^1 manifold over a real finite-dimensional space is C^1-diffeomorphic to a C^∞ manifold.*

Proof. The proof of this result is beyond the scope this monograph. A proof can be found in [77, Chapter 2, Theorem 2.10]. □

Theorem 9.11 states that there is a diffeomorphism $F: X \to X_0$. Recall that the latter means that F is invertible of class C^1 such that also $F^{-1}: X_0 \to X$ is of class C^1.

Unfortunately, there is no corresponding result for C^0 manifolds, that is, C^0 manifolds need not be homeomorphic to C^1 manifolds.

It is unknown to the author whether AC can be avoided in the proof of Theorem 9.11. However, the following variant does not require AC.

Theorem 9.12. *Let X be a C^1 manifold X over a real finite-dimensional space E, and $K \subseteq X$ be compact. Then there is an open neighborhood $U \subseteq X$ of K which is C^1-diffeomorphic to a C^∞ manifold. If K is contained in a component of X, it may be arranged that U is connected.*

Sketch of proof. We just sketch how the above mentioned proof of [77, Chapter 2, Theorem 2.10] has to be modified to avoid Zorn's lemma. By the compactness, K has a finite cover by open sets $U_1, \dots, U_n$, each of which is mapped by a chart c_k diffeomorphically onto an open subset E. Put $U := U_1 \cup \dots \cup U_n$. Then the argument from the proof of [77, Chapter 2, Theorem 2.10] shows that one can inductively define a diffeomorphism $\hat{c}_k$ of U_k onto an open subset of E such that $\hat{c}_k \circ c_j^{-1}$ $(j < k)$ are diffeomorphisms of class C^∞. Then the charts $\hat{c}_k$ $(k = 1, \dots, n)$ turn X into a C^∞ manifold $\hat{X}$, and $\mathrm{id}_X: X \to \hat{X}$ is the required diffeomorphism.

The last assertion follows from the first. Indeed, assume that K is contained in a component M of X. Propositions 2.23 and 8.16 imply that M is open, and M is connected by Proposition 2.17. Proposition 9.1 thus implies that K is contained in a path-connected compact set $K_0 \subseteq M$. By the first assertion there is an open neighborhood $U_0 \subseteq M$ of K_0 which is diffeomorphic to a C^∞ manifold. Then the component U of K_0 in the space U_0 has the required property. Note that U is open by Propositions 2.23 and 8.16 and connected by Proposition 2.17. □

To formulate the famous theorem of Morse–Sard, we recall that a subset of $\mathbb{R}^m$ is called a null set if it has Lebesgue measure zero or, equivalently, if for every $\varepsilon > 0$ it is covered by countably many cuboids of total volume less than ε.

Theorem 9.13 (Morse–Sard). *Let $\Omega \subseteq \mathbb{R}^n$ be open, and $F \in C^r(\mathbb{R}^n, \mathbb{R}^m)$. If $r \geq 1$ and $r > n - m$ then the set of critical values of F is a null set in $\mathbb{R}^m$.*

Proof. The proof of Theorem 9.13 in the general case is rather involved and beyond the scope of this monograph. A relatively simple proof can be found in [1, §15]. □

It turns out that for the Brouwer degree, we will not need Theorem 9.13 in its full generality but only two special cases which are both much simpler to prove: These are the special cases $n \le m$ and $r = \infty$. For the case $r = \infty$, relatively simple proofs can be found in e.g. [77, Chapter 3, Section 1] or [107, Section 3]. We give only the proof for the case $n \le m$ which is the case considered originally by Sard.

Proof of Theorem 9.13 *in case* $n \le m$. Let C denote the set of critical points of F. Note that Ω is second countable by Proposition 9.1 and thus Lindelöf by Proposition 2.61. The family of all open cubes contained in Ω is a cover of Ω and thus has a countable subcover. Since each open cube is a countable union of compact cubes, it follows that Ω can be written as a union of countably many (not disjoint) compact cubes K_k ($k \in \mathbb{N}$). Recall that the countable union of null sets is a null set. Hence, it suffices to show that each $F(K_k \cap C)$ ($k \in K$) is a null set since then also their union $F(C)$ is a null set.

Thus, given a compact cube $K \subseteq \Omega$, we are to show that $F(K \cap C)$ is a null set. Since K is compact, there is a constant $M \in [0, \infty)$ with $\|dF(x)\| \le M$ for all $x \in K$. Moreover, since dF is uniformly continuous on the compact set K, we find for every $\varepsilon > 0$ some $\delta \in (0, 1]$ such that $\|dF(x) - dF(y)\| \le \varepsilon$ whenever $x, y \in K$ satisfy $\|x - y\| \le \delta$. Shrinking $\delta > 0$ if necessary, we can assume that $\ell := \operatorname{diam} K/\delta$ is an integer. We write K as the union of ℓ^n compact cubes Q_k ($k = 1, \dots, \ell^n$) with $\operatorname{diam} Q_k = \delta$.

We are to show that if Q_k contains a point from C then

$$F(Q_k) \text{ is contained in a cuboid of volume } 2^m \delta^m M^{m-1} \varepsilon. \tag{9.1}$$

To see this, let $Q = Q_k$ be fixed and contain some point $x_k \in C$. In order to prove (9.1), we can replace F by $F - F(x_k)$ if necessary and thus assume without loss of generality that $F(x_k) = 0$. Moreover, since $x_k \in C$ and thus $A := dF(x_k)$ is not onto, $\mathrm{R}(A)$ is a subspace of dimension at most $m - 1$. Letting $J \in \operatorname{Iso}(\mathbb{R}^m)$ be an isometry which maps this subspace into $\{0\} \times \mathbb{R}^{m-1}$ and replacing F by $J \circ F$ if necessary, we can assume without loss of generality that $\mathrm{R}(A) = \{0\} \times \mathbb{R}^{m-1}$, that is, writing $F(x) = (F_1(x), \dots, F_m(x))$ that $dF_1(x_k) = 0$. It follows that $\|dF_1(x)\| \le \varepsilon$ for all $x \in Q$, and so the elementary mean value theorem (or Theorem 8.2) implies

$$|F_1(x)| = |F_1(x) - F_1(x_k)| \le \varepsilon \|x - x_k\| \quad \text{for all } x \in Q.$$

In particular, $F(Q)$ is contained in a subset of the stripe $S := [-\varepsilon\delta, \varepsilon\delta] \times \mathbb{R}^{m-1}$.

To find estimates for the other coordinates $F_k(Q)$ ($k = 2, \dots, m$), we use Theorem 8.2 to obtain in view of $\|dF(x)\| \le M$ for all $x \in K$ that

$$\|F(x)\| = \|F(x) - F(x_k)\| \le M\,\|x - x_k\| \quad \text{for all } x \in Q.$$

It follows that $F(Q)$ is contained in $[-\varepsilon\delta, \varepsilon\delta]\times[-M\delta, M\delta]^{m-1}$. This shows (9.1).

Using (9.1), we find that $F(K \cap C)$ is contained in a finite union of cuboids of total volume at most

$$\ell^n 2^m \delta^m M^{m-1}\varepsilon = (\operatorname{diam} K)\delta^{m-n}2^m M^{m-1}\varepsilon \le \varepsilon 2^m M^{m-1} \operatorname{diam} K,$$

because $\delta \in (0, 1]$ and $n \le m$. Since $\varepsilon > 0$ was arbitrary, we obtain that $F(K \cap C)$ is a null set, as required. □

For the Brouwer degree, we will not use Theorem 9.13 directly but only the following simple consequence:

Theorem 9.14 (Morse–Sard). *Let Y denote a manifold over a real finite-dimensional space E_Y. Let Ω be a second countable C^r manifold over a real finite-dimensional space E_X where $r \ge 1$ satisfies $r > \dim E_X - \dim E_Y$, and let U be a second countable C^{r+1} manifold over $\mathbb{R} \times E_X$ with boundary ∂U. Then each map from $C^r(\Omega, Y)$, $C^r(\partial U, Y)$, or $C^{r+1}(U, Y)$ has a dense set of regular values.*

Proof. Let $V \subseteq Y$ be nonempty and open. We have to show that V contains at least one regular value for each map F of the described classes. There is a chart defined on some nonempty open subset of V. Considering the composition of this chart with an isomorphism of E_Y to $\mathbb{R}^m$ ($m := \dim E_Y$), we thus find a nonempty open subset $V_0 \subseteq V$ and a diffeomorphism J (of class C^r or C^{r+1}, respectively) of V_0 onto an open subset of $\mathbb{R}^m$. It suffices to show that $G := F|_{F^{-1}(V_0)}$ has a regular value in V_0. Hence, it suffices to show that $J \circ G$ has a regular value in $J(V_0)$. This proves that we can assume without loss of generality that $Y = \mathbb{R}^m$.

Thus, let $F \in C^r(\Omega, \mathbb{R}^m)$, $F \in C^r(\partial U, \mathbb{R}^m)$, or $F \in C^{r+1}(U, \mathbb{R}^{m+1})$, respectively. Let $\mathcal{U}$ denote the family of all open subsets of $X = \Omega$ or $X = U$, respectively, which are contained in the domain of a chart. Then $\mathcal{U}$ is an open cover of X. Since X is second countable and thus Lindelöf (Proposition 2.61), $\mathcal{U}$ has a countable subcover U_k ($k \in \mathbb{N}$). Let c_k ($k \in \mathbb{N}$) be charts of X defined on U_k, and I be an isomorphism of $\mathbb{R}^n$ or $\mathbb{R}^{n+1}$ onto E_X or $\mathbb{R} \times E_X$, respectively, where $n = \dim E_X$. Then $\Omega_k := I^{-1}(c_k(U_k))$ is open in $\mathbb{R}^n$ or $\mathbb{R}^{n+1}$, and $G_k : \Omega_k \to \mathbb{R}^m$, $G_k := F \circ c_k^{-1} \circ I$ is of class C^r or C^{r+1}, respectively. Thus, Theorem 9.13 implies that the set C_k of critical values of G_k is a null set in $\mathbb{R}^m$.

Note that C_k is exactly the set of critical values of $F|_{U_k}$. Hence, the union C of all sets C_k is the set of all critical values of F. Since the union of countably many null sets is a null set, and since null sets have a dense complement, it follows that $\mathbb{R}^m \setminus C$ is dense. This is exactly the set of regular values of F. □

For the Brouwer degree, we will actually only use the case $\dim E_X = \dim E_Y$ of Theorem 9.14, and only for maps from $C^1(\Omega, Y)$, $C^\infty(\partial U, Y)$, and $C^\infty(U, Y)$, respectively. For the proof of these cases, the earlier mentioned simpler special cases of Theorem 9.13 are sufficient.

Unfortunately, even if we would be willing to apply the most general form of Theorem 9.13, we have in the above situation $\dim E_X = \dim E_Y$ that Theorem 9.14 requires at least maps of class $C^2(U, Y)$ and that U is a C^2 manifold: This is the reason why considering only manifolds and maps of class C^1 is not enough for our arguments and why we will have to use Theorems 9.9 and 9.12.

We will also make use of the classification theorem of compact 1-dimensional manifolds. The following proof is inspired by [107]. We note that an extension of this proof can also be used to characterize even all second countable 1-dimensional manifolds (as those manifolds which consist of at most countably many components being diffeomorphic to circles or intervals), but we will need only the compact case which is slightly simpler.

Theorem 9.15. *Every compact connected* 1*-dimensional* C^1 *manifold* X *(with or without boundary) is diffeomorphic to either the circle* $S^1 := S_1(0)$ *in* $\mathbb{R}^2$ *or to* $[0, 1]$.

Proof. By shrinking the charts if necessary, we can assume that X is covered by open sets, each of which is diffeomorphic to an interval. By the compactness, we have $X = U_1 \cup \cdots \cup U_n$ such that each U_k is diffeomorphic to an interval. Now if X is not diffeomorphic to S^1, we show inductively that there are pairwise different $k_1, \dots, k_j$ and diffeomorphisms f_j of an interval I_j onto $X_j := U_{k_1} \cup \cdots \cup U_{k_j}$. Then we are done for $j = n$.

For the induction start, we put $k_1 := 1$. For the induction step, note that the connectedness of X implies that there is some k_{j+1} different from $k_1, \dots, k_j$ such that $U := U_{k_{j+1}}$ intersects X_j. Let f be a diffeomorphism of an interval I onto U. We put now $M := f_j^{-1}(U \cap U_j)$ and consider

$$\Gamma := \operatorname{graph}(f^{-1} \circ f_j|_M) = \{(t, s) \in I_j \times I : f(s) = f_j(t)\}$$

as a subset of $I_j \times I$. Note that $g = f^{-1} \circ f_j|_M$ is a diffeomorphism (of M onto $f^{-1}(U \cap U_j)$), in particular Γ is the graph of the one-to-one C^1 function g with nonzero derivative. Moreover, if $(t, s) \in \Gamma$ belongs to the interior of $I_j \times I$ then

Γ extends to the left and right of t, that is, Γ cannot end in the interior of the rectangle $I_j \times I$. Since, by the injectivity of g, Γ intersects each side of the rectangle at most once, it follows that there are at most two components of Γ, each starting and ending at different sides of the rectangle. If there is only one component, M is an interval. In this case, it is clear that we can (after reparametrizing f) "concatenate" f_j and f to a homeomorphism of $X_j \cup U$ to an interval. Thus, assume Γ has two components, each connecting two sides of the rectangle. By the injectivity of g, these must be "neighboring" sides, M is the union of two disjoint open in M intervals M_1 and M_2, and by the injectivity of g the intervals $g(M_1)$ and $g(M_2)$ cannot intersect. Filling the "gaps" between the intervals M_1 and M_2 using f_j and between $g(M_1)$ and $g(M_2)$ using f, we see that there is a diffeomorphism h of S^1 onto $X_j \cup U$. Since $h(S^1) = X_j \cup U$ is open and compact and thus closed in X (Proposition 2.45), the connectedness of X implies $X = h(S^1)$. Hence, X is diffeomorphic to S^1 which we had excluded for the induction. □

9.2 Orientation of Continuous Maps and of Manifolds

In this section, we introduce the concept of an oriented continuous map in finite-dimensional manifolds. Although such a concept is already known, in principle, see e.g. [39, Chapter VIII, Exercise 4.10.6], it seems that the concept which we present here is much simpler than other similar concepts which can be found in literature. In fact, it turns out to be a special case of the concept which we introduced in Section 7.3.

Throughout this section, let X and Y be C^1 manifolds over real finite-dimensional spaces E_X and E_Y, respectively, with $\dim E_X = \dim E_Y > 0$.

Note that only in finite dimensions it makes sense to speak about the orientation of continuous maps and generalized homotopies in the following sense.

Definition 9.16. Let I be a topological space, $D_F \subseteq X$ and $D_H \subseteq I \times X$, and $F\colon D_F \to Y$ and $H\colon D_H \to Y$ be continuous. We associate to F and H the maps $A \in \mathcal{L}(D_F, TX, TY)$ and $B \in \mathcal{L}(D_H, TX, TY)$ where for $x \in D_F$ and $(t,x) \in D_H$ the corresponding maps $A(x) \in \mathcal{L}(T_xX, T_{F(x)}Y)$ or $B(t,x) \in \mathcal{L}(T_xX, T_{H(t,x)}Y)$ are defined as the zero maps: $A(x)u := 0$, $B(t,x)u := 0$ $(u \in T_xX)$.

An *orientation* of F or H on a subset $M \subseteq D_F$ or $M \subseteq D_H$ is an orientation σ of A or B on M in the sense of Definition 7.33, respectively.

F and H are called *orientable* (on M) if an orientation exists (on M). The couples (F,σ) or (H,σ) are called *oriented continuous maps*. Notationally, we just write F or H and refer to σ as "the orientation" of F or H.

Remark 9.17. Before we discuss Definition 9.16 in more detail, let us point out that in literature there are also (purely topological) notions of oriented maps which are available even for maps between C^0 manifolds, see for instance [39, Chapter VIII, Exercise 4.10.6].

In contrast, C^1 manifolds are essential for Definition 9.16, since otherwise we could not speak about the tangent bundles. The advantage of Definition 9.16 is that it is much more easier than the purely topological definition of oriented maps. In particular, Definition 9.16 can be verified in practice rather easily if F and σ are given "explicitly". In contrast, orientability in C^0 manifolds is defined by abstract lifting properties which can be very hard to verify. For this reason, we will not discuss the latter and just confine ourselves with the much simpler definition above for which we will show in the following a lot of useful properties.

The disadvantage is of course that we cannot deal with C^0 manifolds in this way which excludes some examples. However, since we will use a differential-topological approach for the degree theory, we will have to exclude C^0 manifolds later, anyway.

Unfortunately, for the case that $F \in C^1(\Omega, Y)$ with open $\Omega \subseteq X$, we have now two different notions of orientation which are not identical: The orientation as a continuous map in the sense of Definition 9.16 and the orientation as a Fredholm operator in the sense of Definition 8.25. The two notions differ in the definition of the associated map A: According to Definition 9.16, the associated map A is always zero while the orientation of Fredholm operators corresponds to the associated map dF. Thus, roughly speaking, from the viewpoint of orientations, "Definition 9.16 acts as if $dF = 0$".

However, this is only a formal difference: It turns out that F is orientable on $M \subseteq \Omega$ as a continuous map if and only if it is orientable on M as a Fredholm map, and if σ_C and σ_{C^1} denote the corresponding orientations, they are related by

$$\sigma_C(x) = dF(x) + \sigma_{C^1}(x) \quad \text{for all } x \in M. \tag{9.2}$$

To see this, since (9.2) holds by definition, we only have to verify that (9.2) is compatible with the additional "continuity" requirements for orientations:

Proposition 9.18. *Let I be a topological space, $\Omega \subseteq X$ be open, $W \subseteq I \times X$ be open, $F \in C^1(\Omega, Y)$ and $H: W \to Y$ be a partial C^1 homotopy. Let σ_C and σ_{C^1} denote the orientations in the sense of Definition* 9.16 *or* 8.25 *(or* 8.26*) pointwise, that is, on each one-element subset of $M \subseteq \Omega$ (or $M \subseteq W$). Then σ_C is an orientation in the sense of Definition* 9.16 *on M if and only if σ_{C^1} is an orientation in the sense of Definition* 8.25 *(or* 8.26*) on M.*

Proof. We show the assertion for H. Let $(t_0, x_0) \in M$, and let c_X and c_Y be charts for x_0 and $H(t_0, x_0)$, and let h_{c_X} and h_{c_Y} be as in Definition 8.20. Putting

$$s_C(t,x) := h_{c_Y}(H(t,x), \cdot)^{-1} \circ \sigma_C(t,x) \circ h_{c_X}(x, \cdot) \subseteq \mathcal{L}(E_X, E_Y)$$

and

$$s_{C^1}(t,x) := h_{c_Y}(H(t,x), \cdot)^{-1} \circ \sigma_{C^1}(t,x) \circ h_{c_X}(x, \cdot) \subseteq \mathcal{L}(E_X, E_Y),$$

we have to show that s_C is lower semicontinuous at (t_0, x_0) if and only if s_{C^1} is lower semicontinuous at (t_0, x_0). Since we have

$$\sigma_C(t,x) = d_X H(t,x) + \sigma_{C^1}(t,x)$$

by (9.2), we have

$$s_C(t,x) = s_{C^1}(t,x) + A(t,x)$$

with

$$A(t,x) := h_{c_Y}(H(t,x), \cdot)^{-1} \circ d_X H(t,x) \circ h_{c_X}(x, \cdot) \subseteq \mathcal{L}(E_X, E_Y).$$

Since the addition is a continuous operation, it thus suffices in view of Proposition 2.94 to show that A is continuous. By (8.8), we have

$$A(t,x) = d(c_Y \circ H(t, \cdot) \circ c_X^{-1})(c_X(x)),$$

and this is continuous, since c_X is continuous and $(t,u) \mapsto d(c_Y \circ H(t, \cdot) \circ c_X^{-1})(u)$ is continuous by the definition of a partial C^1 homotopy. The proof for F is analogous (just omit t in the arguments). □

In order to have a consistent definition of a sign, we thus define in the above case, that is, when $F \in C^1(\Omega, Y)$ and $\sigma = \sigma_c$ denotes the orientation of F in the sense of Definition 9.16:

$$\operatorname{sgn} dF(x) := \begin{cases} 1 & \text{if } dF(x) \in \sigma(x), \\ -1 & \text{if } dF(x) \in \operatorname{Iso}(T_x X, T_{F(x)} Y) \setminus \sigma(x), \\ 0 & \text{if } dF(x) \notin \operatorname{Iso}(T_x X, T_{F(x)} Y). \end{cases}$$

This is of course nothing else than saying that $\operatorname{sgn} dF(x)$ is the sign for the orientation of dF as a Fredholm operator induced by $\sigma_c = \sigma$ in the sense (9.2).

In case of C^1 maps, we can define a corresponding notion of inherited orientation:

Let $\Omega \subseteq X$ be open, $F \in C^1(\Omega, Y)$ be oriented, and $Y_0 \subseteq Y$ be a submanifold of dimension m which is transversal to F on an open subset $\Omega_0 \subseteq \Omega$. Recall that Theorem 8.55 implies that $X_0 := \Omega_0 \cap F^{-1}(Y_0)$ is a submanifold of Ω_0 (hence of X) of dimension m.

Definition 9.19. In the above situation, suppose that F is oriented on $M \subseteq \Omega_0$ in the sense of Definition 9.16. Then the *inherited orientation* of $F_0 := F|_{X_0} \in C^0(X_0, Y_0)$ on $M \cap X_0$ is defined as follows. We orient F as a C^1 map according to (9.2), obtain an inherited orientation on F_0 as a C^1 map according to Definition 8.65, and then, using (9.2) for F_0, we obtain the required orientation for F.

Proposition 9.20. *The inherited orientation is an orientation in the sense of Definition* 9.16.

Proof. According to Proposition 9.18, the orientation of F as a C^1 map is an orientation in the sense of Definition 8.25. By Proposition 8.67, also F_0 becomes oriented in the same sense, and the orientation in the sense of Definition 9.16 follows by applying Proposition 9.18 for F_0 in the opposite direction. □

Similarly, also the notion of the natural orientation of a diffeomorphism can be understood in the sense of Definition 9.16:

Definition 9.21. Let $U \subseteq X$ be open and $J \in C^1(U, V)$ be a diffeomorphism onto an open subset $V \subseteq Y$. The *natural orientation* of J is defined for $x \in U$ pointwise such that $dJ(x) \in \operatorname{Iso}(T_x X, T_y Y)$ represents the orientation of J at x in the sense of Definition 9.16.

Proposition 9.22. *The natural orientation of diffeomorphisms is an orientation in the sense of Definition* 9.16.

Proof. In view of Proposition 9.18, the assertion follows from Proposition 8.37. □

We have the following analogue of parts of Theorem 8.27.

Theorem 9.23. *Let I be a topological space, $\Omega \subseteq X$ and $W \subseteq [0, 1] \times X$ be open, and let $F: \Omega \to Y$ and $H: W \to Y$ be continuous.*

(a) *σ is an orientation for F or H if and only if the (pointwise) opposite orientation is an orientation for F or H, respectively.*

(b) *If $C \subseteq \Omega$ (or $C \subseteq W$) is connected and σ is an orientation for F (or H) on C then σ is uniquely determined if its value is known in one point of C.*

(c) *Let I be locally path-connected. If Ω is simply connected and if in some point of each component of Ω or W an orientation for F or H is given then F and H have a unique corresponding orientation on Ω or W.*

(d) *If* $W = [0,1] \times \Omega$ *then* H *is orientable if and only if* $H(t_0, \cdot)$ *is orientable for some* $t_0 \in [0,1]$. *In this case, the orientation of* H *is uniquely determined by the orientation of* $H(t_0, \cdot)$.

Proof. In view of Proposition 8.16, the proof of the corresponding assertions of Theorem 8.13 carries over word by word. □

For the case of C^1 maps, we also have the following analogues of previous results:

Theorem 9.24. *Let* I *be a topological space,* $\Omega \subseteq X$ *and* $W \subseteq I \times X$ *be open,* $F \in C^1(\Omega, Y)$, *and* $H: W \to Y$ *be a generalized partial* C^1 *homotopy.*

(a) *In case* $X = E_X$ *and* $Y = E_Y$ *and if* $C \subseteq \Omega$ *(or* $C \subseteq W$*) is connected and* $dF|_C$ *(or* $d_X H_C$*) is constant then all orientations of* F *or* H *on* C *are constant.*

(b) *Let* $M := C \times \{x\} \subseteq W$ *with a connected* $C \subseteq I$. *If* $d_X H|_M$ *is constant then all orientations of* H *on* M *are constant.*

(c) *Let* F *(or* H*) be oriented on a connected set* $C \subseteq \Omega$ *(or* $C \subseteq W$*). Suppose that* $dF(x) \subseteq \operatorname{Iso}(T_x X, T_{F(x)} Y)$ *for all* $x \in C$ *(or* $d_X H(t,x) \subseteq \operatorname{Iso}(T_x X, T_{H(t,x)} Y)$ *for all* $(t,x) \in C$*). Then* $\operatorname{sgn} F(x)$ *(or* $\operatorname{sgn} H(t, \cdot)(x)$*) are the same for all* $x \in C$ *(or* $(t,x) \in C$*).*

Proof. Using Proposition 9.18, we can transfer the assertions directly from Propositions 8.14, 8.28, and Theorem 8.27(b), respectively. □

Theorem 9.25. *Let* I *be a topological space,* $W \subseteq I \times \Omega$ *be open, and* $H: W \to Y$ *be a generalized partial* C^1 *homotopy. Let* J_X *be a diffeomorphism of an open subset of* E_X *onto* $U \subseteq X$. *We put* $W_0 := W \cap (I \times U)$. *Let* $V \subseteq Y$ *be a neighborhood of* $H(W_0)$, *and* J_Y *be a diffeomorphism of* V *onto an open subset of* E_Y. *For* $t \in I$, *we put* $h_t := J_Y \circ H(t, \cdot) \circ J_X$.

Let H *be oriented on* $M \subseteq W_0$, *and let* (t_i, x_i) $(i = 0, 1)$ *belong to the same component of* M *and be such that* $h_{t_i}(x_i) \in \operatorname{Iso}(E_X, E_Y)$ $(i = 0, 1)$. *Then we have with* $u_i := J_X^{-1}(x_i)$ $(i = 0, 1)$ *that*

$$\operatorname{sgn} d_X H(t_1, x_1) = \operatorname{sgn} d_X H(t_2, x_2) \iff \operatorname{sgn} \det(dh_{t_1}(x_1)^{-1} dh_{t_2}(x_2)) > 0.$$

Proof. We equip $H|_{W_0}(t, \cdot)$ with the corresponding orientation of Proposition 9.18, that is, we understand $H|_{W_0}$ correspondingly oriented as a Fredholm homotopy in the sense of Definition 8.26. Proposition 8.38 implies that the orientation of $H|_{W_0}$ on M induces an orientation of the Fredholm homotopy

$h(t,x) := h_t(x)$ on $M_0 := \{(t,u) : (t, J_X(u)) \in M\}$ such that

$$\operatorname{sgn} d_X h(t,u) = \operatorname{sgn} d_X H(t, J_X(u)) \quad \text{for all } (t,u) \in M_0.$$

Note that the right-hand side is also the sign in the sense of the assertion. Since Proposition 7.22 implies

$$\operatorname{sgn} d_X h(t_1,x_1) = \operatorname{sgn} d_X h(t_2,x_2) \iff \\ \operatorname{sgn} \det(d_X h(t_1,x_1)^{-1} d_X h(t_2,x_2)) > 0,$$

the assertion follows. □

Definition 9.26. For $i = 1,2$, let X_i and Y_i be C^1 manifolds over real vector spaces E_{X_i} and E_{Y_i}, respectively, such that $\dim E_{X_i} = \dim E_{Y_i} < \infty$. Let $\Omega_i \subseteq X_i$ be open and $F_i \in C(\Omega_i, Y_i)$ have orientations σ_i on $M_i \subseteq \Omega_i$. Then the *product orientation* $\sigma(x_1,x_2)$ of $F_1 \otimes F_2 \in C(\Omega_1 \times \Omega_2, Y_1 \times Y_2)$ is for $(x_1,x_2) \in M_1 \times M_2$ defined as the product orientation of $\sigma_1(x_1)$ and $\sigma_2(x_2)$.

Proposition 9.27. *The above defined product orientation is an orientation of $F_1 \otimes F_2$ on $M_1 \times M_2$.*

Proof. The assertion follows from Proposition 7.43. □

Definition 9.28. Let X, Y, Z be C^1 manifolds over real vector spaces E_X, E_Y, and E_Z with $0 < \dim E_X = \dim E_Y = \dim E_Z < \infty$. Let $M \subseteq X$, $M_0 \subseteq Y$ and let $F\colon M \to Y$ and $G\colon M_0 \to Z$ be continuous and oriented with orientations σ_F and σ_G, respectively, and such that $F(M) \subseteq M_0$.

Then the *composite orientation* $\sigma_{G \circ F}$ of $G \circ F\colon M \to Z$ is defined by

$$\sigma_{G\circ F}(x) := \sigma_G(F(x)) \circ \sigma_F(x). \tag{9.3}$$

Proposition 9.29. *The formula* (9.3) *gives pointwise the composite orientation in the sense of Definition* 7.45 *for the corresponding maps from Definition* 9.16.

The composite orientation is an orientation of $G \circ F$ in the sense of Definition 9.16. *Moreover:*

(a) *For each orientation of $G \circ F$ and each orientation of G there is exactly one orientation of F such that $G \circ F$ carries the corresponding composite orientation.*

(b) *If $F\colon M \to F(M)$ is an oriented homeomorphism then for each orientation of $G \circ F$ there is exactly one orientation of $G|_{F(M)}$ such that $G \circ F$ carries the corresponding composite orientation.*

Proof. Let $A_F(x) \in \mathcal{L}(T_xX, T_{F(x)}Y)$, $A_G(y) \in \mathcal{L}(T_yY, T_{G(y)}Z)$, and $A_{G\circ F}(x) \in \mathcal{L}(T_xX, T_{G(F(x))}Z)$ denote the zero maps. Then (9.3) is the composite orientation of $A_G(F(x))A_F(x)$ on M according to Definition 7.5. This shows the first assertion. Now the remaining assertions follow from Proposition 7.45 with $\tau = F$. □

Corollary 9.30. *Let X, X_0 and Y, Y_0 be manifolds over real vector spaces of the same finite dimension. Let $M_0 \subseteq X$, $M \subseteq X$, J_X an oriented homeomorphism of M_0 onto M, $F \in C(M, Y)$, and $J_Y \in C(F(M), Y_0)$. Then the orientations of $F_0 := J_Y \circ F \circ J_X$ are exactly the composite orientations induced by the orientations of J_Y and J_X and an orientation of F.*

Proof. The assertion follows by applying Proposition 9.29 to the compositions $J_Y \circ F$ and $F \circ J_X$, respectively. □

In the case that J_X and J_Y are diffeomorphisms, we can choose the natural orientations:

Corollary 9.31. *Proposition* 8.38 *holds also when all orientations are understood in the sense of Definition* 9.16 *or Definition* 9.21, *respectively.*

Proof. Equipping J_X and J_Y with the natural orientations, the pointwise formulas for the orientation follow straightforwardly from Corollary 9.30 and Proposition 7.6. The remaining assertions carry over from Proposition 8.38 in view of Proposition 9.22. □

There is also a corresponding variant of Proposition 9.29 for generalized homotopies.

Let X, Y, Z be manifolds over real vector spaces of the same finite dimension. For $M \subseteq [0,1] \times X$ and continuous $H_1 \colon M \to Y$, we define $\hat{H}_1 \colon M \to [0,1] \times Y$ by $\hat{H}_1(t,x) := (t, H_1(t,x))$. Let $M_0 \subseteq [0,1] \times Y$ satisfy $\hat{H}_1(M) \subseteq M_0$, and let $H_2 \colon M_0 \to Z$ be continuous. Then $H := H_2 \circ \hat{H}_1 \colon M \to Z$ is continuous. We define the composite orientation of H pointwise for each fixed t as the composite orientation of the maps $H_1(t,\cdot)$ and $H_2(t,\cdot)$.

Proposition 9.32. *In the above situation, if H_1 and H_2 are oriented, then the composite orientation of $H := H_2 \circ \hat{H}_1$ is an orientation of H. Moreover:*

(a) *If H and H_2 are oriented, then there is a unique orientation of H_1 on M such that $H_2 \circ \hat{H}_1$ carries the composite orientation.*

(b) *If H and H_1 are oriented on M and if $\hat{H}_1 \colon M \to \hat{H}_1(M)$ is a homeomorphism then there is a unique orientation of H_2 on $\hat{H}_1(M)$ such that $H_2 \circ \hat{H}_1$ carries the composite orientation.*

Proof. The assertion follows from Proposition 7.45 with $\tau = \hat{H}_1$. □

Only in finite dimensions, it makes sense to speak about the orientation of manifolds.

Definition 9.33. An *orientation* of X on $M \subseteq X$ is an orientation of the tangent bundle TX on M in the sense of Definition 8.20.

There is a strong relation between oriented maps and orientations of manifolds:

Proposition 9.34. *Let $D_F \subseteq X$, $D_H \subseteq [0,1] \times X$, and let $F: D_F \to Y$ and $H: D_H \to Y$ be continuous. Then the following holds:*

(a) *If Y is oriented on M_Y and X is oriented on M_X then the pointwise induced orientation in the sense of Definition* 7.11 *is an orientation of F and H on $M_X \cap F^{-1}(M_Y)$ or $([0,1] \times M_X) \cap H^{-1}(M_Y)$, respectively.*

(b) *If Y is oriented on M_Y and F is oriented on M_X then the pointwise orientation induced by F and Y in the sense of Proposition* 7.13 *is an orientation of X on $M_X \cap F^{-1}(M_Y)$.*

(c) *If F and X are oriented on M and $F|_M$ is a homeomorphism onto $F(M) \subseteq Y$ then the pointwise orientation induced by F and X in the sense of Proposition* 7.13 *is an orientation of Y on $F(M)$.*

Proof. The assertions follow immediately from the corresponding assertions of Proposition 7.48. □

9.3 The C^r Brouwer Degree

In this section, let X and Y be Banach manifolds of class C^q $(1 \leq q \leq \infty)$ without boundaries over real vector spaces E_X and E_Y with $0 < \dim E_X = \dim E_y < \infty$.

The aim of the following considerations is to establish the Brouwer degree for maps $F \in C^r(\Omega, Y)$ $(0 \leq r \leq q)$ with open $\Omega \subseteq X$.

The Brouwer degree actually comes in two flavors: One for general maps, in which case the Brouwer degree is a number from the group $\mathbb{Z}_2 = \{0, 1\}$ (with the usual addition modulo 2), and one for oriented maps, in which case the Brouwer degree is a number from the group $\mathbb{Z}$.

Definition 9.35. We write $(F, \Omega, y) \in \mathcal{B}^r(X, Y)$ if $\Omega \subseteq X$ is open, $F \in C^r(\Omega, Y)$, $y \in Y$, and $F^{-1}(y)$ is compact.

For the oriented version of the Brouwer degree, we assume in this definition also that F is oriented, that is, we also assume an orientation σ of F in the sense of Definition 9.16, although we do not write σ explicitly. In the oriented case, the set $\mathcal{B}^r(X,Y)$ consists of less maps in general, since not every map is orientable, in general. So, strictly speaking, we should use a different symbol for the oriented case and also write (F,σ) in that case instead of just F. However, we do not do this to treat the two cases (oriented and non-oriented) similar and to simplify notation.

We recall that in the oriented case and if F is of class C^1 in a neighborhood of x then

$$\operatorname{sgn} dF(x) = \begin{cases} 1 & \text{if } dF(x) \in \sigma(x), \\ -1 & \text{if } dF(x) \in \operatorname{Iso}(T_xX, T_{F(x)}Y) \setminus \sigma(x), \\ 0 & \text{if } dF(x) \notin \operatorname{Iso}(T_xX, T_{F(x)}Y). \end{cases}$$

In the non-oriented case, when we calculate only in $\mathbb{Z}_2$ modulo 2, we avoid case distinctions if we put

$$\operatorname{sgn} dF(x) := \begin{cases} 1 & \text{if } dF(x) \in \operatorname{Iso}(T_xX, T_{F(x)}Y), \\ 0 & \text{if } dF(x) \notin \operatorname{Iso}(T_xX, T_{F(x)}Y). \end{cases}$$

Now we come to the Brouwer degree. Roughly speaking, the idea is that $\deg(F,\Omega,y)$ should define an "homotopically invariant count" of the number of solutions of the equation $F(x) = y$ in Ω where "counting" has to be understood with certain multiplicities (where a multiplicity can also be negative, depending on the orientation). For the oriented case, we emphasize once more that we understand F tacitly equipped with an orientation σ, that is, we should write more precisely $\deg((F,\sigma),\Omega,y)$, but we do not do this to simplify notation. However, when we calculate with the degree in the oriented case, we have to specify, of course, which orientation is meant.

The definition which we use is the following. We fix $I_{E_X,E_Y} \in \operatorname{Iso}(E_X, E_Y)$ once and for all.

Definition 9.36. The C^r *Brouwer degree* $\deg = \deg_{(r,X,Y)}$ on X and Y is a map which associates to each $(F,\Omega,y) \in \mathcal{B}^r(X,Y)$ (with F being oriented in the oriented case) a number from $\mathbb{Z}_2$ (resp. from $\mathbb{Z}$ in the oriented case) such that the following holds.

($A_{\mathcal{B}}$) **(Homotopy Invariance).** If $\Omega \subseteq X$ is open and $H\colon [0,1] \times \Omega \to Y$ is an (oriented) partial C^r homotopy with $H^{-1}(y)$ being compact then

$$\deg(H(0,\cdot),\Omega,y) = \deg(H(1,\cdot),\Omega,y).$$

($B_{\mathcal{B}}$) **(Normalization).** Suppose that there are charts c and c_y defined at least on Ω and y, respectively, and that there is $v \in E_Y$ such that $(c_y \circ F \circ c^{-1})(u) = v + I_{E_X,E_Y}u$ for all $u \in c(\Omega)$. Suppose also that there is some (automatically unique) $x \in \Omega$ with $F(x) = y$. Then

$$\deg(F, \Omega, y) = \operatorname{sgn} dF(x).$$

($C_{\mathcal{B}}$) **(Excision).** If $\Omega_0 \subseteq \Omega$ is an open neighborhood of $F^{-1}(y)$ then

$$\deg(F, \Omega_0, y) = \deg(F, \Omega, y).$$

($D_{\mathcal{B}}$) **(Additivity).** If $\Omega = \Omega_1 \cup \Omega_2$ with disjoint open subsets $\Omega_i \subseteq \Omega$ then

$$\deg(F, \Omega, y) = \deg(F, \Omega_1, y) + \deg(F, \Omega_2, y).$$

For the homotopy invariance, we note that the assumptions automatically imply $(H(i,\cdot), \Omega, y) \in \mathcal{B}^r(X, Y)$ $(i = 0, 1)$, since $H(i,\cdot)^{-1}(y)$ is compact by Proposition 2.62.

Concerning the normalization property, we recall that by a chart we mean only a member of the atlas and that we do not assume that the atlas is maximal. In particular, there might be very few charts in which case the normalization property is only a requirement for very few maps F.

Remark 9.37. The formulation of the excision and additivity properties is a bit sloppy:
To be formally correct, we should have written $\deg(F|_{\Omega_i}, \Omega_i, y)$ instead of $\deg(F, \Omega_i, y)$ (for $i = 0, 1, 2$), and in the oriented case, we should have said that we use the restriction of the original orientation.

However, in situations like these, such formal restriction symbols are omitted for better readability.

One aim of the next two sections is to prove:

Theorem 9.38. *For each fixed manifolds X, Y of class C^q $(1 \le q \le \infty)$ and each $0 \le r \le q$ there is exactly one oriented C^r Brouwer degree $\deg_{(r,X,Y)}$. The degrees are compatible in the sense that*

$$\deg_{(R,X,Y)} = \deg_{(r,X,Y)}|_{\mathcal{B}^R(X,Y)} \quad \text{for } 0 \le r \le R \le q. \tag{9.4}$$

The oriented and non-oriented versions $\deg$ *and* $\deg^*$ *are compatible in the sense that for all $(F, \Omega, y) \in \mathcal{B}^r(X, Y)$*

$$\deg(F, \Omega, y) \equiv \deg^*(F, \Omega, y) \pmod 2 \quad \text{if } F \text{ is oriented.} \tag{9.5}$$

In the next two sections, we will also learn a lot of further useful properties of the degree so that these sections should not be skipped completely, even if the reader is not interested in the proof of Theorem 9.38.

Remark 9.39. For the case that $X = E_X$ and $Y = E_Y$ (with id_X and id_Y as the only charts), the normalization property can be replaced by an apparently even weaker property:

($\mathrm{B}_{\mathscr{B}}$) **(Normalization for Vector Spaces)**.

$$\deg(I_{E_X,E_Y}, X, y) = \operatorname{sgn} I_{E_X,E_Y}.$$

We carry out the following proof of Remark 9.39 very detailed to explain how to work with the orientation.

Proposition 9.40. *Let X and Y be fixed. In case $X = E_X$, $Y = E_Y$, the C^r Brouwer degree with the normalization property replaced by that from Remark* 9.39 *satisfies automatically the normalization property of Definition* 9.36.

Proof. Let $\Omega \subseteq X = E_X$ be open and $y, y_0 \in Y$ such that $y \in I_{E_X,E_Y}(\Omega) + y_0$. Putting $F := I_{E_X,E_Y}|_\Omega + y_0$ and $x_0 := I_{E_X,E_Y}^{-1}(y - y_0)$, we have to prove that $\deg(F, \Omega, y) = \operatorname{sgn} dF(x_0)$.

In the oriented case, let σ denote the orientation of $dF(x_0) = I_{E_X,E_Y}$. Let $\Omega_0 \subseteq \Omega$ be open and connected with $x_0 \in \Omega_0$. Note that $dF(x) = I_{E_X,E_Y}$ for every $x \in \Omega_0$. Since Ω_0 is connected, Theorem 9.24(a) thus implies that the orientation of $dF(x)$ is actually independent of $x \in \Omega_0$. We put now $G := I_{E_X,E_Y} + y_0$ where we equip $dG(x) = I_{E_X,E_Y}$ with the same orientation as $dF(x_0) = I_{E_X,E_Y}$. Then the restriction $G|_{\Omega_0}$ has the same orientation as $F|_{\Omega_0}$. Using now the excision property of the degree twice, we find that

$$\deg(F, \Omega, y) = \deg(F, \Omega_0, y) = \deg(G, \Omega_0, y) = \deg(G, X, y).$$

In the non-oriented case, we have of course the same formula, and it would not have been necessary to pass to Ω_0 first, since F and $G|_\Omega$ are identical as maps. However, the orientation of these maps might differ on components of Ω which do not contain Ω_0.

As a final step, we consider the Fredholm homotopy $H(t,x) := (1-t)y_0 + I_{E_X,E_Y}x$. In the oriented case, we equip $d_X H(t,x) = I_{E_X,E_Y}$ for every $(t,x) \in [0,1] \times X$ with the orientation of $dF(x_0) = I_{E_X,E_Y}$. The homotopy invariance implies

$$\deg(G, X, y) = \deg(H(0,\cdot), X, y) = \deg(H(1,\cdot), X, y) = \deg(I_{E_X,E_Y}, X, y).$$

The claim now follows by using the normalization property from Remark 9.39. □

Remark 9.41. A more classical definition of the Brouwer degree does not consider oriented maps but requires instead that the manifolds X and Y be oriented.

Note that when X and Y are oriented then all maps and generalized homotopies obtain an induced orientation according to Proposition 9.34.

If one restricts in the oriented case the Brouwer degree to maps with that induced orientation we speak of the C^r Brouwer degree for oriented manifolds.

Of course, the existence assertion of Theorem 9.38 implies immediately that a C^r Brouwer degree for oriented manifolds exists. However, the uniqueness of that degree does not immediately follow from the uniqueness assertion of Theorem 9.38, since the Brouwer degree for oriented manifolds is only a restriction of the degree considered in Theorem 9.38.

For this reason we point out that actually all proofs and statements in Sections 9.4, and 9.5 hold also with obvious modifications for the case that X and Y are oriented manifolds and that we consider only the orientations of maps induced by X and Y. For instance, in the proof of Proposition 9.40, we need not discuss orientations at all since the induced orientation is automatically the (only) "correct" orientation for which we can apply the homotopy invariance. Similar considerations hold for the subsequent proofs.

In particular, it follows then that (also the uniqueness assertion of) Theorem 9.38 holds in this case.

Remark 9.42. We should briefly mention the advantages that we have (or not have) by considering oriented *maps* instead of considering oriented *manifolds* as in Remark 9.41.

One advantage is evident: There are cases where one can speak about the Brouwer degree with values in $\mathbb{Z}$ in case of oriented maps while the (oriented) degree of Remark 9.41 does not exist. For example, every diffeomorphism of X onto Y is orientable by Proposition 9.22, even if X (and thus necessarily also Y) fails to be orientable. Thus, we really obtain a theory which applies for more maps than the classical degree of Remark 9.41 on oriented manifolds.

On the other hand, the most interesting case for us in the later applications is that $Y = E_Y$ is a (finite-dimensional) normed space. In this case, Y is obviously orientable, and so Proposition 9.34 implies that if there is an orientable $F \in C(X, Y)$ then X is orientable.

However, the existence of such a map is not a requirement for our (oriented) degree theory: Even if X is non-orientable, we can apply the (oriented) degree theory for oriented maps $F \in C(\Omega, Y)$ with open $\Omega \subseteq X$. This implies that Ω is orientable (and thus $\Omega \neq X$), but typically there are a lot of such maps. For instance, if $Y = \mathbb{R}^2$ and X is the Moebius strip, any set $\Omega \subseteq X$ can do which only omits a line which "cuts through the strip".

Of course, one can say that in such a case, one could again apply the degree theory of Remark 9.41 with the oriented manifolds Ω and Y. This is true, but then one cannot easily deduce relations between the degrees of map when the set $\Omega \subseteq X$ changes. In contrast, such a change causes no technical difficulties for the degree theory for oriented *maps*: The generalized homotopy invariance which we will prove later on and which combines the excision property and homotopy invariance in a sense can easily deal with varying sets Ω. In contrast, it would be very hard even to just formulate such a result for the degree of Remark 9.41 if X is non-orientable.

A more practical advantage of the degree theory for oriented *maps* (compared to the degree theory of maps between oriented *manifolds*) will become clear in Chapter 10: For Fredholm maps in infinite-dimensional spaces, we can only speak about orientation of maps, and for certain restrictions of such maps, we obtain naturally an inherited orientation for maps between finite-dimensional submanifolds. For this application, it appears more natural to use directly the degree theory for oriented maps.

Remark 9.43. The classical Brouwer degree on a finite-dimensional normed space E occurs by putting $X = Y = E_X = E_Y = E$ (with only one chart) and considering the same orientation on X and Y. For the normalization in Remark 9.39, one chooses $I_{E,E} = \mathrm{id}_E$ for the classical Brouwer degree, that is, the normalization property becomes simply

$$\deg(\mathrm{id}_E, E, y) = 1.$$

In view of Remark 9.41, our proof shows also the uniqueness of this classical Brouwer degree on E under this very natural normalization property. This special case is of course a well-known result.

Remark 9.44. In the classical definition of the Brouwer degree, one considers only triples (F, Ω, y) with the additional property that $\overline{\Omega}$ is compact and F has a continuous extension to $\overline{\Omega}$, and $y \notin F(\partial\Omega)$. It turns out that the definitions are actually equivalent (by Corollary 9.3 and the excision property), but nevertheless our definition applies to a larger class of maps and is therefore more convenient to apply. There is one important exception, namely the Rouché property of the degree which we discuss later on is a bit more complicated to formulate. However, for the case $Y \neq E_Y$ the classical Rouché property cannot be formulated, anyway.

The reader who is completely unfamiliar with the Brouwer degree in $\mathbb{R}^n$ might want to have a look at Section 9.6 to get a feeling how useful the degree is to prove difficult and interesting assertions. This might help to motivate to go through the details of the rather cumbersome existence and uniqueness proofs of the following sections.

9.4 Uniqueness of the Brouwer Degree

Throughout this section, let X and Y be manifolds of class C^q $(1 \leq q \leq \infty)$ without boundaries over real vector spaces E_X and E_Y with $0 < \dim E_X = \dim E_Y < \infty$, and let $0 \leq r \leq q$.

Proposition 9.45. *The C^r Brouwer degree on fixed X and Y automatically satisfies for every $(F, \Omega, y) \in \mathcal{B}^r(X, Y)$:*

($\mathrm{E}_{\mathcal{B}}$) **(Existence).** *If $\deg(F, \Omega, y) \neq 0$ then $y \in F(\Omega)$.*

($\mathrm{F}_{\mathcal{B}}$) **(Excision-Additivity).** *If $\Omega_i \subseteq \Omega$ $(i \in I)$ is a family of pairwise disjoint open sets with $F^{-1}(y) \subseteq \bigcup_{i \in I} \Omega_i$ such that $F^{-1}(y) \cap \Omega_i$ is compact for all $i \in I$ then*

$$\deg(F, \Omega, y) = \sum_{i \in I} \deg(F, \Omega_i, y),$$

where in the sum at most a finite number of summands is nonzero.

Proof. If $y \notin F(\Omega)$, we can choose $\Omega_0 = \varnothing$ in the excision property and find $\deg(F, \Omega, y) = \deg(F, \varnothing, y)$. Choosing $\Omega_1 = \Omega_2 = \varnothing$ in the additivity property, we obtain in turn that $\deg(F, \varnothing, y) = 0$. Hence, $\deg(F, \Omega, y) = 0$.

For the excision-additivity property, we note that Ω_i $(i \in I)$ constitute an open cover of the compact set $F^{-1}(y)$. Hence, there is a finite subcover $\Omega_{i_1}, \dots, \Omega_{i_n}$. All Ω_i different from these sets satisfy $y \notin F(\Omega_i)$, and so the existence property implies $\deg(F, \Omega_i, y) = 0$ for those i. Now the excision property with $\Omega_0 := \Omega_{i_1} \cup \dots \cup \Omega_{i_n}$ and the additivity applied for $\deg(F, \Omega_0, y)$ give the required formula after a trivial induction. □

The reader be warned that the converse of the existence property does not hold, that is, $\deg(F, \Omega, y) = 0$ does not imply that $y \notin F(\Omega)$. In fact, assume for instance that $F^{-1}(y)$ consists of exactly two points $x_1, x_2 \in \Omega$ such that there are disjoint neighborhoods Ω_i of x_i with $\deg(F, \Omega_i, y) = (-1)^i$. Then we have by the excision-additivity that $\deg(F, \Omega, y) = 0$. However, a similar phenomenon can also occur if $F^{-1}(y)$ consists only of a single point: Intuitively, that point might have a "multiplicity" of 0. An example with $\Omega = X = Y = \mathbb{R}$ and $y = 0$

is the map $F(x) = x^2$. Indeed, the homotopy invariance with $H(t,x) := x^2 + t$ and existence property imply $\deg(F, \Omega, 0) = \deg(F + 1, \Omega, 0) = 0$.

Let us now prove the uniqueness of $\deg(F, \Omega, y)$ for the case that $F \in C^1$ and y is a regular value of F. Note that in this case Remark 8.58 implies that $F^{-1}(y)$ is discrete so that the compactness of $F^{-1}(y)$ implies that $F^{-1}(y)$ is actually a finite set.

The uniqueness (if y is a regular value) now follows from the subsequent regular normalization property:

Proposition 9.46. *The C^r Brouwer degree on fixed X and Y automatically satisfies for every $(F, \Omega, y) \in \mathcal{B}^r(X, Y)$:*

($G_{\mathcal{B}}$) **(Normalization for Diffeomorphisms).** *If F is a diffeomorphism onto an open subset of Y then*

$$\deg(F, \Omega, y) = \begin{cases} 0 & \text{if } y \notin F(\Omega), \\ \operatorname{sgn} dF(F^{-1}(y)) & \text{otherwise.} \end{cases}$$

($H_{\mathcal{B}}$) **(Regular Normalization).** *If F is C^1 in an open neighborhood $\Omega_0 \subseteq \Omega$ of $F^{-1}(y)$ and if y is a regular value of $F|_{\Omega_0}$ then we have a finite sum*

$$\deg(F, \Omega, y) = \sum_{x \in F^{-1}(y)} \operatorname{sgn} dF(x).$$

Recall that we define the empty sum as 0. Hence, the normalization property for diffeomorphisms is just a simple special case of the regular normalization property. The normalization property for diffeomorphisms deserves a special mentioning anyway, since a lot of references call this the normalization property of the degree. However, we point out once more that the normalization property of Definition 9.36 is much easier to verify, since if the atlas of X and Y is small, it involves only a few maps. In fact, if one wants to verify that property, one might use a minimal atlas for X and Y, as we will see.

Proof of the Regular Normalization. In case $F^{-1}(y) = \varnothing$, the existence property implies that we must have $\deg(F, \Omega, y) = 0$. Thus, it remains to discuss the case that $F^{-1}(y)$ contains finitely many points $x_1, \ldots, x_n \in \Omega_0$ $(n \geq 1)$.

By the inverse function Theorem 8.39, there are open neighborhoods $U_k \subseteq \Omega_0$ of x_k $(k = 1, \ldots, n)$ such that $F|_{U_k}$ is a diffeomorphism onto an open subset of Y containing y. Let c_y be a chart of Y for y, and $R > 0$ be such that the range of c_y contains at least the ball $B_R(v) \subseteq E_Y$ with $v := c_y(y)$. Shrinking U_k if necessary, we can assume that the U_k are pairwise disjoint and that there are charts c_k of X defined at least on U_k and that c_y is defined at least on $F(U_k)$.

By Proposition 9.1, there are open neighborhoods $\Omega_k \subseteq X$ of x_k such that $\overline{\Omega}_k \subseteq U_k$ is compact with $c_k(\Omega_k) = B_{r_k}(u_k) \subseteq E_X$ and $c_k(\overline{\Omega}_k) = K_{r_k}(u_k)$ with $u_k := c_k(x_k)$. We obtain by the excision-additivity property that

$$\deg(F, \Omega, y) = \deg(F, \Omega_1, y) + \cdots + \deg(F, \Omega_n, y).$$

Hence, it suffices to prove

$$\deg(F, \Omega_k, y) = \operatorname{sgn} dF(x_k). \tag{9.6}$$

By construction, $G_k := c_y \circ F \circ c_k|_{\overline{U}_k}^{-1}$ is a diffeomorphism of $c_k(U_k)$ onto an open subset of E_Y. Let $A_k := dG_k(u_k) \in \operatorname{Iso}(E_X, E_Y)$. Shrinking r_k if necessary, we can assume by the definition of the derivative

$$\|G_k(u) - v - A_k(u - u_k)\| \le \frac{1}{2\|A_k^{-1}\|}\|u - u_k\| \quad \text{for all } u \in K_{r_k}(u_k). \tag{9.7}$$

We consider now the homotopies $h_k\colon [0, 1] \times K_{r_k}(u_k) \to E_Y$, defined by

$$h_k(t, u) := (1 - t)G(u) + t(v + A_k(u - u_k)).$$

For u sufficiently close to u_k, we have $G(u), v + A_k(u - u_k) \in B_R(v)$. Shrinking r_k if necessary, we can assume without loss of generality that this is the case for all $u \in K_{r_k}(u_k)$. Proposition 3.3 implies then that $h_k([0, 1] \times K_{r_k}(u_k)) \subseteq B_R(v)$. Hence, $H_k(t, x) := c_y^{-1}(h_k(t, c_k(x)))$ defines a partial C^r homotopy on Ω_k which is also continuous on $[0, 1] \times \overline{\Omega}_k$. Note that $H_k(0, \cdot) = c_y^{-1} \circ G \circ c_X = F$. For h_{c_k} and h_{c_y} as in Definition 8.20, we calculate

$$\begin{aligned} dH_k(t, x_k) &= h_{c_y}(y, \cdot) \circ ((1 - t) + t)A_k \circ h_{c_k}^{-1}(x_k, \cdot) \\ &= h_{c_y}(y, \cdot) \circ A_k \circ h_{c_k}^{-1}(x_k, \cdot) = dF(x_k). \end{aligned}$$

We put $F_k := H_k(1, \cdot)$. If F is oriented then Theorem 9.23(d) implies that H_k has a unique orientation such that $H_k(0, \cdot) = F$ has the same orientation as F. Since $d_X H_k(t, x_k) = dF(x_k)$ is independent of $t \in [0, 1]$, we obtain from Theorem 9.24(b) that $H_k(1, \cdot)$ carries on $\{x_k\}$ the same orientation as F. We equip F_k with the orientation of $H_k(1, \cdot)$.

For all $t \in [0, 1]$ and all $u \in K_{r_k}(u_k)$, we calculate by (9.7) that

$$\begin{aligned} \|h_k(t, u) - v\| &= \|(1 - t)G_k(u) + tA_k(u - u_k) - (1 - t)v\| \\ &\ge \|A_k(u - u_k)\| - (1 - t)\frac{\|u - u_k\|}{2\|A_k^{-1}\|}. \end{aligned}$$

Since $(1-t)\|u-u_k\| \leq \|A_k^{-1}\|\|A_k(u-u_k)\|$, we obtain $h_k(t,u) \neq v$ for $u \neq u_k$, in particular for $\|u-u_k\| = r_k$. Hence, $y \notin H_k([0,1] \times \partial\Omega_k)$. It follows that the compact set $H_k^{-1}(y) \subseteq \overline{\Omega}_k$ is contained in Ω_k, and so $H_k|_{[0,1]\times\Omega_k}^{-1}(y)$ is compact. The homotopy invariance of the degree thus implies

$$\begin{aligned}\deg(F,\Omega_k,y) &= \deg(H_k(0,\cdot),\Omega_k,y)\\ &= \deg(H_k(1,\cdot),\Omega_k,y) = \deg(F_k,\Omega_k,y),\end{aligned}$$

where in the oriented case the orientations of F and F_k are such that $dF(x_k) = dF_k(x_k)$ have the same orientation, in particular $\operatorname{sgn} dF(x_k) = \operatorname{sgn} dF_k(x_k)$.

Let now $k \in \{1,\dots,n\}$ be fixed. The subsequent Lemma 9.47 implies that there is a continuous map $\gamma_0\colon [0,1] \to \operatorname{Iso}(E_Y)$ with $\gamma_0(0) := A_k I_{E_X,E_Y}^{-1}$ and $\gamma_0(1) = \operatorname{id}_{E_Y}$ (in the case $\det(A_k I_{E_X,E_Y}^{-1}) > 0$) or $\gamma_0(1) = J$ (in the case $\det(A_k I_{E_X,E_Y}^{-1}) < 0$) where $J \in \operatorname{Iso}(E_Y)$ is an arbitrary fixed map with $\det J < 0$. Then $\gamma\colon [0,1] \to \operatorname{Iso}(E_X,E_Y)$, defined by $\gamma(t) := \gamma_0(t) \circ I_{E_X,E_Y}$, is continuous with $\gamma(0) = A_k$ and $\gamma(1) = I_{E_X,E_Y}$ or $\gamma(1) = JI_{E_X,E_Y}$. We put $h(t,u) := v + \gamma(t)(u-u_k)$ and $H(t,x) := c_y^{-1}(h(t,c_k(x)))$. Shrinking r_k if necessary, we can assume that H is defined on $[0,1] \times \Omega_k$.

Since $H(0,\cdot) = F_k$, Theorem 9.23(d) implies that in the oriented case, we can orient H uniquely on $[0,1] \times \Omega_k$ such that $H(0,\cdot)$ is oriented as F_k. Theorem 9.24(c) implies in this case that

$$\operatorname{sgn} d_X H(1,x_k) = \operatorname{sgn} d_X H(0,x_k) = \operatorname{sgn} dF_k(x_k) = \operatorname{sgn} dF(x_k).$$

The homotopy invariance gives us

$$\deg(F_k,\Omega_k,y) = \deg(H(1,\cdot),\Omega_k,y).$$

In the case $\det(A_k I_{E,Y}^{-1}) > 0$, we have $(c_y \circ H(1,\cdot) \circ c_k^{-1})(u) = v + I_{X,Y}(u-u_k)$, and so the normalization property of the degree implies

$$\deg(H(1,\cdot),\Omega_k,y) = \operatorname{sgn} d_X H(1,x_k) \tag{9.8}$$

which implies (9.6). We have to prove (9.8) also for the remaining k, that is in case $\det(A_k I_{E_X,E_Y}^{-1}) < 0$. In that case, we have $H(1,x) = c_y^{-1}(v + JI_{E_X,E_Y}(c_k(x) - u_k))$ where we are still free to choose $J \in \operatorname{Iso}(E_Y)$ with $\det J < 0$. We fix some $e \in E_Y$ with $0 < \|e\| < R/3$ and $f \in E_Y^*$ with $f(e) = 1$. Note that if we choose e as one vector of the basis in E_Y then $J(z) := \operatorname{id}_{E_Y}(z) - 2f(z)e$ is represented as a diagonal matrix which has -1 in one diagonal entry (corresponding to the basis vector e), and $+1$ in all others. Hence, $\det J = -1$. Shrinking r_k if necessary, we assume $\|I_{E_X,E_Y}\|(1 + \|f\|\|e\|)r_k <$

$R/3$. We fix now $\varepsilon > 0$ so small that $c_k(\Omega_k)$ contains all points from $M_0 := u_k + I_{E_X,E_Y}^{-1}([0,\varepsilon]e)$.

Let $g\colon \mathbb{R} \to (-1,1)$ be a C^r-function which is strictly negative in $(0,\varepsilon)$, strictly positive outside of $[0,\varepsilon]$, and which satisfies $g(s) = -s$ in a neighborhood of 0 and $g(s) = s - \varepsilon$ in a neighborhood of ε. We put $h_0(t,s) := g(s) - s + t$ and define $h_g\colon [0,1] \times E_X \to E_Y$ by

$$h_g(t,u) := v + I_{E_X,E_Y}(u - u_k) + h_0(t, f(I_{E_X,E_Y}(u - u_k)))e.$$

Note that $|h_0(t,s)| < 2 + |s|$ for $t \in [0,1]$, and so

$$\begin{aligned}\|h_g(t,u) - v\| &< \|I_{E_X,E_Y}\|\|u - u_k\| + (2 + \|f\|\|I_{E_X,E_Y}\|\|u - u_k\|)\|e\| \\ &< \|I_{E_X,E_Y}\|(1 + \|f\|\|e\|)r_k + 2\|e\| < R\end{aligned}$$

for $u \in B_{r_k}(u_k)$. Hence, $h_g([0,1] \times B_{r_k}(u_k)) \subseteq B_R(v)$. We observe that $h_g(t,u) = v$ holds if and only if $z := I_{E_X,E_Y}(u - u_k)$ satisfies $z = se$ with $s := -h_0(t, f(z))$. Since $f(e) = 1$, the latter holds if and only if $s = -h_0(t,s)$, that is, if and only if $g(s) + t = 0$. Note that this implies $s \in [0,\varepsilon]$, that is, $z \in [0,\varepsilon]e$ and thus $u \in M_0$. Moreover, $g(s) + t = 0$ has for $t = 1$ no solution and for $t = 0$ exactly the two solutions $s = 0$ and $s = \varepsilon$, corresponding to $u = u_k$ and $u = v_k := u_k + I_{E_X,E_Y}^{-1}(\varepsilon e)$. Hence, if we define a partial C^r homotopy $H_0\colon [0,1] \times \Omega_k \to Y$ by $H_0(t,x) := c_y^{-1}(h_g(t, c_k(x)))$, the closed in $[0,1] \times \Omega_k$ set $H_0^{-1}(y)$ is contained in the compact subset $[0,1] \times c_k^{-1}(M_0) \subseteq [0,1] \times \Omega_k$ and thus compact. Moreover, $H_0(0,x) = y$ for $x \in \Omega_k$ if and only if $x = x_k$ or $x = \hat{x}_k := c_k^{-1}(v_k)$, and $y \notin H_0(\{1\} \times \Omega_k)$. Now by our choice of g we have for a sufficiently small open neighborhood $U_- \subseteq B_{r_k}(u_k)$ of v_k that

$$h_g(t,u) = v + I_{E_X,E_Y}(u - u_k) + (t - \varepsilon)e \quad \text{for all } u \in U_-. \tag{9.9}$$

Similarly, for a sufficiently small neighborhood $U_+ \subseteq B_{r_k}(u_k)$ of u_k, we have

$$\begin{aligned}h_g(0,u) &= v + I_{E_X,E_Y}(u - u_k) - 2f(I_{E_X,E_Y}(u - u_k)) \\ &= v + JI_{E_X,E_Y}(u - u_k) \quad \text{for all } u \in U_+.\end{aligned} \tag{9.10}$$

Then $\Omega_\pm := c_k(U_\pm)$ are neighborhoods of x_k and $\hat{x}_k$, respectively, without loss of generality disjoint. It follows that $H_0(0,\cdot)$ coincides with $H(1,\cdot)$ in Ω_+, in particular $d_X H_0(0,x_k) = d_X H(1,x_k)$.

In the oriented case, since Ω_k is homeomorphic to $B_{r_k}(u_k)$ and thus simply connected, Theorem 9.23(c) implies that we can orient H_0 on $[0,1] \times \Omega_k$ uniquely such that the orientation at $(0,x_k)$ coincides with the orientation of H at $(1,x_k)$. Since $d_X H_0(0,x_k) = d_X H(1,x_k)$, we obtain

$$\operatorname{sgn} d_X H_0(0,x_k) = \operatorname{sgn} d_X H(1,x_k).$$

We have by the excision property of the degree that

$$\deg(H(1,\cdot),\Omega_k,y)=\deg(H(1,\cdot),\Omega_+,y)=\deg(H_0(0,\cdot),\Omega_+),$$

and by the excision-additivity and homotopy invariance of the degree that

$$\begin{aligned}&\deg(H_0(0,\cdot),\Omega_+,y)+\deg(H_0(0,\cdot),\Omega_-,y)\\ &\quad=\deg(H_0(0,\cdot),\Omega_k,y)=\deg(H_0(1,\cdot),\Omega_k,y).\end{aligned}$$

The existence property of the degree (Proposition 9.45) implies that the last number in this formula is 0. Summarizing, we have shown

$$\deg(H(1,\cdot),\Omega_k,y)=-\deg(H_0(0,\cdot),\Omega_-,y). \tag{9.11}$$

In the oriented case, we note that (9.9) and (9.10) imply

$$\det((dh_g(0,\cdot))(v_k)^{-1}dh_g(0,\cdot)(u_k))=\det(I_{E_X,E_Y}^{-1}JI_{E_X,E_Y})=\det J<0.$$

Since u_k and v_k belong to the same component of the connected set Ω_k, we thus obtain from Theorem 9.25 that $\operatorname{sgn} d_X H_0(0,x_k)=-\operatorname{sgn} d_X H_0(0,\hat{x}_k)$. Note that (9.9) and the normalization property of the degree imply

$$\begin{aligned}\deg(H_0(0,\cdot),\Omega_-,y)&=\operatorname{sgn} d_X H_0(0,\hat{x}_k)\\ &=-\operatorname{sgn} d_X H_0(0,x_k)=-\operatorname{sgn} d_X H(1,x_k).\end{aligned}$$

Together with (9.11), this implies (9.8). □

Lemma 9.47. *Let $E\neq\{0\}$ be a finite-dimensional real normed space. Then* $\operatorname{Iso}(E)$ *consists of exactly two components which are path-components, consisting of the operators with positive and negative determinant, respectively.*

Remark 9.48. The proof will also show that the determinant is negative if and only if there is an odd number of eigenvalues in $(-\infty,0)$. Here, eigenvalues are understood as possibly complex eigenvalues, counted with algebraic multiplicity.

Proof. Since $\det\colon \operatorname{Iso}(E)\to\mathbb{R}\setminus\{0\}$ is continuous, it follows that $\operatorname{Iso}(E)$ is the union of the disjoint open sets $D_-:=\det^{-1}((-\infty,0))$ and $D_+:=\det^{-1}((0,\infty))$, and so Corollary 2.18 implies that the components of $\operatorname{Iso}(E)$ are contained in either D_+ or in D_-. It thus remains to show that if $A,B\in\operatorname{Iso}(E)$ belong both to D_+ or both to D_-, that is if $\det(A)/\det(B)>0$, then A and B can be connected by a path in $\operatorname{Iso}(E)$.

We recall that for any $J \in \mathrm{Iso}(E)$ there is a basis in E such that J can be represented with respect to this basis as a matrix in real Jordan normal form that is, as a diagonal block matrix with blocks of the form

$$\begin{pmatrix} \lambda & 1 & & \\ & \ddots & \ddots & \\ & & & 1 \\ & & & \lambda \end{pmatrix} \quad \text{or} \quad \begin{pmatrix} \mathrm{Re}\,\lambda & \mathrm{Im}\,\lambda & 1 & 0 & & & \\ -\mathrm{Im}\,\lambda & \mathrm{Re}\,\lambda & 0 & 1 & & & \\ & & & \ddots & & & \\ & & & & & 1 & 0 \\ & & & & & 0 & 1 \\ & & & & & \mathrm{Re}\,\lambda & \mathrm{Im}\,\lambda \\ & & & & & -\mathrm{Im}\,\lambda & \mathrm{Re}\,\lambda \end{pmatrix},$$

depending on whether the eigenvalue $\lambda \neq 0$ is real or complex. Replacing 1 in the above representation by $1 - t$ $(0 \le t \le 1)$ and considering this as a path in $\mathrm{Iso}(E)$, we see that J belongs to the same path-components as the matrix J_0 where the entries 1 in the above presentation are replaced by 0. Note that J_0 is a diagonal block matrix consisting of blocks of the form

$$(\lambda) \quad \text{or} \quad \begin{pmatrix} \mathrm{Re}\,\lambda & \mathrm{Im}\,\lambda \\ -\mathrm{Im}\,\lambda & \mathrm{Re}\,\lambda \end{pmatrix},$$

depending on whether λ is real or complex. In case $\lambda \in \mathbb{R}$, we replace λ by $(1-t)\lambda + t \operatorname{sgn} \lambda$ $(0 \le t \le 1)$. In case of complex λ with $\mathrm{Re}\,\lambda \neq 0$, we replace $\mathrm{Re}\,\lambda$ by $(1-t)\,\mathrm{Re}\,\lambda + t \operatorname{sgn} \mathrm{Re}\,\lambda$ and $\mathrm{Im}\,\lambda$ by $(1-t)\,\mathrm{Im}\,\lambda$. In the remaining case $\mathrm{Re}\,\lambda = 0$, we replace $\mathrm{Im}\,\lambda$ by $(1-t)\,\mathrm{Im}\,\lambda + t \operatorname{sgn} \mathrm{Im}\,\lambda$. This path shows that J_0 (and thus J) belongs to the same path-component as the matrix J_1 consisting of diagonal blocks of the form

$$(\operatorname{sgn} \lambda) \cdot (1) \quad \text{or} \quad (\operatorname{sgn} \mathrm{Re}\,\lambda) \begin{pmatrix} 1 & 0 \\ 0 & 1 \end{pmatrix} \quad \text{or} \quad (\operatorname{sgn} \mathrm{Im}\,\lambda) \begin{pmatrix} 0 & 1 \\ -1 & 0 \end{pmatrix}.$$

In the last case, we combine with the path

$$\begin{pmatrix} \sin \pi t & \cos \pi t \\ -\cos \pi t & \sin \pi t \end{pmatrix} \qquad (0 \le t \le \frac{1}{2}).$$

All in all, we end with a matrix J_2 in the same path-component of $\mathrm{Iso}(E)$ as J which has only entries ± 1 in the diagonal, where the -1 occurs exactly at those places where $\lambda \in (-\infty, 0)$ in the matrix J. In the same manner, we can replace J_2 by a matrix where two successive diagonal entries have changed their sign: Indeed, on the corresponding 2×2 diagonal block, we just have to replace the block by the path

$$\begin{pmatrix} \sigma_1 \cos \pi t & \sigma_1 \sin \pi t \\ -\sigma_2 \sin \pi t & \sigma_2 \cos \pi t \end{pmatrix} \qquad (0 \le t \le 1)$$

where σ_1, σ_2 are the original diagonal entries. Changing successively the signs of the first two diagonal entries, the second and third diagonal entry, etc, if necessary, it can be arranged that all diagonal entries are 1 except for possibly the last one. The latter holds if and only if the number of negative signs in J_2 is odd which by the construction is the case if and only if the number of complex eigenvalues of J in $(-\infty, 0)$ is odd. This also implies the assertion of Remark 9.48.

To finish the proof of Lemma 9.47, we apply what we have shown to $J := B^{-1}A$. Then $\det J = \det A / \det B > 0$, and by what we have shown there is a continuous path $\gamma\colon [0,1] \to \mathrm{Iso}(E)$ with $\gamma(0) = J$ and $\gamma(1) = \mathrm{id}_E$. It follows that $\gamma_0(t) := B \circ \gamma(t)$ defines a continuous path $\gamma_0\colon [0,1] \to \mathrm{Iso}(E)$ with $\gamma_0(0) = A$ and $\gamma_0(1) = B$. □

Proposition 9.46 is the key to the proof of the uniqueness of the Brouwer degree. It also shows for the existence proof how we have to define the Brouwer degree for regular values. The idea of the proof in the general case is to use Theorem 9.14 to replace y by a "sufficiently close" regular value and to show that the result is independent of the choice of this value. Concerning the uniqueness, this is a relatively simple task, as we will see in a moment.

The reader familiar with the "classical" Brouwer degree in $\mathbb{R}^n$ should be aware that we still have to tackle two difficulties.

The first of these difficulties is more or less formal: We have to reduce the situation to compact sets, since only on such sets one can formulate a continuous dependency of $\deg(F, \Omega, y)$ with respect to y.

The other difficulty is more serious: Our axioms of Definition 9.36 do not include any continuous dependency with respect to y. For the Brouwer degree in $\mathbb{R}^n$ many text books allow in the definition of the homotopy invariance also that y depends (continuously) on t. With such a requirement, the uniqueness would be very easy to solve. However, such a requirement is not very natural and in practice not easy to verify. Without such a requirement we have to *prove* such a continuous dependence.

We will simultaneously solve another problem: It makes no sense at all to speak about regular values of a C^0 function F, and Theorem 9.14 is not directly applicable to F. Therefore, we not only have to approximate y but also F (by a C^1 function, using Theorem 9.9), and we need a result that the degree is independent of the approximation.

Thus, our next aim is to prove a result which states, roughly speaking, that

$$\deg(F, \Omega, y) = \deg(F_0, \Omega, y_0)$$

whenever (F_0, y_0) is "sufficiently close" to (F, y). Once we have this, we obtain immediately the uniqueness of the degree from the regular normalization property. The reader who is familiar with the degree theory mentioned in Remark 9.44 will have no doubt about such a "stability" of the degree with respect to y.

Unfortunately, things are not as simple in our case for several reasons:

Example 9.49. Let $X = Y = \mathbb{R}$, and $F(x) = x/(1+x)$. Then the regular normalization property implies the two examples

$$\deg(\mathrm{id}_X, (0,1), y) = \deg(F, (0,\infty), y) = \begin{cases} 1 & \text{if } y \in (0,1), \\ 0 & \text{if } y \in \mathbb{R} \setminus (0,1). \end{cases}$$

Note that both degrees are defined for every $y \in \mathbb{R}$.

In the first of these examples the degree is defined for $y = 0$ or $y = 1$ in our sense but not in the sense of Remark 9.44, because $1 \in \mathrm{id}_X(\partial(0,1))$. We will see that the second example breaks down for $y = 1$ in the sense of Remark 9.44 due to the noncompactness of $\overline{(0,\infty)}$ (closure in X).

Thus, for a precise formulation, we have to reduce the general situation to the case that $\overline{\Omega}$ is compact. However, we have to tackle further difficulties: If Y is a manifold, we have to specify a notion of "uniform closeness" of F and F_0 in Y, since the topology in Y alone cannot define such a notion. Moreover, we must guarantee that we can compare the orientations of F and approximating maps F_0. To achieve all this, the subsequent notion of a Rouché triple turns out to be useful.

We recall that X and Y are assumed to be of class C^q $(1 \le q \le \infty)$.

Definition 9.50. (Ω_0, V, J) is a *Rouché triple* for $(F, \Omega, y) \in \mathscr{B}^r(X, Y)$ of class C^p $(1 \le p \le q)$ if the following holds:

(a) $V \subseteq Y$ is an open neighborhood of y.

(b) J is a C^p-diffeomorphism of V onto an open convex subset of E_Y.

(c) $\Omega_0 \subseteq \Omega$ is open with $F^{-1}(y) \subseteq \Omega_0$.

(d) $\overline{\Omega_0}$ is compact.

(e) F has a continuous extension $F\colon \overline{\Omega_0} \to Y$.

(f) $F(\overline{\Omega_0}) \subseteq V$, $y \notin F(\partial\Omega_0)$.

This Rouché triple induces on V the metric

$$d_J(x, y) := \|J(x) - J(y)\|.$$

Remark 9.51. Since Y is Hausdorff, the continuous extension of F is unique by Lemma 2.55, and so there is no ambiguity when we denote this extension also by F.

Proposition 9.52. *For each $y \in Y$ there are (V, J) such that, whenever $(F, \Omega, y) \in \mathcal{B}^r(X, Y)$, there is a Rouché triple (Ω_0, V, J) of class C^q with $\overline{\Omega_0} \subseteq \Omega$.*

Proof. Let $c\colon U \to E_Y$ be a chart for y. Then $v := c(y)$ is an interior point of the open set $c(U)$, and so there is some $r > 0$ with $B := B_r(v) \subseteq c(U)$. Since B is convex by Proposition 3.3, the set $V := c^{-1}(B)$ is an open neighborhood of y. Now if $(F, \Omega, y) \in \mathcal{B}^r(X, Y)$, the set $\Omega_1 := F^{-1}(V)$ is open with $K := F^{-1}(y) \subseteq \Omega_1$. Corollary 9.3 implies that there is an open set $\Omega_0 \subseteq X$ with compact $\overline{\Omega_0} \subseteq \Omega_1 \subseteq \Omega$ and $K \subseteq \Omega_0$. Hence, (Ω, V, c) is a Rouché triple. □

Proposition 9.53. *Let (Ω_0, V, J) be a Rouché triple for $(F, \Omega, y) \in \mathcal{B}^r(X, Y)$ of class C^p $(1 \le p \le q)$. Then for each $\varepsilon > 0$ there are functions $F_0 \in C^p(\Omega, Y)$ and a regular value $y_0 \in V$ of $F_0|_{\Omega_0}$ such that F_0 has a continuous extension to $\overline{\Omega_0}$ with $F_0(\overline{\Omega_0}) \subseteq V$ and*

$$d_J(y, y_0) < \varepsilon \quad \textit{and} \quad d_J(F(x), F_0(x)) < \varepsilon \quad \textit{for all } x \in \overline{\Omega_0}.$$

Proof. We put $G := J \circ F|_{\overline{\Omega_0}}$ and observe that $K := F(\overline{\Omega_0}) \cup \{y\}$ is compact by Proposition 2.100 and disjoint from the closed set $A := E_Y \setminus J(V)$. Corollary 3.14 implies that there is some $\rho > 0$ with $B_\rho(K) \subseteq J(V)$. Without loss of generality, we assume $\varepsilon < \rho$. Proposition 9.1 implies that there is a relatively compact, hence second countable, open neighborhood $U \subseteq X$ of the compact set $\overline{\Omega_0}$. By Theorem 9.9, there is $G_0 \in C^q(U, Y)$ satisfying

$$\|G(x) - G_0(x)\| < \varepsilon \quad \text{for all } x \in \overline{\Omega_0}.$$

Note that $B_\varepsilon(K) \subseteq J(V)$ thus implies $G_0(\overline{\Omega_0}) \subseteq V$. Then $F_0 := J^{-1} \circ G_0|_{\Omega_0}$ is a function with the required properties: The existence of a regular value y_0 of $F_0|_{\Omega_0}$ in the neighborhood $J^{-1}(B_\varepsilon(J(y)))$ of y follows from Theorem 9.14. □

Proposition 9.54. *The C^r Brouwer degree for fixed X and Y automatically satisfies the following property for $(F, \Omega, y) \in \mathcal{B}^r(X, Y)$ and $0 \le r \le p \le q$, $p \ge 1$:*

($I_{\mathcal{B}}$) **(Rouché Property).** *Let* (Ω_0, V, J) *be a Rouché triple for* (F, Ω, y) *of class* C^p, *and* $F_0 \in C^r(\Omega_0, Y)$ *have a continuous extension to* $\overline{\Omega}_0$ *with* $F_0(\overline{\Omega}_0) \subseteq V$ *and*

$$d_J(F(x), F_0(x)) \neq d_J(F(x), y) + d_J(F_0(x), y) \quad \text{for all } x \in \partial\Omega_0. \tag{9.12}$$

Then $(F_0, \Omega_0, y) \in \mathcal{B}^r(X, Y)$, (Ω_0, V, d) *is a Rouché triple for* (F_0, Ω_0, y), *and*

$$\deg(F, \Omega, y) = \deg(F, \Omega_0, y) = \deg(F_0, \Omega_0, y).$$

The orientation of F_0 *on* Ω_0 *is described below.*

($J_{\mathcal{B}}$) **(Stability).** *Let* (Ω_0, V, J) *be a Rouché triple for* (F, Ω, y) *of class* C^p. *Then there is some* $\varepsilon > 0$ *such that if* $y_0 \in V$, $F_0 \in C^r(\Omega_0, Y)$ *has a continuous extension to* $\overline{\Omega}_0$ *with* $F_0(\overline{\Omega}_0) \subseteq V$,

$$d_J(y, y_0) \leq \varepsilon \quad \text{and} \quad d_J(F(x), F_0(x)) \leq \varepsilon \quad \text{for all } x \in \partial\Omega_0,$$

then $(F_0, \Omega_0, y_0), (F, \Omega_0, y_0) \in \mathcal{B}^r(X, Y)$, (Ω_0, V, J) *is a Rouché triple for* (F_0, Ω_0, y_0) *and for* (F, Ω_0, y_0), *and*

$$\deg(F, \Omega, y) = \deg(F, \Omega_0, y_0) = \deg(F_0, \Omega_0, y) = \deg(F_0, \Omega_0, y_0). \tag{9.13}$$

The orientation of F_0 *on* Ω_0 *is described below.*

($K_{\mathcal{B}}$) **(Local Constantness in** y**).** *Let* (Ω_0, V, J) *be a Rouché triple for* (F, Ω, y), *and* $M \subseteq V$ *be such that* $F^{-1}(M) \subseteq \Omega_0$. *Then we have for any* $z \in M$ *that* $(F, \Omega, z) \in \mathcal{B}^r(X, Y)$, (Ω_0, V, J) *is a Rouché triple for* (F, Ω, z), *and*

$$\deg(F, \Omega, \cdot) \text{ is constant on the components of } M.$$

In the oriented case, the orientation of F_0 on Ω_0 is defined as follows. First, the diffeomorphism J with the natural orientation induces an orientation of $J \circ F$, see Corollary 9.31. For a fixed orientation of E_Y, the orientation of $J \circ F$ induces an orientation on Ω_0 according to Proposition 9.34. A further application of Proposition 9.34 shows that the orientations on Ω_0 and E_Y induce an orientation of $J \circ F_0$ on Ω_0. According to Corollary 9.31 (with the natural orientation of J), we obtain an induced orientation of F_0 on Ω_0. The orientations on the maps are well-defined, since if the opposite orientation is chosen in E_Y, then one obtains the opposite orientation of Ω_0 and thus the same orientation of the maps $J \circ F_0$ and F_0 on Ω_0.

Proof. By definition, there is an isometry J of V onto an open convex subset $W \subseteq E_Y$ and a continuous extension $F\colon \overline{\Omega}_0 \to V$ of F. We put $v := J(y)$ and $G := J \circ F$. In the oriented case, we equip J with the natural orientation and fix an orientation on E_Y and a corresponding orientation on Ω_0 as described above.

We show first the Rouché property. Condition (9.12) means that $G_0 := J \circ F_0$ satisfy

$$\|G(x) - G_0(x)\| \neq \|G(x) - v\| + \|v - G_0(x)\| \quad \text{for all } x \in \partial\Omega_0. \tag{9.14}$$

We consider the homotopy $h\colon [0,1] \times \overline{\Omega}_0 \to W$,

$$h(t,x) := tG_0(x) + (1-t)G(x),$$

and claim that $v \notin h([0,1] \times \partial\Omega_0)$. Indeed, assume by contradiction that there are $(t,x) \in [0,1] \times \partial\Omega_0$ with $h(t,x) = v$. Then

$$t\|G(x) - G_0(x)\| = \|G(x) - h(t,x)\| = \|G(x) - v\|$$

and

$$(1-t)\|G(x) - G_0(x)\| = \|h(t,x) - G_0(x)\| = \|v - G_0(x)\|,$$

and adding both equations, we obtain a contradiction to (9.14). We thus obtain that the set $h^{-1}(v)$ is contained in $[0,1] \times \Omega_0$. Note that the continuity of h implies that $h^{-1}(v)$ is closed in $\overline{\Omega}$ and thus compact by Proposition 2.29. Hence, if we define the partial C^r homotopy $H\colon [0,1] \times \Omega_0 \to Y$ by $H := J^{-1} \circ h|_{[0,1]\times\Omega_0}$, it follows that $H^{-1}(y)$ is compact. In the oriented case, we orient h on $[0,1] \times \Omega_0$ according to the orientations on E_Y and Ω_0 by Proposition 9.34, and since $h = J \circ H$, we obtain an induced orientation on H according to Corollary 9.31. It follows that $F_0|_{\Omega_0} = H(1,\cdot)$ is equipped with the orientation as described above. The homotopy invariance of the degree implies

$$\deg(F, \Omega_0, y) = \deg(F_0, \Omega_0, y),$$

and so the assertion follows from the excision property of the degree.

To prove the stability property, we note that $\partial\Omega_0$ is closed in the compact set $\overline{\Omega}_0$ and thus compact. Proposition 2.100 implies that $B := G(\partial\Omega_0)$ is compact. Corollary 3.14 implies in view of $v \notin B$ that $\rho := \operatorname{dist}(v, B) > 0$ (in case $\partial\Omega_0 = \emptyset$, we fix $\rho \in (0,\infty)$ arbitrary). We show first the auxiliary claim:

Whenever F_0 and y_0 are as in the stability property with $\varepsilon \in (0, \rho/2)$, then $(F, \Omega_0, y_0), (F_0, \Omega_0, y_0) \in \mathcal{B}^r(X, Y)$; moreover, (Ω_0, V, J) is a Rouché triple for (F_0, Ω_0, y_0) and for (F, Ω_0, y_0), and

$$\deg(F_0, \Omega_0, y_0) = \deg(F, \Omega_0, y_0).$$

Indeed, we have for all $x \in \partial\Omega_0$ that

$$d_J(F(x), y_0) \geq d_J(F(x), y) - \varepsilon \geq \rho - \varepsilon > 0.$$

In particular, the compact set $F^{-1}(y_0) \cap \overline{\Omega}_0$ is contained in Ω_0, and (Ω_0, V, J) is a Rouché triple for $(F, \Omega_0, y_0) \in \mathcal{B}^r(X, Y)$. Moreover, we have for every $x \in \partial\Omega_0$ the estimate

$$\begin{aligned} d_J(F(x), y_0) + d_J(F_0(x), y_0) &\geq 2d_J(F(x), y_0) - \varepsilon \geq 2d_J(F(x), y) - 3\varepsilon \\ &\geq 2\rho - 3\varepsilon > \varepsilon \geq d_J(F(x), F_0(x)), \end{aligned}$$

so that the auxiliary claim now follows by applying the already proved Rouché property with $(F, \Omega_0, y_0) \in \mathcal{B}^r(X, Y)$.

Since $A := E_Y \setminus W$ is closed and disjoint from the compact set $K := G(\overline{\Omega}_0)$, Corollary 3.14 implies the there is some $r > 0$ with $B_r(K) \subseteq W$.

We show now that the stability property holds with $\varepsilon \in (0, \rho/4)$ if $\varepsilon \leq r/2$. Thus, let F_0 and y_0 be as in the stability property. Note that the auxiliary claim already implies that (Ω_0, V, J) is a Rouché triple for $(F_0, \Omega_0, y_0), (F, \Omega_0, y_0) \in \mathcal{B}^r(X, Y)$. By Proposition 9.53, we find that there are $F_1 \in C(\overline{\Omega}_0, Y)$ and a regular value $y_1 \in V$ of $F_1|_{\Omega_0} \in C^p(\Omega_0, Y)$ with

$$d_J(y_0, y_1) < \varepsilon \quad \text{and} \quad d_J(F(x), F_1(x)) < \varepsilon \quad \text{for all } x \in \overline{\Omega}_0.$$

Putting $G_1 := J \circ F_1$ and $G_2 := J(y_0) - J(y_1) + G_1$, we have for every $x \in \overline{\Omega}_0$ that $\|G_2(x) - G(x)\| < 2\varepsilon$, and so $G_2(\overline{\Omega}_0) \subseteq B_{2\varepsilon}(K) \subseteq W = J(V)$. Hence, we can define $F_2 := J^{-1} \circ G_2 \in C(\overline{\Omega}_0, Y)$, and we have $d_J(F(x), F_2(x)) < 2\varepsilon < \rho/2$ for all $x \in \partial\Omega_0$. In the oriented case, we orient F_1 and F_2 analogously to F_0 on Ω_0, induced by the orientations on Ω_0 and E_Y as described earlier. Applying the auxiliary claim twice, we find on the one hand that

$$\deg(F_0, \Omega_0, y_0) = \deg(F, \Omega_0, y_0) = \deg(F_2, \Omega_0, y_0).$$

Using the excision property and the auxiliary claim, we have on the other hand

$$\deg(F, \Omega, y) = \deg(F, \Omega_0, y) = \deg(F_1, \Omega_0, y_1).$$

Hence, if we can show that

$$\deg(F_2, \Omega_0, y_0) = \deg(F_1, \Omega_0, y_1), \tag{9.15}$$

we obtain two of the equalities in (9.13); applying these with the particular choice $F_0 = F$, we obtain also the remaining equality in (9.13).

To see (9.15), we note that the definition of G_2 implies that for $x \in \Omega_0$ the equality $F_2(x) = y_0$ is equivalent to the equality $F_1(x) = y_1$. Moreover, since the definition of G_2 implies also that $dG_2 = dG_1$, we find by the chain rule of manifolds (Proposition 8.24) for every such x in view of $G_i = J \circ F_i$ $(i = 1, 2)$ that $dJ(y_0)dF_2(x) = dJ(y_1)dF_1(x)$ are invertible, because y_1 is a regular value of $F_1|_{\Omega_0}$. In the oriented case, the above formula shows in view of the definition of the orientation also that $\operatorname{sgn} dF_2(x) = \operatorname{sgn} dF_1(x)$. Now the regular normalization property (Proposition 9.46) yields

$$\deg(F_2, \Omega_0, y_0) = \sum_{x \in \Omega_0 \cap F_2^{-1}(y_0)} \operatorname{sgn} dF_2(x) = \sum_{x \in \Omega_0 \cap F_1^{-1}(y_1)} \operatorname{sgn} dF_1(x). \tag{9.16}$$

Using the regular normalization property once more, we obtain (9.15). Hence, the stability property is established.

The local constantness follows from the stability property. Indeed, for any $z \in M$, the set $F^{-1}(z)$ is a closed subset of the compact set $\overline{\Omega}_0$ and thus compact. Since $F^{-1}(z) \subseteq \Omega_0 \subseteq \Omega$, it follows that $(F, \Omega, z) \in \mathcal{B}^r(X, Y)$ and that (Ω_0, V, J) is a Rouché triple for (F, Ω, z). By the excision property, we have

$$\deg(F, \Omega, z) = \deg(F, \Omega_0, z) \quad \text{for all } z \in M.$$

By the stability property, the right-hand side, considered as a function of $z \in M$, is locally constant, and so Proposition 2.19 implies that it is constant on the components of M. □

The Rouché property of Proposition 9.54 should be understood as a more quantitative formulation of the stability property. Its condition (9.12) deserves some remarks, since it is weaker than the condition which the reader knows perhaps from other text books.

Note that in view of the triangle inequality of d_J the condition (9.12) can be equivalently rewritten as

$$d_J(F(x), F_0(x)) < d_J(F(x), y) + d_J(F_0(x), y) \quad \text{for all } x \in \partial\Omega_0.$$

For the particular case that $X = Y = E_X = E_Y = \mathbb{C}$, considered as a real vector space (that is, considered as $\mathbb{R}^2$), one can choose $V = Y$ and $J = \mathrm{id}_Y$, and so this condition becomes in case $y = 0$:

$$|F(z) - F_0(z)| < |F(z)| + |F_0(z)| \qquad \text{for all } z \in \partial\Omega.$$

In this case the famous theorem of Rouché from complex analysis states that if F and F_0 are holomorphic in a bounded open set $\Omega \subseteq X$ then F and F_0 have the same numbers of zeroes in Ω (counted with multiplicity in the usual sense of complex analysis). This is exactly a special case of the Rouché property of Proposition 9.54 (with $\Omega_0 = \Omega$) since for holomorphic F the number $\deg(F, \Omega, 0)$ is the number of zeroes of F (for the natural orientation of F), see Proposition 9.55. This explains our choice of the name "Rouché" in this connection.

Actually, most text books on complex analysis only treat special cases of the theorem of Rouché, like e.g. when Ω is connected or even a disk. Moreover, usually, the more restrictive condition

$$|F(z) - F_0(z)| < |F(z)| \qquad \text{for all } z \in \partial\Omega$$

is required. Therefore it is worth mentioning that our hypothesis (9.12) is weaker than the hypothesis which would be analogous to this special case

$$d_J(F(x), F_0(x)) < d_J(F(x), y) \quad \text{for all } x \in \partial\Omega_0.$$

Our condition (9.12) not only has the advantage that it is weaker than the above condition. It also has the advantage that it is symmetric with respect to F and F_0 which in some cases can simplify considerations. To complete the above remark, let us show the following result:

Proposition 9.55. *Let $X = Y = E_X = E_Y = \mathbb{C}$, considered as a real vector space, equipped with some fixed orientation (the same orientation for X as for Y). Let $\Omega \subseteq X$ be open and $F\colon \Omega \to Y$ be holomorphic and equipped with the induced orientation. If $y \in Y$ is such that $F^{-1}(y)$ is compact then $(F, \Omega, y) \in \mathcal{B}^\infty(X, Y)$, and $\deg(F, \Omega, y)$ is exactly the number of zeroes of $F - y$, counted according to multiplicity in the sense of complex analysis.*

Proof. By Proposition 9.1 there is an open set $\Omega_0 \subseteq Y$ containing $F^{-1}(y)$ with compact $\overline{\Omega_0} \subseteq Y$. Then $(\Omega_0, Y, \mathrm{id}_Y)$ is a Rouché triple for (F, Ω, y). The classical theorem of Rouché of complex analysis implies that there is a neighborhood $V \subseteq Y$ of y such that the number of zeroes of $F - y_0$ in Ω_0 is the same for all $y_0 \in V$. By the stability property of Proposition 9.54 also $\deg(F, \Omega_0, y_0) = \deg(F, \Omega, y)$ for all $y_0 \in V$ sufficiently close to y. By Theorem 9.14, we find some such y_0 which is a regular value of $F|_{\Omega_0}$. It follows that all zeroes of $F - y_0$ in Ω_0 are simple (in the sense of complex analysis), and the regular normalization property 9.46 implies

$$\deg(F, \Omega, y) = \deg(F, \Omega_0, y_0) = \sum_{z \in \Omega_0 \cap F^{-1}(y_0)} \operatorname{sgn} dF(z). \tag{9.17}$$

Note that the Cauchy–Riemann equations imply $\det dF(z) = |F'(z)|^2 \geq 0$, where F' denotes the complex derivative of F. Hence, $dF(z)$ is orientation preserving by Proposition 7.12, and so Corollary 7.14 implies $\operatorname{sgn} dF(z) = 1$ for all $z \in \Omega_0 \cap F^{-1}(y_0)$. The latter is exactly the set of zeroes of $F - y_0$ on Ω_0 which, as observed above, has exactly as many elements as the number of zeroes of F (according to multiplicity). Thus, the assertion follows from (9.17). □

The reader might have observed that the independence of the right-hand side of (9.16) of the Brouwer degree already implies the uniqueness of the degree. For completeness, we write down this fact explicitly, giving a slightly different proof.

Theorem 9.56. *Let $1 \leq q \leq \infty$ and $0 \leq r \leq q$. On C^q manifolds X and Y over E_X and E_Y with $0 < \dim E_X = \dim E_Y < \infty$ there is at most one C^r Brouwer degree. Moreover:*

(a) *If X and Y are oriented, and if in the definition of the C^r Brouwer degree only the induced orientations of Proposition 9.34 are considered, then there is at most one such degree.*

(b) *If in the definition of the C^r Brouwer degree in the non-oriented case only orientable maps and orientable homotopies are allowed, then there is at most one such degree.*

Proof. Let $(F, \Omega, y) \in \mathcal{B}^r(X, Y)$. By Proposition 9.52, there is a Rouché triple (Ω_0, V, J) for (F, Ω, y). Let $\varepsilon > 0$ be as in the stability property of Proposition 9.54. Choosing a corresponding F_0 and y_0 as in Proposition 9.53, we obtain by the stability property and the regular normalization (Proposition 9.46) that

$$\deg(F, \Omega, y) = \deg(F_0, \Omega_0, y_0) = \sum_{x \in \Omega_0 \cap F_0^{-1}(y)} \operatorname{sgn} dF_0(x),$$

and the last expression does not depend on deg.

For the other two assertions, we note that Proposition 9.46 and 9.54 hold also for the correspondingly defined degrees in view of Remark 9.41, and since, if we started from an oriented map, the degree has only been used for oriented maps in the proofs. □

Corollary 9.57. *The Brouwer degrees satisfy the compatibility conditions* (9.4) *and* (9.5).

Proof. For $0 \le r \le R \le q$ (and $q \ge 1$) the map $\deg_{(r,X,Y)}|_{\mathcal{B}^R(X,Y)}$ has all properties of the C^r Brouwer degree. Hence, the uniqueness of the latter implies (9.4). Similarly, the degree

$$\deg^*(F,\Omega,y) := \deg(F,\Omega,y) \bmod 2$$

for oriented maps has all properties of the C^r Brouwer degree of non-oriented (but orientable) maps. Hence, (9.5) follows from Theorem 9.56(b). □

We note that a uniqueness proof of the degree in a similar spirit can also be found in [16].

The uniqueness implies very important properties of the Brouwer degree.

Proposition 9.58. *The Brouwer degree has the following property for every* $(F,\Omega,y) \in \mathcal{B}^r(X,Y)$.

($L_{\mathcal{B}}$) **(Topological Invariance).** *Let X_0 and Y_0 be manifolds of class C^1 over real vector spaces E_{X_0} and E_{Y_0} with $0 < \dim E_{X_0} = \dim E_{Y_0} < \infty$. Let J_1 and J_2 be (oriented) homeomorphisms of an open subset of X_0 onto Ω or of an open neighborhood $U \subseteq Y$ of $F(\Omega) \cup \{y\}$ onto an open subsets of Y_0, respectively. Then*

$$\deg_{(r,X,Y)}(F,\Omega,y) = \deg_{(0,X_0,Y_0)}(J_2 \circ F \circ J_1, J_1^{-1}(\Omega), J_2(y)).$$

In the oriented case, the orientation of $J_2 \circ F \circ J_1$ is the composite orientation of Corollary 9.30.

($M_{\mathcal{B}}$) **(Diffeomorphic Invariance).** *Let X_0 and Y_0 be manifolds of class C^s $(1 \le s \le r)$ over real vector spaces E_{X_0} and E_{Y_0} with $0 < \dim E_{X_0} = \dim E_{Y_0} < \infty$. Let J_1 and J_2 be C^s-diffeomorphisms of an open subset of X_0 onto Ω or of an open neighborhood $U \subseteq Y$ of $F(\Omega) \cup \{y\}$ onto an open subsets of Y_0, respectively. Then*

$$\deg_{(r,X,Y)}(F,\Omega,y) = \deg_{(s,X_0,Y_0)}(J_2 \circ F \circ J_1, J_1^{-1}(\Omega), J_2(y)).$$

In the oriented case, the orientation of $J_2 \circ F \circ J_1$ is understood in the sense of Corollary 9.31.

($N_{\mathcal{B}}$) **(Restriction).** *Let $X_0 \subseteq X$ and $Y_0 \subseteq Y$ be open. Then*

$$\deg_{(r,X_0,Y_0)} = \deg_{(r,X,Y)}|_{\mathcal{B}^r(X_0,Y_0)}. \tag{9.18}$$

Proof. For the restriction property, we note that it can be verified straightforwardly that the right-hand side of (9.18) defines a C^r Brouwer degree on the manifolds X_0 and Y_0. Hence, (9.18) follows from the uniqueness of such a degree (Theorem 9.56).

To prove the topological invariance, we can consider Ω and U as manifolds. On these manifolds, we can define a C^r Brouwer degree by putting

$$\deg_{(r,\Omega,U)}(\hat{F},\hat{\Omega},\hat{y}) := \deg_{(0,X_0,Y_0)}(J_2 \circ \hat{F} \circ J_1, J_1^{-1}(\hat{\Omega}), J_2(\hat{y})).$$

Indeed, since $\deg_{(0,X_0,Y_0)}$ satisfies the normalization property for diffeomorphisms (Proposition 9.46), it follows that $\deg_{(r,\Omega,U)}$ satisfies the normalization property. Since $\deg_{(0,X_0,Y_0)}$ satisfies the homotopy invariance, additivity and excision properties, it follows straightforwardly that also $\deg_{(r,\Omega,U)}$ has these properties (for the orientation of homotopies, recall Proposition 9.32). Hence, $\deg_{(r,\Omega,U)}$ is actually the (unique) C^r Brouwer degree on the manifolds Ω and U, and so the restriction property implies

$$\deg_{(r,X,Y)}(F,\Omega,y) = \deg_{(r,\Omega,U)}(F,\Omega,y).$$

Combining this formula with our above definition of $\deg_{(r,\Omega,U)}$, we obtain the topological invariance. In view of the compatibility (9.4), the diffeomorphic invariance is the special case of the topological invariance when the diffeomorphisms J_1 and J_2 carry the natural orientation. □

Remark 9.59. If we change an atlas of a manifold such that the charts of the new atlas are diffeomorphisms with respect to the original atlas then the two manifolds are diffeomorphic. Proposition 9.58 implies in particular that the degree does not change under such a change of the atlas. This has two important consequences:

(a) We can always enlarge or shrink the atlas (by compatible charts) without changing the degree.

(b) We can always reduce considerations to manifolds over $E_X = E_Y = \mathbb{R}^n$.

Both was not clear from the very beginning: In fact, the choice of the atlas apparently modifies the meaning of the normalization property. Also that one can choose $E_X = E_Y = \mathbb{R}^n$ is evident only once Proposition 9.58 and the existence of the degree are established.

If X and Y are oriented, and the orientations are understood in the sense of Remark 9.41, the orientations of the maps $J_2 \circ F \circ J_1$ in Proposition 9.58 should be understood anyway as described in Proposition 9.58. This means that the orientation of $J_2 \circ F \circ J_1$ is induced by the orientations of $J_1^{-1}(\Omega)$ and $J_2(U)$ which in turn is induced by the orientations of J_1 and J_2.

This may be somewhat confusing if the manifold X_0 and Y_0 are open subsets of X and Y, respectively, since in this case, it can mean that $J_2 \circ F \circ J_1$ is not necessarily equipped with the orientations inherited from X and Y, since $J_1^{-1}(\Omega)$ and $J_2(U)$ can carry different orientations. As a trivial example of this observation, consider for instance $X = Y = \mathbb{R}^n$ with some fixed orientation, and $J_1, J_2 \in \mathrm{Iso}(\mathbb{R}^n)$. Then the diffeomorphic invariance means

$$\begin{aligned}\deg_{(r,X,Y)}(F, \Omega, y) &= (\operatorname{sgn}\det J_1)(\operatorname{sgn}\det J_2)\\ &\quad\times \deg_{(r,X,Y)}(J_2 \circ F \circ J_1, J_1^{-1}(\Omega), J_2(y)).\end{aligned}$$

Here, the change in the signs comes from the fact that $J_1(\Omega) \subseteq X$ and $J_2(V) \subseteq Y$ carry the opposite orientation than X and Y in case $\operatorname{sgn}\det J_1 < 0$ or $\operatorname{sgn}\det J_2 < 0$, respectively.

9.5 Existence of the Brouwer Degree

Throughout this section, let X and Y be manifolds of class C^q $(1 \le q \le \infty)$ without boundaries over real vector spaces E_X and E_Y with $0 < \dim E_X = \dim E_Y < \infty$, and let $0 \le r \le q$.

Concerning the existence of the C^r Brouwer degree, the uniqueness proof of Section 9.4 indicated how one *must* define $\deg(F, \Omega, y)$: Namely as the sum

$$\deg(F, \Omega, y) := \sum_{\Omega_0 \cap F_0^{-1}(y_0)} \operatorname{sgn} dF_0(x), \tag{9.19}$$

where (Ω_0, V, J) is a Rouché triple for F and (F_0, y_0) are "sufficiently close" to (F, y), and y_0 is a regular value for $F_0|_{\Omega_0}$.

One of the major technical difficulties if one wants to define the degree in this way is that one has to show the stability property for this definition: Note that one has to show that this (the only possible) definition of the degree is actually well-defined, that is, independent of the particular choice of F_0 and y_0. For the case $r = 1$, that is if $F \in C^1$, one might at a first glance try to use only $F_0 = F$ so that the only difficulty here is to prove the stability with respect to y_0. Unfortunately, we will be able to prove this stability only in a very cumbersome way and only in the case that $F \in C^2$ so that for $r < 2$ we still have to use an approximating function F_0.

The above remark is not completely true: We *are* able to prove the stability with respect to y_0 also in a simple manner and for $F \in C^1$, but only under the additional condition that y is a regular value of F as the following result shows. The difficulty is to get rid of this additional condition.

Lemma 9.60. *Let $(F, \Omega, y) \in \mathcal{B}^1(X, Y)$ be such that y is a regular value of F. Then for each Rouché triple (Ω_0, V, J) for (F, Ω, y), there is a neighborhood $M \subseteq Y$ of y such that every $\hat{y} \in M$ is a regular value of $F|_{\Omega_0}$, and*

$$\sum_{x \in F^{-1}(y)} \operatorname{sgn} dF(x) = \sum_{x \in \Omega_0 \cap F^{-1}(\hat{y})} \operatorname{sgn} dF(x).$$

Proof. The compact set $K := F^{-1}(y)$ is discrete by Remark 8.58 and thus consists of finitely many points $x_1, \ldots, x_n$. By the inverse function Theorem 8.39 and Proposition 9.1 there are disjoint connected open neighborhoods $\Omega_k \subseteq \Omega_0$ of x_k $(k = 1, \ldots, n)$ such that F is a diffeomorphism of Ω_k onto an open neighborhood $Y_k \subseteq Y$ of y. Note that

$$K := \overline{\Omega}_0 \setminus (\Omega_1 \cup \cdots \cup \Omega_n)$$

is closed in $\overline{\Omega}_0$ and thus compact by Proposition 2.29. Since Y is Hausdorff, Proposition 2.45 implies that y and the compact set $F(K)$ have disjoint open neighborhoods $M, M_0 \subseteq Y$. Shrinking M if necessary, we can assume that $M \subseteq Y_1 \cap \ldots \cap Y_n$. For every $\hat{y} \in M$, we have $\hat{y} \notin M_0 \supseteq F(K)$, and so $F^{-1}(\hat{y})$ is disjoint from K, that is $\Omega_0 \cap F^{-1}(\hat{y}) \subseteq \Omega_1 \cup \cdots \cup \Omega_n$. Since F is a diffeomorphism of Ω_k onto Y_k, it follows that for each $\hat{y} \in M$ the set $\Omega_0 \cap F^{-1}(\hat{y})$ contains exactly one point $\hat{x}_k$ from each Ω_k $(k = 1, \ldots, n)$ and no other points. Hence, $\hat{y}$ is a regular value of $F|_{\Omega_0}$. Moreover, since Ω_k is connected and $dF(x) \in \operatorname{Iso}(T_x, T_{F(x)}Y)$ is regular for every $x \in \Omega_k$, Theorem 8.27(b) implies in the oriented case that $\operatorname{sgn} dF(x_k) = \operatorname{sgn} dF(\hat{x}_k)$. This holds of course also in the non-oriented case. It follows that

$$\sum_{k=1}^{n} \operatorname{sgn} dF(x_k) = \sum_{k=1}^{n} \operatorname{sgn} dF(\hat{x}_k),$$

which is a reformulation of the assertion. □

Remark 9.61. Using Lemma 9.60, one could prove the existence of a "small" Brouwer degree for $r \geq 1$ which is only defined for $(F, \Omega, y) \in \mathcal{B}^r(X, Y)$ when y is a regular value of F. For this "small" degree, it would not be too hard to prove all required properties except for the homotopy invariance which could be proved easily only under the additional hypothesis that y is a regular value of $H(t, \cdot)$ for every $t \in [0, 1]$.

One might be less ambitious and consider only such a "small" degree theory, hoping that in most applications the additional requirement that y be a regular value of F is not very restrictive. Unfortunately, the requirement that y is a regular value of $H(t,\cdot)$ makes this degree theory almost useless: For most Fredholm homotopies h such that y is a regular value of $h(i,\cdot)$ $(i=0,1)$ and $h^{-1}(y)$ is compact there exists typically no Fredholm homotopy H with $H(i,\cdot)=h(i,\cdot)$ $(i=0,1)$ and compact $H^{-1}(y)$ which satisfies this additional requirement that y is a regular value for every $H(t,\cdot)$ $(0\le t\le 1)$. Thus, even if one wants to use "adapted" homotopies, one can usually not apply the "small" degree theory which is therefore rather useless.

The major obstacle in the existence proof of the degree is that there is no method known how to prove an analogue of Lemma 9.60 directly for the case that y is not a regular value of F. All known proofs use an approximation argument by smooth (at least C^2) maps and then some technical lemma showing that certain integrals or sums vanish. (For alternative topological approaches by using homology theory, one also implicitly approximates the given maps by "simplicial" maps of one type or another.) Therefore, early attempts to generalize the approach to an infinite-dimensional setting also required C^2 smoothness. Only in finite dimensions, it is easy to relax this smoothness by an approximation argument.

We use an approach which also gives us the so-called bordism invariance of the degree which we explain in a moment. For this approach, the "technical lemma" mentioned above is some sort of bordism invariance for the quantity

$$\sum_{x\in F^{-1}(\hat{y})} \operatorname{sgn} dF(x).$$

It is hard to trace back who used a lemma of such a type first for the definition of the degree, but it seems that it goes back to Pontryagin. We use some ideas of the presentation in [107] in the following.

To formulate the lemma and the notion of bordism invariance, we need a notion of orientation of the boundary of an oriented manifold.

Let W be an oriented manifold over $\mathbb{R}\times E$ with boundary ∂W. Recall that for $x\in\partial W$ there is a chart c which maps an open neighborhood of x onto an open subset of a closed halfspace

$$\{(t,u)\in\mathbb{R}\times E : f(t,u)\ge 0\}$$

where $f\in(\mathbb{R}\times E)^*$ and, since $x\in\partial W$, we can assume that $f^*\ne\{0\}$ and that $f(c(x))=0$. There is a vector $e\in\mathbb{R}\times E$ with $f(e)<0$. With h_c from Definition 8.20, this vector corresponds by the chart c with a vector $\hat{e}:=h_c(x,e)\in T_xW$ which, geometrically speaking, points from $x\in\partial W$ "outside" of W. There is also a basis $e_1, \dots, e_n$ of $\mathrm{N}(f)$ ($n = \dim E$). The corresponding vectors $\hat{e}_k := h_c(x, e_k) \in T_x W$ are tangential to the manifold ∂W, that is, they are a basis of $T_x \partial W$. Choosing the opposite of one of these vectors, if necessary, we can assume that $(\hat{e}, \hat{e}_1, \dots, \hat{e}_n)$ belongs to the orientation of $T_x W$. Then we define the orientation of $T_x \partial W$ such that it is represented by $(\hat{e}_1, \dots, \hat{e}_n)$. In this way, ∂W becomes an oriented manifold.

Definition 9.62. The bordism invariance is the following property of the Brouwer degree.

($\mathrm{O}_{\mathcal{B}}$) **(Bordism Invariance for C^q Manifolds).** Let W be a C^q manifold over $\mathbb{R} \times E_X$ with boundary ∂W. Let $\Omega_0, \Omega_1 \subseteq \partial W$ be open in ∂W and disjoint, $H \in C^r(W, Y)$, and $y \in Y$ be such that $H^{-1}(y)$ is compact and $H^{-1}(y) \cap \partial W \subseteq \Omega_0 \cup \Omega_1$. Then we have for $i = 0, 1$ with $F_i := H|_{\Omega_i} \in C^r(\Omega_i, Y)$ that $(F_i, \Omega_i, y) \in \mathcal{B}^r(\Omega_i, Y)$ for $i = 0, 1$, and

$$\deg_{(r, \Omega_0, Y)}(F_0, \Omega_0, y) = \deg_{(r, \Omega_1, Y)}(F_1, \Omega_1, y).$$

It is admissible that $\Omega_i = \varnothing$ in which case the corresponding degree in this formula is considered as 0. The orientation of F_i in the oriented case is described below.

In the oriented case, we assume that Y is oriented in a neighborhood V of $H(W) \cup \{y\}$, that W is oriented, and that ∂W is oriented as described above. We assume that F_0 is oriented according to the orientation of $\Omega_0 \subseteq \partial W$ and V in the sense of Proposition 9.34. Concerning F_1, we assume that it has the opposite orientation as that which is induced by the orientations of $\Omega_1 \subseteq \partial W$ and V.

Note that $(F_i, \Omega_i, y) \in \mathcal{B}^r(\Omega_i, Y)$ for $i = 0, 1$ is actually automatic, since $F_i^{-1}(y) = H^{-1}(y) \cap (\partial W \setminus \Omega_{1-i})$ is a closed subset of the compact set $H^{-1}(y)$ and thus closed by Proposition 2.29.

Now the "technical lemma" we announced states that the bordism invariance holds in case of regular values for the only possible definition of the degree.

Lemma 9.63. *In the situation of Definition* 9.62, *suppose that y is simultaneously a regular value of H, of F_0, and of F_1. Then*

$$\sum_{x \in F_0^{-1}(y)} \operatorname{sgn} dF_0(x) = \sum_{x \in F_1^{-1}(y)} \operatorname{sgn} dF_1(x).$$

Proof. For the proof, it is more convenient for us to put $\Omega := \Omega_0 \cup \Omega_1$ and to define $F \in C^r(\Omega, Y)$ by $F(x) = F_0(x)$ for $x \in \Omega_0$ with the corresponding orientation, and $F(x) = F_1(x)$ for $x \in \Omega_1$ with the opposite orientation. Then the assertion means

$$\sum_{x \in F^{-1}(y)} \operatorname{sgn} dF(x) = 0. \tag{9.20}$$

In the oriented case, we have by Corollary 7.14 for all $x \in F^{-1}(y)$:

$$\operatorname{sgn} dF(x) > 0 \iff dF(x) \in \operatorname{Iso}(T_x \partial W, T_y Y) \text{ is orientation preserving.}$$

Now we prove (9.20). Corollary 8.57 implies that $W_0 := H^{-1}(y)$ is a compact submanifold of W of dimension 1 with boundary $\partial W_0 = W_0 \cap \partial W \subseteq \Omega$ and that $T_x W_0 = \mathrm{N}(dH(x))$ $(x \in W_0)$. Since $\Omega \subseteq \partial W$, we have $\partial W_0 = W_0 \cap \Omega$. Proposition 9.1 implies for the manifold W_0 and $K := M := W_0$ that K intersects at most finitely many components of M, that is, W_0 consists of only finitely many components $C_1, \dots, C_m$ which are closed (Proposition 2.17) and thus compact. By Theorem 9.15, each C_k is diffeomorphic to either a circle (without boundary points) or to $[0, 1]$ (with two boundary points corresponding to $\{0, 1\}$). Note that the sum in (9.20) is over the points from $W_0 \cap \Omega = \partial W_0$. Hence, if C_k is diffeomorphic to a circle then it does not contribute to the sum, and if C_k is diffeomorphic to $[0, 1]$, only the two boundary points contribute to the sum. We show that the sum for these two boundary points is zero, and so (9.20) follows. In the non-oriented case, this is trivial since $1 + 1 = 0$ in $\mathbb{Z}_2$. It remains to verify that if x_0 and x_1 denotes the two boundary points of C_k then exactly one of the maps $dF(x_i) \in \operatorname{Iso}(T_x \partial W, T_y Y)$ $(i = 0, 1)$ is orientation preserving, so that $\operatorname{sgn} dF(x_0) = -\operatorname{sgn} dF(x_1)$.

To see this, we define an orientation along the path C_k from x_0 to x_1 as follows. For $x \in C_k$, let $v \in T_x C_k \setminus \{0\}$. Note that, since y is a regular value of H, $dH(x) \in \mathcal{L}(T_x W, T_y Y)$ has rank n where $n := \dim T_y Y = \dim T_x W - 1$. Since $v \in T_x W_0 = \mathrm{N}(dH(x))$, it follows that if we extend v to some basis $(v, v_1, \dots, v_n)$ of $T_x W$, without loss of generality representing the orientation of $T_x W$ then $(dH(x)v_1, \dots, dH(x)v_n)$ is a basis of Y. If this represents the orientation of Y, then v should represent the orientation of $T_x C_k$. Otherwise, $-v$ will represent the orientation of $T_x C_k$ in the described sense.

By the continuity of the representations, it is clear that C_k becomes oriented this way. Obviously, there are only two possibilities to orient a path C_k at all: The orientations at the endpoints x_0 and x_1 must be opposite. At the endpoints x_i $(i = 0, 1)$, we can consider a basis $(v, v_1, \dots, v_n)$ of $T_x W$ with $v_1, \dots, v_n \subseteq T_{x_i} \partial W$ and $v \in T_{x_i} C_k$. It follows from the definitions that $v \in T_{x_i} C_k$ represents the orientation of the path if and only if $dF(x_i)$ is orientation preserving. Since the orientations at the endpoints of the path is opposite, we obtain that exactly one of the maps $dF(x_i)$ is orientation preserving, as claimed. □

We obtain that an analogue of Lemma 9.63 holds even if y is not a regular value of H but under the additional hypothesis that H is C^∞. In fact, by the latter hypothesis, we can apply Sard's Theorem 9.14 in the following proof.

Lemma 9.64. *In the situation of Definition* 9.62*, suppose that* y *is simultaneously a regular value of* F_0 *and of* F_1*, and that* W *and* H *are of class* C^∞*.*

Then there is a neighborhood $M \subseteq Y$ *of* y *such that every* $y_1, y_2 \in M$ *is a regular value of* F_i *with*

$$\sum_{x \in F_i^{-1}(y)} \operatorname{sgn} dF_i(x) = \sum_{x \in \Omega_0 \cap F_0^{-1}(y_1)} \operatorname{sgn} dF_0(x) = \sum_{x \in \Omega_1 \cap F_1^{-1}(y_2)} \operatorname{sgn} dF_1(x)$$

for $i = 0$ *and* $i = 1$*.*

Proof. Lemma 9.60 implies that there is a neighborhood $M \subseteq Y$ of y such that every $y_0 \in M$ is a regular value of F_i with

$$\sum_{x \in F_i^{-1}(y)} \operatorname{sgn} dF_i(x) = \sum_{x \in \Omega_i \cap F_i^{-1}(y_0)} \operatorname{sgn} dF_i(x) \tag{9.21}$$

for $i = 0, 1$. Putting $K := H^{-1}(y) \cup \bigcup_{i=1,2} \{i\} \times \overline{\Omega}_i$, we obtain from Proposition 9.1 that there is some open neighborhood $W_0 \subseteq W$ of $K \subseteq W_0$ with compact $\overline{W}_0 \subseteq W$. Replacing Ω_i by $W_0 \cap \Omega_i$ if necessary, we can assume without loss of generality that $\Omega_i \subseteq W_0$ for $i = 0, 1$. By Theorem 9.14 there is some $y_3 \in M$ which is a regular value of $H|_{W_0}$. Lemma 9.65, applied with $H|_{W_0}$, thus gives

$$\sum_{x \in \Omega_0 \cap F_0^{-1}(y_3)} \operatorname{sgn} dF_0(x) = \sum_{x \in \Omega_1 \cap F_1^{-1}(y_3)} \operatorname{sgn} dF_1(x),$$

and the assertion follows by combining this equality with the equalities coming from (9.21) with the choices $y_0 \in \{y_1, y_2, y_3\}$ and $i \in \{0, 1\}$. □

We show now that the bordism invariance is actually a generalization of the homotopy invariance. In particular, we are now in a position to prove the required analogue of Lemma 9.60 for the case that y is not a regular value.

Lemma 9.65. *Let* $W \subseteq [0,1] \times X$ *be open,* $H \colon W \to Y$ *be a generalized oriented homotopy, and* $y \in Y$*. Then there is an orientation on an open neighborhood* $V \subseteq Y$ *of* y *and an orientation on* $W_0 := H^{-1}(V)$ *such that the orientation of* $F_i := H(i, \cdot)$ *on* $\Omega_i := \{x : (i,x) \in W_0\}$ *is induced as in Definition* 9.62 *when we replace* W *by* W_0*.*

Proof. Let $V \subseteq Y$ be an open neighborhood of y such that there is a chart $c\colon V \to E_Y$. We claim that the assertion follows with $W_0 := H^{-1}(V)$. We must find corresponding orientations of V and W_0. We start by fixing some orientation on V. Such an orientation exists, since we can fix an orientation of E_Y and let the chart c induce an orientation on V by Proposition 9.34. It remains to define the orientation of W_0.

To this end, we observe that for each $(t,x) \in W_0$ the orientation of $H(t,\cdot)$ at x induces by the orientation of $T_{H(t,x)}Y$ an orientation on T_xX according to Proposition 7.13. Let $(e_1,\dots,e_n)$ represent the orientation of T_xX. Then $((1,0),(0,e_1),\dots,(0,e_n))$ represents an orientation of $\mathbb{R}\times T_xX = T_{(t,x)}([0,1]\times X) = T_{(t,x)}W_0$ (the first equality is understood by the canonical identification of product manifolds). According to the definition of the orientation induced on ∂W_0, we find that F_0 carries on Ω_0 the orientation induced by ∂W_0 and V in the sense of Proposition 9.34, and that F_1 carries on Ω_1 the opposite of the orientation induced by ∂W_0 and V. □

Corollary 9.66. *The bordism invariance for q-manifolds ($q \ge r$), the excision property, and the restriction property of the Brouwer degree together imply:*

($P_{\mathcal{B}}$) **(Generalized Homotopy Invariance with Constant y).** *If $W \subseteq [0,1]\times X$ is open, $H\colon W \to Y$ a generalized (oriented) partial C^r homotopy, and $y \in Y$ are such that $H^{-1}(y)$ is compact then we have with $\Omega_i := \{x : (i,x) \in W\}$ $(i = 0,1)$ that*

$$\deg_{(r,X,Y)}(H(0,\cdot),\Omega_0,y) = \deg_{(r,X,Y)}(H(1,\cdot),\Omega_1,y).$$

Note that $(H(i,\cdot),\Omega_i,y) \in \mathcal{B}^r(X,Y)$ is automatic, since $H(i,\cdot)^{-1}(y) \subseteq \Omega_i$ is compact by Proposition 2.62.

Proof. Let W_0 and V be as in Lemma 9.65. By the excision property, it is no loss of generality to replace W by W_0. Hence, the bordism invariance implies

$$\deg_{(r,\Omega_0,Y)}(H(0,\cdot),\Omega_0,y) = \deg_{(r,\Omega_1,Y)}(H(1,\cdot),\Omega_1,y).$$

The assertion follows now from the restriction property. □

Corollary 9.67. *Let X be a C^∞ manifold, $W \subseteq [0,1]\times X$ be open, and $H \in C^\infty(W,Y)$ be a generalized (oriented) partial C^1 homotopy. Let $y \in Y$ be a regular value of $F_i := H(i,\cdot)$ $(i = 0,1)$ and $H^{-1}(y)$ be compact. For $i = 0,1$, put $W_i := \{x : (i,x) \in W\}$ and $K_i := \{x : H(t,x) = y\}$. Let $\Omega_i \subseteq X$ be open with $K_i \subseteq \Omega_i$ and compact $\overline{\Omega_i} \subseteq W_i$.*

Then there is a neighborhood $M \subseteq Y$ of Y such that every $y_1, y_2 \in M$ is a regular value of F_i with

$$\sum_{x \in F_i^{-1}(y)} \operatorname{sgn} dF_i(x) = \sum_{x \in \Omega_0 \cap F_0^{-1}(y_1)} \operatorname{sgn} dF_0(x) = \sum_{x \in \Omega_1 \cap F_1^{-1}(y_2)} \operatorname{sgn} dF_1(x)$$

for $i = 0$ and $i = 1$.

Proof. Let W_0 and V be as in Lemma 9.65. Then the assertion follows from Lemma 9.64 with W replaced by W_0. □

Using Corollary 9.67, we finally obtain the existence of the degree for C^∞ manifolds by an approximation argument with C^∞ maps.

We note that we could replace C^∞ by C^r in Lemma 9.64, Corollary 9.67, and Lemma 9.68 (and in their proofs) if we would be willing to use Theorem 9.13 in C^r with $m = \dim E_Y$ and $n = m + 1$ (recall that the proof of the latter is much more involved than in the case $r = \infty$). Unfortunately, since a hypothesis of Theorem 9.13 is $r > n - m$, we must have $r \geq 2$ in any case so that the best possible result which we could obtain in this way is a result for C^2 manifolds.

Lemma 9.68. *Let X be a C^∞ manifold and $Y = E_Y$. Then for each $0 \leq r \leq \infty$ there is a C^r Brouwer degree for X and Y which satisfies the bordism invariance for C^∞ manifolds.*

Before we go into the details of the proof, let us point out that the idea of the proof is to use (9.19) for a smooth approximation of F. We have the additional technical difficulty here in the choice of the set Ω_0 which determines how good the approximation must be: We must also prove the independence of this definition from Ω_0. The trick we use here is to consider *two* sets $\Omega_0 \subseteq U_0$ and to define the closeness with respect to both sets.

Proof. We define the degree for $T := (F, \Omega, y) \in \mathscr{B}^r(X, Y)$ as follows. By Corollary 9.3, there is an open neighborhood $U_0 \subseteq X$ of $F^{-1}(y)$ with compact $\overline{U}_0 \subseteq \Omega$. For each open neighborhood $\Omega_0 \subseteq U_0$ of $F^{-1}(y)$, we can define by Proposition 9.4 a number $r_T(\Omega_0, U_0) > 0$ as the supremum of all $\varepsilon \in (0, 1]$ such that whenever $\overline{\Omega}_0 \subseteq Z \subseteq \overline{U}_0$ and $H\colon [0,1] \times Z \to V$ is continuous with

$$\|H(t,x) - F(x)\| \leq 3\varepsilon \quad \text{for all } (t,x) \in [0,1] \times Z,$$

and $y_0 \in B_\varepsilon(y)$, then the set $H^{-1}(y_0)$ is contained in $[0,1] \times \Omega_0$ and compact. It follows that r_T is monotone in the sense that $\hat{\Omega}_0 \subseteq \Omega_0$ and $U_0 \subseteq \hat{U}_0$ imply $r_T(\hat{\Omega}_0, \hat{U}_0) \leq r_T(\Omega_0, U_0)$.

By Proposition 2.45 there is a neighborhood $\Omega_0 \subseteq X$ of $F^{-1}(y)$ with $\overline{\Omega_0} \subseteq U_0$. Then $(\Omega_0, Y, \mathrm{id}_Y)$ and (U_0, Y, id_Y) are both Rouché triples for (F, Ω, y). By Proposition 9.53 there is $F_0 \in C^\infty(U_0, V)$ and a regular value y_0 of F_0 such that

$$\|y - y_0\| < r_T(\Omega_0, U_0) \quad \text{and} \quad \sup_{x \in U_0} \|F(x) - F_0(x)\| < r_T(\Omega_0, U_0).$$

We define

$$\deg(F, \Omega, y) := \sum_{x \in \Omega_0 \cap F_0^{-1}(y_0)} \operatorname{sgn} dF_0(x). \tag{9.22}$$

In the oriented case, we fix some orientation of $Y = E_Y$, and note that in the sense of Proposition 9.34 the orientation of F then induces an orientation of U_0 which in turn induces an orientation of F_0. (Clearly, this orientation is independent of the choice of the orientation of $Y = E_Y$, since for the opposite orientation, we obtain the opposite orientation of U_0 and thus the same orientation of F_0.) We understand F_0 equipped with this orientation.

We must show that (9.22) is well-defined, that is, independent of the particular choice of U_0, Ω_0, F_0, and y_0. Thus, let U_1, Ω_1, F_1, and y_1 be corresponding possibly different choices, and we have to show

$$\sum_{x \in \Omega_0 \cap F_0^{-1}(y_0)} \operatorname{sgn} dF_0(x) = \sum_{x \in \Omega_1 \cap F_1^{-1}(y_1)} \operatorname{sgn} dF_1(x). \tag{9.23}$$

We show this first for the special case $U_0 = U_1$ and $\Omega_0 = \Omega_1$. To this end, we define $H\colon [0,1] \times U_0 \to E_Y$ by

$$H(t,x) := t(F_1(x) + (y_0 - y_1)) + (1-t)F_0(x).$$

Since $\|y_0 - y_1\| < 2r_T(\Omega, U_0)$, we have

$$\begin{aligned} &\|H(t,x) - F(x)\| \\ &\quad = \|t(F_1(x) - F(x) + (y_0 - y_1)) + (1-t)(F_0(x) - F(x))\| \le 3\varepsilon \end{aligned} \tag{9.24}$$

for some $\varepsilon < r_T(\Omega_0, U_0)$. In the oriented case, we fix as above an orientation of Y and note that F induces an orientation of U_0 which in turn induces an orientation of the partial C^∞ homotopy H. This orientation is actually independent of the choice of the orientation of Y, and $d_X H(i, \cdot) = dF_i$ have the same orientations for $i = 0, 1$. By (9.24) and the definition of $\varepsilon < r_T(\Omega_0, U)$, we obtain that $H^{-1}(y_0)$ is a compact subset of $[0,1] \times \Omega_0$. Since y_i are regular values for F_i $(i = 0, 1)$ it follows that y_0 is a regular value for $H(0, \cdot)$ and for $H(1, \cdot)$. Corollary 9.67 thus implies that

$$\sum_{x \in \Omega_0 \cap F_0^{-1}(y_0)} \operatorname{sgn} dF_0(x) = \sum_{x \in \Omega_0 \cap F_1^{-1}(y_0)} \operatorname{sgn} dF_1(x).$$

By the definition of H, this means exactly (9.23).

Now we show (9.23) in the general case. Let $U_2 := U_0 \cup U_1$, $\Omega_2 := \Omega_0 \cap \Omega_1$, and let (F_2, y_2) be correspondingly as in the definition of the degree. By the monotonicity of r_T, we have $r_T(\Omega_2, U_2) \le r_T(\Omega_i, U_i)$ $(i = 0, 1)$. Hence, the choices $(F_2|_{U_i}, y_2)$ are actually also admissible for our definition of the degree when we replace (Ω_2, U_2) by (Ω_i, U_i) with $i = 0$ or $i = 1$. It follows that the special case we proved so far implies

$$\sum_{x \in \Omega_i \cap F_i^{-1}(y_i)} \operatorname{sgn} dF_i(x) = \sum_{x \in \Omega_i \cap F_2^{-1}(y_2)} \operatorname{sgn} dF_2(x) \quad \text{for } i = 0, 1.$$

By definition of $r_T(\Omega_2, U_2)$, we have $F^{-1}(y_2) \subseteq \Omega_2$, that is, the sum on the right-hand side is actually over all $x \in F_2^{-1}(y_2)$ and thus independent of $i \in \{0, 1\}$. Hence, also the left-hand side is independent of $i \in \{0, 1\}$ which means (9.23).

Summarizing, we have shown that the degree is well-defined by (9.22). We have to show that it has all properties required for the C^r Brouwer degree. The normalization property (even the regular normalization property) follows immediately from the definition. For the excision property, suppose that $\hat{\Omega} \subseteq \Omega$ is open and contains $F^{-1}(y)$. Then we can choose the set Ω_0 in the definition of the degree even such that additionally $\overline{\Omega_0} \subseteq \hat{\Omega}$, and then the independence of our definition of the particular choice of Ω_0 implies that

$$\deg(F, \Omega, y) = \deg(F, \hat{\Omega}, y).$$

For the additivity, let Ω be the union of disjoint open subsets Ω_i $(i = 1, 2)$. We note that we can choose (F_0, y_0) in the above definition of $\deg(F, \Omega, y)$ such that the same choice holds also for the definition of $\deg(F, \Omega_i, y)$ $(i = 1, 2)$. Then the additivity is obvious from (9.22). The restriction property of the degree is immediately clear from its definition. Hence, to prove the homotopy invariance, it suffices by Corollary 9.66 to prove the bordism invariance.

Thus, let W be a C^∞ manifold over $\mathbb{R} \times E_X$ with boundary ∂W, $W_0, W_1 \subseteq \partial W$ open and disjoint, $H \in C^r(W, Y)$ be of class C^r, and $y \in Y$ with compact $H^{-1}(y)$ and $H^{-1}(y) \cap \partial W \subseteq W_0 \cup W_1$. For $i = 0, 1$, let $F_i := H|_{W_i} \in C^r(W_i, Y)$. In the oriented case, we assume that Y is oriented on an open neighborhood V of $H(W) \cup \{y\}$, W is oriented, and F_i are oriented as described in Definition 9.62.

By Proposition 9.1, there is an open set $U \subseteq W$ with $H^{-1}(y) \subseteq U$ and compact $\overline{U} \subseteq W$. For $i = 0, 1$, we put $U_i := W_i \cap U$. We let $\Omega_i \subseteq X$ be open neighborhoods of $F_i^{-1}(y)$ with $\overline{\Omega_i} \subseteq U_i$, put $T_i := (F_i, W_i, y)$, and let $\varepsilon > 0$ be strictly less than $r_{T_i}(\Omega_i, U_i)$ for $i = 0, 1$. By the compactness of $C := H(\overline{U}) \cup \{y\} \subseteq V$, we may assume in view of Corollary 3.14 that

$B_\varepsilon(C) \subseteq V$. By Proposition 9.1, the compact set $\overline{U}$ has a relatively compact open neighborhood $X_1 \subseteq W$ which is second countable. By Theorem 9.9 there is a map $h \in C^\infty(X_1, E_Y)$ with

$$\|h(t,x) - H(t,x)\| \leq \varepsilon \quad \text{for all } (t,x) \in \overline{U}.$$

Then $F_{i,0} := h|_{U_i} \in C^\infty(U_i, E_Y)$ $(i = 0,1)$, even $F_{i,0}(U_i) \subseteq V$, and so we can equip $F_{i,0}$ with the orientations analogously to F_i. Now we note that Theorem 9.14 implies that the map $h|_{\partial U}$ has a regular value y_0 in $B_\varepsilon(y)$. Note that this implies in view of $\partial U \subseteq \partial W$ that y_0 is a regular value for $F|_{i,0}$, simultaneously for $i = 0$ and $i = 1$. The definition of the degree now implies

$$\deg_{(r,W_i,Y)}(F_i, W_i, y) = \sum_{x \in \Omega_i \cap F_{i,0}^{-1}(y_0)} \operatorname{sgn} dF_{i,0}(x) \quad \text{for } i = 0,1.$$

Lemma 9.64 shows that the right-hand side is actually independent of $i \in \{0,1\}$. Hence also the left-hand side is independent of $i \in \{0,1\}$ which means that the degree satisfies the bordism invariance for C^∞ manifolds. □

Lemma 9.69. *Let X and Y be of class C^q $(1 \leq q \leq \infty)$. For each $0 \leq r \leq q$ there is a C^r Brouwer degree on X and Y satisfying the bordism invariance for manifolds of class C^q.*

Proof. We define the degree for $(F, \Omega, y) \in \mathcal{B}^r(X,Y)$ as follows. By Theorem 9.12, there is an open neighborhood $U \subseteq X$ of $F^{-1}(y)$ and a diffeomorphism J_1 of a C^∞ manifold X_0 onto U. There is a chart $J_2 := c$ of y which maps a neighborhood $V \subseteq Y$ of y diffeomorphically onto an open subset of E_Y. Replacing U by $U \cap F^{-1}(V)$ if necessary, we can assume that $F(U) \subseteq V$. Then we define

$$\deg_{(r,X,Y)}(F, \Omega, y) := \deg_{(r,X_0,E_Y)}(J_2 \circ F \circ J_1, J_1^{-1}(U), J_2(y)). \tag{9.25}$$

The degree on the right-hand side of (9.25) exists by Lemma 9.68. In the oriented case, the orientation of $J_2 \circ F \circ J_1^{-1}$ is induced in the sense of Corollary 9.31. The diffeomorphic invariance of the degree (Proposition 9.58) implies that we obtain the same value also for different choices of J_1, X_0, and J_2, and moreover, that we obtain the same values also for smaller sets U and V. It follows that the above definition is actually independent of the particular choice of the auxiliary data used for the definition.

It remains to verify that the thus defined degree has all required properties. Since $\deg_{(r,X_0,E_Y)}$ satisfies the normalization property for diffeomorphisms (Proposiion 9.46), it follows that $\deg_{(r,X,Y)}$ satisfies the normalization property. The excision property follows from the very definition. The same holds for the restriction property. The additivity follows straightforwardly from the additivity of $\deg_{(r,X_0,E_Y)}$. In view of Corollary 9.66, it thus suffices to show the bordism invariance.

To prove the bordism invariance, let W be a manifold of class C^q over $\mathbb{R} \times E_X$ with boundary ∂W, $W_0, W_1 \subseteq \partial W$ open and disjoint, $H \in C^r(W, Y)$ be of class C^r, and $y \in Y$ with compact $K := H^{-1}(y)$ and $K \cap \partial W \subseteq W_0 \cup W_1$. For $i = 0, 1$, let $F_i := H|_{W_i} \in C^r(W_i, Y)$. In the oriented case, we assume that Y is oriented on an open neighborhood V of $H(W) \cup \{y\}$, W is oriented, and F_i are oriented as described in Definition 9.62. Let $J_2 := c$ be a chart of y which maps an open neighborhood $V_0 \subseteq V$ of y diffeomorphically onto an open subset of E_Y. By Theorem 9.12, there is an open neighborhood $U \subseteq H^{-1}(V_0)$ of K and a diffeomorphism J_1 of a C^∞ manifold M onto U. For $i = 0, 1$, we put $U_i := W_i \cap U$, $\Omega_i := J_1^{-1}(U_i)$, and note that $J_{1,i} = J_1|_{\Omega_i}$ is a diffeomorphism of Ω_i onto U_i. In particular, the above definition of the degree implies

$$\deg_{(r,W_i,Y)}(F_i, W_i, y) = \deg_{(r,\Omega_i,E_Y)}(J_2 \circ F_i \circ J_{1,i}, \Omega_i, J_2(y)). \tag{9.26}$$

Note that $H(U) \subseteq V$ implies that we can define $H_0 := J_2 \circ H \circ J_1$. Then $J_2 \circ F_i \circ J_{1,i}$ is the restriction of H_0 to $\Omega_i \subseteq \partial M = J_1^{-1}(\partial U) = J_1^{-1}(U \cap \partial W)$. In the oriented case, we equip J_1 with the natural orientation, and M with the orientation induced by J_1 from W in the sense of Proposition 9.34. Now the bordism invariance of the degree of Lemma 9.68 implies that the right-hand side of (9.26) is independent of $i \in \{0, 1\}$. □

The proof of Theorem 9.38 is now complete: The assertions have been proved in Theorem 9.56, Corollary 9.57, and Lemma 9.69.

The proof of Lemma 9.69 shows why it is so useful to have the uniqueness assertion of Theorem 9.38 or at least the diffeomorphic invariance of the degree: Without knowing the diffeomorphic invariance, we could not have defined the degree by (9.25) so easily.

The proof of Lemma 9.69 shows also why it is useful to have the bordism invariance of the degree or at least the *generalized* homotopy invariance with constant y: If we would have attempted to prove the homotopy invariance directly for $H: W \to Y$ with $W := [0, 1] \times \Omega$ and $W_i := \{0, 1\}$ $(i = 0, 1)$, we would have run into the problem that H_0 in the above proof is not necessarily defined on a set of the form $[0, 1] \times M_0$, since $H^{-1}(V_0)$ does in general not contain a set of such a form.

Therefore, we emphasize that the bordism invariance and the generalized homotopy invariance is a side result of our proof. For completeness, let us remark that it is actually not necessary that y be constant for the generalized homotopy invariance. This is a somewhat surprising complement to the stability property which can be formulated without Rouché triples:

Theorem 9.70. *The C^r Brouwer degree satisfies the bordism invariance for C^q manifolds ($0 \le r \le q \le \infty$, $q \ge 1$) and the following extension of the generalized homotopy invariance with constant y:*

($Q_{\mathcal{B}}$) **(Generalized Homotopy Invariance).** *Let $W \subseteq [0,1] \times X$ be open, $H: W \to Y$ be a generalized (oriented) partial C^r homotopy, and $y \in C([0,1], Y)$.*
If $\{(t,x) \in W : H(t,x) = y(t)\}$ is compact then we have with $W_t := \{x : (t,x) \in W\}$ that $(H(t,\cdot), W_t, y(t))) \in \mathcal{B}^r(X,Y)$ for all $t \in [0,1]$, and

$$\deg(H(t,\cdot), W_t, y(t)) \text{ is independent of } t \in [0,1].$$

Proof. For the assertion $(H(t,\cdot), W_t, y(t)) \in \mathcal{B}^r(X,Y)$, it suffices to note that $H(t,\cdot)^{-1}(y(t)) \subseteq W_t$ is compact by Proposition 2.62.

By Lemma 9.69, we know that there is a C^r Brouwer degree which satisfies the bordism invariance. By the uniqueness, there is no other C^r Brouwer degree. Hence, the C^r Brouwer degree satisfies the bordism invariance. Corollary 9.66 implies that it also satisfies the generalized homotopy invariance with constant y.

We have to show that even in case $y \in C([0,1], Y)$ the number

$$d(t) := \deg(H(t,\cdot), W_t, y(t))$$

is constant. Since $[0,1]$ is connected by Proposition 2.14, Proposition 2.19 implies that it suffices to prove that d is locally constant. Thus, let $t_0 \in [0,1]$. Let $V \subseteq Y$ be an open neighborhood of $y_0 := y(t_0)$, and let $c: V \to E_Y$ be a chart. Shrinking W if necessary (which we can do by the excision property), we can assume without loss of generality that there is a neighborhood $I_0 \subseteq [0,1]$ of t_0 such that for each $t \in I_0$ and $x \in W_t$, we have $H(t,x), y(t) \in V \cap J_t^{-1}(c(V))$ with

$$J_t(x) := c(x) + c(y(t_0)) - c(y(t)).$$

Note that the diffeomorphic invariance implies

$$d(t) = \deg_{(r,X,E_Y)}(J_t(H(t,\cdot)), W_t, J_t(y(t))),$$

and that $J_t(y(t))$ is actually independent of $t \in I_0$. Hence, replacing H by $(t,x) \mapsto J_t(H(t,x))$ and y by $t \mapsto J_t(y(t))$ if necessary, we can assume without loss of generality that $y(t) = y_0$ is constant on I_0. We have to show that d is constant on I_0.

Thus, let $t_1, t_2 \in I_0$, $t_1 < t_2$. We consider the continuous map $g\colon [0,1] \times X \to [t_1, t_2] \times X$, $g(s,x) := (st_1 + (1-s)t_2, x)$, and apply the generalized homotopy invariance with constant $y = y_0$ with $\hat{W} := g^{-1}(W)$ and the generalized partial C^r homotopy $\hat{H} := H \circ g\colon \hat{W} \to W$. Note that, since $h := g|_{\hat{W}}$ is a homeomorphism onto $W_{t_1,t_2} := W \cap ([t_1, t_2] \times X)$ and $M := H^{-1}(y) \cap ([t_1, t_2] \times X)$ is closed in W and thus compact, it follows that $\hat{H}^{-1}(y) = h^{-1}(M)$ is indeed compact. In the oriented case, if σ denotes the orientation of H, we define the orientation of $\hat{H}$ by $\hat{\sigma}(s,x) = \sigma(g(s,x))$. The homotopy invariance with constant y now implies with $\hat{W}_s := \{x : (s,x) \in \hat{W}\}$ that

$$d(t_2) = \deg(\hat{H}(1,\cdot), \hat{W}_1, y_0) = \deg(\hat{H}(0,\cdot), \hat{W}_0, y_0) = d(t_1),$$

and so d is indeed constant on I_0. □

Corollary 9.71. *The C^r Brouwer degree satisfies the following property.*

($R_{\mathcal{B}}$) **(Elimination of** y). *If $Y = E_Y$ then $(F, \Omega, y) \in \mathcal{B}^r(X,Y)$ is equivalent to $(F - y, \Omega, 0) \in \mathcal{B}^r(X,Y)$, and in this case*

$$\deg_{(r,X,E_Y)}(F, \Omega, y) = \deg_{(r,X,E_Y)}(F - y, \Omega, 0).$$

Proof. The assertion follows by applying the generalized homotopy invariance with $y(t) = ty_0$ and $H(t, y_0) := F(x) - y(t)$. □

Remark 9.72. By using an approach by homology theory instead of differential topology, it is possible to prove the existence of a C^0 Brouwer degree also if X and Y are only C^0 manifolds. Of course, orientations of maps and manifolds have to be defined differently for C^0 manifolds: As we have mentioned already in Remark 9.17, this is a much more complicated definition than in the case of C^1 manifolds.

Such an approach for degree theory on oriented C^0 manifolds can be found in [39, Chapter VIII, §4], and the extension to oriented maps (on not necessarily orientable manifolds) in [39, Chapter VIII, Exercise 4.10.6]. By definition, that degree satisfies the topological invariance. Moreover, by [134, Section V.14] that degree also satisfies the bordism invariance (for C^0 manifolds!).

However, it is unknown to the author whether the such defined degree must be unique in case of C^0 manifolds, i.e., whether it is already completely determined by the properties of Definition 9.36.

Remark 9.73. For the case $r \geq 1$, that is, for the C^1 Brouwer degree, we might of course speak throughout about the orientation in the sense of Definition 8.25

instead of Definition 9.16: This is more or less just a convention, since the orientations can be transformed into each other by means of (9.2) (and Proposition 9.18). When we choose the orientation in the sense of Definition 8.25 for the C^1 Brouwer degree, we obtain a special case of the Benevieri–Furi degree which we study in Section 10.2.

9.6 Some Classical Applications of the Brouwer Degree

Mainly for the reader who is unfamiliar with the classical Brouwer degree on a finite-dimensional normed space E, we provide now some applications of that degree which demonstrate what kind of results one can obtain by using degree theory.

Many of these examples are based on the so-called continuation principle:

Proposition 9.74 (Continuation Principle). *Let X and Y be C^1 manifolds over real finite-dimensional vector spaces of the same finite nonzero dimension, $\Omega \subseteq X$ be open, and $H\colon [0,1] \times \Omega \to Y$ be a continuous (oriented or non-oriented) homotopy with*

$$\deg(H(0,\,\cdot\,), \Omega, y) \neq 0. \tag{9.27}$$

If $H^{-1}(y)$ is compact then the equation $H(1,x) = y$ has a solution $x \in \Omega$.

Proof. By the homotopy invariance

$$\deg(H(1,\,\cdot\,), \Omega, y) = \deg(H(0,\,\cdot\,), \Omega, y) \neq 0,$$

so the assertion follows from the existence property of the degree. □

Remark 9.75. The easiest case of this principle is if $H(0,\,\cdot\,)$ is a diffeomorphism of Ω onto an open subset of E with $y \in H(\{0\} \times \Omega)$. In this case, (9.27) holds by the diffeomorphic normalization property of the degree.

As a special case, we prove Brouwer's fixed point theorem. The reader should be aware that all known elementary proofs of Brouwer's fixed point theorem are relatively cumbersome; although meanwhile some are not that complicated, there is no really short proof. The proofs which are less cumbersome make use of some heavy machinery: algebraic topology or differential geometry (Gauß' theorem) or, in our case, degree theory.

Theorem 9.76 (Brouwer Fixed Point). *Let M be a nonempty closed convex subset of a finite-dimensional normed vector space E, and $\varphi \in C(M,M)$ be such that $\varphi(M)$ is bounded. Then φ has a fixed point.*

Proof. By Theorem 4.36, M is an CE_M for the class of T_4 spaces. Since φ is compact, we can thus extend φ to some compact $f \in C(E, M)$. We consider the map $H \in C([0,1] \times E, E)$, $H(t,x) := x - tf(x)$. Then $H^{-1}(0)$ is closed and bounded in $[0,1] \times E$ and thus compact. The continuation principle (Remark 9.75) implies that there is some $x \in \Omega := E$ satisfying $H(1,x) = 0$, that is, $x = f(x) \in M$. Since $x \in M$, x is a fixed point of the original map $\varphi = f|_M$. □

The underlying intuitive idea in the above application of the continuation principle is that during the homotopy $H(t,x) = x - tf(x)$ all zeroes of $H(t,\cdot)$ depend in a sense "continuously" on t and so the zero of $H(0,\cdot) = \mathrm{id}_E$ may not vanish suddenly if it remains in a bounded (compact) set.

This intuition is not true if one counts zeroes of $H(t,\cdot)$ in the ordinary way, but if one understands the number of zeroes as the degree, this argument is precisely the homotopy invariance of the degree.

Since we actually only need that during the homotopy the zeroes lie in a compact subset of Ω, we obtain immediately a generalization of Brouwer's fixed point theorem by relaxing the hypothesis $f(M) \subseteq M$:

Theorem 9.77 (Leray–Schauder Alternative). *Let Ω be an open subset of a finite-dimensional normed space E. Let $\varphi \in C(\overline{\Omega}, E)$. Then at least one of the following holds:*

(a) *φ has a fixed point in $\overline{\Omega}$.*

(b) *For each $x_0 \in \Omega$ the set*

$$\bigcup_{\lambda > 1} \{x \in \overline{\Omega} : \varphi(x) - x_0 = \lambda(x - x_0)\} \tag{9.28}$$

is unbounded or intersects $\partial\Omega$.

Proof. Assume by contradiction that both properties fail. Then there is $x_0 \in \Omega$ such that (9.28) is a bounded subset of Ω. We define $H \in C([0,1] \times \overline{\Omega}, E)$ by $H(t,x) := x - t(\varphi(x) - x_0)$. Since φ has no fixed point in $\overline{\Omega}$, we have $H(1,\cdot)^{-1}(x_0) = \varnothing$, and so the assumption implies that $H^{-1}(x_0) \subseteq [0,1] \times \Omega$ is bounded and closed in $[0,1] \times \overline{\Omega}$ and thus compact. Hence, the continuation principle with $y = x_0$ implies in view of Remark 9.75 that there is some $x \in \Omega$ with $H(1,x) = x_0$, contradicting $H(1,\cdot)^{-1}(x_0) = \varnothing$. □

We note that Theorem 9.77 actually contains Brouwer's fixed point theorem for sets of the form $M = \overline{\Omega}$ as a special case. In fact, we need to verify the hypothesis only on $\partial\Omega$:

Corollary 9.78 (Rothe's Fixed Point Theorem). *Let Ω be a nonempty open convex subset of a finite-dimensional normed space E. Let $\varphi \in C(\overline{\Omega}, E)$ be such that $\varphi(\Omega)$ is bounded and $\varphi(\partial\Omega) \subseteq \overline{\Omega}$. Then φ has a fixed point.*

Proof. Let $x_0 \in \Omega$. Since $\varphi(\overline{\Omega}) \subseteq \overline{\varphi(\Omega)}$ is bounded, it follows that the set (9.28) is bounded. We claim that this set is disjoint from $\partial\Omega$. Assume by contradiction that there is some $x \in \partial\Omega$ and some $\lambda > 1$ such that $\varphi(x) - x_0 = \lambda(x - x_0)$. Then $\varphi(x) \in \overline{\Omega}$, and so Lemma 4.41 with $t := 1 - \lambda^{-1} \in (0, 1)$ implies that $x = (1 - t)\varphi(x) + tx_0 \in \Omega$ which is a contradiction. □

There is a "dual" variant of the continuation principle:

Proposition 9.79 (Inverse Continuation Principle). *Let X and Y be C^1 manifolds over real finite-dimensional vector spaces of the same finite nonzero dimension, $\Omega \subseteq X$ be open, and $H\colon [0, 1] \times \overline{\Omega} \to Y$ be a continuous (oriented or non-oriented) homotopy such that the following degrees are defined and satisfy*

$$\deg(H(0, \cdot), \Omega, y) \neq \deg(H(1, \cdot), \Omega, y).$$

Then the set $H^{-1}(y)$ fails to be compact. In particular, if $\overline{\Omega}$ is compact and H has an extension to a continuous map $H_0\colon [0, 1] \times \overline{\Omega} \to Y$ then $y \in H_0([0, 1] \times \partial\Omega)$.

Proof. The first assertion is a reformulation of the homotopy invariance. If the second assertion is false then, since $\{y\}$ is closed, $H_0^{-1}(\{y\}) = H^{-1}(y)$ is a closed subset of the compact set $[0, 1] \times \overline{\Omega}$ and thus compact. □

The inverse continuation principle typically implies the existence of "nonlinear eigenvalues" for certain classes of functions. Here is a simple result in that direction:

Theorem 9.80. *Let E be a real normed space of finite dimension, $\Omega \subseteq E$ a bounded open neighborhood of 0, and $f \in C(\overline{\Omega}, E)$ satisfy $0 \notin f(\overline{\Omega})$. Then there are $\lambda_+ > 0 > \lambda_-$ and $x_\pm \in \partial\Omega$ with $f(x_\pm) = \lambda_\pm x_\pm$.*

Proof. Since $f^{-1}(0) = \emptyset$, the existence and normalization properties of the degree imply

$$\deg(f, \Omega, 0) = 0 \neq \deg(\mathrm{id}_\Omega, \Omega, 0).$$

Hence, we can apply the inverse continuation principle with the homotopy $H(t, x) := tx + (1 - t)f(x)$ which implies in view of $f(x) \neq 0$ that there are $\lambda_- < 0$ and $x_- \in \partial\Omega$ with $f(x_-) = \lambda_- x_-$. The existence of λ_+ and x_+ follows by applying this assertion with $-f$ in place of f. □

Remark 9.81. Our above proofs show that Brouwer's fixed point theorem and Theorem 9.80 both follow straightforwardly from degree theory. It is interesting to observe that one can see elementary that actually both assertions are equivalent, at least in the most important case $\Omega = B_1(0)$ and $M = \overline{\Omega}$:

In fact, if Brouwer's fixed point theorem holds and $f \in C(M, \mathbb{R}^n)$ satisfies $0 \notin f(M)$ then the maps $F_\pm(x) := \pm f(x)/\|f(x)\|$ have fixed points $x_\pm \in M$ which satisfy $x_\pm = F_\pm(x_\pm) \in \partial\Omega$ and $f_\pm(x_\pm) = \lambda_\pm x_\pm$ with $\lambda_\pm := \pm\|f_\pm(x_\pm)\| \neq 0$.

Conversely, if the assertion of Theorem 9.80 holds and $f \in C(M, M)$ would have no fixed point then $F(x) := f(x) - x$ would satisfy $0 \notin F(\overline{\Omega})$, and so there are $\lambda_+ > 0$ and $x_+ \in \partial\Omega$ with $F(x_+) = \lambda_+ x_+$. Then $f(x_+) = (1 + \lambda_+)x_+$ would imply $\|f(x_+)\| = 1 + \lambda_+ > 1$ which is a contradiction.

For a different kind of application of the inverse continuation principle, we equip $X = Y = E = \mathbb{R}^n$ with some orientation (the same on X and Y). With the induced orientation on the maps, the diffeomorphic normalization property implies for every open neighborhood $\Omega \subseteq E$ of 0 that

$$\deg(-\operatorname{id}_\Omega, \Omega, 0) = \operatorname{sgn}(-\operatorname{id}_E) = \operatorname{sgn}\det(-\operatorname{id}_E) = (-1)^n.$$

For odd n this differs from

$$\deg(\operatorname{id}_\Omega, \Omega, 0) = \operatorname{sgn}(\operatorname{id}_E) = 1.$$

By this observation the inverse continuation principle now implies the existence of "nonlinear eigenvalues" in spaces of odd dimension without the hypothesis $0 \notin f(\overline{\Omega})$:

Theorem 9.82. *Let E be a real normed space of odd finite dimension, $\Omega \subseteq E$ be an open bounded neighborhood of 0, and $f \in C(\overline{\Omega}, E)$. Then there are $\lambda \in \mathbb{R}$ and $x \in \partial\Omega$ with $f(x) = \lambda x$.*

Proof. We fix an orientation on $X = Y = E$ and consider the homotopies $H_0(t,x) = tx + (1-t)f(x)$ and $H_1(t,x) = t(-x) + (1-t)f(x)$ with the induced orientations. We can assume that $f(x) \neq 0$ for all $x \in \partial\Omega$, since otherwise we are done. Then the following degrees are defined:

$$\deg(H_0(0,\cdot), \Omega, 0) = \deg(f, \Omega, 0) = \deg(H_1(0,\cdot), \Omega, 0).$$

Since we have calculated above that

$$\deg(H_0(1,\cdot), \Omega, 0) = 1 \neq -1 = \deg(H_1(1,\cdot), \Omega, 0),$$

the inverse continuation principle applies for at least one of the maps H_0 and H_1. Hence, there are $t \in [0,1]$ and $x \in \partial\Omega$ with $H_i(t,x) = 0$ for $i = 0$ or $i = 1$ which implies the assertion. □

For Ω being the open unit ball in $\mathbb{R}^n$, we obtain the famous "hedgehog theorem" which states that it is not possible to "comb" a hedgehog. We use for $x = (x_1, \dots, x_n), y = (y_1, \dots, y_n) \in \mathbb{R}^n$ the scalar product

$$\langle x, y\rangle := x_1 y_1 + \cdots + x_n y_n.$$

Corollary 9.83 (Hedgehog Theorem). *For the unit sphere $S = S_1(0)$ in $\mathbb{R}^n$ with odd n there is no continuous tangential field $f \in C(S, \mathbb{R}^n \setminus \{0\})$, that is $\langle x, f(x)\rangle = 0$ for all $x \in S$.*

Proof. Otherwise, we put $\Omega := B_1(0)$ and use Tietze's extension theorem to extend f to a map $f \in C(\overline{\Omega}, \mathbb{R}^n)$. Theorem 9.82 implies that there are $x \in S$ and $\lambda \in \mathbb{R}$ with $f(x) = \lambda x$. Then $0 = \langle x, f(x)\rangle = \lambda\langle x, x\rangle$ implies $\lambda = 0$, that is $f(x) = 0$, contradicting the hypothesis. □

We note that Corollary 9.83 (and thus also Theorem 9.82) fails for even dimension n, since then a continuous tangential field is given by

$$f(x_1, \dots, x_n) = (x_2, -x_1, x_4, -x_3, \dots, x_n, -x_{n-1}).$$

A different kind of application of degree theory uses Borsuk's theorem that odd maps have an odd degree on neighborhoods of zero:

Theorem 9.84 (Borsuk). *Let $X = Y = E$ be a real finite-dimensional vector space and $\Omega \subseteq X$ be open with $-\Omega = \Omega$. Let $F \in C(\Omega, Y)$ be odd, that is $F(-x) = -F(x)$ for all $x \in \Omega$. If $F^{-1}(0)$ is compact then*

$$\deg(F, \Omega, 0) \text{ is } \begin{cases} \textit{odd} & \textit{if } 0 \in \Omega, \\ \textit{even} & \textit{if } 0 \notin \Omega. \end{cases} \tag{9.29}$$

Proof. In view of (9.5), it suffices to consider the non-oriented degree. For the case that F is of class C^1 and that 0 is a regular value of F, the regular normalization property implies that $\deg(F, \Omega, 0)$ is odd/even if and only if the finite number of elements in $Z = F^{-1}(0)$ is odd/even. Since $-Z = Z$ and $0 \in Z$ if and only if $0 \in \Omega$, the assertion follows.

The idea of the proof is to reduce the general case to this special case. This is somewhat technical since the straightforward idea to approximate F by a smooth function and then to apply Sard's lemma does not work, because $F - c$ fails to be odd for $c \neq 0$. However, the idea works when we construct the approximating function coordinatewise appropriately.

Therefore, we assume without loss of generality that $X = Y = E = \mathbb{R}^n$. By the excision property of the degree, we can shrink Ω without changing the assertion, as long as the shrinked set Ω satisfies $-\Omega = \Omega$ and contains $F^{-1}(0)$. Hence, without loss of generality, we can assume in view of Proposition 9.1 that Ω is bounded, that F has a continuous extension to $\overline{\Omega}$, and in case $0 \notin \Omega$ that there is some $\varepsilon > 0$ such that $K := [-\varepsilon, \varepsilon]^n$ is disjoint from Ω. In case $0 \in \Omega$, we assume conversely that $K \subseteq \Omega$.

By construction, $(\Omega, Y, \mathrm{id}_E)$ is a Rouché triple for $(F, \Omega, 0)$. The stability property of the degree (Proposition 9.54) implies that there is some $\delta_F > 0$ such that any function $G \in C(\overline{\Omega}, Y)$ satisfies the implication

$$\max_{x \in \partial\Omega} \|F(x) - G(x)\| < \delta_F \implies \deg(F, \Omega, 0) = \deg(G, \Omega, 0). \tag{9.30}$$

By Theorem 9.9 there is $G \in C^\infty(X, Y)$ satisfying the left-hand side of (9.30). In case $0 \in \Omega$, we can assume that $G|_K = \mathrm{id}_K$, because by Theorem 9.8 there is $G_0 \in C^\infty(X, Y)$ satisfying $G_0|_{\partial\Omega} = G|_{\partial\Omega}$ and $G_0|_K = \mathrm{id}_K$, and we can replace G by G_0 if necessary. Moreover, replacing G by

$$\hat{G}(x) := \frac{1}{2}(G(x) - G(-x))$$

if necessary, we can assume without loss of generality that G is odd. In view of (9.30), it suffices to show the assertion for G in place of F.

This argument shows that, without loss of generality, we can assume from the very beginning that F is the restriction of a function from $C^\infty(X, Y)$, and in case $0 \in \Omega$ that $F|_K = \mathrm{id}_K$.

We fix now an odd $\varphi\colon \mathbb{R} \to \mathbb{R}$ of class C^∞ satisfying $\varphi^{-1}(0) = [-\varepsilon, \varepsilon]$; for instance, with the function g of Lemma 9.5, we can put $\varphi(t) := tg(t^2 - \varepsilon^2)$. For $k = 1, \dots, n$, we put

$$\Omega_k := \{(x_1, \dots, x_n) \in \Omega : |x_k| > \varepsilon\}.$$

We put $F_0 := F$. By induction, we define for $k = 1, \dots, n$ functions $G_k\colon \Omega_k \to Y$ by

$$G_k(x_1, \dots, x_n) := \frac{F_{k-1}(x_1, \dots, x_n)}{\varphi(x_k)}, \tag{9.31}$$

and by the Lemma of Sard there is a regular value c_k of G_k such that $\|c_k\|$ is so small that, in view of (9.30) the function $F_k\colon \overline{\Omega} \to Y$,

$$F_k(x_1,\dots,x_n) := F(x_1,\dots,x_n) - c_1\varphi(x_1) - \dots - c_k\varphi(x_k) \tag{9.32}$$

satisfies $\deg(F_k,\Omega,0) = \deg(F_{k-1},\Omega,0)$.

Then F_n is an odd function satisfying $\deg(F_n,\Omega,0) = \deg(F,\Omega,0)$. Hence, we are done if we can show that 0 is a regular value of F_n. To see this, we note first that $F_n|_K = \mathrm{id}_K$ so that we only have to show that 0 is a regular value of $F_n|_{\Omega\setminus K}$. Since $\Omega \setminus K = \Omega_1 \cup \dots \cup \Omega_n$, it suffices to show by induction on $k = 1,\dots,n$ that 0 is a regular value for the restriction of F_k to $\Omega_1 \cup \dots \cup \Omega_k$.

Thus, let $x = (x_1,\dots,x_n) \in F_k^{-1}(0)$ satisfy $x \in \Omega_1 \cup \dots \cup \Omega_k$. We have to show that $dF_k(x)$ is invertible. In case $x \notin \Omega_k$, that is if $|x_k| \le \varepsilon$, we have $\varphi'(x_k) = 0$ and $x \in \Omega_1 \cup \dots \cup \Omega_{k-1}$. In this case, it follows that $dF_k(x) = dF_{k-1}(x)$ is invertible by induction hypothesis. It remains to consider the case $x \in \Omega_k$. We have $0 = F_k(x) = F_{k-1}(x) - c_k\varphi(x_k)$ which implies

$$c_k\varphi(x_k) = F_{k-1}(x). \tag{9.33}$$

In particular, we obtain $G_k(x) = c_k$. Since c_k is a regular value of G_k by construction, we find that $dG_k(x)$ is invertible. We will show that

$$dF_k(x) = \varphi(x_k)dG_k(x). \tag{9.34}$$

Since $x \in \Omega_k$ implies $\varphi(x_k) \neq 0$, we then obtain from (9.34) that $dF_k(x)$ is invertible, as required. In order to show (9.34), it suffices by Proposition 8.1 to compare the partial derivatives, that is, we have to show that

$$d_jF_k(x) = \varphi(x_k)d_jG_k(x) \tag{9.35}$$

for all $j = 1,\dots,n$. This is clear in case $j \neq k$, since in that case the function F_k has by (9.32) the same j-th partial derivatives as F_{k-1}, and by (9.31) the function G_k has the same j-the partial derivatives as F_{k-1} up to the factor $1/\varphi(x_k)$, hence,

$$d_jG_k(x) = \frac{d_jF_{k-1}(x)}{\varphi(x_k)} = \frac{d_jF_k(x)}{\varphi(x_k)},$$

which means (9.35). For $j = k$, we use the quotient rule for (scalar) derivatives in (9.31) and obtain in view of (9.33) that

$$\begin{aligned} d_kG_k(x) &= \frac{d_kF_{k-1}(x)\varphi(x_k) - \varphi'(x_k)F_{k-1}(x)}{\varphi(x_k)^2} \\ &= \frac{d_kF_{k-1}(x) - c_k\varphi'(x_k)}{\varphi(x_k)} = \frac{d_kF_k(x)}{\varphi(x_k)}, \end{aligned}$$

hence, (9.35) holds also for $j = k$. □

If Ω in Theorem 9.84 is bounded and F has a continuous extension to $\overline{\Omega}$, it suffices to verify that $F|_{\partial\Omega}$ is odd:

Corollary 9.85 (Borsuk Fixed Point I). *Let E be a finite-dimensional normed vector space, $\Omega \subseteq E$ be open and bounded with $-\Omega = \Omega$, and $F \in C(\overline{\Omega}, E)$ be such that $0 \notin F(\partial\Omega)$ and $F|_{\partial\Omega}$ is odd. Then* (9.29) *holds, and in case $0 \in \Omega$, we have $0 \in F(\Omega)$.*

Proof. The function $G(x) := \frac{1}{2}(F(x) - F(-x))$ is odd and satisfies $G|_{\partial\Omega} = F|_{\partial\Omega}$. Since $(\Omega, E, \mathrm{id}_E)$ is a Rouché triple for F, the stability property of the degree implies $\deg(F, \Omega, 0) = \deg(G, \Omega, 0)$. Theorem 9.84 implies (9.29). In case $0 \in \Omega$, we have $0 \in F(\Omega)$ by the existence property of the degree. □

Using the homotopy invariance of the degree, we show now that Corollary 9.85 holds also if $F|_{\partial\Omega}$ is not necessarily odd, but if only for all $x \in \partial\Omega$ the (nonzero) vectors $F(x)$ and $F(-x)$ do not point into the same direction, that is, if

$$\|F(-x)\|F(x) \neq \|F(x)\|F(-x) \quad \text{for all } x \in \partial\Omega. \tag{9.36}$$

This is of course satisfied if $0 \notin F(\partial\Omega)$ and $F|_{\partial\Omega}$ is odd. Hence, Corollary 9.85 is actually a special case of the subsequent Corollary 9.86.

Corollary 9.86 is also applicable for unbounded Ω or if F does not necessarily have a continuous extension to $\overline{\Omega}$.

Corollary 9.86 (Borsuk Fixed Point II). *Let E be a finite-dimensional normed space, $\Omega \subseteq E$ be open with $-\Omega = \Omega$, and $F \in C(\Omega, E)$ such that*

$$C := \{x \in \Omega : \|F(-x)\|F(x) = \|F(x)\|F(-x)\}$$

is compact; this holds in particular if Ω is bounded and $F \in C(\overline{\Omega}, E)$ satisfies (9.36). *Then* (9.29) *holds, and in case $0 \in \Omega$, we have $0 \in F(\Omega)$.*

Proof. In case (9.36), we even have

$$C = \{x \in \overline{\Omega} : \|F(-x)\|F(x) = \|F(x)\|F(-x)\},$$

and by the continuity of F and of the norm, the set on the right-hand side is closed in E and bounded (if Ω is bounded) and thus compact.

To prove the main assertion, we define $H\colon [0,1] \times \Omega \to E$ by

$$H(t, x) := F(x) - tF(-x).$$

We have $H(t,x) = 0$ if and only if $F(x) = tF(-x)$. In case $F(x) \neq 0$, this implies that $F(x)$ is a positive multiple of $F(-x)$, and so $x \in C$. In case $F(x) = 0$, we have trivially $x \in C$. Hence, $H^{-1}(0)$ is a closed subset of the compact set $[0,1] \times C$ and thus compact. The homotopy invariance of the degree implies that

$$\deg(F,\Omega,0) = \deg(H(0,\,\cdot),\Omega,0) = \deg(H(1,\,\cdot),\Omega,0).$$

Since $H(1,\,\cdot)$ is an odd map, Borsuk's Theorem 9.84 implies that the latter degree satisfies (9.29), and so $0 \in F(\Omega)$ follows in case $0 \in \Omega$ from the existence property of the degree. □

Remark 9.87. In case $E = \mathbb{R}$, Corollary 9.86 is nothing else than the intermediate value theorem of continuous functions. Indeed, if $F \in C([-a,a],\mathbb{R})$ is such that $F(-a)$ and $F(a)$ have opposite (nonzero) signs then (9.36) holds with $\Omega = (-a,a)$, and so Corollary 9.86 implies that there is some $x \in (-a,a)$ with $F(x) = 0$.

In this sense, degree theory and in particular Theorem 9.84 might be considered as a higher-dimensional generalization of the intermediate value theorem.

Remark 9.88. The reader might be surprised why we call Corollaries 9.85 and 9.86 "fixed point theorems". The reason is that they become indeed fixed point theorems for φ if one applies the results with $F := \mathrm{id}_\Omega - \varphi$. Note that F is odd if and only if φ is odd. This fixed point formulation might appear artificial in the moment but is actually more natural in the setting of (infinite-dimensional) Banach spaces, as we will see in Section 13.3: Theorem 13.24 will be an infinite-dimensional generalization of Corollary 9.86.

Here are the most important consequence of Theorem 9.84:

Theorem 9.89. *The following statements hold.*

(a) **(Borsuk Fixed Point Theorem on Balls).** *Let $K = K_1(0) \subseteq \mathbb{R}^n$ and $F \in C(K,\mathbb{R}^n)$. If $F|_{S_1(0)}$ is odd then $0 \in F(K)$.*

(b) **(Borsuk-Ulam Theorem).** *Let $S = S_1(0) \subseteq \mathbb{R}^n$ and $F \in C(S,\mathbb{R}^m)$ with $m < n$. Then there is some $x \in S$ with $F(x) = F(-x)$.*

(c) **(Ljusternik-Schnirel'man Theorem).** *Let $S = S_1(0) \subseteq \mathbb{R}^n$. For all closed sets $A_1,\dots,A_n \subseteq \mathbb{R}^n$ with $S \subseteq A_1 \cup \dots \cup A_n$ there is some k such that $S \cap A_k$ contains an antipodal pair, that is, there is $x \in S$ with $x \in A_k$ and $-x \in A_k$.*

Remark 9.90. As in Remark 9.88, we can reformulate assertion (a) of Theorem 9.89 equivalently in "fixed point form" (with $F = \mathrm{id}_K - \varphi$):

(a) (**Borsuk Fixed Point Theorem on Balls**). Let $K = K_1(0) \subseteq \mathbb{R}^n$ and $\varphi \in C(K, \mathbb{R}^n)$. If $\varphi|_{S_1(0)}$ is odd then φ has a fixed point in K.

Proof. The assertion (a) is the special case of Corollary 9.85 with $\Omega = B_1(0)$.

To prove (b), we put $\Omega := B_1(0)$ and extend F by Tietze–Urysohn (Corollary 2.67) to some $F \in C(\overline{\Omega}, \mathbb{R}^m)$. Assuming by contradiction that there is no $x \in S$ with $F(x) = F(-x)$, we obtain that $G \in C(\overline{\Omega}, \mathbb{R}^m)$, $G(x) := (F(x) - F(-x), 0)$, satisfies $G^{-1}(0) \subseteq \Omega$. Hence, $G^{-1}(0)$ is compact. Since G is odd, we obtain by Borsuk's Theorem 9.84 that $\deg(G, \Omega, 0)$ is odd, in particular nonzero. Since $(\Omega, E, \mathrm{id}_E)$ is a Rouché triple for $(G, \Omega, 0)$, we obtain by the stability property of the degree (Proposition 9.54) that also $\deg(G, \Omega, y) \neq 0$ for all y in some neighborhood of 0. The existence property implies that $y \in G(\Omega)$ for all y in a neighborhood of 0, contradicting the fact that by definition of G, all elements from $G(\Omega)$ have 0 in their last coordinate.

To prove (c), we assume by contradiction that none of the sets $S_k := S \cap A_k$ $(k = 1, \dots, n)$ contains an antipodal pair. Then

$$S \subseteq \bigcup_{k=1}^{n} S_k \subseteq \bigcup_{k=1}^{n-1} (S_k \cap (-S_k)), \tag{9.37}$$

because for every $x \in S_n$, we have $-x \notin S_n$ and thus $-x \in S_k$ for some $k < n$. Since S_k and $-S_k$ are closed and disjoint, we find by Urysohn's lemma functions $g_k \in C(\mathbb{R}^n, [-1, 1])$ satisfying $g_k(S_k) = \{1\}$ and $g_k(S_k) = \{-1\}$. We put $\Omega := B_1(0) \subseteq \mathbb{R}^n$ and define $G\colon \overline{\Omega} \to \mathbb{R}^n$ by

$$G(x) := (g_1(x), \dots, g_{n-1}(x), 1).$$

Then $F(x) := \frac{1}{2}(G(x) - G(-x))$ is an odd function. Moreover, the homotopy

$$H(t, x) := tG(x) + (1 - t)F(x)$$

satisfies $0 \notin H([0, 1] \times \partial\Omega)$, because for every $x \in S$ some of the first $n - 1$ coordinates of $H(0, x) = F(x)$ is ± 1 by (9.37), and because for $t > 0$ the last coordinate of $H(t, x)$ is $t > 0$. Hence, $H^{-1}(0)$ is a compact subset of $[0, 1] \times \Omega$. Then $\deg(H(1, \cdot), \Omega, 0) = \deg(F, \Omega, 0)$ is odd, in particular nonzero, by Borsuk's Theorem 9.84. It follows from the continuation principle that $0 \in H(\{0\} \times \Omega) = G(\Omega)$ which is a contradiction, since the definition of G implies that the last coordinate of every $y \in G(\Omega)$ is 1. □

The assertions (b) and (c) of Theorem 9.89 have famous real-world interpretations, which we discuss now.

For $n = 3$ and $m = 2$ the assertion (b) means that at each fixed time there is an antipodal pair on the earth surface with the same temperature and the same pressure: Indeed, for every function $F \in C(S, \mathbb{R}^2)$ which associates to each point x of the earth surface S temperature and pressure at a given time there is some $x \in S$ with $F(x) = F(-x)$.

Another famous interpretation of (b) for $n = 3$ is that it is possible to divide a piece of bread, an egg, and a ham simultaneously into two halves with one straight cut:

Corollary 9.91 (Egg–Ham–Sandwich). *Let $M_1, \ldots, M_n \subseteq \mathbb{R}^n$ be Lebesgue measurable with finite measure. Then there is a hyperplane which cuts each M_k into two sets of equal measure.*

Proof. Let $S = S_1(0) \subseteq \mathbb{R}^{n+1}$. For $x = (x_1, \ldots, x_n, t) \in S$, we define the hyperplane

$$H_x := \{y = (y_1, \ldots, y_n) \in \mathbb{R}^n : x_1 y_1 + \cdots + x_n y_n = t\},$$

and the corresponding "upper halfspace"

$$H_x^+ := \{y = (y_1, \ldots, y_n) \in \mathbb{R}^n : x_1 y_1 + \cdots + x_n y_n > t\}.$$

Note that the corresponding "lower halfspace" is H_{-x}^+. Hence, it suffices to prove that there is some $x \in S$ satisfying

$$\operatorname{mes}(A_k \cap H_x^+) = \operatorname{mes}(A_k \cap H_{-x}^+) \quad \text{for } k = 1, \ldots, n.$$

Defining $F\colon S \to \mathbb{R}^n$ by

$$F(x) := (\operatorname{mes}(A_1 \cap H_x^+), \ldots, \operatorname{mes}(A_n \cap H_x^+)),$$

we thus have to prove that there is some $x \in S$ satisfying $F(x) = F(-x)$. This is the assertion of Theorem 9.89(b) if F is continuous. The continuity of F follows by the representation

$$\operatorname{mes}(A_k \cap H_x^+) = \int_{\mathbb{R}^n} \chi_{A_k \cap H_x^+}(y)\, dy, \tag{9.38}$$

where χ_M denotes the characteristic function of M, that is

$$\chi_M(x) := \begin{cases} 1 & \text{if } x \in M, \\ 0 & \text{otherwise.} \end{cases}$$

Indeed, if $x_j \in S$ converges to $x \in S$, we have $\chi_{A_k \cap H_{x_j}^+} \to \chi_{A_k \cap H_x^+}$ almost everywhere. Since $0 \le \chi_{A_k \cap H_{x_j}^+} \le \chi_{A_k}$ and the latter function is integrable (since A_k has finite measure), we obtain by Lebesgue's dominated convergence theorem and (9.38) that $\operatorname{mes}(A_k \cap H_{x_j}^+) \to \operatorname{mes}(A_k \cap H_x)$. □

For $n = 3$, the assertion (c) of Theorem 9.89 implies that for peeling a potato by straight cuts, one needs at least 4 cuts. More generally, the following result holds:

Corollary 9.92. *If a bounded set $M \subseteq \mathbb{R}^n$ with an interior point is "bordered" by hyperplanes, then at least $n + 1$ such hyperplanes are used.*

Proof. By scaling and shifting M, we may assume without loss of generality that $0 \in \overset{\circ}{M}$ and $M \subseteq B_1(0)$. We assume by contradiction that n hyperplanes are sufficient to "border" M. The k-th hyperplane divides $\mathbb{R}^n$ into two halfspaces. Let A_k denote that (closed) halfspace which lies on the opposite side of 0. Then $S_1(0) \subseteq A_1 \cup \cdots \cup A_n$, and no $S \cap A_k$ contains an antipodal pair. This is a contradiction to Theorem 9.89(c). □

Remark 9.93. Our above proof shows that all assertions of Theorem 9.89 follow straightforwardly from degree theory. It is interesting to observe that one can see elementary that actually all assertions of Theorem 9.89 are equivalent.

To see the equivalence of (a) and (b), we assume that $\mathbb{R}^n$ and $\mathbb{R}^{n+1}$ are equipped with the Euclidean norm, and we introduce the following notation. Let $K := K_1(0) \subseteq \mathbb{R}^n$, $S := S_1(0) \subseteq \mathbb{R}^n$, and

$$S^+ := \{(x,t) \in \mathbb{R}^{n+1} : \|x\|^2 + t^2 = 1 \text{ and } t \ge 0\}.$$

Then $S_0 := S^+ \cup (-S^+)$ is the unit sphere in $\mathbb{R}^{n+1}$.

To prove (b)⇒(a), we assume now that $F \in C(K, \mathbb{R}^n)$ is such that $F|_S$ is odd. We define $G \colon S_0 \to \mathbb{R}^n$ by

$$G(x,t) := \begin{cases} F(x) & \text{if } (x,t) \in S^+, \\ -F(-x) & \text{if } (x,t) \in -S^+. \end{cases}$$

This is well-defined, since for $(x,t) \in S^+ \cap (-S^+)$, we have $x \in S$, and so $F(-x) = -F(x)$. The glueing lemma (Lemma 2.93) implies that G is continuous. We find by (b) some $(x,t) \in S_0$ with $G(x,t) = G(-x,-t)$, without loss of generality $(x,t) \in S^+$. Then $F(x) = -F(x)$, hence $F(x) = 0$.

To prove the converse implication (a)$\Rightarrow$(b), let $F \in C(S_0, \mathbb{R}^m)$ be given with some $m \le n$. We define $G \in C(K, \mathbb{R}^n)$ by

$$G(x) := (F(x, \sqrt{1 - \|x\|^2}) - F(-x, -\sqrt{1 - \|x\|^2}), 0).$$

Then $G|_S$ is odd, and so (a) implies that there is $x \in K$ with $G(x) = 0$. Putting $t := \sqrt{1 - \|x\|^2}$, we obtain $F(x, t) = F(-x, -t)$.

The following proof of the equivalence of (c) and (b) works with any choice of the norm. To prove the implication (b)$\Rightarrow$(c), we assume by contradiction that (c) fails, that is, S is covered by closed sets $A_1, \dots, A_n \in \mathbb{R}^n$ without antipodal pairs. Then we find as in the proof of Theorem 9.89 that (9.37) holds with $S_k := S \cap A_k$. We define $F \in C(S, \mathbb{R}^{n-1})$ by

$$F(x) := (\operatorname{dist}(x, S_1), \dots, \operatorname{dist}(x, S_{n-1})).$$

By (b), there is $x \in S$ with $F(x) = F(-x)$, that is $d_k = \operatorname{dist}(x, S_k) = \operatorname{dist}(x, -S_k)$ for $k = 1, \dots, n-1$. We must have $d_k \ne 0$, since otherwise $x \in S_k \cap (-S_k)$ which is a contradiction. It follows that $x \notin S_k \cup (-S_k)$ for $K = 1, \dots, n-1$, and so $x \notin S$ by (9.37) which is also a contradiction.

To prove the remaining implication (c)$\Rightarrow$(b), we assume by contradiction that (b) fails, that is, there is $F \in C(S, \mathbb{R}^m)$ with $m < n$ satisfying $F(x) \ne F(-x)$ for all $x \in S$. Then $G(x) := F(x) - F(-x)$ is odd and satisfies $0 \notin G(S)$. The compactness of $G(S)$ and Corollary 3.14 imply $\operatorname{dist}(0, G(S)) > 0$. Hence, with $G = (G_1, \dots, G_m)$ there is $\delta > 0$ such that $\max\{|G_k(x)| : k = 1, \dots, m\} > \delta$ for each $x \in S$. Since all G_k are odd, it follows that the closed set

$$A := \{x \in S : G_1(x), \dots, G_m(x) \in (-\infty, \delta]\}$$

contains no antipodal pair. Since G_k are odd, also the closed sets

$$A_k := \{x \in S : G_k(x) \ge \delta\} \qquad (k = 1, \dots, m)$$

contain no antipodal pair. However, the $m + 1 \le n$ sets $A_1, \dots, A_m, A$ cover S by their definition. This contradicts (c).

Borsuk's fixed point theorem on balls (or any of the other equivalent assertions of Theorem 9.89) is a surprisingly powerful result. In fact, one can show without using degree theory that it actually implies the most general form of Borsuk's fixed point theorem (Corollary 9.86 without the assertion about the degree), and also that it implies Brouwer's fixed point theorem, and that the latter implies the Leray–Schauder alternative.

If one looks at our proofs of Brouwer's fixed point theorem and the Leray–Schauder alternative, it is not surprising that from a very abstract point of view, these are in a sense special cases of Borsuk's result: The latter two results were essentially consequences of the assertion $\deg(\mathrm{id}_E, E, 0) = 1$, but actually only $\deg(\mathrm{id}, E, 0) \neq 0$ is needed for the proof. The latter in turn is a trivial special case of Borsuk's Theorem 9.84. This heuristic is of course not a proof that already the special case Theorem 9.89(a) implies all the other assertions in an elementary manner, but a rigorous proof can be given using the theory of so-called epi maps (sometimes also called essential maps) which we will not cover in this monograph: For a proof of the mentioned implications without using degree theory, we refer to [7].

We come to another important application of Borsuk's theorem. We call a map $F: X \to Y$ *locally one-to-one* if every $x \in X$ has a neighborhood U such that $F|_U$ is one-to-one.

Theorem 9.94 (Invariance of Domain). *Let E be a finite-dimensional normed space, and $\Omega \subseteq E$ be open. If $F \in C(\Omega, E)$ is locally one-to-one then $F(\Omega)$ is open in E.*

Proof. We have to show that every $y_0 \in F(\Omega)$ is an interior point of $F(\Omega)$. Let $x_0 \in \Omega$ satisfy $y_0 = F(x_0)$, and let $r > 0$ be such that $\Omega_0 := B_r(x_0)$ satisfies $\overline{\Omega_0} \subseteq \Omega$ and such that $F|_{\Omega_0}$ is one-to-one. Without loss of generality, we can assume that $y_0 = 0$ and $x_0 = 0$. We define $H: [0,1] \times \Omega_0 \to E$ by

$$H(t, x) := F(x) - F(-tx).$$

Since $F|_{\Omega_0}$ is one-to-one, we have $H(t, x) = 0$ if and only if $x = -tx$ which holds if and only if $x = 0$. Hence, $H^{-1}(0) = [0,1] \times \{0\}$ is compact. By the homotopy invariance of the degree, we obtain

$$\deg(F, \Omega_0, 0) = \deg(H(0, \cdot), \Omega_0, 0) = \deg(H(1, \cdot), \Omega_0, 0).$$

Since $H(1, \cdot)$ is odd, the latter degree is odd by Borsuk's Theorem 9.84, in particular nonzero. Note that $(\Omega_0, E, \mathrm{id}_E)$ is a Rouché triple for $(F, \Omega_0, 0)$. We obtain by the stability property of the degree (Proposition 9.54) that also $\deg(F, \Omega_0, y) \neq 0$ for all y in some neighborhood U of 0. The existence property of the degree thus implies $U \subseteq F(\Omega_0) \subseteq F(\Omega)$. □

Also Theorem 9.94 could have been reduced to Borsuk's fixed point theorem on balls [7].

Corollary 9.95 (Dimension Invariance). *Let nonempty open sets $\Omega_1 \subseteq \mathbb{R}^n$ and $\Omega_2 \subseteq \mathbb{R}^m$ be homeomorphic. Then $n = m$.*

Proof. Assume by contradiction without loss of generality that $m < n$. By hypothesis, there is a one-to-one map $H \in C(\Omega_1, \Omega_2)$. Then we can define a one-to-one map $F \in C(\Omega_1, \mathbb{R}^n)$ by putting $F(x) := (H(x), 0)$. Theorem 9.94 implies that $F(\Omega_1)$ is open. This is a contradiction, since all $y \in F(\Omega_1)$ have 0 as their last coordinate. □

Chapter 10

The Benevieri–Furi Degrees

The aim of this chapter is to define a degree theory for (oriented) Fredholm maps F which act from subsets of a Banach manifold X into a Banach manifold Y. Moreover, also a coincidence degree for pairs of maps (F, φ) where φ is continuous and compact should be developed. Similarly as the Brouwer degree, this degree should "count" in a homotopically invariant manner the number of solutions of the equation $F(x) = y$ or $F(x) = \varphi(x)$, respectively.

Of course, the coincidence degree is more general. Unfortunately, it is an open problem of whether the coincidence degree exists in this general setting: Our approach can prove its existence only for the case that Y is a Banach space.

Therefore, we present two degree theories: one for the case that Y is a Banach manifold, and the coincidence degree when Y is a Banach space. These degrees go back to [17] and [19]. In [19] only the case that X is a Banach space is covered, but we will develop the approach also if X is a Banach manifold.

Some historical notes are in order. There were several attempts to define a degree theory for Fredholm maps. The work initiating the main research in this direction is due to S. Smale [131], although similar results were obtained much earlier by R. Caccioppoli [31]. In these early attempts only a degree with values in $\mathbb{Z}_2$ (without orientation) was considered. Using various definitions of orientation or, more general, of something corresponding to $\operatorname{sgn} dF(x)$ in our setting, it was observed in several subsequent research papers that essentially the same approach as for the Brouwer degree works if one considers C^2 maps, see e.g. [51], [54]. At a first glance, it appears that one cannot do better than C^2 this way, since Corollary 9.67 (or similar results for other approaches) require at least C^2 smoothness. The first successful attempts to avoid C^2 smoothness have been made by Ju. I. Sapronov [26] and then later in [125], [126]. However, these approaches appear not so simple than that from [19], so we will use the latter approach.

The crucial idea of [17] by which C^1 maps can be treated is to *use* the Brouwer degree. This has a drawback: Even if one is only interested in Banach spaces X and Y, it is necessary for this approach to consider submanifolds and to use the Brouwer degree of manifolds.

We point out that our definition for both degrees differs from the corresponding definition in [17] and [19]: We have chosen the so-called reduction property as the key property for the definition of the degrees. This is the property which makes the connection with the Brouwer degree most obvious. This approach by the reduction property has the advantage that we can use the same idea later also in the more general setting of function triples: The latter possibility was observed only in [142].

We need to discuss this reduction property first in the setting of the Brouwer degree.

10.1 Further Properties of the Brouwer Degree

Due to the compatibility property (9.4) of the Brouwer degree, all properties which can be proved for the C^0 Brouwer degree also hold for the C^r Brouwer degree for $r \geq 1$ if they can be formulated for that degree. Therefore, we do not formulate the subsequent properties for smoother maps than C^1, although this would be possible (but gives nothing new). For this reason, we also omit the index r in the definition of deg, that is, from now on we write just $\deg_{(X,Y)}$ instead of $\deg_{(r,X,Y)}$.

Throughout this section, let X and Y be manifolds without boundary of class C^1 over real vector spaces E_X and E_Y with $\dim E_X = \dim E_Y < \infty$.

The assertion of the subsequent reduction property is in the simplest case that if $\deg_{(X,Y)}(G, \Omega, y)$ is defined for some C^1 map G, and if $Y_0 \subseteq Y$ is a submanifold which is transversal to G and satisfies $y \in Y_0$ then this degree can be calculated by considering the restriction of G to $X_0 := G^{-1}(Y_0)$. Note that this makes sense since by the transversality theorem (Theorem 8.55) the restriction $G_0 = G|_{X_0}: X_0 \to Y_0$ acts between manifolds of the same dimension, and therefore one can speak about $\deg_{(X_0,Y_0)}(G_0, X_0, y)$.

However, for the coincidence degree which we develop later, we need a similar assertion also when we replace G by a map F which is a "continuous perturbation of G along Y_0". In the case that $Y = E_Y$ is a vector space and Y_0 is a linear subspace, we mean by this that F has the form $F = G - \varphi$ with $\varphi \in C(\Omega, Y_0)$. If Y is only a manifold (and not a vector space), we cannot define $G - \varphi$ directly, and therefore the formulation of the general case is a bit clumsy. An additional technical difficulty in the oriented case is that Definition 9.19 only provides us an orientation for G_0 if G is oriented, but we are only given an orientation of F and need to define an orientation for the restriction F_0 of F.

Theorem 10.1. *The Brouwer degree has the following property:*

($S_{\mathcal{B}}$) (C^0 **Reduction**). *Let $(F, \Omega, y) \in \mathcal{B}^0(X, Y)$, and let J be a diffeomorphism of an open neighborhood $V \subseteq Y$ of y onto an open subset of E_Y. Let $\Omega_0 \subseteq \Omega$ be an open neighborhood containing $F^{-1}(y)$ with $F(\Omega_0) \subseteq V$, and let $G \in C^1(\Omega_0, V)$ and $\varphi \in C(\Omega_0, V)$ be such that*

$$J \circ F = (J \circ G) - (J \circ \varphi). \tag{10.1}$$

Let $Z \subseteq E_Y$ be a linear subspace such that $Y_0 := J^{-1}(Z)$ satisfies:

(a) *$\varphi(\Omega_0) \subseteq Y_0$ and $y \in Y_0$.*
(b) *Y_0 is transversal to G on Ω_0.*

Then $X_0 := \Omega_0 \cap G^{-1}(Y_0)$ is a submanifold of X of the same dimension as Z, and if this dimension is positive, the map $F_0 := F|_{X_0} \in C(X_0, Y_0)$ satisfies $(F_0, X_0, y) \in \mathcal{B}^0(X_0, Y_0)$ and

$$\deg_{(X,Y)}(F, \Omega, y) = \deg_{(X_0,Y_0)}(F_0, X_0, y).$$

The orientation of F_0 for the oriented case is described below.

In the oriented case, we fix some orientation of E_Y and equip J with the natural orientation. Then J induces an orientation on V by Proposition 9.34. The given orientation of F induces a corresponding orientation on Ω_0. The orientations on Ω_0 and V induce an orientation on G. Note that the latter orientation is independent of the choice of the orientation of E_Y, for if we choose the opposite orientation of E_Y then the orientations on V and Ω_0 are opposite, but the orientation of G stays the same. We equip $G_0 := G|_{X_0} \in C^1(X_0, Y_0)$ with the inherited orientation in the sense of Definition 9.19. We fix now some orientation of Z, and equip the diffeomorphism $J_0 := J|_{X_0} \in C^1(Y_0, Z)$ with the natural orientation. Then J_0 induces an orientation on Y_0, G induces a corresponding orientation on X_0, and these two orientations induce the required orientation on F_0. As above, this orientation is actually independent of the choice of the orientation of Z.

Proof. We note first that for each $x \in \Omega_0$ with $F(x) \in Y_0$ we have $J(G(x)) = J(F(x)) + J(\varphi(x)) \in Z + Z \subseteq Z$, and so $x \in X_0$. Hence,

$$\Omega_0 \cap F^{-1}(Y_0) \subseteq X_0, \qquad \text{in particular} \quad F^{-1}(y) = F_0^{-1}(y). \tag{10.2}$$

Thus, $F_0 \in C(X_0, Y_0)$, and $(F_0, X_0, y) \in \mathcal{B}^0(X_0, Y_0)$ follows from $(F, \Omega, y) \in \mathcal{B}^0(X, Y)$. By the excision property, it remains to show that

$$\deg_{(X,Y)}(F, \Omega_0, y) = \deg_{(X_0,Y_0)}(F_0, X_0, y). \tag{10.3}$$

Note that by Corollary 9.3 there is some open neighborhood $U \subseteq X$ containing $F^{-1}(y)$ with compact $\overline{U} \subseteq \Omega_0$. In view of the excision property, it suffices to prove the assertion when we replace Ω_0 by U. Hence, without loss of generality, we can assume that (Ω_0, J, V) is a Rouché triple for (F, Ω_0, y), (X_0, J_0, Y_0) is a Rouché triple for (F_0, X_0, y), and that, moreover, also G and φ have continuous extensions to $\overline{\Omega}_0$ satisfying $G(\overline{\Omega}_0) \cup \varphi(\overline{\Omega}_0) \subseteq V$. Since $\overline{\Omega}_0$ is compact and $Y \setminus J(V)$ is closed, we find by Corollary 3.14 some $\varepsilon > 0$ such that

$$B_\varepsilon((J \circ \varphi)(\overline{\Omega}_0) \cup (J \circ F)(\overline{\Omega}_0)) \subseteq J(V).$$

In particular, whenever $H \in C(\overline{\Omega}_0, Z)$ satisfies

$$\max_{x \in \overline{\Omega}_0} \|H(x) - (J \circ \varphi)\| < \varepsilon, \tag{10.4}$$

then $\hat{\varphi} := J^{-1} \circ H \in C(\overline{\Omega}_0, Y_0)$ and

$$\hat{F} := J^{-1} \circ ((J \circ G) - H) \in C(\overline{\Omega}_0, Y_0)$$

are defined, and $\hat{F}_0 := \hat{F}|_{X_0} \in C(X_0, Y_0)$. The stability property of the degree (Proposition 9.54) implies that if $\varepsilon > 0$ is chosen sufficiently small then none of the two sides in (10.3) changes if we replace F by $\hat{F}$ and F_0 by $\hat{F}_0$, respectively.

Using Theorem 9.9, we find some $H \in C^1(\Omega, Z)$ satisfying (10.4). Since then $H|_{X_0} \in C^1(X_0, Z)$, we can assume by Theorem 9.14, replacing H by $H + z$ with some small $z \in Z$ if necessary, that $J(y)$ is a regular value of $J \circ G_0 - H|_{X_0} \in C^1(X_0, Z)$ (we put $G_0 := G|_{X_0} \in C^1(X_0, Y_0)$ as above). By the above argument, it suffices to prove the assertion when we replace φ by $\hat{\varphi}$ and correspondingly F by $\hat{F}$ and F_0 by $\hat{F}_0$.

Hence, we have shown: It suffices to prove the reduction property for the case that $\varphi \in C^1(\Omega_0, Y_0 \cap V)$ and that y is a regular value of $F_0 \in C^1(X_0, Y_0)$.

We will show that it follows also that y is a regular value of $F|_{\Omega_0}$. To see this, we observe that by (10.2) a point $x \in X$ belongs to $F^{-1}(y)$ if and only if it belongs to $F_0^{-1}(y)$. We have to show for every such x that $dF(x) \in \operatorname{Iso}(T_xX, T_yY)$, that is, $A := d(J \circ F)(x) \in \operatorname{Iso}(T_xX, Z)$. Since $x \in X_0$ and y is a regular value of F_0, we know that $dF_0(x) \in \operatorname{Iso}(T_xX_0, T_yY_0)$, that is, $A_0 := d(J_0 \circ F_0)(x) \in \operatorname{Iso}(T_xX_0, Z)$.

Since Y_0 is transversal to G on Ω_0, we have that Z is transversal to $J \circ G$ on Ω_0. Putting $B := d(J \circ G)(x) \in \mathcal{L}(T_xX, Y)$ and $B_0 := d(J \circ G_0)(x) \in \mathcal{L}(T_xX_0, Z)$, we thus obtain from Theorem 8.55 that $T_xX_0 = B^{-1}(Z)$ has the same dimension as Z, and $B_0 = B|_{T_xX_0}$. Let $E_1 \subseteq T_xX$ be some subspace with $T_xX = T_xX_0 \oplus E_1$. Since $T_xX_0 = B^{-1}(Z)$, we have $\mathrm{R}(B) = Z \oplus B(E_1)$. On the other hand Z is transversal to $J \circ G$ on Ω_0; since $x \in \Omega_0$, this implies by definition that Z is transversal to B. Hence, $Z + \mathrm{R}(B) = Y$. Since $\mathrm{R}(B) = Z \oplus B(E_1)$, it follows with $Z_1 := B(E_1)$ that $Y = Z \oplus Z_1$. Since $\dim T_x X = \dim Y$ and $\dim T_x X_0 = \dim Z$, it follows that $\dim Z_1 = \dim E_1$, and so $B_1 := B|_{E_1} \in \mathrm{Iso}(E_1, Z_1)$.

Next, we note that $(J \circ \varphi)(\Omega_0) \subseteq Z$ implies that $C := d(J \circ \varphi)(x) \in \mathcal{L}(T_x X, Z)$. Let $P \in \mathcal{L}(T_x X)$ denote the projection onto $T_x X_0$ with $\mathrm{N}(P) = E_1$, and $Q := \mathrm{id}_{T_x X} - P$ the projection onto E_1 with kernel $T_x X_0$ (Proposition 6.18). Then $C = CP + CQ \in \mathcal{L}(T_x X, Z)$. Finally, we note that (10.1) implies $A = B - C$. Hence, according to the splitting $T_x X = T_x X_0 \oplus E_1$ and $Y = Z \oplus Z_1$, we can write $A = B - C$ in matrix form

$$A = \begin{pmatrix} B_0 & 0 \\ 0 & B_1 \end{pmatrix} - \begin{pmatrix} CP & CQ \\ 0 & 0 \end{pmatrix} = \begin{pmatrix} B_0 - CP & -CQ \\ 0 & B_1 \end{pmatrix} = \begin{pmatrix} A_0 & -CQ \\ 0 & B_1 \end{pmatrix}.$$

Since A_0 and B_1 are isomorphisms, it follows that A is an isomorphism, as required. Hence y is a regular value of $F|_{\Omega_0}$. Applying the regular normalization property (Proposition 9.46) to both sides of (10.3), we thus obtain the assertion if we can show that even $\operatorname{sgn} dF(x) = \operatorname{sgn} dF_0(x)$. This is clear in the non-oriented case.

In the oriented case, we fix orientations on E_Y and Z, and let Ω_0 and X_0 (hence $T_x X$ and $T_x X_0$) be oriented as described earlier. We have to show that $\operatorname{sgn} A = \operatorname{sgn} A_0$ with the corresponding induced orientations. To this end, we observe that $\operatorname{sgn} A_0$ is determined by the inherited orientation of

$$B = \begin{pmatrix} B_0 & 0 \\ 0 & B_1 \end{pmatrix}.$$

Note that Corollary 7.17 implies $\operatorname{sgn} B_0 = \operatorname{sgn} B$. We have

$$\det(A^{-1}B) = \det \begin{pmatrix} A_0^{-1}B_0 & A_0^{-1}CQ \\ 0 & \mathrm{id}_{E_1} \end{pmatrix} = \det(A_0^{-1}B_0).$$

If this number is positive (negative), then A and B and A_0 and B belong to the same (opposite) class of correctors of the zero operator, and so $\operatorname{sgn} A = \operatorname{sgn} B$, $\operatorname{sgn} A_0 = \operatorname{sgn} B_0$ (or $\operatorname{sgn} A = -\operatorname{sgn} B$, $\operatorname{sgn} A_0 = -\operatorname{sgn} B_0$). In both cases, we obtain from $\operatorname{sgn} B_0 = \operatorname{sgn} B$ that $\operatorname{sgn} A_0 = \operatorname{sgn} A$. Hence, $\operatorname{sgn} dF(x) = \operatorname{sgn} dF_0(x)$, and the reduction property is proved. □

It seems that Theorem 10.1 was never observed before, although the special case $Y = E_Y$ with $J = \mathrm{id}_Y$ was mentioned without proof in [19].

The reader familiar with the fixed point index will see an analogue of Theorem 10.1 with the reduction property of the fixed point index. If $X = Y = E_X =$

E_Y with some fixed (the same) orientation and if $(\mathrm{id}_X - \varphi, \Omega, 0) \in \mathscr{B}^0(X, Y)$ then we define the *fixed point index* on X by the formula

$$\mathrm{ind}_X(\varphi, \Omega) := \deg_{(X,X)}(\mathrm{id}_X - \varphi, \Omega, 0),$$

where the orientation is understood as the induced orientation. For the case that $X_0 \subseteq X$ is a linear subspace and $\varphi(\Omega) \subseteq X_0$, the so-called reduction property of the fixed point index states that

$$\mathrm{ind}_X(\varphi, \Omega) = \mathrm{ind}_{X_0}(\varphi, \Omega \cap X_0).$$

This is in fact nothing else than the special case of Theorem 10.1 with $G = \mathrm{id}_X$.

In fact, our proof of Theorem 10.1 imitates the proof of this classical special case. This strategy for the proof is rather natural: Using the stability of the degree, one reduces the situation to the regular case and then one obtains the result by direct calculation, using just basic linear algebra. Since the details in the above result were a bit involved, let us give an example where the same strategy can be applied more straightforwardly.

Theorem 10.2. *The Brouwer degree has the following property:*

($\mathrm{T}_{\mathscr{B}}$) **(Cartesian Product).** *For $i = 1, 2$, let X_i and Y_i be manifolds without boundary of class C^1 over real vector spaces E_{X_i} and E_{Y_i}, respectively, with $0 < \dim E_{X_i} = \dim E_{Y_i} < \infty$. Let $\Omega_i \subseteq X_i$ be open, $F_i \in C(\Omega_i, Y_i)$, and $y_i \in Y_i$ ($i = 1, 2$). We put $X := X_1 \times X_2$, $\Omega := \Omega_1 \times \Omega_2$, $Y := Y_1 \times Y_2$, $F := F_1 \otimes F_2$, and $y := (y_1, y_2)$. Then $(F, \Omega, y) \in \mathscr{B}^0(X, Y)$ if and only if $(F_i, \Omega_i, y_i) \in \mathscr{B}^0(X_i, Y_i)$ for $i = 1, 2$ or if $F_i^{-1}(y_i) = \emptyset$ for $i = 1$ or $i = 2$. In the former case*

$$\deg_{(X,Y)}(F, \Omega, y) = \deg_{(X_1,Y_1)}(F_1, \Omega_1, y_1) \deg_{(X_2,Y_2)}(F_2, \Omega_2, y_2). \tag{10.5}$$

In the oriented case F is equipped with the product orientation.

Proof. We have

$$F^{-1}(y) = F_1^{-1}(y_1) \times F_2^{-1}(y_2), \tag{10.6}$$

and so Theorem 2.63 and Corollary 2.101 imply that $F^{-1}(y)$ is compact if and only if $F_i^{-1}(y_i)$ ($i = 1, 2$) are compact. Hence, $(F, \Omega, y) \in \mathscr{B}^0(X, Y)$ is equivalent to $(F_i, \Omega_i, y_i) \in \mathscr{B}^0(X_i, Y_i)$ for $i = 1, 2$.

If the latter holds, we find by Proposition 9.52 Rouché triples $(\Omega_{0,i}, V_i, J_i)$ for (Ω_i, F_i, y_i) ($i = 1, 2$). Putting $\Omega_0 := \Omega_{0,1} \times \Omega_{0,2}$, $V := V_1 \otimes V_2$, and $J := J_1 \otimes J_2$, we find by Theorem 2.63 that (Ω_0, V, J) is a Rouché triple for (Ω, F, y).

Using the stability property (Proposition 9.54), we find some $\varepsilon > 0$ such that for every $F_{0,i} \in C^1(\Omega_{0,i}, V_i)$ and every $y_{0,i} \in V_i$ with

$$d_{J_i}(y_i, y_{0,i}) \leq \varepsilon \quad \text{and} \quad d_{J_i}(F_i(x), F_{0,i}(x)) \leq \varepsilon \quad \text{for all } x \in \overline{\Omega}_{0,i},$$

we have $(F_{0,i}, \Omega_{0,i}, y_{0,i}) \in \mathcal{B}^0(X_i, Y_i)$ and

$$\deg_{(X_i,Y_i)}(F_i, \Omega_i, y_i) = \deg_{(X_i,Y_i)}(F_{0,i}, \Omega_{0,i}, y_{0,i})$$

for $i = 1, 2$ and, moreover, for every $F \in C^1(\Omega_0, V)$ and every $y_0 \in V$ with

$$d_J(y, y_0) \leq 2\varepsilon \quad \text{and} \quad d_J(F(x), F_0(x)) \leq 2\varepsilon \quad \text{for all } x \in \overline{\Omega}_0,$$

we have $(F_0, \Omega_0, y_0) \in \mathcal{B}^0(X_0, Y_0)$ and

$$\deg_{(X,Y)}(F, \Omega, y) = \deg(F_0, \Omega_0, y_0).$$

(In the oriented case the orientation must be chosen as described after Proposition 9.54.) By Proposition 9.53, there are $F_{0,i} \in C^1(\Omega_i, V_i)$ and regular values $y_{0,i} \in V$ for $F_{0,i}$ satisfying the above requirement. It follows that also $F_0 := F_{0,1} \otimes F_{0,2}$ and $y_0 := (y_{0,1}, y_{0,2})$ satisfy the above requirement. Moreover, in the oriented case, since F is equipped with the product orientation, it follows that the product orientation of F_0 corresponds to the orientation described after Proposition 9.53.

The above argument shows that it suffices to prove the result for the special case that F_i are functions of class C^1 and that y_i is a regular value of F_i for $i = 1, 2$. In this case, also F is of class C^1, and moreover, y is a regular value of F. To see the latter, recall that by (10.6), we have for $x = (x_1, x_2)$ that $F(x) = y$ if and only if $F_i(x_i) = y_i$ ($i = 1, 2$). Since $T_x X = T_{x_1} X_1 \times T_{x_2} X_2$, $T_y Y = T_{y_1} Y_1 \times T_{y_2} Y_2$, and $dF(x) = dF(x_1) \otimes dF(x_2)$, it follows that $dF(x)$ is onto if (and only if) $dF(x_i)$ is onto for $i = 1, 2$. Hence y is a regular value of F. Noting that Proposition 7.10 implies for $x = (x_1, x_2) \in X$ that

$$\operatorname{sgn} dF(x) = \operatorname{sgn} dF(x_1) \operatorname{sgn} dF(x_2),$$

we calculate in view of (10.6) that

$$\begin{aligned}
\deg(F, \Omega, y) &= \sum_{x \in F^{-1}(y)} \operatorname{sgn} dF(x) = \sum_{(x_1,x_2) \in F^{-1}(y)} \operatorname{sgn} dF_1(x_1) \operatorname{sgn} dF_2(x_2) \\
&= \sum_{\substack{x_1 \in F_1^{-1}(y_1) \\ x_2 \in F_2^{-1}(y_2)}} \operatorname{sgn} dF_1(x_1) \operatorname{sgn} dF_2(x_2) \\
&= \Big(\sum_{x_1 \in F_1^{-1}(y_1)} \operatorname{sgn} dF_1(x_1) \Big) \Big(\sum_{x_2 \in F_2^{-1}(y_2)} \operatorname{sgn} dF_2(x_2) \Big).
\end{aligned}$$

By the regular normalization property of the degree (Proposition 9.46) this means exactly (10.5). □

In the special case of the fixed point index, we show now that the Cartesian product property implies the remarkable commutativity property of the fixed point index. The latter means that, roughly speaking, the fixed point index of $F \circ G$ is the same as that of $G \circ F$. This becomes reasonable if one considers the diagram

$$X \underset{G}{\overset{F}{\rightleftarrows}} Y$$

where the arrows should be understood such that F and G are not necessarily defined on the whole space X and Y but only on a subset. This diagram implies that each fixed point x of $G \circ F$ becomes mapped by F into a fixed point y of $F \circ G$ which then by G is mapped back to x. Conversely, each fixed point y of $F \circ G$ is mapped by G into a fixed point of $G \circ F$ which then by F is mapped back into y. If follows that F and G map the corresponding fixed point sets onto each other and are inverse to each other on these sets, that is, their restriction to the fixed point sets is a bijection to the respective other fixed point set with the other map as inverse. In particular, the cardinality of the fixed points of $F \circ G$ and of $G \circ F$ is the same. Thus, it is perhaps not too surprising that also their "homological count", the fixed point index, is the same. The perhaps surprising fact is that we do not even need for the latter that X and Y have the same dimension.

Theorem 10.3 (Commutativity of the Fixed Point Index). *Let $X = E_X$ and $Y = E_Y$ be finite-dimensional real normed spaces (not necessarily of the same dimension). For open subsets $\Omega_X \subseteq X$ and $\Omega_Y \subseteq Y$, let $F\colon \Omega_X \to Y$ and $G\colon \Omega_Y \to X$ be continuous. Then*

$$\operatorname{ind}_X(G \circ F, \Omega_X \cap F^{-1}(\Omega_Y)) = \operatorname{ind}_Y(F \circ G, \Omega_Y \cap G^{-1}(\Omega_X)), \qquad (10.7)$$

in the strong sense that if one of the two indices is defined then so is the other.

Proof. Let $K_X \subseteq \Omega_1 := \Omega_X \cap F^{-1}(\Omega_Y)$ and $K_Y \subseteq \Omega_2 := \Omega_Y \cap G^{-1}(\Omega_X)$ denote the fixed point set of $G \circ F|_{\Omega_1}$ or $F \circ G|_{\Omega_2}$, respectively. As we have seen above, $F|_{K_X}$ is a bijection onto K_Y with inverse $G|_{K_Y}$. In particular, if K_X is compact then so is $F(K_X) = K_Y \subseteq \Omega_Y \cap F^{-1}(\Omega_X)$ and if K_Y is compact then so is $G(K_Y) = K_X \subseteq \Omega_X \cap G^{-1}(\Omega_Y)$. Thus, if one of the fixed point indices in (10.7) is defined then so is the other. Assume now that this is the case. In case $K_X = \varnothing$ or, equivalently, $K_Y = \varnothing$, both indicesin (10.7) vanish. Hence, assume that there is some $(x_0, y_0) \in K_1 \times K_2$. We consider the homotopies

$$\begin{aligned}
&H_1: \Omega_1 \times \Omega_2 \to X \times Y, && H_1(t,(x,y)) := ((1-t)G(y) + tG(F(x)), F(x))\\
&H_2: \Omega_1 \times \Omega_2 \to X \times Y, && H_2(t,(x,y)) := (G(y), (1-t)F(x) + tF(G(y)))\\
&H_3: \Omega_1 \times Y \to X \times Y, && H_3(t,(x,y)) := (G(F(x)), (1-t)F(x) + ty_0)\\
&H_4: X \times \Omega_2 \to X \times Y, && H_4(t,(x,y)) := ((1-t)G(y) + tx_0, F(G(y)))
\end{aligned}$$

and put $h_i(t,(x,y)) := (x,y) - H_i(t,(x,y))$ $(i = 1,2,3,4)$. If (x,y) is a fixed point of $H_i(t,\cdot)$ for $i = 1$ or $i = 2$ then $F(x) = y$ and $G(y) = x$, hence the fixed point set of $H_i(t,\cdot)$ $(i = 1,2)$ is contained in the set

$$K := \{(x,y) \in K_X \times K_Y : F(x) = y\} = \{(x,y) \in K_X \times K_Y : G(y) = x\}.$$

which is a closed subset of $K_X \times K_Y$ and thus compact. Conversely, every point from K is a fixed point of $H_i(t,\cdot)$ for $i = 1,2$, and so $h_i^{-1}(0) = [0,1] \times K$ is compact for $i = 1,2$. Similarly,

$$\begin{aligned}
h_3^{-1}(0) &= \{(t,x,y) \in [0,1] \times K_X \times Y : y = (1-t)F(x) + ty_0\},\\
h_4^{-1}(0) &= \{(t,x,y) \in [0,1] \times X \times K_Y : x = (1-t)G(y) + tx_0\}.
\end{aligned}$$

Since $F(K_X) = K_Y$ and $G(K_Y) = K_X$, it follows that these are closed subsets of $[0,1] \times K_X \times (\overline{\operatorname{conv}}\, K_Y)$ and $[0,1] \times (\overline{\operatorname{conv}}\, K_X) \times K_Y$, respectively. The latter sets are compact by Corollary 3.62 and Theorem 2.63, and so $h_3^{-1}(0)$ and $h_4^{-1}(0)$ are compact. Moreover, we have

$$\begin{aligned}
h_3(0,\cdot)^{-1}(0) &= \{(x,y) \in K_X \times Y : y = F(x)\} \subseteq K_X \times K_Y \subseteq \Omega_X \times \Omega_Y,\\
h_4(0,\cdot)^{-1}(0) &= \{(x,y) \in X \times K_Y : x = G(y)\} \subseteq K_X \times K_Y \subseteq \Omega_X \times \Omega_Y.
\end{aligned}$$

Applying the homotopy invariance and excision property, we thus calculate

$$\begin{aligned}
&\operatorname{ind}_{X\times Y}(H_3(1,\cdot), \Omega_1 \times Y) = \operatorname{ind}_{X\times Y}(H_3(0,\cdot), \Omega_1 \times Y)\\
&\quad = \operatorname{ind}_{X\times Y}(H_1(1,\cdot), \Omega_1 \times \Omega_2) = \operatorname{ind}_{X\times Y}(H_1(0,\cdot), \Omega_1 \times \Omega_2)\\
&\quad = \operatorname{ind}_{X\times Y}(H_2(0,\cdot), \Omega_1 \times \Omega_2) = \operatorname{ind}_{X\times Y}(H_2(1,\cdot), \Omega_1 \times \Omega_2)\\
&\quad = \operatorname{ind}_{X\times Y}(H_4(0,\cdot), X \times \Omega_2) = \operatorname{ind}_{X\times Y}(H_4(1,\cdot), X \times \Omega_2).
\end{aligned}$$

Considering the constant maps $c_Y(y) := y_0$ and $c_X(x) := x_0$, we obtain by two further applications of the excision property that

$$\operatorname{ind}_{X\times Y}((G \circ F) \otimes c_Y, \Omega_1 \times \Omega_2) = \operatorname{ind}_{X\times Y}(c_X \otimes (F \circ G), \Omega_1 \times \Omega_2).$$

Since the normalization of the degree implies that $\operatorname{ind}_X(c_X, \Omega_1) = 1$ and $\operatorname{ind}_Y(c_Y, \Omega_2) = 1$, we obtain (10.7) from the Cartesian product property. □

The above proof uses ideas from [147].

Note that we only used the Cartesian product property of the degree in the proof (besides the homotopy invariance and normalization property). Note also that the reduction property of the fixed point index is a special case of the commutativity with $Y := X_0$ and $G: X_0 \to X$ being the inclusion map. Hence, for the fixed point index, the reduction property can be considered as a special case of the Cartesian product property. However, it is unknown to the author whether this is also the case in the more general setting of the Brouwer degree on manifolds. At least, it seems that if the reduction property can be proved by means of the Cartesian product property in the general case, then this proof cannot be much simpler than the more direct proof of Theorem 10.1 which we had given.

Anyway, the above considerations show that the Cartesian product property of the degree is an important property which is worth to be established also for the more advanced degree theories which we discuss in this monograph.

10.2 The Benevieri–Furi C^1 Degree

Throughout this section, let X and Y be C^1 Banach manifolds without boundary over real Banach spaces E_X and E_Y, respectively. In order to avoid trivialities, we assume throughout:

$$E_X^* \neq \{0\},\ E_Y \neq \{0\},\ \text{and at least one of } E_X^* \text{ or } E_Y^* \text{ has full support.} \quad (10.8)$$

If one assumes AC, the hypothesis (10.8) just means $E_X \neq \{0\}$ and $E_Y \neq \{0\}$ by the Hahn–Banach extension theorem (Corollary 6.25).

Similarly as the Brouwer degree, the Benevieri–Furi degree comes in a non-oriented and in an oriented flavor: In the non-oriented case, it assumes values in $\mathbb{Z}_2$, and in the oriented case, it assumes values in $\mathbb{Z}$.

We denote by $\mathcal{F}_{\mathrm{BF}}(X,Y)$ the system of all (F,Ω,y) where $\Omega \subseteq X$ is open, $F:\Omega \to Y$, $y \in Y$, $F^{-1}(y)$ is compact, and such that there is an open neighborhood $\Omega_0 \subseteq \Omega$ of $F^{-1}(y)$ with $F \in \mathcal{F}_0(\Omega_0, Y)$. For the oriented version of the degree, we assume that F is oriented on Ω_0.

Definition 10.4. The *Benevieri–Furi C^1 degree* is the map $\deg = \deg_{(X,Y)}$ which associates to each $(F,\Omega,y) \in \mathcal{F}_{\mathrm{BF}}(X,Y)$ a number from $\mathbb{Z}_2$ (or $\mathbb{Z}$ in the oriented case) such that the following property holds:

($\mathrm{A}_{\mathcal{F}_{\mathrm{BF}}}$) ($C^1$ **Reduction**). Let $Y_0 \subseteq Y$ be a finite-dimensional submanifold with $y \in Y_0$ which is transversal to F on an open neighborhood $\Omega_0 \subseteq \Omega$ of $F^{-1}(y)$ with $F \in \mathcal{F}_0(\Omega_0, Y)$ (and F being oriented on Ω_0). For each such Ω_0 and Y_0 the set $X_0 := \Omega_0 \cap F^{-1}(Y_0)$ is either empty or a submanifold of the same dimension as Y_0, and if the dimension of Y_0 is positive, the map $F_0 := F|_{X_0} \in C^1(X_0, Y_0)$ satisfies

$$\deg_{(X,Y)}(F, \Omega, y) = \deg_{(X_0,Y_0)}(F_0, X_0, y), \tag{10.9}$$

where the right-hand side denotes the C^1 Brouwer degree. In case $X_0 = \varnothing$ the right-hand side is defined as zero. In the oriented case, the orientation of F_0 is the inherited orientation according to Definition 8.65.

Note that the inherited orientation is an orientation in the sense of Definition 8.25, that is, we understand the C^1 Brouwer degree according to Remark 9.73. (Equivalently, we can convert the orientation into an orientation of C^0 maps by means of (9.2) and Proposition 9.18, and use the C^0 Brouwer degree.)

Theorem 10.5. *For fixed Banach manifolds X and Y there is exactly one C^1 Benevieri–Furi degree. This degree satisfies the following properties for every $(F, \Omega, y) \in \mathcal{F}_{\mathrm{BF}}(X, Y)$.*

($\mathrm{B}_{\mathcal{F}_{\mathrm{BF}}}$) **(Regular Normalization).** *If y is a regular value of F then we have a finite sum*

$$\deg(F, \Omega, y) = \sum_{x \in F^{-1}(y)} \operatorname{sgn} dF(x).$$

($\mathrm{C}_{\mathcal{F}_{\mathrm{BF}}}$) **(Excision).** *If $\Omega_0 \subseteq \Omega$ is an open neighborhood of $F^{-1}(y)$ then*

$$\deg(F, \Omega, y) = \deg(F, \Omega_0, y).$$

($\mathrm{D}_{\mathcal{F}_{\mathrm{BF}}}$) **(Compatibility with the Brouwer Degree).** *In case $\dim E_X = \dim E_Y < \infty$ the C^1 Benevieri–Furi degree is the same as the C^1 Brouwer degree.*

($\mathrm{E}_{\mathcal{F}_{\mathrm{BF}}}$) **(Compatibility with the Non-oriented Case).** *The degrees for the oriented and non-oriented case are the same modulo 2 (if the oriented case applies).*

Proof. Let $(F, \Omega, y) \in \mathcal{F}_{\mathrm{BF}}(X, Y)$. We note first that (10.8) implies that there is some submanifold Y_0 of finite positive dimension with $y \in Y_0$ which is transversal to F on an open neighborhood $\Omega_0 \subseteq \Omega$ of $F^{-1}(K)$ where $F \in \mathcal{F}_0(\Omega_0, Y)$. Indeed, this follows from Corollary 8.72 (and Remark 8.71) with $H(t, x)$

$:= F(x)$, $K := F^{-1}(y)$, and a chart of $y = F(K)$ as the required diffeomorphism onto an open subset of E_Y. Theorem 8.55 implies that $X_0 := F|_{\Omega_0}^{-1}(Y_0)$ is a submanifold of Ω_0 (and thus a submanifold of X) of the same dimension as Y_0. We put $F_0 := F|_{X_0}$. If $\deg = \deg_{(X,Y)}$ is a C^1 Benevieri–Furi degree, the C^1 reduction property implies that (10.9) holds, and so $\deg_{(X,Y)}(F, \Omega, y)$ is uniquely determined. This shows the uniqueness.

For the existence, we let Y_0 and Ω_0 be as above, define X_0, F_0 as above, and define $\deg(F, \Omega, y)$ by (10.9). We have to show that this is well-defined, that is, independent of the particular choice of Ω_0 and Y_0. Let us first show that this is the case if $y \in Y$ is a regular value of F. By Proposition 8.68 the latter is equivalent to y being a regular value of F_0. Moreover, in view of (8.13) and $F^{-1}(y) = F_0^{-1}(y)$, we obtain

$$\deg_{(X_0,Y_0)}(F_0, X_0, y) = \sum_{x \in F_0^{-1}(y)} \operatorname{sgn} dF_0(x) = \sum_{x \in F^{-1}(y)} \operatorname{sgn} dF(x).$$

The latter expression is independent of the particular choice of Y_0 and Ω_0, and so $\deg(F, \Omega, y)$ is well-defined in this case. Moreover, we have also proved the regular normalization property of the C^1 Benevieri–Furi degree. This in turn implies the excision property for the case that y is a regular value of F.

To see that $\deg(F, \Omega, y)$ is well-defined also if y is not a regular value of F, let Ω_0, Y_0, X_0, and F_0 be as in the definition of the C^1 reduction property. By Proposition 9.52, there is a Rouché triple (Ω_1, V, J) for (F_0, X_0, y). By definition of the inherited topology there is an open set $\hat{\Omega} \subseteq \Omega$ with $\Omega_1 = X_0 \cap \hat{\Omega}$. Proposition 9.54 implies that there is an open neighborhood $V_0 \subseteq V$ of y such that

$$\deg_{(X_0,Y_0)}(F_0, X_0, y) = \deg_{(X_0,Y_0)}(F_0, \Omega_1, z)$$

for all $z \in V_1$. By Proposition 9.53, there is a set $V_2 \subseteq V_1$ of regular values of F_0 with $y \in \overline{V_2}$. By the special case we proved before, each $z \in V_2$ is a regular value of F, and so

$$\deg_{(X_0,Y_0)}(F_0, X_0, y) = \deg(F_0, \hat{\Omega}, z) \quad \text{for all } z \in V_2,$$

where we already know that the degree on the right-hand side is well-defined and satisfies the excision property. The latter implies that the right-hand side is actually independent of the particular choice of Ω_0 and Y_0, and so also the left-hand side is independent of that choice which means that $\deg(F, \Omega, y)$ is well-defined. By definition, this degree satisfies the C^1 reduction property. Hence, the existence is established. Since the degree is well-defined, the excision property is contained in the definition of the C^1 reduction property. The compatibility with the Brouwer degree follows with the choices $Y_0 := Y$ and $\Omega_0 := \Omega$ in the C^1 reduction property. The compatibility of the oriented and non-oriented case follows from the C^1 reduction property and the corresponding property of the Brouwer degree. □

Theorem 10.6. *The C^1 Benevieri–Furi degree satisfies the following properties for every $(F, \Omega, y) \in \mathcal{F}_{\mathrm{BF}}(X, Y)$.*

($\mathrm{F}_{\mathcal{F}_{\mathrm{BF}}}$) **(Generalized Homotopy Invariance).** *Let $W \subseteq [0,1] \times X$ be open, $H: W \to Y$ and $y \in C([0,1], Y)$. Suppose that the coincidence set $K := \{(t,x) \in [0,1] \times X : H(t,x) = y(t)\}$ is compact and that there is an open neighborhood $U \subseteq W$ such that $H|_U$ is a generalized (oriented) Fredholm homotopy of index 0. Then $W_t := \{x : (t,x) \in W\}$ satisfies $(H(t, \cdot),$
$W_t, y(t)) \in \mathcal{F}_{\mathrm{BF}}(X, Y)$ $(t \in [0,1])$, and*

$$\deg(H(t, \cdot), W_t, y(t)) \textit{ is independent of } t \in [0,1].$$

($\mathrm{G}_{\mathcal{F}_{\mathrm{BF}}}$) **(Additivity).** *If $\Omega = \Omega_1 \cup \Omega_2$ with disjoint open subsets $\Omega_1, \Omega_2 \subseteq \Omega$ then*

$$\deg(F, \Omega, y) = \deg(F, \Omega_1, y) + \deg(F, \Omega_2, y).$$

($\mathrm{H}_{\mathcal{F}_{\mathrm{BF}}}$) **(Normalization for Diffeomorphisms).** *If F is a diffeomorphism onto an open subset of Y then*

$$\deg(F, \Omega, y) = \begin{cases} 0 & \textit{if } y \notin F(\Omega), \\ \operatorname{sgn} dF(F^{-1}(y)) & \textit{otherwise.} \end{cases}$$

($\mathrm{I}_{\mathcal{F}_{\mathrm{BF}}}$) **(Existence).** *If $\deg(F, \Omega, y) \neq 0$ then $y \in F(\Omega)$.*

($\mathrm{J}_{\mathcal{F}_{\mathrm{BF}}}$) **(Excision-Additivity).** *If $\Omega_i \subseteq \Omega$ $(i \in I)$ is a family of pairwise disjoint open sets with $F^{-1}(y) \subseteq \bigcup_{i \in I} \Omega_i$ such that $\Omega_i \cap F^{-1}(y)$ is compact for all $i \in I$, then*

$$\deg(F, \Omega, y) = \sum_{i \in I} \deg(F, \Omega_i, y),$$

where in the sum at most a finite number of summands is nonzero.

($\mathrm{K}_{\mathcal{F}_{\mathrm{BF}}}$) **(Diffeomorphic Invariance).** *Let J_1 and J_2 be diffeomorphisms of an open subset of a Banach manifold X_0 onto Ω or of an open neighborhood*

$U \subseteq Y$ of $F(\Omega) \cup \{y\}$ onto an open subsets of a Banach manifold Y_0, respectively. Then

$$\deg_{(X,Y)}(F, \Omega, y) = \deg_{(X_0,Y_0)}(J_2 \circ F \circ J_1, J_1^{-1}(\Omega), J_2(y)).$$

In the oriented case, the orientation of $J_2 \circ F \circ J_1$ is understood in the sense of Proposition 8.38.

($L_{\mathcal{F}_{BF}}$) **(Restriction)**. *Let $X_0 \subseteq X$ and $Y_0 \subseteq Y$ be open. Then*

$$\deg_{(X_0,Y_0)} = \deg_{(X,Y)} |_{\mathcal{F}_{BF}(X_0,Y_0)}.$$

($M_{\mathcal{F}_{BF}}$) **(Cartesian Product)**. *For $i = 1, 2$, let X_i and Y_i be manifolds without boundary of class C^1 over Banach spaces E_{X_i} and E_{Y_i}, respectively. Let $\Omega_i \subseteq X_i$ be open, $F_i \in \mathcal{F}_0(\Omega_i, Y_i)$, and $y_i \in Y_i$ ($i = 1, 2$). We put $X := X_1 \times X_2$, $Y := Y_1 \times Y_2$, $\Omega := \Omega_1 \times \Omega_2$, $F := F_1 \otimes F_2$, and $y := (y_1, y_2)$. Then $(F, \Omega, y) \in \mathcal{F}_{BF}(X, Y)$ if and only if $(F_i, \Omega_i, y_i) \in \mathcal{F}_{BF}(X_i, Y_i)$ for $i = 1, 2$ or if $F_i^{-1}(y_i) = \varnothing$ for $i = 1$ or $i = 2$. In the former case*

$$\deg_{(X,Y)}(F, \Omega, y) = \deg_{(X_1,Y_1)}(F_1, \Omega_1, y_1) \deg_{(X_2,Y_2)}(F_2, \Omega_2, y_2).$$

In the oriented case F is equipped with the product orientation.

($N_{\mathcal{F}_{BF}}$) **(Elimination of** y**)**. *If $Y = E_Y$ is a Banach space then $(F, \Omega, y) \in \mathcal{F}_{BF}(X, Y)$ is equivalent to $(F - y, \Omega, 0) \in \mathcal{F}_{BF}(X, Y)$, and in this case*

$$\deg_{(X,E_Y)}(F, \Omega, y) = \deg_{(X,E_Y)}(F - y, \Omega, 0).$$

Proof. The normalization for diffeomorphism and the existence properties are immediate consequences of the regular normalization. The additivity, restriction, diffeomorphic invariance, and Cartesian product properties are immediate consequences of the definition and the corresponding properties of the Brouwer degree: For example, for the Cartesian product property, we note that if $Y_{i,0} \subseteq Y_i$ are finite-dimensional submanifolds which are transversal to F_i on $\Omega_{i,0}$ and satisfy $y_i \in Y_{i,0}$ ($i = 1, 2$) then $Y_0 := Y_{1,0} \times Y_{2,0}$ is a submanifold of Y which is transversal to F on Ω and satisfies $y \in Y_0$.

The excision-additivity follows from the excision and additivity properties; the reasoning here is completely analogously to the Brouwer degree.

To prove the generalized homotopy invariance, we note first that Proposition 2.62 implies by the compactness of K that $K_t := \{x : H(t, x) = y(t)\}$ is compact for every $t \in [0, 1]$, and so $(H(t, \cdot), W_t, y(t)) \in \mathcal{F}_{\mathrm{BF}}(X, Y)$. By Proposition 2.19, it suffices to show that

$$d(t) := \deg(H(t, \cdot), W_t, y(t))$$

is locally constant on $[0, 1]$. Thus, let $t_0 \in [0, 1]$, and we must show that d is constant in a neighborhood of t_0.

Let $c\colon V \to E_Y$ be a chart of Y with some open neighborhood $V \subseteq Y$ of $y(t_0)$. Shrinking W if necessary (which we can do by the excision property) we can assume without loss of generality that there is a neighborhood $I_0 \subseteq [0, 1]$ of t_0 such that for each $t \in I_0$ and $x \in W_t$, we have $H(t, x), y(t) \in V \cap J_t^{-1}(c(V))$ with

$$J_t(x) := c(x) + c(y(t_0)) - c(y(t)).$$

Note that the diffeomorphic invariance implies

$$d(t) = \deg_{(X,E_Y)}(J_t(H(t, \cdot)), W_t, J_t(y(t))),$$

and that $J_t(y(t))$ is actually independent of $t \in I_0$. Hence, replacing H by $(t, x) \mapsto J_t(H(t, x))$ and y by $J_t(y(t))$ if necessary, we can assume in the following without loss of generality that $y(t) = y_0$ for all $t \in I_0$, and that $Y = E_Y$ (because we will only consider $t \in I_0$ in the following). We can also assume that I_0 is a compact interval.

By Proposition 8.70 and Remark 8.71, there is a finite-dimensional subspace $Y_0 \subseteq Y$ with $y_0 \in Y_0$, $\dim Y_0 \geq 1$, and an open neighborhood $U \subseteq W$ of $K_0 := K \cap (I_0 \times X)$ such that Y_0 is transversal to H on U. Corollary 8.63 implies that $Z := (I_0 \times X) \cap U \cap H^{-1}(Y_0)$ is a finite-dimensional manifold (with a boundary) of class C^0. By Proposition 9.1, the space Z contains an open neighborhood $U_0 \subseteq Z$ of K_0 whose closure C in Z is compact. By definition of the subspace topology, there is an open set $W_0 \subseteq W$ with $U_0 = Z \cap W_0$. Note that $\partial_Z U_0 = C \setminus U_0$ is closed in C and thus compact. Hence, $H(\partial_Z U_0)$ is compact and thus closed. In particular, there is an open ball $V_0 \subseteq Y_0$ around y_0 which is disjoint from $H(\partial_Z U_0)$. Note that $\Omega_t := \{x : (t, x) \in U_0\}$ has a compact closure in $X_t := \{x : (t, x) \in Z\}$ with $\{t\} \times \partial_{X_t}\Omega_t \subseteq \partial_Z U_0$. Putting $F_t := H(t, \cdot)|_{X_t} \in \mathcal{F}_0(X_t, Y_0)$, we have by definition of the Benevieri–Furi degree and the excision property of the Brouwer degree that

$$d(t) = \deg_{(X_t,Y_0)}(F_t, \Omega_t, y_0) \quad \text{for all } t \in I_0.$$

Since $(\Omega_t, Y_0, \mathrm{id}_{Y_0})$ is a Rouché triple for (F_t, Ω_t, y_0) and $\partial_{X_t}\Omega_t \cap H^{-1}(V_0) = \emptyset$, we obtain by the local constantness in y (Proposition 9.54) that

$$d(t) = \deg_{(X_t,Y_0)}(F_t, \Omega_t, z) \quad \text{for all } z \in V_0, t \in I_0. \tag{10.10}$$

By Theorem 9.14, we may assume that z is a regular value of $F_{t_0}|_{\Omega_{t_0}}$. Put $H_0 := H|_{W_0}$. Proposition 8.68 and the normalization property of the Benevieri–Furi degree imply that z is a regular value for $H_0(t,\cdot)$, and so

$$d(t_0) = \sum_{x \in H_0(t_0,\cdot)^{-1}(z)} \operatorname{sgn} d_X H_0(t_0, z).$$

Note that $H_0(t,\cdot)^{-1}(z)$ consists for $t = t_0$ of at most finitely many points $x_1(t_0), \dots, x_n(t_0)$. By the partial implicit function theorem (Theorem 8.40), there exist disjoint open neighborhoods $U_1, \dots, U_n \subseteq X$ of these points such that also for all t in a neighborhood of t_0, the set U_k contains exactly one point $x_k(t)$ of $H_0(t,\cdot)^{-1}(z)$, and that this is a regular point of $H_0(t,\cdot)|_{U_k}$ and depends continuously on t.

The remaining set $C_0 := \{(t,x) \in C : x \notin U_1 \cup \cdots \cup U_n\}$ is compact. Hence, the closed subset $M := C_0 \cap H_0^{-1}(z)$ is compact by Proposition 2.29. Putting $p(t,x) := t$, we obtain that $p(M) \subseteq I_0$ is compact and thus closed in I_0. Since $t_0 \notin p(M)$ we can assume, shrinking I_0 if necessary, that $I_0 \cap p(M) = \varnothing$, that is, $H_0^{-1}(z)$ consists precisely of the at most finitely many points $x_1(t), \dots, x_n(t)$. It follows that z is a regular value of $H_0(t,\cdot)$. The regular normalization property of the Benevieri–Furi degree implies together with (10.10)

$$d(t) = \sum_{k=1}^{n} \operatorname{sgn} d_X H_0(t, x_k(t)).$$

Since I_0 is an interval, the set $\{(t, x_k(t)) : t \in I_0\}$ is path-connected and thus connected (Proposition 2.14). Theorem 8.27(b) thus implies that d is constant on I_0.

We thus have proved the generalized homotopy invariance. This property in turn implies the elimination of y by considering $y(t) := ty_0$ and $H(t,y) := F(x) - y(t)$. □

Our above proof of the homotopy invariance, Theorem 10.6($\mathrm{F}_{\mathscr{F}_{\mathrm{BF}}}$), was completely based on the regular normalization property of the Benevieri–Furi C^1 degree. We will discuss two alternative strategies for a proof in Remark 10.11.

We point out that the uniqueness of the Benevieri–Furi C^1 degree can also be obtained without any reduction property by using only the natural properties ($\mathrm{F}_{\mathscr{F}_{\mathrm{BF}}}$) (with $W = [0,1] \times \Omega$), ($\mathrm{G}_{\mathscr{F}_{\mathrm{BF}}}$), and ($\mathrm{H}_{\mathscr{F}_{\mathrm{BF}}}$) of Theorem 10.6, see [16].

10.3 The Benevieri–Furi Coincidence Degree

Finding solutions of equations $F(x) = \varphi(x)$ when φ is compact and F is a Fredholm map is rather important in the theory of PDEs. For the case that F is a linear Fredholm map, such a so-called coincidence degree was first developed by Mawhin [66], [104] in case of Fredholm index 0. The case of positive index was first studied by Nirenberg [113], [114]. The Benevieri–Furi degree which we will discuss now treats the case of nonlinear Fredholm maps with index 0.

So it is a natural question whether it might also be possible to treat the case of nonlinear Fredholm maps of positive index. This seems indeed possible, see [149] (see also [131]), but the corresponding definitions are much more involved than the relatively simple notion of orientability which one can use for the Benevieri–Furi degree. Moreover, the degree for the case of positive index is not a number but only an element of a homotopy group of spheres (or even just a cobordism) and thus much more complicated to deal with. In fact, the homotopy groups of spheres are still not completely understood, although meanwhile they can be tackled computationally.

For *fixed linear* Fredholm operators of positive index there is no need to define some higher-dimensional analogue of an orientation, and so the problem is somewhat easier. In fact, even a corresponding degree theory for function triples has been developed by Kryszewski [93] (see also e.g. [63]).

In this monograph, we only consider the case of index 0 in order to have a conveniently accessible approach which is general enough to treat almost all related applications: For nonlinear Fredholm operators, this is the Benevieri–Furi coincidence degree. The generalization to function triples will be developed later in this monograph.

We confine ourselves in this section to the case that the range of φ is contained in a finite-dimensional subspace instead of considering locally compact maps φ. Although the latter was done for function pairs in [19], it was observed in [142] that if one wants to obtain similar results for multivalued maps (i.e. for function triples), one should *first* extend the degree to function triples and only *afterwards* to (locally) compact maps. We will do this in the later sections, and therefore we do not need to repeat essentially the same procedure already in this section: The degree developed later will cover of course the single-valued case (of function pairs) as well as we will discuss in Theorems 13.14 and 13.19. For this reason, we will study in this section only maps φ with values in finite-dimensional subspaces.

A more severe restriction might lie in the nature of things: We already mentioned in the beginning of Chapter 10 that we will be forced to work with the case that the image space Y is a Banach space, and we cannot treat the case of a Banach manifold. In fact, it appears that the linear structure of Y is crucial for the whole approach. For instance, we will not be able to prove a diffeomorphic invariance of the corresponding degree but only an "isomorphic" invariance (with respect to Y). The reason is that we cannot work with submanifolds Y_0 of Y to define the degree by means of the reduction property but that we must use a very restricted class of submanifolds (we use here subspaces): The underlying problem is that the union of two finite-dimensional submanifolds is in general (for "badly" situated submanifolds) not contained in a finite-dimensional submanifold, and so the reduction property of Theorem 10.1 cannot be used in some larger submanifold to prove that the reduction property will give the same degree for both manifolds. By using linear subspaces instead of submanifolds, this problem vanishes, since the union of two finite-dimensional subspaces is always contained in a finite-dimensional subspace, namely in the sum of the two subspaces. It might be possible to prove an extension of the Benevieri–Furi degree for Banach manifolds Y and certain subsets of admissible function pairs, if one considers another subclass of finite-dimensional submanifolds with the property that each union of two submanifolds of that class is contained in a submanifold of that class. Such an extension might lead to a corresponding form of a "diffeomorphic invariance". However, since currently there seems to be no particular use for such an extended theory and since it seems not possible to cover general diffeomorphisms this way, we confine ourselves for the rest of this monograph to the simple situation that Y is a Banach space.

Throughout this section, we assume that X is a Banach manifold over a real Banach space $E = E_X$ and that $Y = E_Y$ is a real Banach space. Moreover, we make throughout the non-degeneracy hypotheses (10.8) which in the presence of AC just means $E \neq \{0\}$ and $Y \neq \{0\}$.

We write $(F, \varphi, \Omega) \in \mathcal{P}_{\mathrm{BF}}(X, Y)$ if $\Omega \subseteq X$ is open and $F, \varphi \colon \Omega \to Y$ are such that

$$\mathrm{coin}_{\Omega}(F, \varphi) = \{x \in \Omega : F(x) = \varphi(x)\}$$

is compact and there is an open neighborhood $\Omega_0 \subseteq \Omega$ of $\mathrm{coin}_{\Omega}(F, \varphi)$ with $F|_{\Omega_0} \in \mathcal{F}_0(\Omega_0, Y)$ and $\varphi|_{\Omega_0} \in C(\Omega_0, Y_0)$ for some finite-dimensional subspace $Y_0 \subseteq Y$. For the oriented version of the degree, we assume that F is oriented on Ω_0.

Definition 10.7. The *Benevieri-Furi coincidence degree* is the map $\deg = \deg_{(X,Y)}$ which associates to each $(F, \varphi, \Omega) \in \mathcal{P}_{\mathrm{BF}}(X, Y)$ a number from $\mathbb{Z}_2$ (or $\mathbb{Z}$ in the oriented case) such that the following property holds:

($\mathrm{A}_{\mathcal{P}_{\mathrm{BF}}}$) **(Reduction).** Let $(F, \varphi, \Omega) \in \mathcal{P}_{\mathrm{BF}}(X, Y)$, $\Omega_0 \subseteq \Omega$ be an open neighborhood of $\mathrm{coin}_{\Omega}(F, \varphi)$ with $F|_{\Omega_0} \in \mathcal{F}_0(\Omega_0, Y)$, and $Y_0 \neq \{0\}$ a finite-dimensional subspace of Y with $\varphi|_{\Omega_0} \in C(\Omega_0, Y_0)$ and such that Y_0 is transversal to F on Ω_0. Then $X_0 := \Omega_0 \cap F^{-1}(Y_0)$ is empty or a submanifold of X of the same dimension as Y_0, and the map $G := (F - \varphi)|_{X_0} \in C(X_0, Y_0)$ satisfies $(G, X_0, 0) \in \mathcal{B}^0(X_0, Y_0)$ and

$$\deg_{(X,Y)}(F, \varphi, \Omega) = \deg_{(X_0,Y_0)}(G, X_0, 0), \qquad (10.11)$$

where the right-hand side denotes the C^0 Brouwer degree. In case $X_0 = \emptyset$, we define the right-hand side of (10.11) as 0. The orientation of G for the oriented case is described below.

In the oriented case, we equip $F_0 := F|_{X_0} \in \mathcal{F}_0(X_0, Y_0)$ with the inherited orientation. We use (9.2) (and Proposition 9.18) to understand this as an orientation of a continuous map in the sense of Definition 9.16. Fixing some orientation of the subspace Y_0, we obtain an induced orientation on X_0. These two orientations in turn induce an orientation on G. This orientation is independent of the choice of the orientation of Y_0, since for the opposite orientation of Y_0, we obtain the opposite orientation of X_0 and thus the same orientation of G.

Theorem 10.8. *For every fixed Banach manifold X and fixed Banach space Y there is exactly one Benevieri–Furi coincidence degree.*

Proof. Shrinking Ω_0 in the reduction property if necessary, we find by Proposition 8.70 in view of Remark 8.71 that for each $(F, \varphi, \Omega) \in \mathcal{P}_{\mathrm{BF}}(X, Y)$ there actually is a finite-dimensional subspace $Y_0 \subseteq Y$ which is transversal to F on Ω_0. Hence, the reduction property implies that (10.11) holds, and so $\deg_{(X,Y)}(F, \varphi, \Omega)$ is uniquely determined. Thus, the uniqueness is established. Concerning the existence, we use (10.11) to define the degree, and we have to show that this is well-defined, that is, independent of the particular choice of Ω_0 and Y_0.

Thus, let Ω_1 and Y_1 be possibly different choices. We find by Proposition 8.70 and Remark 8.71 a finite-dimensional subspace $Y_2 \subseteq Y$ containing $Y_0 + Y_1$ which is transversal to F on some open neighborhood $\Omega_2 \subseteq \Omega$ of $\mathrm{coin}_\Omega(F, \varphi)$. Then in particular $Y_i \subseteq Y_2$ for $i = 0, 1$. Replacing Ω_2 by $\Omega_0 \cap \Omega_1 \cap \Omega_2$ if necessary, we can assume that $\Omega_2 \subseteq \Omega_i$ $(i = 0, 1)$. For $i = 0, 1, 2$, we put $X_i := \Omega_i \cap F^{-1}(Y_i)$ and $G_i := (F - \varphi)|_{X_i} \in C(X_i, Y_i)$. In the oriented case, we orient G_i and $\hat{G}_i$ according to the orientation of F as described after Definition 10.7. For $i = 0, 1$, the C^0 reduction property of the degree (Theorem 10.1) implies with $\hat{X}_i := \Omega_2 \cap X_i$ that we have for the Brouwer degree

$$\deg_{(X_2,Y_2)}(G_2, X_2, 0) = \deg_{(\hat{X}_i,Y_i)}(G_i, \hat{X}_i, 0).$$

By the restriction and excision property of the Brouwer degree, we find

$$\deg_{(\hat{X}_i,Y_i)}(G_i, \hat{X}_i, 0) = \deg_{(X_i,Y_i)}(G_i, \Omega_2 \cap X_i, 0) = \deg_{(X_i,Y_i)}(G_i, X_i, 0)$$

for $i = 0, 1$. Summarizing, we have shown that

$$\deg_{(X_0,Y_0)}(G_0, X_0, 0) = \deg_{(X_2,Y_2)}(G_2, X_2, 0) = \deg_{(X_1,Y_1)}(G_1, X_1, 0).$$

This shows that the right-hand side of (10.11) is independent of the particular choice of Y_0 and Ω_0, and so the Benevieri–Furi coincidence degree exists. □

We point out once more the reason why we had to consider subspaces instead of submanifolds in the above proof, and why we require that Y is a Banach space: For any two finite-dimensional subspaces $Y_0, Y_1 \subseteq Y$ there is a finite-dimensional subspace $Y_2 \subseteq Y$ containing Y_0 and Y_1 (and transversal to F on some Ω_2). A corresponding statement does not hold if one replaces "subspace" by "submanifold". In addition, for the application of Theorem 10.1 in the above proof for the case that Y is only a Banach manifold, we would also have to require that $F(\Omega_0) \cup \varphi(\Omega_0)$ is contained in a set which is diffeomorphic to a Banach manifold. Since we can consider only subspaces and not submanifolds, it is not clear whether a corresponding definition would be independent of the choice of the diffeomorphism (which in general only sends subspaces to submanifolds).

Theorem 10.9. *The Benevieri–Furi coincidence degree satisfies the following properties for every* $(F, \varphi, \Omega) \in \mathcal{P}_{\mathrm{BF}}(X, Y)$.

($\mathrm{B}_{\mathcal{P}_{\mathrm{BF}}}$) **(Generalized Homotopy Invariance).** *Let* $W \subseteq [0,1] \times X$ *be open,* $H: W \to Y$, *and* $h: W \to Y$ *be such that* $\mathrm{coin}_W(H, h)$ *is compact and that there is an open neighborhood* $U \subseteq W$ *of* $\mathrm{coin}_W(H, h)$ *such that* $H|_U$ *is a generalized (oriented) Fredholm homotopy of index* 0 *and that* $h|_U \in C(U, Y_0)$ *for some finite-dimensional subspace* $Y_0 \subseteq Y$. *Then* $W_t := \{x : (t, x) \in W\}$ *satisfies* $(H(t, \cdot), h(t, \cdot), W_t) \in \mathcal{P}_{\mathrm{BF}}(X, Y)$ $(t \in [0,1])$ *and*

$$\deg(H(t, \cdot), h(t, \cdot), W_t) \text{ is independent of } t \in [0,1].$$

($\mathrm{C}_{\mathcal{P}_{\mathrm{BF}}}$) **(Excision).** *If* $\Omega_0 \subseteq \Omega$ *is an open neighborhood of* $\mathrm{coin}_\Omega(F, \varphi)$ *then*

$$\deg(F, \varphi, \Omega) = \deg(F, \varphi, \Omega_0).$$

($\mathrm{D}_{\mathcal{P}_{\mathrm{BF}}}$) **(Additivity).** *If* $\Omega = \Omega_1 \cup \Omega_2$ *with disjoint open subsets* $\Omega_1, \Omega_2 \subseteq \Omega$ *then*

$$\deg(F, \varphi, \Omega) = \deg(F, \varphi, \Omega_1) + \deg(F, \varphi, \Omega_2).$$

($\mathrm{E}_{\mathcal{P}_{\mathrm{BF}}}$) **(Compatibility with the Brouwer Degree).** *If* $0 < \dim E_X = \dim E_Y < \infty$ *then*

$$\deg(F, \varphi, \Omega) = \deg(F - \varphi, \Omega, 0),$$

where the right-hand side denotes the Brouwer degree. The orientation of $F - \varphi$ *in the oriented case is described below.*

($F_{\mathscr{P}_{\mathrm{BF}}}$) **(Compatibility with the C^1 Degree I)**. *If $\varphi(x) = y$ for all $x \in \Omega$ then*

$$\deg(F, \varphi, \Omega) = \deg(F, \Omega, y),$$

where the right-hand side denotes the Benevieri–Furi C^1 degree.

($G_{\mathscr{P}_{\mathrm{BF}}}$) **(Compatibility with the C^1 Degree II)**. *If $\varphi \in C^1(\Omega, y)$ then*

$$\deg(F, \varphi, \Omega) = \deg(F - \varphi, \Omega, 0),$$

where the right-hand side denotes the Benevieri–Furi C^1 degree. The orientation of $F - \varphi$ is described below.

($H_{\mathscr{P}_{\mathrm{BF}}}$) **(Compatibility with the Non-Oriented Case)**. *The degrees for the oriented and non-oriented case are the same modulo 2 (if the oriented case applies).*

($I_{\mathscr{P}_{\mathrm{BF}}}$) **(Existence)**. *If* $\deg(F, \varphi, \Omega) \neq 0$ *then* $\mathrm{coin}_\Omega(F, \varphi) \neq \varnothing$.

($J_{\mathscr{P}_{\mathrm{BF}}}$) **(Excision-Additivity)**. *If $\Omega_i \subseteq \Omega$ $(i \in I)$ is a family of pairwise disjoint open sets with* $\mathrm{coin}_\Omega(F, \varphi) \subseteq \bigcup_{i \in I} \Omega_i$ *such that* $\mathrm{coin}_{\Omega_i}(F, \varphi)$ *is compact for all $i \in I$, then*

$$\deg(F, \varphi, \Omega) = \sum_{i \in I} \deg(F, \varphi, \Omega_i),$$

where in the sum at most a finite number of summands is nonzero.

($K_{\mathscr{P}_{\mathrm{BF}}}$) **(Diffeomorphic-Isomorphic Invariance)**. *Let J_1 be a diffeomorphism of an open subset of a Banach manifold X_0 onto Ω and J_2 an isomorphism of a subspace containing $F(\Omega) \cup \varphi(\Omega)$ onto a Banach space Y_0. Then*

$$\deg_{(X,Y)}(F, \varphi, \Omega) = \deg_{(X_0,Y_0)}(J_2 \circ F \circ J_1, J_2 \circ \varphi \circ J_1, J_1^{-1}(\Omega)).$$

In the oriented case, the orientation of $J_2 \circ F \circ J_1$ is understood in the sense of Proposition 8.38.

($L_{\mathscr{P}_{\mathrm{BF}}}$) **(Restriction)**. *Let $X_0 \subseteq X$ be open, and $Y_0 \subseteq Y$ a subspace. Then*

$$\deg_{(X_0,Y_0)} = \deg_{(X,Y)}|_{\mathscr{P}_{\mathrm{BF}}(X_0,Y_0)}.$$

($M_{\mathscr{P}_{\mathrm{BF}}}$) **(Cartesian Product)**. *For $i = 1, 2$, let X_i be a Banach manifold without boundary of class C^1 over the Banach space E_{X_i}, and Y_i be a Banach space. Let $\Omega_i \subseteq X_i$ be open, $F_i \in \mathscr{F}_0(\Omega_i, Y_i)$, and $\varphi_i \in C(\Omega_i, Y_i)$. Put*

$X := X_1 \times X_2$, $Y := Y_1 \times Y_2$, $\Omega := \Omega_1 \times \Omega_2$, $F := F_1 \otimes F_2$, and $\varphi := \varphi_1 \otimes \varphi_2$. Then $(F, \varphi, \Omega) \in \mathcal{P}_{\mathrm{BF}}(X, Y)$ if and only if $(F_i, \varphi_i, \Omega_i) \in \mathcal{P}_{\mathrm{BF}}(X_i, Y_i)$ for $i = 1, 2$ or if $\operatorname{coin}(F_i, \varphi_i) = \varnothing$ for $i = 1$ or $i = 2$. In the former case

$$\deg_{(X,Y)}(F, \varphi, \Omega) = \deg_{(X_1,Y_1)}(F_1, \varphi_1, \Omega_1) \deg_{(X_2,Y_2)}(F_2, \varphi_2, \Omega_2).$$

In the oriented case F is equipped with the product orientation.

For the compatibility with the Brouwer degree, we define the orientation as follows: We understand the orientation of F as an orientation of F as a continuous map by means of (9.2) (recall Proposition 9.18). We fix an orientation for Y, and equip Ω with the orientation induced by the orientations of F in the sense of Proposition 9.34, and $F - \varphi$ with the orientations induced by the orientations of Ω and Y. This definition is independent of the choice of the orientation of Y, since for the opposite orientation on Y, we obtain the opposite orientation on Ω and the same orientation on $F - \varphi$.

For the compatibility with the C^1 degree, note that there is some open neighborhood $\Omega_0 \subseteq \Omega$ of $\operatorname{coin}_{\Omega}(F, \varphi)$ such that $F \in \mathcal{F}_0(\Omega_0, Y)$ and $\varphi \in C(\Omega_0, Y_0)$ for some finite-dimensional subspace $Y_0 \subseteq Y$. Theorem 6.40 implies that $(F - \varphi)|_{\Omega_0} \in \mathcal{F}_0(\Omega_0, Y)$, and so $(F - \varphi, \Omega, 0) \in \mathcal{F}_{\mathrm{BF}}(X, Y)$. Moreover, by Theorem 6.40 the map $H(t, x) := F(x) - t\varphi(x)$ is a Fredholm homotopy of index 0. By Theorem 8.27(e), it has a unique orientation such that $H(0, \cdot) = F$ is oriented as F. The orientation required in ($\mathrm{G}_{\mathcal{P}_{\mathrm{BF}}}$) is that of $H(1, \cdot) = F - \varphi$.

Proof. With the exception of the homotopy invariance and the compatibility with the C^1 degree, the properties follow straightforwardly from the corresponding properties of the Brouwer degree. For example, for the Cartesian product property, we can argue as in the proof of Theorem 10.6, only replacing "subspace" by "submanifold". For the diffeomorphic-isomorphic invariance, we note that if $Y_1 \subseteq Y_0$ is a finite-dimensional subspace which is transversal to $J_2 \circ F \circ J_1$ on Ω_0 then $J_2^{-1}(Y_1) \subseteq Y$ is a finite-dimensional subspace which is transversal to F on $J_1(\Omega_0)$ and thus can serve as the space which is needed in Definition 10.7. The existence and excision-additivity properties follow from the excision and additivity properties by the same arguments as for the Brouwer degree.

To prove the generalized homotopy invariance, we note first that Proposition 2.62 implies by the compactness of K that $K_t := \{x : H(t, x) = h(t, x)\}$ is compact for every $t \in [0, 1]$, and so $(H(t, \cdot), h(t, \cdot, W_t) \in \mathcal{P}_{\mathrm{BF}}(X, Y)$. By Proposition 8.70 and Remark 8.71, we can assume that Y_0 is transversal to H on U. In view of the excision property, we can replace W by U if necessary and thus assume without loss of generality that H is a generalized (oriented) Fredholm homotopy. It suffices to show by Proposition 2.19, that

$$d(t) := \deg(H(t,\,\cdot), h(t,\,\cdot), W_t)$$

is locally constant on $[0,1]$. Thus, let $t_0 \in [0,1]$, and we have to show that d is constant in some neighborhood of t_0.

Putting $X_t := H(t,\,\cdot)^{-1}(Y_0)$ and $Z_t := \{t\} \times X_t$, we find by Corollary 8.63 that $Z = \bigcup_{t\in[0,1]} Z_t$ is a partial C^1 manifold. In particular Z is a finite-dimensional C^0 manifold (with boundary), and so we find by Proposition 9.1 that there is some open neighborhood $V \subseteq Z$ of $\operatorname{coin}_W(H,h)$ whose closure C in Z is compact. The function $H-h$ is nonzero on the compact relative boundary $\partial_Z V = C \setminus V$. Hence, Corollary 3.14 implies $\varepsilon = \operatorname{dist}(0,(H-h)(\partial_Z V)) > 0$. By Remark 9.10, there is a partial C^1 map $h_0\colon Z \to Y_0$ with $\|h(t,x) - h_0(t,x)\| < \varepsilon/2$ for all $(t,x) \in \partial_Z V$. For $(t,x) \in Z$, we put $H_0(t,x) := H(t,x) - h_0(t,x)$. By Theorem 9.14 there is a regular value of $H_0(t_0,\,\cdot) \in C^1(Z_{t_0}, Y_0)$ in $B_{\varepsilon/2}(0) \subseteq Y_0$. Adding such a regular value if necessary, we have $\|H_0(t,x)\| < \varepsilon$ for all $(t,x) \in \partial_Z V$, and 0 is a regular value of $H_0(t_0,\,\cdot)$. We put $V_t := \{x : (t,x) \in V\}$. Then (V_t, Y, id_Y) is a Rouché triple for $(H(t,\,\cdot) - h(t,\,\cdot), X_t, 0)$, and so the Rouché property of the Brouwer degree (Proposition 9.54) implies

$$d(t) = \deg_{(X_t,Y_0)}(H(t,\,\cdot) - h(t,\,\cdot), X_t, 0) = \deg_{(X_t,Y_0)}(H_0(t,\,\cdot), V_t, 0). \tag{10.12}$$

In the oriented case, the orientation $H(t,\,\cdot) - h(t,\,\cdot)$ is defined as described after Definition 10.7 and the orientation of $H_0(t,\,\cdot)$ is defined correspondingly as described after Proposition 9.54.

Since 0 is a regular value of $H_0(t_0,\,\cdot)$, the set $H_0(t,\,\cdot)^{-1}(0)$ consists for $t = t_0$ of at most finitely many points $x_1(t_0), \ldots, x_n(t_0) \in V_{t_0}$. By the partial implicit function theorem (Theorem 8.40), there exist open neighborhoods $U_1, \ldots, U_n \subseteq X$ of these points such that also for all t in a neighborhood of t_0, the set U_k contains exactly one point $x_k(t)$ of $H_0(t,\,\cdot)^{-1}(0)$, and that this is a regular point of $H_0(t,\,\cdot)|_{U_k}$ and depends continuously on t.

The remaining set $C_0 := V \setminus (U_1 \cup \cdots \cup U_n)$ is compact. Hence, the closed subset $M := C_0 \cap H_0^{-1}(0)$ is compact by Proposition 2.29. Putting $p(t,x) := t$, we obtain that $p(M) \subseteq [0,1]$ is compact and thus closed. Since $t_0 \notin p(M)$ we can assume there is a neighborhood $I_0 \subseteq [0,1]$ of t_0 with $I_0 \cap p(M) = \varnothing$, that is, $H_0^{-1}(z)$ consists for $t \in I_0$ precisely of the at most finitely many points $x_1(t), \ldots, x_n(t)$. It follows that z is a regular value of $H_0(t,\,\cdot)$. The regular normalization property of the Brouwer degree implies together with (10.12)

$$d(t) = \sum_{k=1}^{n} \operatorname{sgn} d_X H_0(t, x_k(t)).$$

Note that the orientation of $d_X H_0$ is actually an orientation when we consider it as a map of the partial tangent bundle $T_X Z$ into Y_0, according to Proposition 8.67. We can assume that I_0 is an interval. Then $\{(t, x_k(t)) : t \in I_0\}$ is path-connected and thus connected (Proposition 2.14). Proposition 7.36 thus implies that d is constant on I_0. Hence, (B$_{\mathcal{P}_{\mathrm{BF}}}$) is established.

Next, we prove the special case $\varphi = 0$ of property (F$_{\mathcal{P}_{\mathrm{BF}}}$). We note that Proposition 8.70 and Remark 8.71 implies that there is an open neighborhood $\Omega_0 \subseteq \Omega$ of $\mathrm{coin}_\Omega(F, \varphi) = F^{-1}(0)$ and a subspace $Y_0 \subseteq Y$ which is transversal to $dF(x)$. It follows from the definition of the Benevieri–Furi degrees with $X_0 := F^{-1}(Y_0)$ that

$$\deg_{(X,Y)}(F, \varphi, \Omega) = \deg_{(X_0,Y_0)}(F, \Omega, 0) = \deg_{(X,Y)}(F, \Omega, 0).$$

To prove (G$_{\mathcal{P}_{\mathrm{BF}}}$), we apply the homotopy invariance with $H(t, x) := F(x) - t\varphi(x)$ and $h(t, x) := t\varphi(x)$. We obtain

$$\deg_{(X,Y)}(F, \varphi, \Omega) = \deg_{(X,Y)}(F - \varphi, 0, \Omega).$$

By the special case $y = 0$ of (F$_{\mathcal{P}_{\mathrm{BF}}}$) which we had proved before, we obtain (G$_{\mathcal{P}_{\mathrm{BF}}}$). For $\varphi(x) = y$, we obtain from (G$_{\mathcal{P}_{\mathrm{BF}}}$) that

$$\deg_{(X,Y)}(F, \varphi, \Omega) = \deg_{(X,Y)}(F - y, \Omega, 0),$$

which implies (F$_{\mathcal{P}_{\mathrm{BF}}}$) by the "elimination of y" property of the C^1 degree. □

Remark 10.10. It is unknown to the author whether the Benevieri–Furi coincidence degree satisfies a diffeomorphic invariance property: The argument used to prove the diffeomorphic-isomorphic invariance in Theorem 10.9 fails if J_2 is only a diffeomorphism, because then $J_2^{-1}(Y_1)$ is only a submanifold and thus cannot be used directly in the definition of the degree.

If the diffeomorphic invariance property can be proved, it is possible to extend the definition of the Benevieri–Furi degree also to the case that Y is a Banach manifold over a Banach space E_Y, at least with the additional restriction that only those $(F, \varphi, \Omega) \in \mathcal{P}_{\mathrm{BF}}(X, Y)$ are considered which have the additional property that there is an open neighborhood $\Omega_0 \subseteq \Omega$ of $\mathrm{coin}_\Omega(F, \varphi)$ such that $F(\Omega_0) \cup \varphi(\Omega_0)$ has a neighborhood which is diffeomorphic to an open subset of E_Y: By the diffeomorphic invariance, the corresponding definition of a degree would be independent of the choice of the diffeomorphism. However, we emphasize once more that the diffeomorphic invariance is currently unknown.

We point out that the homotopy invariance in Theorem 10.9(B$_{\mathcal{P}_{\mathrm{BF}}}$) in this strong (and natural) form is a new result, to the author's knowledge:

In [19], the homotopy invariance was established only under the additional hypothesis $H \in C^1(U, Y)$ in which case things are much easier, as we will see in moment. Also in [126] it is required that $H \in C^1(U, Y)$.

According to personal communication, this stronger result was known to the authors of [19], but in this case it is not clear whether one can define the degree in terms of the map $F - \varphi$ as in [19] (instead of in terms of the couple (F, φ) which we consider, so that this difficulty does not arise in our context).

Although our above proof might appear to be a rather direct transfer of the proof of Theorem 10.6($\mathrm{F}_{\mathscr{F}_{\mathrm{BF}}}$), this is actually not the case. In fact, we needed to introduce the involved machinery of partial tangent bundles and their orientation (like e.g. Proposition 8.67) to deal with this more general setting. (Our arguments in the proof of Theorem 10.6($\mathrm{F}_{\mathscr{F}_{\mathrm{BF}}}$) avoided this machinery.) The underlying difficulty in the proof is that one cannot approximate continuous functions by C^1 functions in Banach spaces, in general: Hence, we had to approximate only the restriction of h to Z, and for this, we had to introduce an appropriate notion of "continuity" for the orientation of that restriction which is not that simple, since Z need not be a C^1 manifold. If we would have assumed that $H \in C^1(U, Y)$, we could have avoided this difficulty.

The argument given in [19] under the assumption $H \in C^1(U, Y)$ is actually completely different: Since this argument is in a sense the most natural way to prove the homotopy invariance, we do not hesitate to give a second proof of Theorem 10.9($\mathrm{B}_{\mathscr{P}_{\mathrm{BF}}}$) by using this argument.

Alternative proof of Theorem 10.9($\mathrm{B}_{\mathscr{P}_{\mathrm{BF}}}$). Let $I := [s_0, s_1] \subseteq [0, 1]$ and $n = \dim Y_0$. Replacing W by U if necessary (using the excision property), and using Proposition 8.70 and Remark 8.71, we can assume that $n \geq 1$ and that Y_0 is transversal to H on $W = U$. By the additional assumption $H \in C^1(W, Y)$, we can apply Corollary 8.63 to find that the space $Z := H^{-1}(Y_0)$ is empty or a submanifold (with boundary) of $I \times X$ of dimension $n+1$ of class C^1. More precisely, putting $X_t := H(t, \cdot)^{-1}(Y_0)$ and $Z_t := \{t\} \times X_t$, we have $Z = \bigcup_{t \in I} Z_t$, $\partial Z = Z_0 \cup Z_1$, and by Theorem 8.60 the charts $c \in C^1(U, \mathbb{R} \times \mathbb{R}^n)$ of Z can be chosen such that $Z_t = U \cap c^{-1}(\{t\} \times \mathbb{R}^n)$. We consider first the oriented case. By Proposition 8.67, the inherited orientation on $H_0 := H|_Z \colon Z \to Y_0$ is an orientation of the partial tangent bundle $T_X Z$. We aim now to apply the bordism invariance of the Brouwer degree. To this end, we fix an orientation of Y_0, and we define a corresponding orientation on Z induced by the inherited orientation on H_0 similarly as in the proof of Lemma 9.65: By Proposition 7.48, the inherited orientation of H_0 and the fixed orientation of Y_0 together induce an orientation of $T_X Z$ as a vector bundle by Proposition 7.13. In particular, for each fixed $(t_0, x_0) \in Z$, that is $x_0 \in X_{t_0}$, we have an orientation of $T_{x_0} X_{t_0}$. Let $(e_1, \dots, e_n) \in T_{x_0} X_{t_0}$

be a basis representing the corresponding orientation. Since $Z_{t_0} = \{t_0\} \times X_{t_0}$ is the same as X_{t_0} under the canonical identification of x with (t_0, x), we also have $T_{x_0} X_{t_0} = T_{(t_0,x_0)} Z_{t_0}$ in a canonical sense. In this canonical sense, we can understand $(e_1, \dots, e_n)$ as a basis of $T_{(t_0,x_0)} Z_{t_0} \subseteq T_{(t_0,x_0)} Z$. Roughly speaking, we will extend this to a basis $(e_0, e_1, \dots, e_n)$ of $T_{(t_0,x_0)} Z$ where the remaining basis vector e_0 is chosen "in direction of t".

More precisely, we choose some chart $c \in C^1(U, \mathbb{R} \times \mathbb{R}^n)$ of Z with $(t_0, x_0) \in U$ such that $Z_t = U \cap c^{-1}(\{t\} \times \mathbb{R}^n)$. Note that $c_0 := c|_{Z_{t_0}} \in C^1(Z_{t_0}, \{t_0\} \times \mathbb{R}^n)$ satisfies $dc_0(t_0, x_0) \in \mathcal{L}(T_{(t_0,x_0)} Z_{t_0}, \{0\} \times \mathbb{R}^n)$. We put $(0, \hat{e}_k) := dc_0(t_0, x_0) e_k$ $(k = 1, \dots, n)$ where e_k is understood as an element of $T_{(t_0,x_0)} Z_{t_0}$ under the canonical identification. Then $((1, 0), (0, \hat{e}_1), \dots, (0, \hat{e}_n))$ represents a basis on $\mathbb{R}^{n+1}$, and $dc(t_0, x_0)^{-1}$ transforms that basis into a basis of $T_{(t_0,x_0)} Z$ of the form $(e_0, e_1, \dots, e_n)$. Since $(1, 0) \in T_{(t_0,x_0)} Z_{t_0}$ points "in direction of t", we can interpret also e_0 as pointing "in direction of t".

Now this extended basis $(e_0, e_1, \dots, e_n)$ represents an orientation of $T_{(t_0,x_0)} Z$, and doing this for every $(t_0, x_0) \in Z$, we obtain an orientation of Z. By construction, the orientation of Z_t for $t \in \{s_0, s_1\}$ is the orientation as the boundary of the manifold Z. We put $F_i := (H - h)|_{Z_{s_i}} \in C(Z_{s_i}, Y_0)$ $(i = 0, 1)$. Equipping F_0 with the orientations induced by Z_{s_0} and Y, and F_1 with the opposite of the orientation induced by Z_{s_1} and Y, we obtain by the bordism invariance of the Brouwer degree that

$$\deg_{(Z_{s_0}, Y_0)}(F_0, Z_{s_0}, 0) = \deg_{(Z_{s_1}, Y_0)}(F_1, Z_{s_1}, 0).$$

The same formula holds of course also in the non-oriented case. Since $Z_{s_i} = X_{s_i}$ up to a canonical identification, in particular, up to diffeomorphisms, the diffeomorphic invariance of the Brouwer degree implies

$$\begin{aligned}&\deg_{(X_{s_0}, Y_0)}(H(s_0, \cdot) - h(s_0, \cdot), X_{s_0}, 0)\\ &\quad = \deg_{(X_{s_1}, Y_0)}(H(s_1, \cdot) - h(s_1, \cdot), X_{s_1}, 0).\end{aligned}$$

By definition of the Benevieri–Furi coincidence degree, this means exactly

$$\deg_{(X,Y)}(H(s_0, \cdot), h(s_0, \cdot), \Omega) = \deg_{(X,Y)}(H(s_1, \cdot), h(s_1, \cdot), \Omega).$$

Hence, the homotopy invariance is established. □

In the author's opinion, this second proof is much more natural and also shows that the bordism invariance of the Brouwer degree is a very useful tool. The disadvantage of this second proof is that it requires the additional hypothesis $H \in C^1(U, Y)$. It might be possible to use the same idea of proof also without this hypothesis, but this is rather hard work in the oriented case:

Instead of orienting the C^1 manifold $H^{-1}(Y_0)$, one would have to orient the C^0 manifold $H^{-1}(Y_0)$ (we use here that Corollary 8.63 holds also with $q = 0$) and to use a C^0 bordism invariance of the Brouwer degree: The latter is true for the Brouwer degree on C^0 manifolds by Remark 9.72. Since on the boundaries, we still have C^1 manifolds and thus uniqueness of the Brouwer degree, the degree of Remark 9.72 must be the same as our Brouwer degree, and so actually our Brouwer degree satisfies a C^0 bordism invariance. Unfortunately, the notion of orientation in C^0 manifolds (which we avoided to discuss in this monograph) is much more involved than the orientation in C^1 manifolds. In particular, it is not completely obvious how the relatively simple construction of the orientation which we used above for the C^1 case should be carried out in the C^0 case, although this can probably be done.

Remark 10.11. After Theorem 10.6, we announced two alternative strategies for the proof of the homotopy invariance. These should be clear after the above discussions. Indeed, by the argument used in the proof of Theorem 10.6, one can reduce the situation to the case that Y is a Banach space. Then the two alternative proofs of Theorem 10.9(B$_{\mathcal{P}_{\mathrm{BF}}}$) can be used with $h(t, x) = y(t)$. Note that, although our first proof of Theorem 10.9(B$_{\mathcal{P}_{\mathrm{BF}}}$) looks formally similar to our proof of Theorem 10.6, it is actually rather different, since the main argument takes place in the finite-dimensional partial manifold Z instead of the original Banach manifold X.

Part III

Degree Theory for Function Triples

Chapter 11

Function Triples

Recall that the Benevieri–Furi coincidence degree is a "count" for the number of solutions of the equation

$$F(x) = \varphi(x) \qquad (x \in \Omega),$$

where F is (oriented) Fredholm of index 0 and $\varphi(\Omega)$ is finite-dimensional. Our aim is to obtain a corresponding notion also for the case that φ is multivalued and then, later, to relax the assumption that $\varphi(\Omega)$ be finite-dimensional to some rather mild compactness hypotheses.

More precisely, we will study inclusions of the type

$$F(x) \in \varphi(\Phi(x)) \qquad (x \in \Omega), \tag{11.1}$$

where Φ is an acyclic* map. Before we describe our approach, we give a brief historical overview.

Homotopical invariants for inclusions with multivalued maps have a long history, although usually only the fixed point case (that is, $F = \mathrm{id}_\Omega$ and usually also $\varphi = \mathrm{id}_{\Phi(\Omega)}$) has been treated. It is almost impossible to give a complete bibliographical overview over the fixed point case, and therefore we just confine ourselves to a view remarks. The first fixed point theorem for multivalued acyclic maps appeared in [48], a Lefschetz number for the case of maps with UV^∞ values was developed in [25]. The general Lefschetz number theory for (11.1) with $F = \mathrm{id}_\Omega$ and acyclic Φ is due to Górniewicz, see e.g. [70]. A degree theory for maps with convex values was developed in [102] (for the noncompact case see e.g. [56]). A degree theory for acyclic maps was developed independently in various forms by many authors. We essentially refer to the surveys [21], [23] and the monographs [5], [71], [83] and their enormous number of references. The following list of particular contributions is by no means complete!

For fixed points of noncompact acyclic maps, we mention [53], [55], [80]. One of the earliest approaches for (11.1) with $F = \mathrm{id}_\Omega$ and acyclic Φ is by means of a so-called coincidence index [73], [94], [95]. Actually, this coincidence index is a count for the fixed points of $\varphi \circ \Phi$ (not of the coincidence points of φ and Φ); in this sense the notion is actually a misnomer. For the treatment of (11.1) with $F = \mathrm{id}_\Omega$ in the noncompact case, see e.g. [120], [148].

All the early approaches which can treat acyclic Φ have the disadvantage that it is unknown whether the corresponding notion is unique or at least topologically invariant. In [130] a topologically invariant degree was developed which works for acyclic Φ. This theory was extended to the noncompact case [47], [58], [59], [144]. However, that theory is extremely technical, and it is an open problem whether it can be extended for the case $F \neq \mathrm{id}_\Omega$.

Moreover, uniqueness of the degree can perhaps only be established if one requires that Φ is acyclic*. For the case that Φ has UV^∞ values, a degree theory for (11.1) with $F = \mathrm{id}_\Omega$ can be based on approximating such Φ by single-valued maps [72]; see also [11] and [3], [4]. This approximation approach can be also used for the case that F is a Fredholm map: The non-oriented case can be found in [121], and an oriented variant in [146] (but the notion of orientation is rather different than ours).

For the case of acyclic Φ with $\dim \Phi(\Omega) < \infty$ the uniqueness (and existence) of the degree for (11.1) with $F = \mathrm{id}_\Omega$ was established by Kryszewski [91] by means of the homotopic Vietoris Theorem 5.25. Kryszewski observed in [93] (see also [62], [63]) that a similar approach can be taken if F is a linear Fredholm map. Kryszewski's original approach still required deep results from infinite-dimensional homotopy theory.

We will take the route to employ the homotopy Vietoris theorem which has the advantage that it works for both, maps with UV^∞ values and acyclic maps with $\dim \Phi(\Omega) < \infty$ (that is, for acyclic* maps in our terminology): Instead of using approximations, we use the homotopic Vietoris theorem to obtain a homotopy which leads to the single-valued situation (we will call this a *simplifier* later on). Thus, from the very nature of the approach, this will lead to a degree theory for which we can prove the uniqueness.

More precisely, in the paper [140] a general scheme was developed by which any coincidence degree theory (for function pairs (F, φ) like the Benevieri-Furi degree) can be extended uniquely to a degree theory for function triples (F, Φ, φ) in the finite-dimensional situation by means of the homotopic Vietoris theorem. In [142], it was shown how such a degree theory for function triples can be extended to a degree theory for the infinite-dimensional situation if $\varphi(\Phi(\Omega))$ is relatively compact. Finally, in [141], it was discussed how such a theory in turn can be extended uniquely to a degree theory under much weaker compactness hypotheses.

We will not present these general axiomatic approaches here. Instead, we will confine ourselves to the extension of the Benevieri-Furi coincidence degree to a corresponding degree for function triples. In this chapter, we discuss function triples in general and how the homotopic Vietoris theorem is related with the existence of homotopies which turn function triples (with a possibly multivalued function) into function pairs (not multivalued).

11.1 Function Triples and Their Equivalences

Recall that our aim is to define a degree for inclusions of the type (11.1). This degree will not only depend on the functions F and $\varphi \circ \Phi$ but actually on the full function triple (F, Φ, φ) where, however, we will see that certain modifications of the couple (Φ, φ) are "harmless" in the sense that they will produce the same degree.

The aim of this section is to make precise for which objects our degree will be defined.

Definition 11.1. We call (F, Φ, φ) or, more verbosely, $(F, \Phi, \varphi, \Omega, Y, \Gamma)$ a *function triple* if $F\colon \Omega \to Y$, $\Phi\colon \Omega \multimap \Gamma$, and $\varphi\colon \Gamma \to Y$.

We can visualize a function triple as diagrams as in Figure 11.1.

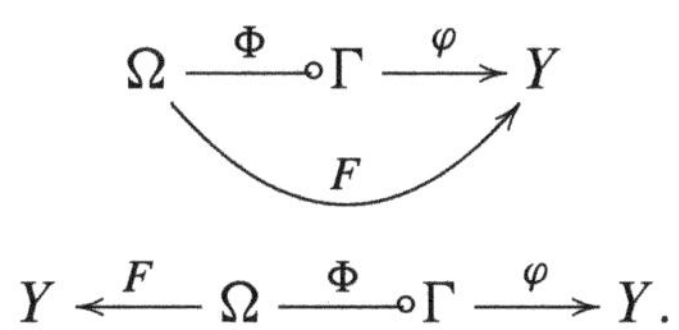

Figure 11.1. Two diagrams for the function triple $(F, \Phi, \varphi, \Omega, Y, \Gamma)$.

Later on, the map F will be an (oriented) Fredholm map of index 0 or, in the finite-dimensional setting, an (oriented) continuous map. For the moment, it is most convenient to consider an arbitrary map F with a closed graph (recall Example 2.120).

Definition 11.2. A function triple (F, Φ, φ) is called *closed* if the following holds:

(a) Ω and Γ are Hausdorff spaces, Y is a topological space.

(b) $\mathrm{graph}(F)$ is closed in $\Omega \times Y$.

(c) $\Phi\colon \Omega \multimap \Gamma$ is upper semicontinuous with compact values $\Phi(x)$.

(d) $\varphi \in C(\Gamma, Y)$.

We call (F, Φ, φ) *acyclic* or *acyclic** if Φ is additionally an acyclic or acyclic* map, respectively.

The point to consider compositions $\varphi \circ \Phi$ in (11.1) is that we do *not* want to require that the composition $\varphi \circ \Phi$ is acyclic. In fact, this is typically not the case:

Example 11.3. Let $\Phi(x) := [0, 2\pi]$ for all $x \in \Omega$ and $\varphi: [0, 2\pi] \to \mathbb{R}^2$ be defined by $\varphi(t) := (\cos t, \sin t)$. Then Φ is acyclic (even acyclic*) but $\varphi \circ \Phi$ fails to be acyclic.

Our degree should "count" (with appropriate multiplicities) the number of solutions of (11.1). These solutions will be denoted by $\mathrm{coin}(F, \Phi, \varphi)$:

Definition 11.4. For a function triple $(F, \Phi, \varphi, \Omega, Y, \Gamma)$ and $M \subseteq \Omega$, we put

$$\mathrm{coin}_M(F, \Phi, \varphi) := \{x \in M : F(x) \in \varphi(\Phi(x))\}.$$

In case $M = \Omega$, we sometimes omit the index M and just write

$$\mathrm{coin}(F, \Phi, \varphi) := \{x \in \Omega : F(x) \in \varphi(\Phi(x))\} = \{x \in \Omega : x \in (F^{-1} \circ \varphi \circ \Phi)(x)\}.$$

If this set is compact, we call (F, Φ, φ) a *proper triple*.

We point out that, although formally $\mathrm{coin}_M(F, \Phi, \varphi)$ is a fixed point set of the multivalued map $F^{-1} \circ \varphi \circ \Phi$, the map F^{-1} typically assumes empty values, so that classical fixed point theory of multivalued maps cannot be used.

In some cases, it is more convenient to "multiply" F^{-1} to the right, that is, to consider $\varphi \circ \Phi \circ F^{-1}$ instead:

Definition 11.5. For a function triple $(F, \Phi, \varphi, \Omega, Y, \Gamma)$ and a set $M \subseteq \Omega$, we put

$$\begin{aligned}\mathrm{fix}_M(F, \Phi, \varphi) &:= \{y \in F(M) : y \in (\varphi \circ \Phi \circ F^{-1})(y)\}\\ &= \{y \in Y : \text{there is } x \in M \text{ with } y = F(x) \in \varphi(\Phi(x))\}\\ &= F(\mathrm{coin}_M(F, \Phi, \varphi)).\end{aligned}$$

In case $M = \Omega$, we sometimes omit the index M and write

$$\mathrm{fix}(F, \Phi, \varphi) = F(\mathrm{coin}(F, \Phi, \varphi)).$$

This formula implies $\mathrm{coin}(F, \Phi, \varphi) \neq \emptyset$ if and only if $\mathrm{fix}(F, \Phi, \varphi) \neq \emptyset$, but the latter set has less elements, in general (unless F is one-to-one). Note also that in general $\operatorname{coin}(F, \Phi, \varphi) \subseteq \Omega$ while $\operatorname{fix}(F, \Phi, \varphi) \subseteq Y$. The set $\operatorname{fix}(F, \Phi, \varphi)$ will play a particular role in Chapter 14.

For the particular case that F is a Vietoris map, that is, F^{-1} is acyclic, then the map $\varphi \circ \Phi \circ F^{-1}$ has a form for which a lot of the earlier mentioned degree theories apply, see e.g. [23], [47], [53], [59], [71], [83], [91], [95], [120], [130], [144], [148]. However, these theories "count" the elements of $\operatorname{fix}(F, \Phi, \varphi)$ (for instance, they satisfy the excision property and additivity on the domain Y).

In contrast, the degree which we intend to define is of a rather different nature: It will "count" the elements of $\operatorname{coin}(F, \Phi, \varphi)$ (that is, it will satisfy the excision property and additivity on the domain Ω). Moreover, we do *not* require that F is onto or that F^{-1} is acyclic. We will require instead that F is (oriented) Fredholm of index 0. In the finite-dimensional setting even arbitrary continuous (oriented) maps F will be allowed.

Proposition 11.6. *Let (F, Φ, φ) be a closed function triple. Then $\operatorname{coin}(F, \Phi, \varphi)$ is closed in Ω. Hence,*

$$\operatorname{coin}_M(F, \Phi, \varphi) = M \cap \operatorname{coin}(F, \Phi, \varphi)$$

is closed in M for every $M \subseteq \Omega$. In particular, (F, Φ, φ) is proper if and only if $\operatorname{coin}(F, \Phi, \varphi)$ is relatively compact in Ω. In this case $\operatorname{fix}_M(F, \Phi, \varphi)$ is compact for any $M \supseteq \operatorname{coin}(F, \Phi, \varphi)$.

Proof. Putting $\Psi := \varphi \circ \Phi$ and $\widetilde{\Phi}(x) := \{F(x)\}$, we have $\operatorname{coin}(F, \Phi, \varphi) = \operatorname{coin}(\widetilde{\Phi}, \Psi)$, so that the assertion follows from Corollary 2.119, since $\operatorname{graph}(\widetilde{\Phi}) = \operatorname{graph}(F)$ is closed, Ψ is upper semicontinuous by Proposition 2.94, and $\Psi(x)$ is compact for every $x \in \Omega$. □

Of course, the degree we intend to define will be invariant under a certain class of homotopies. However, it is somewhat important that the degree will also be invariant under certain modifications of Φ, φ, and Γ. For instance, it would be rather strange if the degree would change when we replace Γ by a homeomorphic copy of Γ or if we shrink or enlarge Γ (as long as it contains $\Phi(\Omega)$). Actually, we will see that we are even allowed to shrink or enlarge the values $\Phi(x)$ in a certain sense.

For this reason, we intend to introduce an equivalence relation on the class of function triples which allows to exchange (Φ, φ, Γ) in a certain way. The degree will then only depend on the equivalence class. This has not only practical advantages for the calculation of the degree, but we will also see that it is crucial for our construction of the degree.

We will now introduce the announced equivalence relation. To define the equivalence relation on the class of acyclic function triples, let us first give some heuristic motivation for the definition in terms of the inclusion (11.1): Assume that the continuous function φ is the composition of $\widetilde{\varphi} \in C(\widetilde{\Gamma}, Y)$ and $J \in C(\Gamma, \widetilde{\Gamma})$, that is, we have $\varphi = \widetilde{\varphi} \circ J$. In particular, we can write (11.1) in the form

$$F(x) \in (\widetilde{\varphi} \circ J)(\Phi(x)). \tag{11.2}$$

The function triple associated to this inclusion is $(F, \Phi, \varphi) = (F, \Phi, \widetilde{\varphi} \circ J)$. On the other hand, we can rewrite (11.2) equivalently as

$$F(x) \in \widetilde{\varphi}((J \circ \Phi)(x))$$

which is associated with the triple $(F, J \circ \Phi, \widetilde{\varphi})$. We would like to consider such triples as equivalent and write

$$(F, \Phi, \widetilde{\varphi} \circ J) \precsim (F, J \circ \Phi, \widetilde{\varphi}).$$

If J is onto or a homeomorphism, respectively, we use the symbols $\preceq$ or $\approx$ instead. Note that unless J is a homeomorphism the construction is not reversible, so that the relations $\precsim$ and $\preceq$ are not symmetric and thus not equivalence relations. Of course, this is only a formal difficulty, since we can define a corresponding equivalence relation by passing to the transitive symmetric closure.

However, in the class of acyclic (or acyclic*) triples, this construction has another problem: If Φ is an acyclic map, the composition $J \circ \Phi$ is not acyclic, in general (recall Example 11.3). Therefore, to enlarge the equivalence classes, we additionally want to allow that $J \circ \Phi$ is replaced by an acyclic map $\widetilde{\Phi}$ with possibly larger sets as values, and in this case, we will write

$$(F, \Phi, \widetilde{\varphi} \circ J) \sqsubseteq (F, \widetilde{\Phi}, \widetilde{\varphi},).$$

Of course, also this construction is irreversible, and we have to define a corresponding equivalence relation by passing to the transitive symmetric closure. We already point out that for the definition to the degree, we have to impose further conditions on the considered triples (e.g. that the triple is proper), and therefore it will be important that actually also the triple $(F, \widetilde{\Phi}, \widetilde{\varphi})$ will satisfy these conditions, that is, the extension $\widetilde{\Phi}$ will actually not be as arbitrary as it is in the moment.

Let us now turn the above heuristics into precise definitions:

Definition 11.7. A function triple $(F, \Phi, \varphi, \Omega, Y, \Gamma)$ is *embedded* into a function triple $(F, \widetilde{\Phi}, \widetilde{\varphi}, \Omega, Y, \widetilde{\Gamma})$ if there is $J \in C(\Gamma, \widetilde{\Gamma})$ with

$$J(\Phi(x)) \subseteq \widetilde{\Phi}(x) \text{ for all } x \in \Omega, \text{ and } \widetilde{\varphi}(J(z)) = \varphi(z) \text{ for all } z \subset \Gamma. \tag{11.3}$$

We write in this case $(F, \Phi, \varphi) \sqsubseteq (F, \widetilde{\Phi}, \widetilde{\varphi})$. If J satisfies the stronger relation

$$J(\Phi(x)) = \widetilde{\Phi}(x) \text{ for all } x \in \Omega, \text{ and } \widetilde{\varphi}(J(z)) = \varphi(z) \text{ for all } z \in \Gamma, \tag{11.4}$$

then we write $(F, \Phi, \varphi) \precsim (F, \widetilde{\Phi}, \widetilde{\varphi})$. If additionally J is onto, we write $(F, \Phi, \varphi) \preceq (F, \widetilde{\Phi}, \widetilde{\varphi})$. If additionally J is a homeomorphism, we write $(F, \Phi, \varphi) \approx (F, \widetilde{\Phi}, \widetilde{\varphi})$.

By definition, the requirements become successively stronger, that is, we have the chain of implications

$$A \approx \widetilde{A} \implies A \preceq \widetilde{A} \implies A \precsim \widetilde{A} \implies A \sqsubseteq \widetilde{A}.$$

Proposition 11.8. *If* $(F, \Phi, \varphi) \sqsubseteq (F, \widetilde{\Phi}, \widetilde{\varphi})$ *then*

$$\varphi(\Phi(x)) \subseteq \widetilde{\varphi}(\widetilde{\Phi}(x)) \qquad \text{for all } x \in \Omega, \tag{11.5}$$

and thus in particular

$$\operatorname{coin}(F, \Phi, \varphi) \subseteq \operatorname{coin}(F, \widetilde{\Phi}, \widetilde{\varphi}) \quad \text{and} \quad \operatorname{fix}(F, \Phi, \varphi) \subseteq \operatorname{fix}(F, \widetilde{\Phi}, \widetilde{\varphi}). \tag{11.6}$$

If $(F, \Phi, \varphi) \precsim (F, \widetilde{\Phi}, \widetilde{\varphi})$ *then*

$$\widetilde{\varphi}(\widetilde{\Phi}(x)) = \varphi(\Phi(x)) \qquad \text{for all } x \in \Omega \tag{11.7}$$

and thus in particular

$$\operatorname{coin}(F, \Phi, \varphi) = \operatorname{coin}(F, \widetilde{\Phi}, \widetilde{\varphi}) \quad \text{and} \quad \operatorname{fix}(F, \Phi, \varphi) = \operatorname{fix}(F, \widetilde{\Phi}, \widetilde{\varphi}). \tag{11.8}$$

Proof. The relation (11.3) implies for every $x \in \Omega$ that

$$\varphi(\Phi(x)) = \widetilde{\varphi}(J(\Phi(x))) \subseteq \widetilde{\varphi}(\widetilde{\Phi}(x))$$

which means (11.5). Hence, (11.6) follows from the definition of coin and fix. The proof of the second assertion is analogous: Instead of an inclusion we have in the above calculation an equality by (11.4). This shows (11.7), and (11.8) follows from the definition of coin and fix. □

Corollary 11.9. *If $(F, \Phi, \varphi) \precsim (F, \widetilde{\Phi}, \widetilde{\varphi})$ then (F, Φ, φ) is proper if and only if $(F, \widetilde{\Phi}, \widetilde{\varphi})$ is proper.*

If $(F, \Phi, \varphi) \sqsubseteq (F, \widetilde{\Phi}, \widetilde{\varphi})$, (F, Φ, φ) is closed and $(F, \widetilde{\Phi}, \widetilde{\varphi})$ is proper then also (F, Φ, φ) is proper.

Proof. The first assertion follows from (11.8). For the second assertion, note that if $(F, \widetilde{\Phi}, \widetilde{\varphi})$ is proper, then (11.6) implies that $\operatorname{coin}(F, \Phi, \varphi)$ is relatively compact in Ω. Proposition 11.6 thus implies that (F, Φ, φ) is proper. □

The notation is justified by the following observation:

Proposition 11.10. *The relation $\sqsubseteq$ is reflexive and transitive, that is, for all function triples we have*

(a) $(F, \Phi, \varphi) \sqsubseteq (F, \Phi, \varphi)$.

(b) *If $(F, \Phi, \varphi) \sqsubseteq (F, \Phi_0, \varphi_0) \sqsubseteq (F, \widetilde{\Phi}, \widetilde{\varphi})$ then $(F, \Phi, \varphi) \sqsubseteq (F, \widetilde{\Phi}, \widetilde{\varphi})$.*

In the same sense $\precsim$ and $\preceq$ are reflexive and transitive, and $\approx$ is even an equivalence relation, that is, $\approx$ is also symmetric.

Proof. The reflexivity follows with the choice $J = \mathrm{id}_\Gamma$ in Definition 11.7. For the transitivity, note that if there are corresponding continuous functions J_0 and $\widetilde{J}$ such that

$$J_0 \circ \Phi \subseteq \Phi_0, \widetilde{J} \circ \Phi_0 \subseteq \widetilde{\Phi}, \varphi = \varphi_0 \circ J_0, \varphi_0 = \widetilde{\varphi} \circ \widetilde{J},$$

then $\widetilde{J} \circ J_0 \circ \Phi \subseteq \widetilde{\Phi}$ and $\varphi = \widetilde{\varphi} \circ \widetilde{J} \circ J_0$, that is, the continuous function $J := \widetilde{J} \circ J_0$ satisfies (11.3). □

Thus, there are several natural ways to obtain an equivalence relation on a class of function triples: We can take $\approx$, or we can either restrict or extend some of $\sqsubseteq$, $\precsim$, or $\preceq$ to make it symmetric. Obviously, the strongest of all these equivalence relations is $\approx$. However, we are interested in obtaining an equivalence relation which is as weak as possible, hence, we choose the way of extending the weakest of the above relations by passing to the symmetric transitive closure. If we are interested only in triples of a certain class, it is important that we consider also the symmetric transitive closure only within this class. Hence, the equivalence relation which we define actually depends on a given class:

Definition 11.11. Let $\mathcal{T}$ denote a class of function triples. Then two triples A and $\widetilde{A}$ from $\mathcal{T}$ are *equivalent* if there is a finite sequence $A = A_0, A_1, \dots, A_n = \widetilde{A}$, all belonging to the class $\mathcal{T}$, such that for each $k = 1, \dots, n$ at least one of the relations $A_{k-1} \sqsubseteq A_k$ or $A_k \sqsubseteq A_{k-1}$ holds true. In this case, we write $A \sim_{\mathcal{T}} \widetilde{A}$.

We point out that also all "intermediate" function triples $A_1, \dots, A_{n-1}$ are required to belong to $\mathcal{T}$. Thus the definition of the relation $\sim_{\mathcal{T}}$ really depends on the class $\mathcal{T}$. In particular, in order to verify the equivalence, it is not sufficient to find corresponding maps $J_k = J$ satisfying (11.3) for the corresponding functions, but one also has to verify that the corresponding function triples indeed belong to the class $\mathcal{T}$.

An example which shows that it is useful that one can vary the direction in Definition 11.11 will be given in Proposition 11.29.

Proposition 11.12. *The relation $\sim_{\mathcal{T}}$ is an equivalence relation in the class $\mathcal{T}$. For $A, \widetilde{A} \in \mathcal{T}$, we have the implications*

$$A \approx \widetilde{A} \implies A \preceq \widetilde{A} \implies A \precsim \widetilde{A} \implies A \sqsubseteq \widetilde{A} \implies A \sim_{\mathcal{T}} \widetilde{A}.$$

Proof. The symmetry and transitivity of $\sim_{\mathcal{T}}$ follows from its definition. Since $\sqsubseteq$ is reflexive (Proposition 11.10), also $\sim_{\mathcal{T}}$ must be reflexive. The last assertion is obvious. □

Proposition 11.13. *We have for each function triple $(F, \Phi, \varphi, \Omega, Y, \Gamma)$ that*

$$(F, \Phi, \varphi|_{\Phi(\Omega)}, \Omega, Y, \Phi(\Omega)) \precsim (F, \Phi, \varphi, \Omega, Y, \Gamma).$$

Proof. Put $J(z) := z$ for all $z \in \Phi(\Omega)$ in Definition 11.7. □

Remark 11.14. Proposition 11.13 implies that under the equivalence relation $\sim_{\mathcal{T}}$ and actually even under the more restrictive relation $\precsim$ the set Γ plays no role; only the restriction of φ to $\Phi(\Omega)$ (and the topology inherited from Γ) will be used.

We will use this observation later to suppress Γ in the notation. If Y is clear, we will later also suppress Y from the notation.

Only under this independence from Γ the following definition deserves its name:

Definition 11.15. The *restriction of a function triple* $(F, \Phi, \varphi, \Omega, Y, \Gamma)$ to $\Omega_0 \subseteq \Omega$ is defined as the function triple $(F, \Phi, \varphi, \Omega_0, Y, \Phi(\Omega_0))$.

Formally more correct, the last expression should read

$$(F|_{\Omega_0}, \Phi|_{\Omega_0}, \varphi|_{\Phi(\Omega_0)}, \Omega_0, Y, \Phi(\Omega_0)),$$

but we used here the convention of omitting obvious restriction symbols for better readability (cf. Remark 9.37).

Proposition 11.16. *If* $(F, \Phi, \varphi) \precsim (F, \widetilde{\Phi}, \widetilde{\varphi})$ *or* $(F, \Phi, \varphi) \sqsubseteq (F, \widetilde{\Phi}, \widetilde{\varphi})$ *then the same relation holds for the corresponding restrictions.*

Proof. The restriction of the map J from Definition 11.7 is the required map. □

Now we come to the crucial definition which relates our objects with Vietoris maps.

Definition 11.17. A function triple (F, Φ, φ) is *in standard form* if there is a continuous map $p\colon \Gamma \to \Omega$ with $\Phi = p^{-1}$, that is, if $\Phi(\Omega) = \Gamma$ and Φ^{-1} is single-valued and continuous.

We can visualize a function triple in standard form as diagrams as in Figure 11.2.

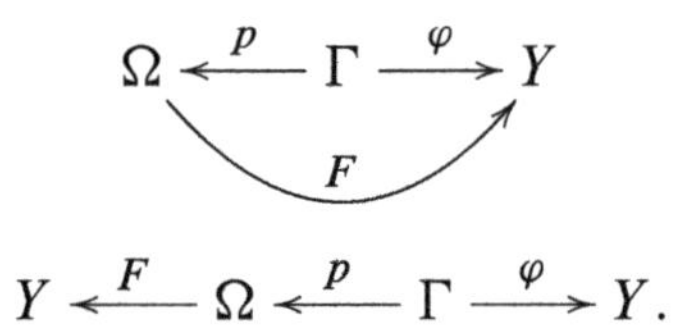

Figure 11.2. Two diagrams for a function triple in standard form.

Proposition 11.18. *If* (F, Φ, φ) *is a closed function triple in standard form then the above map* p *is automatically closed and proper. If the triple is acyclic (acyclic*), then* p *is Vietoris (Vietoris*).*

Proof. Corollary 2.106 implies that p is closed and proper. The second assertion follows from $p^{-1}(x) = \Phi(x)$. □

Formally, the degree which we intend to consider later can only be defined for a certain class $\mathcal{T}$ of acyclic function triples in standard form. Of course, "most" acyclic function triples are not in standard form. Therefore, it is important for us to consider the equivalence relation $\sim_{\mathcal{T}}$ and to show:

(a) Each equivalence class contains a triple in standard form, and

(b) the corresponding degree is independent of the choice of that triple in the equivalence class.

This idea is not new, of course. It can be traced back at least to the work of L. Gorniewicz [70] who defined so-called *admissible pairs* which are essentially acyclic function triples in standard form in case $\Omega \subseteq Y$ and F being the inclusion map. In a slightly more general terminology, W. Kryszewski defined in [93] so-called co-triads which in our terminology correspond to acyclic function triples in standard form. Also in [140]–[142], essentially only acyclic function triples in standard form were considered. This has the technical advantage that the multivalued terminology vanishes from the notations, and so all topological considerations can be carried out with single-valued maps. However, it has the disadvantage that for the application of the theory to a problem in which a multivalued map occurs, one first has to "translate" the problem into a standard form which can be notationally rather inconvenient.

Therefore, from the viewpoint of applications, it is more natural to use multivalued terminology. Note that this multivalued terminology has not only a formal advantage: For the equivalence relation of Definition 11.11 it is not only unnecessary that the two considered triples A and $\widetilde{A}$ are in standard form, but also the "intermediate" triples $A_1, \dots, A_{n-1}$ are not required to be in standard form. The corresponding more restrictive (at a first glance) definition when all triples are required to be in standard form is in view of the subsequent Proposition 11.19 the equivalence relation considered in [140]. Hence, at least formally, our equivalence relation is easier to verify than that from [140].

Let us now equivalently reformulate the above relations for the case that the function triples are in standard form.

Proposition 11.19. *Let $(F, \Phi, \varphi, \Omega, Y, \Gamma)$ and $(F, \widetilde{\Phi}, \widetilde{\varphi}, \Omega, Y, \widetilde{\Gamma})$ be two function triples in standard form, $\Phi = p^{-1}$ and $\widetilde{\Phi} = \widetilde{p}^{-1}$. Then the following statements are equivalent.*

(a) $(F, \Phi, \varphi) \sqsubseteq (F, \widetilde{\Phi}, \widetilde{\varphi})$.

(b) *There is a continuous map $J\colon \Gamma \to \widetilde{\Gamma}$ with*

$$\widetilde{p} \circ J = p \quad \text{and} \quad \widetilde{\varphi} \circ J = \varphi. \tag{11.9}$$

The corresponding maps J are exactly those from Definition 11.7, *that is, actually* (11.3) *and* (11.9) *are equivalent.*

Proof. For $x \in \Omega$ and $J\colon \Gamma \to \widetilde{\Gamma}$ the relation

$$J(\Phi(x)) \subseteq \widetilde{\Phi}(x)$$

is in view of $\widetilde{\Phi}(x) = \widetilde{p}^{-1}(x)$ equivalent to

$$\widetilde{p}(J(\Phi(x))) \subseteq \{x\}.$$

This means that for all $z \in p^{-1}(x) = \Phi(x)$ the equality

$$\widetilde{p}(J(z)) = p(z)$$

holds. Since p^{-1} is onto, it follows that $J \circ \Phi \subseteq \widetilde{\Phi}$ is equivalent to $\widetilde{p} \circ J = p$. Hence, (11.3) and (11.9) are equivalent which implies the claim. □

Corollary 11.20. *For triples in standard form the following statements are equivalent. (We use the notation of Proposition* 11.19*).*

(a) $(F, \Phi, \varphi) \precsim (F, \widetilde{\Phi}, \widetilde{\varphi})$.

(b) $(F, \Phi, \varphi) \preceq (F, \widetilde{\Phi}, \widetilde{\varphi})$.

(c) *There is a continuous map* $J : \Gamma \to \widetilde{\Gamma}$ *with*

$$\widetilde{p}^{-1} \subseteq J \circ p^{-1}, \quad \widetilde{p} \circ J = p, \quad \textit{and} \quad \widetilde{\varphi} \circ J = q. \tag{11.10}$$

Proof. The first inclusion in (11.10) is exactly the inclusion which is missing from (11.3) to yield (11.4). Since $\widetilde{p}^{-1}$ is onto, this inclusion implies that J is onto. □

Corollary 11.21. *For triples in standard form the following statements are equivalent. (We use the notation of Proposition* 11.19*).*

(a) $(F, \Phi, \varphi) \approx (F, \widetilde{\Phi}, \widetilde{\varphi})$.

(b) *There is a homeomorphism* $J : \Gamma \to \widetilde{\Gamma}$ *with* (11.10).

Proof. The claim follows from the same argument as in Corollary 11.20. □

Corollary 11.21 shows that the relation $\approx$ corresponds to the strong equivalence relation introduced in [93, Definition 4.39] (which in turn is based on a definition from [89]).

From the viewpoint of applications of degree theory, it is a big advantage that we will be able to show that already the weaker equivalence relation $A \sim_{\mathcal{T}} \widetilde{A}$ implies that the corresponding degrees of A and $\widetilde{A}$ are equal. For this reason, it is worth to discuss now briefly how our equivalence relation $\sim_{\mathcal{T}}$ compares to a more popular equivalence relation which can be found in literature.

There is a notion of equivalence classes for the earlier mentioned admissible pairs from [70]. In our more general setting this corresponds to the notion of equivalence for co-triads which can be found e.g. in [93, Definition 4.6]: In our terminology, the latter means in view of Proposition 11.19 that two triples A and $\widetilde{A}$ in standard form are equivalent (in the sense of [93, Definition 4.6]) if there is a third triple A_0 in standard form with $A_0 \sqsubseteq A$ and $A_0 \sqsubseteq \widetilde{A}$. Note that if A and $\widetilde{A}$ are proper and A_0 is closed then A_0 is automatically proper by Corollary 11.9.

In particular, if $\mathcal{T}$ denotes the class of proper acyclic function triples and A and $\widetilde{A}$ belong to this class and are in standard form and equivalent in the sense of [93, Definition 4.6], then $A \sim_{\mathcal{T}} \widetilde{A}$ in our sense. However, we do not know whether also the converse holds, that is, our equivalence classes are perhaps even *strictly* larger than the equivalence classes considered in [93, Definition 4.6].

In this sense, our notion $\sim_{\mathcal{T}}$ of equivalence is weaker than all related popular notions of equivalences which can be found in literature. Note that this means that our claim that the degree depends only on the equivalence class improves all related statements in literature.

The equivalence relation introduced in this section has not only practical advantages for the calculation of the degree, but it is also crucial for our construction of the degree in two respects:

(a) Strictly speaking, we can define the degree only if the triple is in standard form. Only by passing to an appropriate equivalent triple, we will be able to produce this standard form.

(b) When we want to interpret homotopies of function triples as function triples, we will need the equivalence relation to obtain a natural meaning for the homotopy "at time t".

The first of these properties will be discussed just now, the second in Section 11.3.

Definition 11.22. Let (F, Φ, φ), more precisely $(F, \Phi, \varphi, \Omega, Y, \Gamma)$, be a function triple. The *corresponding standard form* is the function triple $(F, \widetilde{\Phi}, \widetilde{\varphi}, \Omega, Y, \widetilde{\Gamma})$, defined as follows.

(a) $\widetilde{\Gamma} := \operatorname{graph}(\Phi)$

(b) $\widetilde{\Phi} := p^{-1}$ where $p\colon \widetilde{\Gamma} \to \Omega$ is defined by $p(x, y) := x$.

(c) $\widetilde{\varphi} := \varphi \circ q$ where $q\colon \widetilde{\Gamma} \to \Gamma$ is defined by $q(x, z) := z$.

The idea of this construction is well-known and has already been used in the first paper on topological fixed point theory for acyclic maps [48]. The following result implies that we obtain indeed a standard form in the corresponding equivalence class:

Theorem 11.23. *If $(F, \widetilde{\Phi}, \widetilde{\varphi})$ is the standard form of (F, Φ, φ), then $(F, \widetilde{\Phi}, \widetilde{\varphi})$ is in standard form satisfying*

$$\widetilde{\Phi}(x) = \{x\} \times \Phi(x) \text{ (which is homeomorphic to } \Phi(x)) \tag{11.11}$$

for all x, and

$$(F, \widetilde{\Phi}, \widetilde{\varphi}) \precsim (F, \Phi, \varphi). \tag{11.12}$$

Moreover, if Φ is upper semicontinuous with compact values then $\widetilde{\Phi}$ has the same property.

In particular, (11.7) *and* (11.8) *hold, and if (F, Φ, φ) is closed, proper, acyclic or acyclic* then $(F, \widetilde{\Phi}, \widetilde{\varphi})$ has the same property.*

Proof. Since p in Definition 11.22 is continuous, we obtain immediately that $(F, \widetilde{\Phi}, \widetilde{\varphi}, \Omega, Y, \widetilde{\Gamma})$ is in standard form. Definition 11.22 implies (11.11), and the map $J := q$ from Definition 11.22 proves (11.12). Indeed, by (11.11), we have $\Phi = q \circ p^{-1} = J \circ \widetilde{\Phi}$ and $\widetilde{\varphi} = \varphi \circ q = \varphi \circ J$, hence (11.4) holds.

If Φ is upper semicontinuous with compact values, then Theorem 2.111 implies that the map p of Definition 11.22 is closed. Proposition 2.104 implies that $\widetilde{\Phi} = p^{-1}$ is upper semicontinuous. The remaining assertions follow from Corollary 11.9 and (11.11). □

The following result implies that we can suppress Γ in the notation when we deal with the standard form:

Proposition 11.24. *The standard form of $(F, \Phi, \varphi, \Omega, Y, \Gamma)$ is independent of Γ in the sense that it depends only on $(F, \Phi, \varphi|_{\Phi(\Omega)})$.*

Proof. The function $\widetilde{\varphi}$ of Definition 11.22 depends only on $\varphi|_{\Phi(\Omega)}$. □

Having the above independence from Γ in mind, we recall that it makes sense to define a restriction (Definition 11.15). This operation commutes with the passage to the standard form

Proposition 11.25. *The standard form of the restriction is the restriction of its standard form.*

Proof. Let $A = (F, \Phi, \varphi, \Omega, Y, \Gamma)$ be some function triple, and A_0 denote the restriction of A to some $\Omega_0 \subseteq \Omega$. The standard form of A is then $\widetilde{A} = (F, \widetilde{\Phi}, \widetilde{\varphi}, \Omega, Y, \widetilde{\Gamma})$ with $\widetilde{\Gamma} = \operatorname{graph}(\Phi)$, $\widetilde{\Phi}(x) := \{x\} \times \Phi(x)$, and $\widetilde{\varphi}(x, z) := \varphi(z)$. The standard form of A_0 is similarly $(F|_{\Omega_0}, \Phi_0, \varphi_0, \Omega_0, Y, \Gamma_0)$ with $\Gamma_0 = \{(x, y) \in \operatorname{graph}(\Phi) : x \in \Omega_0\}$, $\Phi_0(x) := \{x\} \times \Phi(x)$, and $\varphi_0(x, z) := \varphi(z)$. Since $\Phi_0(\Omega_0) = \Gamma_0$, it follows that this is the restriction of $\widetilde{A}$ to Ω_0. □

Formally, even if a triple is in standard form, its corresponding standard form is a different object. Fortunately, this difference is covered by our equivalence relation, even by the strongest equivalence relation which we consider.

Proposition 11.26. *Let (F, Φ, φ) be in standard form, and let $(F, \widetilde{\Phi}, \widetilde{\varphi})$ be the corresponding standard form. Then*

$$(F, \widetilde{\Phi}, \widetilde{\varphi}) \approx (F, \Phi, \varphi). \tag{11.13}$$

Proof. Using the notation of the proof of Theorem 11.23, we have to show that $J := q : \widetilde{\Gamma} \to \Gamma$ is a homeomorphism. Since $\Phi = p_0^{-1}$ for some continuous map $p_0 : \Gamma \to \Omega$, we have for any $z \in \Gamma$ that $z \in p_0^{-1}(p_0(z)) = \Phi(p_0(z))$, hence $q_0(z) := (p_0(z), z) \in \widetilde{\Gamma} = \operatorname{graph}(\Phi)$. Thus, q_0 defines a map $q_0 : \Gamma \to \widetilde{\Gamma}$. Since p_0 is continuous, also q_0 is continuous. For every $(x, z) \in \widetilde{\Gamma}$, we have $z \in \Phi(x) = p_0^{-1}(x)$, hence $p_0(z) = x$, and so $(x, z) = q_0(z) = q_0(q(x, z))$. This shows that $q_0 \circ q$ is the identity map on $\widetilde{\Gamma}$. Since also $q \circ q_0$ is the identity map on Γ, it follows that $J = q$ is invertible with the continuous inverse q_0. □

It is a technical difficulty for us that Proposition 11.26 fails if (F, Φ, φ) is not in standard form:

Example 11.27. Let $\Omega = Y = \Gamma = [0, 1]$, $\Phi(x) := \Gamma$, and $F = \varphi = \mathrm{id}_\Omega$. Then the corresponding standard form is $(F, \widetilde{\Phi}, \widetilde{\varphi}, \Omega, Y, \widetilde{\Gamma})$ with $\widetilde{\Gamma} = [0, 1] \times [0, 1]$. Then (11.13) fails, because $\widetilde{\Gamma}$ is not homeomorphic to Γ. The latter can be seen by using the dimension invariance result of Corollary 9.95 or also elementary by observing that $\Gamma \setminus \{\frac{1}{2}\}$ is disconnected while $\widetilde{\Gamma} \setminus \{z\}$ is connected for every $z \in \widetilde{\Gamma}$.

Example 11.27 shows that for our purposes the equivalence relation $\approx$ would be too restrictive. However, the equivalence relation $\sim_{\mathcal{T}}$ will be sufficient in view of (11.12).

Since we define the degree first only on triples in standard form, there is a technical difficulty when we want to show that the degree depends only on the equivalence class: We need also that the "intermediate" triples in Definition 11.11 can be chosen in standard form. It seems that this does not follow from the previous results, so we have to show it:

Theorem 11.28. *If $A \sqsubseteq A_0$, $A \precsim A_0$, $A \preceq A_0$, or $A \approx A_0$ then the same relation holds for the corresponding standard forms of A and A_0.*

Proof. Let $A = (F, \Phi, \varphi, \Omega, Y, \Gamma)$ and $A_0 = (F, \Phi_0, \varphi_0, \Omega, Y, \Gamma_0)$, and let $\widetilde{A} = (F, \widetilde{\Phi}, \widetilde{\varphi}, Y, \widetilde{\Gamma})$ and $\widetilde{A}_0 = (F, \widetilde{\Phi}_0, \widetilde{\varphi}_0, Y, \widetilde{\Gamma}_0)$ denote the respective corresponding standard forms. If $A \sqsubseteq A_0$, there is a continuous map $J : \Gamma \to \Gamma_0$ with

$$J \circ \Phi \subseteq \Phi_0 \quad \text{and} \quad \varphi = \varphi_0 \circ J.$$

Since $\widetilde{\Gamma} = \text{graph}(\Phi)$ and $\text{graph}(J \circ \Phi) \subseteq \text{graph}(\Phi_0) = \widetilde{\Gamma}_0$, we can define a map $J_0 \colon \widetilde{\Gamma} \to \widetilde{\Gamma}_0$ by $J_0(x, z) := (x, J(z))$. We show that this is the map required for Definition 11.7. Indeed, J_0 is continuous and satisfies for all $x \in \Omega$

$$J_0(\{x\} \times \Phi(x)) = \{x\} \times J(\Phi(x)) \subseteq \{x\} \times \Phi_0(x)$$

which means $J_0 \circ \widetilde{\Phi} \subseteq \widetilde{\Phi}_0$. If q (or q_0) denote the canonical projections of $\text{graph}(\Phi)$ (or $\text{graph}(\Phi_0)$) to Γ (or Γ_0, respectively), then our definition of J_0 means for all $(x, z) \in \Gamma$ that

$$q_0(J_0(x, z)) = J(z) = J(q(x, z)),$$

that is $q_0 \circ J_0 = J \circ q$, and so

$$\widetilde{\varphi} = \varphi \circ q = \varphi_0 \circ J \circ q = \varphi_0 \circ q_0 \circ J_0 = \widetilde{\varphi}_0 \circ J_0.$$

Hence, J_0 has all required properties to show that $\widetilde{A} \sqsubseteq \widetilde{A}_0$.

The proof of the other assertions is similar: In the above arguments, one just has to observe that one can replace throughout $\subseteq$ by $=$, and that in this case, if J is onto or a homeomorphism, also J_0 is onto or a homeomorphism, respectively. □

The following result will be useful to establish a relation of the degree for function triples with the ordinary degree in the single-valued case. This is also an example which shows why it is useful to consider such a general equivalence relation as in Definition 11.11:

Proposition 11.29. *Let $(F, \Phi, \varphi, \Omega, Y, \Gamma)$ be a function triple such that $\varphi \circ \Phi$ is single-valued on Ω. If $(F, \widetilde{\Phi}, \widetilde{\varphi}, \Omega, Y, \widetilde{\Gamma})$ denotes the standard form of (F, Φ, φ) we have*

$$(F, \Phi, \varphi) \gtrsim (F, \widetilde{\Phi}, \widetilde{\varphi}) \lesssim (F, \text{id}_\Omega, \varphi \circ \Phi). \tag{11.14}$$

In particular, if all three of the above triples belong to $\mathcal{T}$ then

$$(F, \Phi, \varphi) \sim_{\mathcal{T}} (F, \text{id}_\Omega, \varphi \circ \Phi).$$

Proof. The first relation in (11.14) is (11.12) of Theorem 11.23. For the second relation, we recall that (11.11) means $\widetilde{\Phi}(x) = \{x\} \times \Phi(x)$ and $\widetilde{\varphi}(x, z) = \varphi(z)$. In particular, $\text{dom}(\widetilde{\Phi}) = \text{dom}(\Phi) = \Omega$, since $\text{dom}(\varphi \circ \Phi) = \Omega$. Recall that $\widetilde{\Phi}$ is in standard form, i.e. $p = \widetilde{\Phi}^{-1}$ is single-valued and continuous. Since $\text{dom}(\widetilde{\Phi}) = \Omega$, it follows that $p(\Gamma) = \Omega$. Putting $J := p$, we calculate

$$J \circ \widetilde{\Phi} = p \circ p^{-1} = \text{id}_\Omega,$$

and for all $(x, z) \in \widetilde{\Gamma} = \operatorname{graph}(\Phi)$, we have in view of $z \in \Phi(x)$, since $\varphi \circ \Phi$ is single-valued that

$$(\varphi \circ \Phi)(J(x, z)) = (\varphi \circ \Phi)(x) = \varphi(\Phi(x)) = \varphi(z) = \widetilde{\varphi}(x, z).$$

The map J in Definition 11.7 thus proves that $(F, \widetilde{\Phi}, \widetilde{\varphi}) \precsim (F, \mathrm{id}_\Omega, \varphi \circ \Phi)$. □

11.2 The Simplifier Property

Now we discuss a homotopic property of the standard form which is the key to our approach to deal with the multivalued map for the degree.

The reason why the standard form is so crucial for our considerations is that only in the standard form the following notion makes sense:

Definition 11.30. Let $(F, \Phi, \varphi, \Omega, Y, \Gamma)$ be a function triple in standard form, that is, $\Phi = p^{-1}$. Let $M \subseteq \Omega$. Then $f \in C(M, Y)$ is called a *simplifier* on M for (F, Φ, φ) if $\varphi(z) = f(p(z))$ for all $z \in \Phi(M)$. In case $M = \Omega$, we omit M and call f just a *simplifier* for (F, Φ, φ).

It may be more intuitive to consider this situation in the form of the commutative diagram in Figure 11.3.

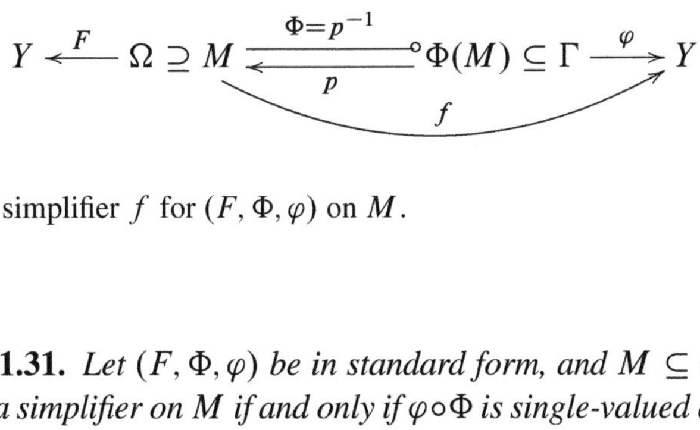

Figure 11.3. A simplifier f for (F, Φ, φ) on M.

Proposition 11.31. *Let (F, Φ, φ) be in standard form, and $M \subseteq \operatorname{dom}(\Phi)$. Then (F, Φ, φ) has a simplifier on M if and only if $\varphi \circ \Phi$ is single-valued and continuous on M, and in this case $f = \varphi \circ \Phi$ is the only simplifier on M.*

Proof. Let $\Phi = p^{-1}$. The relation $\varphi = f \circ p$ on $\Phi(M)$ implies $\varphi \circ \Phi = f \circ p \circ \Phi = f \circ p \circ p^{-1} = f$ on M. Hence, $f = \varphi \circ \Phi$ is the only candidate for a simplifier, and it is a simplifier if and only if f is continuous. □

The idea is that the single-valued function f can on M replace the couple (Φ, φ) in a sense. This idea makes sense according to our equivalence relation:

Proposition 11.32. *If f is a simplifier for $(F, \Phi, \varphi, \Omega, Y, \Gamma)$ then*

$$(F, \mathrm{id}_\Omega, f) \preceq (F, \Phi, \varphi).$$

In particular, $\mathrm{coin}(F,f) = \mathrm{coin}(F, \mathrm{id}_\Omega, f) = \mathrm{coin}(F, \Phi, \varphi)$ *and* $\mathrm{fix}(F, \mathrm{id}_\Omega, f) = \mathrm{fix}(F, \Phi, \varphi)$.

Proof. We have $\Phi = p^{-1}$. Then the map $J := p$ has the properties required for Definition 11.7. Indeed, $J = \varphi \circ p = \varphi \circ J$ holds by hypothesis, and $J \circ p^{-1} = p \circ p^{-1} = \mathrm{id}_\Omega$ since p is onto. □

Not every function triple in standard form has a simplifier, of course. However, concerning the degree, it will be sufficient to allow certain homotopies with respect to φ, that is, it will be sufficient to consider simplifiers which are admissible in the following homotopic sense.

Definition 11.33. Let $(F, \Phi, \varphi, \Omega, Y, \Gamma)$ be a function triple in standard form, that is, $\Phi = p^{-1}$. Let $M_0 \subseteq M \subseteq \Omega$ and $Y_0 \subseteq Y$. Then $f \in C(M, Y_0)$ is called a (M_0, Y_0)-*admissible simplifier* on M if there is a continuous function $h\colon [0,1] \times \Phi(M) \to Y_0$ with $h(0,\cdot) = \varphi$ and $\mathrm{coin}_{M_0}(F, \Phi, h(t,\cdot)) = \varnothing$ for all $t \in [0,1]$ such that f is a simplifier for $(F, \Phi, h(1,\cdot))$ on M, that is, $h(1,\cdot) = f \circ p|_{\Phi(M)}$.

We attempt to visualize Definition 11.33 by a commutative diagram in Figure 11.4.

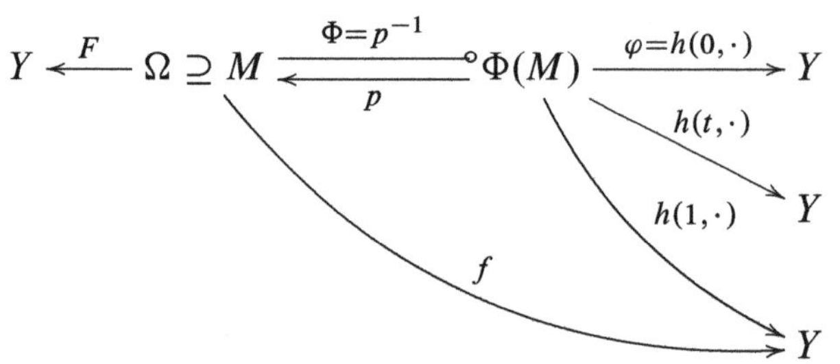

Figure 11.4. An admissible simplifier f on M.

A crucial (but invisible) point of Figure 11.4 is that $\mathrm{coin}_{M_0}(F, \Phi, h(t,\cdot)) = \varnothing$ for all $t \in [0,1]$, that is that for each $x \in M_0 \subseteq M$ the corresponding value $F(x)$ "to the left" in that diagram does not lie in any of the corresponding sets $h(t, \Phi(x))$ "to the right" of the diagram.

Actually, we would like to have not only the existence of an admissible simplifier but also uniqueness in a homotopic sense:

Definition 11.34. Two maps $f_0, f_1 \in C(M, Y_0)$ are (F, M_0, Y_0)-*homotopic* on M if there is a continuous function $h: [0,1] \times M \to Y_0$ with $h(i, \cdot) = f_i$ $(i = 0, 1)$ and $\operatorname{coin}_{M_0}(F, h(t, \cdot)) = \varnothing$ for all $t \in [0,1]$.

Now we come to the point why we consider acyclic* function triples: For acyclic* function triples, we can prove the existence (and uniqueness up to homotopy) of admissible simplifiers.

In our applications for the degree, one should in the above definitions have the following case in mind: Ω is an open set in a manifold, $M \subseteq \Omega$ the intersection with a finite-dimensional submanifold, and Y_0 a finite-dimensional normed space with $F(M) \subseteq Y_0$. By modifying φ appropriately, we will be able to arrange that $\varphi(\Phi(M)) \subseteq Y_0$. In our application, the set $M_0 \subseteq M$ will be compact with $\operatorname{coin}_{M_0}(F, \Phi, \varphi) = \varnothing$; for instance, M_0 can be the boundary in M of a small neighborhood neighborhood containing $\operatorname{coin}_M(F, \Phi, \varphi)$ (but we will also need some other cases).

The following example shows that even in this setting, admissible simplifiers need not always exist for acyclic function triples:

Example 11.35. Let $Y_0 = Y = \mathbb{R}^4$, and $M_0 = M = \Omega = \{x_0\}$. Recall that by Example 4.58 there exists an acyclic metric compact space Γ such that there is a map $\varphi \in C(\Gamma, S^3)$ which fails to be homotopic to a constant map. We consider φ as a map from Γ into $Y_0 \setminus \{0\}$ and put $F(x) := 0$ and $\Phi(x) := \Gamma$ for all $x \in \Omega$ (that is, for $x = x_0$).

Then (F, Φ, φ) is an acyclic function triple in standard form and satisfies $\operatorname{coin}(F, \Phi, \varphi) = \varnothing$. However, (F, Φ, φ) has no (M_0, Y_0)-admissible simplifier on M. Indeed, otherwise there would exist a continuous map $h: [0,1] \times \Gamma \to Y_0$ with $h(0, \cdot) = \varphi$ and $\operatorname{coin}_{M_0}(F, \Phi, h(t, \cdot)) = \varnothing$ for all $t \in [0,1]$ such that $h(1, \cdot) = f \circ p$ for a map $f: M \to Y_0$. Since $M = \{x_0\}$, the latter means that $h(1, \cdot)$ is constant, and $\operatorname{coin}_{M_0}(F, \Phi, h(t, \cdot)) = \varnothing$ means that $0 \notin h([0,1] \times \Gamma)$. Hence, denoting by $\rho: Y_0 \setminus \{0\} \to S^3$ the radial retraction $\rho(x) = x/\|x\|$, we found that the map $H := \rho \circ h: [0,1] \times \Gamma \to S^3$ is continuous with $H(0, \cdot) = \varphi$ and constant $H(1, \cdot)$. We thus have shown that $\varphi \in C(\Gamma, S^3)$ is homotopic to a constant map which contradicts the choice of φ.

The map Φ in Example 11.35 is acyclic, but not acyclic*. The latter is not accidental. In fact, the following result implies for acyclic* function triples the existence of a (M_0, Y_0)-admissible simplifier for any compact subset $M_0 \subseteq M$

for the setting sketched above. Moreover, this result also implies the uniqueness of the simplifier (up to homotopies).

This result is the key property for the degree for function triples and will be used many times. Since it is so important, we try to formulate it as general as possible.

Recall that a topological group is a group endowed with a topology such that the group operation and the forming of the inverse are continuous maps.

Theorem 11.36 (Unique Existence of Simplifiers). *Let $(F, \Phi, \varphi, \Omega, Y, \Gamma)$ be an acyclic* function triple in standard form. Let $M_0 \subseteq M \subseteq \Omega$ and $\varphi(\Phi(M)) \subseteq Y_0 \subseteq Y$. Put $M_1 := M_0 \cap F^{-1}(Y_0)$, and suppose that the following holds:*

(a) $\operatorname{coin}_{M_1}(F, \Phi, \varphi) = \varnothing$.

(b) $\operatorname{Ind} M_1 < \infty$.

(c) *Φ is finite-dimensional acyclic or M_1 is T_6.*

(d) *There is $y_0 \in Y_0$ such that $Y_0 \setminus \{y_0\}$ is paracompact and homotopy equivalent to an* ANR.

(e) *Y_0 is an* AE *for $[0, 1] \times \Phi(M)$ and for $[0, 1] \times M$.*

(f) *Y_0 is homeomorphic to a topological Hausdorff group.*

(g) *M_1 is compact.*

Then there exists a (M_0, Y_0)-admissible simplifier f for (F, Φ, φ) on M. Moreover, every such simplifier g is (F, M_0, Y_0)-homotopic to f on M.

Remark 11.37. Theorem 11.36 holds even when we replace (f) by the following weaker hypothesis:

(f) Y_0 is Hausdorff, and there is $H_{y_0} \in C(F(M_1) \times Y_0, Y_0)$ with $H_{y_0}(y, y) = y_0$ for every $y \in F(M_1)$ such that $H_{y_0}(y, \cdot)$ is a homeomorphism on Y_0 for every $y \in F(M_1)$ and such that its inverse $h_{y_0}(y, \cdot)$ satisfies $h_{y_0} \in C(F(M_1) \times Y_0, Y_0)$.

If g is a homeomorphism of Y_0 onto a Hausdorff topological group, this condition is satisfied with $H_{y_0}(y, z) := g^{-1}(g(y_0) \cdot g(y) \cdot g(z)^{-1})$.

Remark 11.38. (AC). (Recall Remark 5.26).

Theorem 11.36 and Remark 11.37 hold also when we replace (g) by the following hypotheses:

(g) (1) M_1 and $\Phi(M_1)$ are paracompact,

(2) M_1 is closed in M, and $\Phi(M_1)$ is closed in $\Phi(M)$.

(3) At least one of the following holds:

   (i) $\Phi(M_1)$ is compact.
   (ii) $Y_0 \setminus \{y_0\}$ is homotopy equivalent to a compact ANR.
   (iii) $\Phi(x)$ is UV^∞ for every $x \in M_1$.

Theorem 11.36 is indeed a special case: If M_1 is compact then $\Phi(M_1)$ is compact by Proposition 2.100, and so M_1 and $\Phi(M_1)$ are paracompact and closed in every Hausdorff space; hence the above assumptions (1)–(3) are automatically satisfied for compact M_1.

Remark 11.39. If we are only interested in the existence assertion of Theorem 11.36 (or Remarks 11.37 or 11.38), that is, in the existence of an (M_0, Y_0)-admissible simplifier, we may relax the requirement that Y_0 be an AE for $[0,1]\times M$ to the requirement that Y_0 be an AE for M.

Proof of Theorem 11.36 *and of Remarks* 11.37, 11.38, *and* 11.39. We put $p = \Phi^{-1}$. For $z \in \Gamma_0 := \Phi(M_1)$, we have $p(z) \in M_1$ and thus $F(p(z)) \in Y_0$ and hence, we can define

$$G(z) := H_{y_0}(F(p(z)), \varphi(z)). \tag{11.15}$$

For $z \in \Gamma_0$, we have $x := p(z) \in M_1$ and $z \in p^{-1}(x)$. The hypothesis $\operatorname{coin}_{M_0}(F, \Phi, \varphi) = \varnothing$ thus implies $F(x) \neq \varphi(z)$, and so $G(z) \neq y_0$. Hence, $G: \Gamma_0 \to Z := Y_0 \setminus \{y_0\}$.

Note that $p|_{\Gamma_0}: \Gamma_0 \to M_1$ is Vietoris* by Proposition 11.18 (even UV^∞-Vietoris if $\Phi(x) = p^{-1}(x)$ is UV^∞ for every $x \in M_1$), and that M_1 and Γ_0 are paracompact. The assumptions (also in Remark 11.38) are such that we can apply the homotopic version of the Vietoris theorem (Theorem 5.25) with the map $p|_{\Gamma_0}$ and the space Z. By this result, the map $p|_{\Gamma_0}$ thus induces a bijection between the homotopy classes of $[M_1, Z]$ and of $[\Gamma_0, Z]$. In particular, there is $f_0 \in C(M_1, Z)$ such that $G \in C(\Gamma_0, Z)$ is homotopic to $f_0 \circ p|_{\Gamma_0} \in C(\Gamma_0, Z)$, and moreover, if $f_1 \in C(M_1, Z)$ is another map with this property then f_0 and f_1 are homotopic to each other.

Let $h_0: [0,1]\times\Gamma_0 \to Z$ be a homotopy connecting $h_0(0,\cdot) = G$ with $h_0(1,\cdot) = f_0 \circ p|_{\Gamma_0}$. We define $H_0: [0,1] \times \Gamma_0 \to Z$ by

$$H_0(t,z) := h_{y_0}(F(p(z)), h_0(t,z)).$$

Then $H_0(0,z) = h_{y_0}(F(p(z)), G(z)) = \varphi(z)$ by (11.15), and $H_0(1,\cdot) = f \circ p|_{\Gamma_0}$ with $f: M_1 \to Y_0$ being defined by

$$f(x) := h_{y_0}(F(x), f_0(x)). \tag{11.16}$$

For $(t, z) \in [0,1] \times \Gamma_0$, we have $h_0(t,z) \neq y_0$ and thus $H_0(t,z) \neq F(p(z))$. In particular, if $t \in [0,1]$ and $x \in M_1$, we have for any $z \in \Phi(x)$ in view of $z \in \Gamma_0$ and $p(z) = x$ that $H_0(t,z) \neq F(x)$. Hence, $F(x) \notin H_0(t, \Phi(x))$. We thus have $\mathrm{coin}_{M_1}(F, \Phi, H_0(t,\,\cdot\,)) = \varnothing$.

Since Y_0 is an AE for M and $M_1 \subseteq M$ is closed, we can extend f to $f \in C(M, Y_0)$ $(i = 0, 1)$. Recall that Y_0 is an AE for $[0,1] \times \Phi(M)$, and that $\Gamma_0 \subseteq \Phi(M)$ is closed. By the both-sided homotopy extension theorem (Theorem 4.45), we can extend H_0 to a homotopy $H_0\colon [0,1] \times \Phi(M) \to Y_0$ such that $H_0(0,\,\cdot\,) = \varphi|_{\Phi(M)}$ and $H_0(1,\,\cdot\,) = f \circ p|_{\Phi(M)}$. Then H_0 is the homotopy showing that f is a (M_0, Y_0)-admissible simplifier of (F, Φ, φ) on M. Indeed, since H_0 assumes its values in Y_0, we have

$$\mathrm{coin}_{M_0}(F, \Phi, H_0(t,\,\cdot\,)) = \mathrm{coin}_{M_1}(F, \Phi, H_0(t,\,\cdot\,)) = \varnothing.$$

To prove the assertion about the uniqueness, let $g\colon M \to Y_0$ be a further (M_0, Y_0)-admissible simplifier on M, that is, there is a continuous map $H_1\colon [0,1] \times \Phi(M) \to Y_0$ with $H_1(0,\,\cdot\,) = \varphi$, $H_1(1,\,\cdot\,) = g \circ p|_M$, and such that $\mathrm{coin}_{M_0}(F, \Phi, H_1(t,\,\cdot\,)) = \varnothing$ for all $t \in [0,1]$. We define $h_1\colon [0,1] \times \Gamma_0 \to Y_0$ by

$$h_1(t,z) := H_{y_0}(F(p(z)), H_1(t,z)).$$

If $t \in [0,1]$ and $z \in \Gamma_0$, we have $z \in \Phi(x)$ for some $x \in M_1$. Then $p(z) = x$, and $\mathrm{coin}_{M_0}(F, \Phi, H_1(t,\,\cdot\,)) = \varnothing$ implies $H_1(t,z) \neq F(x) = F(p(z))$. It follows that we have actually $h_1(t,z) \neq y_0$, that is, $h_1\colon [0,1] \times \Gamma_0 \to Z$. Hence, h_1 is a homotopy connecting $h_1(0,\,\cdot\,) = G\colon \Gamma_0 \to Z$ with the map $h_1(1,\,\cdot\,) = f_1 \circ p|_{\Gamma_0}\colon \Gamma_0 \to Z$ where $f_1\colon M_1 \to Y$ is defined by

$$f_1(x) := H_{y_0}(F(x), g(x)). \tag{11.17}$$

Since $p(\Gamma_0) = M_1$ and $h_1(\{1\} \times \Gamma_0) \subseteq Z$, we have actually $f_1\colon M_1 \to Z$. By what we had already observed from the homotopic version of the Vietoris theorem, it follows that $f_0, f_1\colon M_1 \to Z$ are homotopic, that is, there is a continuous map $h_2\colon [0,1] \times M_1 \to Z$ with $h_2(i,\,\cdot\,) = f_i$ for $i = 0, 1$. Now we define $H_2\colon [0,1] \times M_1 \to Y_0$ by

$$H_2(t,x) := h_{y_0}(F(x), h_2(t,x)).$$

For $i = 0, 1$, we have $H_2(i,x) = h_{y_0}(F(x), f_i(x))$, and so (11.16) and (11.17), imply $H_2(0,\,\cdot\,) = f|_{M_1}$, $H_2(1,\,\cdot\,) = g|_{M_1}$. Since M_1 is closed and Y_0 is an AE for $[0,1] \times M$, we can use the both-sided homotopy extension theorem (Theorem 4.45) to extend H_2 to a homotopy $H_2\colon [0,1] \times M \to Y_0$ satisfying $H_2(0,\,\cdot\,) = f$ and $H_2(1,\,\cdot\,) = g$. Then H_2 shows that f and g are (F, M_0, Y_0)-homotopic on

M. Indeed, assume by contradiction that there are $(t, x) \in [0,1] \times M_0$ satisfying $F(x) = H_2(t, x) \in Y_0$. Then $x \in F^{-1}(Y_0)$, and so $x \in M_1$. We obtain

$$F(x) = H_2(t, x) = h_{y_0}(F(x), h_2(t, x))$$

and thus $h_2(t, x) = H_{y_0}(F(x), F(x)) = y_0$, contradicting the fact that h_2 assumes its values in Z. □

Theorem 11.36 will be our key tool to reduce the multivalued case to a setting with single-valued maps concerning degree theory. However, essentially the same idea works not only for degree theory but also for even finer and more sophisticated tools from homotopy theory, see e.g. [64], [65].

11.3 Homotopies of Triples

Now we intend to define generalized homotopies between function triples.

Definition 11.40. A function triple (G, H, h) or, more verbosely, $(G, H, h, W, X, Y, \Gamma)$ is called a *generalized homotopy triple* for the family $(G_t, H_t, h_t, W_t, Y, \Gamma_t)$ $(t \in I)$ if the following holds:

(a) I, X, and Γ are Hausdorff spaces, Y is a topological space.

(b) $W \subseteq I \times X$, $W_t = \{x \in X : (t, x) \in W\}$.

(c) $\Gamma_t \subseteq \Gamma$ $(t \in I)$. We equip $\widetilde{\Gamma} := \bigcup_{t \in I}(\{t\} \times \Gamma_t)$ with the topology inherited from the product topology of $I \times \Gamma$.

(d) $G\colon W \to Y$, $\operatorname{graph}(G)$ is closed in $W \times Y$, $G_t = G(t, \cdot)\colon W_t \to Y$.

(e) $H\colon W \multimap \Gamma$ is upper semicontinuous with compact values $H(t, x)$, and $H_t = H(t, \cdot)\colon \Omega \multimap \Gamma_t$.

(f) $h \in C(\widetilde{\Gamma}, Y)$, $h_t = h(t, \cdot)\colon \Gamma_t \to Y$.

If additionally H is acyclic or acyclic*, we call the generalized homotopy triple *acyclic* or *acyclic**. If

$$\bigcup_{t \in I}(\{t\} \times \operatorname{coin}(G_t, H_t, h_t)) = \{(t, x) \in W : G_t(x) \in h_t(H_t(x))\} \tag{11.18}$$

is compact in W, we call the generalized homotopy triple *proper*.

These are indeed homotopies for closed function triples. Moreover, in a canonical manner, we can interpret the homotopy triples as closed function triples.

Proposition 11.41. *We put in the above setting* $\widetilde{H}(t,x) := \{t\} \times H(t,x)$. *Then* $(G,\widetilde{H},h,W,Y,\widetilde{\Gamma})$ *is a closed function triple which is acyclic, acyclic*, or proper, respectively. The set* (11.18) *is* $\mathrm{coin}(G,\widetilde{H},h)$ *and closed in* W. *The family* (G_t,H_t,h_t) $(t \in I)$ *consists automatically of closed function triples which are acyclic, acyclic*, or proper, respectively. Moreover,*

$$(G(t,\cdot),\widetilde{H}(t,\cdot),h|_{\{t\}\times\Gamma_t},W_t,Y,\{t\}\times\Gamma_t) \approx (G_t,H_t,h_t,W_t,Y,\Gamma_t). \quad (11.19)$$

Proof. The upper semicontinuity of $\widetilde{H}$ follows from Proposition 2.99. The upper semicontinuity of H_t follows from Proposition 2.90. Since $\widetilde{H}(t,x)$ is homeomorphic to $H(t,x)$ for every $(t,x) \in W$, we find that $\widetilde{H}$ is acyclic or acyclic* if and only if H has this property. Since $(G,\widetilde{H},h)$ is closed, Proposition 2.29 yields that $\mathrm{coin}(G,\widetilde{H},h)$ is closed in W. If this set is compact, the compactness of $\mathrm{coin}(G_t,H_t,h_t)$ follows from Proposition 2.62. The relation (11.19) follows by considering the canonical homeomorphism $J\colon \{t\}\times\Gamma_t \to \Gamma_t$, $J(t,x) := x$ in Definition 11.7. □

Proposition 11.42. *Let* $(G,\widetilde{H},h,W,Y,\widetilde{\Gamma})$ *be as in Proposition* 11.41. *Denote the corresponding standard form by* $(G,\hat{H},\hat{h},W,Y,\hat{\Gamma})$, *and the standard form of* (G_t,H_t,h_t) *by* $A_t := (G_t,\widetilde{H}_t,\widetilde{h}_t,W_t,Y,\widetilde{\Gamma}_t)$. *Then* $\widetilde{\hat{H}}(t,x) = \{t\}\times\hat{H}(t,x)$ *where* $(G,\hat{H},\hat{h})$ *is a generalized homotopy triple for a family*

$$(G_t,\hat{H}_t,\hat{h}_t,W_t,Y,\hat{\Gamma}_t) \approx A_t \qquad (t \in I).$$

Proof. We have

$$\hat{\Gamma} = \mathrm{graph}(\widetilde{H}) = \{(t,x,t,z) : (t,x) \in W, z \in H(t,x)\},$$

$\widetilde{\hat{H}}(t,x) = \{(t,x,t)\}\times H(t,x)$, and $\hat{h}(t,x,t,z) = h(t,z)$. Hence, $\hat{H}(t,x) = \{(x,t)\}\times H(t,x)$. We find that we can choose $\hat{H}_t(x) := \{(x,t)\}\times H(t,x)$,

$$\hat{\Gamma}_t := \{(x,t,z) : x \in W_t, z \in H(t,x)\},$$

$\hat{h}_t(x,t,z) := h(t,z)$. On the other hand, we have

$$\widetilde{\Gamma}_t = \{(x,z) : x \in W_t, z \in H(t,x)\},$$

$\widetilde{H}_t(t,x) = \{x\}\times H(t,x)$, and $h_t(x,z) = h(t,z)$. Comparison shows that we can choose the canonical homeomorphism $J_t\colon \hat{\Gamma}_t \to \widetilde{\Gamma}_t$, $J_t(x,t,z) := (x,z)$, in Definition 11.7. □

Definition 11.40 is probably what the reader might intuitively expect about a generalized homotopy of function triples: Roughly speaking, it should be a generalized homotopy in each of the three functions. However, in order to interpret the homotopy triple itself as a function triple, we had to introducc an auxiliary map, and with this auxiliary map the natural correspondence (11.19) "at time t" holds only after a canonical identification. This is another reason why it is important for us that the degree does not change under such identifications.

Although Definition 11.40 might appear rather natural, it might also appear unnecessarily restrictive: Although it is natural to require that W is a subset of a product space, it is perhaps not so natural to expect the same for $\widetilde{\Gamma} \subseteq I \times \Gamma$. In fact, for the admissible pairs in [70], the co-triads in [93], and also in the general approach from [140], another notion of homotopies is used where an arbitrary Hausdorff space $\widetilde{\Gamma}$ is used. If we also drop the requirement that W be a subset of a product space, this corresponds to the following definition.

Definition 11.43. A function triple (G, H, h) or, more verbosely, (G, H, h, W, Y, Γ) is called an *abstract homotopy triple* for the family $(G_t, H_t, h_t, W_t, Y, \Gamma_t)$ $(t \in I)$ if the following holds:

(a) X and Γ are Hausdorff spaces Y is a topological space.

(b) $W_t \subseteq W$ is closed in W for each $t \in I$.

(c) $\Gamma_t \subseteq \Gamma$ $(t \in I)$.

(d) $G\colon W \to Y$, $\operatorname{graph}(G)$ is closed in $W \times Y$, $G_t = G|_{W_t}\colon W_t \to Y$.

(e) $H\colon W \multimap \Gamma$ is upper semicontinuous with compact values $H(x)$, and $H_t := H|_{W_t}\colon \Omega \multimap \Gamma_t$.

(f) $h \in C(\Gamma, Y)$, $h_t = h|_{\Gamma_t}\colon \Gamma_t \to Y$.

If additionally H is acyclic or acyclic*, we call the abstract homotopy triple *acyclic* or *acyclic**. If $\operatorname{coin}(G, H, h)$ is relatively compact in W, we call the abstract homotopy triple *proper*.

Proposition 11.44. *In the above setting, the triple (G, H, h, W, Y, Γ) is a closed function triple which is acyclic, acyclic*, or proper, respectively. The family $(G_t, H_t, h_t, W_t, Y, \Gamma_t)$ $(t \in I)$ consists automatically of closed function triples which are acyclic, acyclic*, or proper, respectively.*

Proof. The first assertion is an immediate consequence of the definitions and Proposition 11.6. The upper semicontinuity of H_t follows from Proposition 2.90. Finally, if (G, H, h) is proper then $\operatorname{coin}(G_t, H_t, h_t) = W_t \cap \operatorname{coin}(G, H, h)$ is compact by Proposition 2.29, since W_t is closed. □

It may be somewhat surprising that in the most important case $W \subseteq I \times X$ (with $W_t = \{x : (t,x) \in W\}$) we actually do *not* obtain more general results by considering abstract homotopies instead of generalized homotopies.

In fact, both notions are equivalent in the sense that to each generalized homotopy triple, we can canonically associate an abstract homotopy triple and vice versa, corresponding to the same families up to some natural identifications:

Proposition 11.45. *Let I and X be Hausdorff spaces, $W \subseteq I \times X$, $W_t = \{x : (t,x) \in W\}$, and let $J_t \colon W_t \to \{t\} \times W_t$ be the canonical homeomorphism $J_t(x) := (t,x)$.*

(a) *Whenever (G,H,h,W,Y,Γ) is a generalized homotopy triple for the family $(G_t,H_t,h_t,W_t,Y,\Gamma_t)$ $(t \in I)$, then $(G,\widetilde{H},h,W,Y,\widetilde{\Gamma})$ with $\widetilde{H}$ and $\widetilde{\Gamma}$ as in Proposition 11.41 is an abstract homotopy triple for the family*

$$(G|_{\{t\}\times W_t}, \widetilde{H}|_{\{t\}\times W_t}, h|_{\{t\}\times\Gamma_t}, \{t\}\times W_t, Y, \{t\}\times\Gamma_t) \\ \approx (G_t \circ J_t, H_t \circ J_t, h_t, W_t, Y, \Gamma_t).$$

Moreover, (G,H,h,W,Y,Γ) is acyclic, acyclic, or proper if and only if $(G,\widetilde{H},h,W,Y,\widetilde{\Gamma})$ has this property.*

(b) *Whenever (G,H,h,W,Y,Γ) is an abstract homotopy triple for the family $(G_t,H_t,h_t,\{t\}\times W_t,Y,\Gamma_t)$ $(t \in I)$ then $(G,H,\widetilde{h},W,Y,\widetilde{\Gamma})$ with*

$$\widetilde{\Gamma} := \bigcup_{t\in I}(\{t\}\times\Gamma_t) \subseteq I \times \Gamma \quad \text{and} \quad \widetilde{h}(t,z) := h(z)$$

is a generalized homotopy triple for the family

$$(G(t,\cdot\,), H(t,\cdot\,), h_t, W_t, Y, \Gamma_t) = (G_t \circ J_t^{-1}, H_t \circ J_t^{-1}, h_t, \{t\}\times W_t, Y, \Gamma_t)$$

Moreover, (G,H,h,W,Y,Γ) is acyclic, acyclic, or proper if and only if $(G,H,\widetilde{h},W,Y,\widetilde{\Gamma})$ has this property.*

Proof. The assertions follow straightforwardly from the definitions or Proposition 11.41. □

Proposition 11.45 states that, roughly speaking, in case $W \subseteq I \times X$ it is just a matter of taste and notation whether one considers abstract homotopy triples or generalized homotopy triples. Typically, the former is more convenient for the theory while the latter is more convenient for applications.

However, for the case that W is not a subset of a space of the form $I \times X$ (like e.g. when $I = \{0,1\}$ and we want to prove a bordism invariance of the degree) it is not even possible to formulate the corresponding assertions by using generalized homotopy triples: In such a case it cannot be avoided to work with abstract homotopy triples.

Abstract homotopy triples satisfy an analogue of Proposition 11.42. It is not even necessary to consider an equivalence relation for the corresponding statement.

Proposition 11.46. *Let (G, H, h) be an abstract homotopy triple for the family $(G_t, H_t, h_t, W_t, Y, \Gamma_t)$ $(t \in I)$. Then the standard form of (G, H, h) is an abstract homotopy triple for the standard forms of (G_t, H_t, h_t) $(t \in I)$.*

Proof. Let $(G, \widetilde{H}, \widetilde{h}, W, Y, \Gamma)$ and $(G_t, \widetilde{H}_t, \widetilde{h}_t, W, Y, \widetilde{\Gamma}_t)$ denote the corresponding standard forms. We have $\Gamma = \operatorname{graph}(H) = \{(x,z) : z \in H(y)\}$ and $\widetilde{\Gamma}_t = \operatorname{graph}(H_t) = \operatorname{graph}(H) \cap (W_t \times Y)$. Hence, $\widetilde{H}(x) = \{x\} \times H(x)$ and $\widetilde{h}(x,z) = h(z)$ while $\widetilde{H}_t(x) = \{x\} \times H(x)$ and $\widetilde{h}_t(x,z) = h(z)$. It follows that $\widetilde{H}_t = \widetilde{H}|_{W_t}$ and $\widetilde{h}_t = \widetilde{h}|_{\widetilde{\Gamma}_t}$. □

11.4 Locally Normal Triples

The following class of function triples will play a particular role in Chapter 14 for the case $I = [0, 1]$.

Definition 11.47. Let I be a Hausdorff space. We call a function triple $(F, \Phi, \varphi, \Omega, Y, \Gamma)$ *locally I-normal* (or *locally normal*) if there is an open neighborhood $\Omega_0 \subseteq \Omega$ of $\operatorname{coin}_\Omega(F, \Phi, \varphi)$ such that for any open subset $\Omega_1 \subseteq \Omega_0$ and any compact $C \subseteq \Phi(\Omega_1)$ there is a neighborhood Z_0 of C in $\Phi(\Omega_1)$ such that $I \times Z_0$ (or Z_0) is normal.

We recall that even in case $I = [0, 1]$, it is slightly more restrictive to require that $I \times Z_0$ is normal than to require that Z_0 is normal, see Corollary 2.76 and the subsequent remarks.

Clearly, $(F, \Phi, \varphi, \Omega, Y, \Gamma)$ is locally normal if Γ is completely normal. However, in most cases it suffices that Γ is normal, and actually even less suffices as we will show now.

Note that if (F, Φ, φ) is in standard form, that is, $\Phi = p^{-1}$ with a continuous map p then Φ sends open sets onto open sets. The following result implies in particular that in such a situation, it suffices that $I \times \Gamma$ is normal.

Proposition 11.48. *Let I be a Hausdorff space. Then $(F, \Phi, \varphi, \Omega, Y, \Gamma)$ is locally I-normal (or locally normal) if one of the following holds:*

(a) $\operatorname{coin}_{\Omega}(F, \Phi, \varphi) = \varnothing$.

(b) *There is some open neighborhood $\Omega_0 \subseteq \Omega$ of $\operatorname{coin}_{\Omega}(F, \Phi, \varphi)$ such that for any $Z \subseteq \Phi(\Omega_0)$ the space $I \times Z$ (or Z) is normal. This holds in particular, if $I \times \Phi(\Omega_0)$ (or $\Phi(\Omega_0)$) is completely normal.*

(c) *There is some open neighborhood $\Omega_0 \subseteq \Omega$ of $\operatorname{coin}_{\Omega}(F, \Phi, \varphi)$ such that $\Phi(\Omega_0)$ is contained in a Hausdorff space Z such that $I \times Z$ (or Z) is T_4, and for any open $\Omega_1 \subseteq \Omega_0$ the set $\Phi(\Omega_1)$ is open in Z.*

Proof. In case (a), we choose $\Omega_0 = \varnothing$ in Definition 11.47.

Let $\Omega_1 \subseteq \Omega_0$ be open, and $C \subseteq \Phi(\Omega_1)$ be compact. In case (b), we put $Z_0 := Z := \Phi(\Omega_1)$ and obtain by hypothesis that $I \times Z_0$ is normal.

Suppose now that $\Phi(\Omega_0)$ is contained in a Hausdorff space Z such that $I \times Z$ is T_4, hence normal, and $\Phi(\Omega_1)$ is open in Z. Note that Z is regular by Theorem 2.36, and so Corollary 2.48 implies that the compact set $C \subseteq \Phi(\Omega_1)$ has a closed neighborhood $Z_0 \subseteq Z$ which is contained in the open set $\Phi(\Omega_1)$. Since $I \times Z_0$ is a closed subset of the normal space $I \times Z$, Proposition 2.44 implies that $I \times Z_0$ is normal. □

Chapter 12

The Degree for Finite-Dimensional Fredholm Triples

12.1 The Triple Variant of the Brouwer Degree

Throughout this section, let X be a C^1 Banach manifold without boundary over a real normed space $E = E_X \neq \{0\}$, and let $Y = E_Y$ be a real normed space with $0 < \dim E_X = \dim E_Y < \infty$.

Definition 12.1. We write $(F, \Phi, \varphi, \Omega) \in \mathcal{T}_{\mathrm{Br}}(X, Y)$ if $\Omega \subseteq X$ is open, $F \in C(\Omega, Y)$, and if (F, Φ, φ) is a proper acyclic* function triple.

Note that Corollary 2.117 implies automatically that $\operatorname{graph}(F)$ is closed.

The purpose of this section is to define a degree on the class $\mathcal{T}_{\mathrm{Br}}(X, Y)$ which "counts" $\operatorname{coin}_\Omega(F, \Phi, \varphi)$.

According to Remark 11.14 and in view of the subsequent weak equivalence invariance of the Brouwer degree, we can omit the set Γ from the notation of function triples (although the topology on $\Phi(\Omega)$ inherited from Γ will be considered given). We also omit Y from the notation.

The degree will assume its values in $\mathbb{Z}_2$. If we want to obtain a degree which assumes its values in $\mathbb{Z}$, we require in addition that F is oriented in the sense of Definition 9.16. As in the discussion of the Brouwer and Benevieri–Furi degrees, we will notationally not distinguish between these cases, although formally we should write the couple (F, σ) (where σ denotes an orientation of F) instead of F in the oriented case.

Definition 12.2. The *Brouwer triple degree* is an operator $\deg = \deg_{(X,Y)}$ which associates to each $(F, \Phi, \varphi, \Omega) \in \mathcal{T}_{\mathrm{Br}}(X, Y)$ a number from $\mathbb{Z}_2$ (or from $\mathbb{Z}$ in the oriented case) such that the following holds for each $(F, \Phi, \varphi, \Omega) \in \mathcal{T}_{\mathrm{Br}}(X, Y)$:

($A_{\mathcal{T}_{\mathrm{Br}}}$) **(Homotopy Invariance in the Last Function).** If $h \in C([0,1] \times \Phi(\Omega), Y)$ is such that

$$\operatorname{coin}_{[0,1]\times\Omega}(F, \Phi, h) = \{(t, x) \in [0,1] \times \Omega : F(x) \in h(t, \Phi(x))\} \tag{12.1}$$

is compact then

$$\deg(F, \Phi, h(0, \cdot), \Omega) = \deg(F, \Phi, h(1, \cdot), \Omega). \tag{12.2}$$

($B_{\mathcal{T}_{\mathrm{Br}}}$) **(Normalization).** If $\Phi = \mathrm{id}_\Omega$ then

$$\deg(F, \mathrm{id}_\Omega, \varphi, \Omega) = \deg(F - \varphi, \Omega, 0),$$

where the right-hand side denotes the Brouwer degree. The orientation of $F - \varphi$ for the oriented case is described below.

($C_{\mathcal{T}_{\mathrm{Br}}}$) **(Excision).** If $\Omega_0 \subseteq \Omega$ is open and contains $\mathrm{coin}(F, \Phi, \varphi)$ then

$$\deg(F, \Phi, \varphi, \Omega_0) = \deg(F, \Phi, \varphi, \Omega).$$

($D_{\mathcal{T}_{\mathrm{Br}}}$) **(Weak Equivalence Invariance).** If $(F, \widetilde{\Phi}, \widetilde{\varphi})$ is an acyclic* function triple then

$$(F, \widetilde{\Phi}, \widetilde{\varphi}) \precsim (F, \Phi, \varphi) \implies \deg(F, \widetilde{\Phi}, \widetilde{\varphi}, \Omega) = \deg(F, \Phi, \varphi, \Omega).$$

Note for the weak equivalence invariance that $(F, \widetilde{\Phi}, \widetilde{\varphi}) \in \mathcal{T}_{\mathrm{Br}}(X, Y)$ is automatic by (11.8).

In the oriented case, the orientation of $F - \varphi$ in the normalization property is defined as follows. We fix some orientation of Y. The orientation of Y and of F induces an orientation on Ω in the sense of Proposition 9.34. The orientations on Ω and Y in turn induce the required orientation on $F - \varphi \in C(\Omega, Y)$. This orientation is well-defined, since for the opposite orientation of Y, we obtain the opposite orientation of Ω and thus the same orientation of $F - \varphi$.

Remark 12.3. The notation (12.1) coincides with our previous definition of coin, if we consider (F, Φ, h) as a function triple, that is, if we consider F and Φ as functions defined on $[0, 1] \times \Omega$ (although they are actually independent of the first argument). We will tacitly make such identifications from now on when we use the notation coin or fix.

Theorem 12.4. *For each fixed (X, Y) there is exactly one Brouwer triple degree* $\deg = \deg_{(X,Y)}$. *Its restriction to triples in standard form is uniquely determined by the corresponding restriction of the properties of Definition* 12.2 *to triples in standard form.*

Moreover, the Brouwer triple degree satisfies the following properties for all $(F, \Phi, \varphi, \Omega) \in \mathcal{T}_{\mathrm{Br}}(X, Y)$:

($E_{\mathcal{T}_{\mathrm{Br}}}$) **(Equivalence Invariance).** *With $\mathcal{T} = \mathcal{T}_{\mathrm{Br}}(X, Y)$, we have*

$$(F, \Phi, \varphi) \sim_{\mathcal{T}} (F, \widetilde{\Phi}, \widetilde{\varphi}) \implies \deg(F, \Phi, \varphi, \Omega) = \deg(F, \widetilde{\Phi}, \widetilde{\varphi}, \Omega).$$

($F_{\mathcal{T}_{\mathrm{Br}}}$) **(Single-Valued Normalization).** *If $\varphi \circ \Phi$ is single-valued then*

$$\deg(F, \Phi, \varphi, \Omega) = \deg(F, \mathrm{id}_\Omega, \varphi \circ \Phi, \Omega) = \deg(F - (\varphi \circ \Phi), \Omega, 0), \tag{12.3}$$

where the right-hand side denotes the Brouwer degree. The orientation of $F - (\varphi \circ \Phi)$ in the oriented case is analogous to the normalization property.

($G_{\mathcal{T}_{\mathrm{Br}}}$) **(Compatibility with the Non-oriented Case).** *The degrees for the oriented and non-oriented case are the same modulo 2 (if the oriented case applies).*

Proof. Let $\mathcal{T}_0$ denote the class of all those $(F, \Phi, \varphi, \Omega) \in \mathcal{T}_{\mathrm{Br}}(X, Y)$ which are in standard form. We prove the assertions first for the restriction of the degree to $\mathcal{T}_0$. By Proposition 9.1, there is an open neighborhood $\Omega_0 \subseteq \Omega$ of $K := \mathrm{coin}(F, \Phi, \varphi)$ with compact $M := \overline{\Omega_0} \subseteq \Omega$. Then $M_0 := \partial\Omega_0 = \overline{\Omega_0} \setminus \Omega_0$ is compact, and $\mathrm{coin}_{M_0}(F, \Phi, \varphi) = \varnothing$. Moreover, M_0 is metrizable (and thus T_6) and satisfies $\dim M_0 \le \mathrm{Ind}\, M_0 < \infty$ by Proposition 9.1. Hence, by Theorem 11.36, there is an (M_0, Y)-admissible simplifier f for (F, Φ, φ) on M.

To prove the uniqueness of the degree, we recall that this means that there is $H \in C([0,1] \times \Phi(M), Y)$ with $H(0, \cdot) = \varphi$ and $\mathrm{coin}_{M_0}(F, \Phi, H(t, \cdot)) = \varnothing$ for all $t \in [0,1]$ such that f is a simplifier for $(F, \Phi, H(1, \cdot))$ on M.

Using the convention of Remark 12.3, we note that Proposition 11.6 implies that $\mathrm{coin}_{[0,1]\times\overline{\Omega_0}}(F, \Phi, H)$ is a closed hence compact subset of $[0,1] \times \overline{\Omega_0}$ and actually contained in $[0,1] \times \Omega_0$. Hence, we can calculate

$$\begin{aligned}\deg(F, \Phi, \varphi, \Omega) &= \deg(F, \Phi, \varphi, \Omega_0)\\ &= \deg(F, \Phi, H(0, \cdot), \Omega_0) = \deg(F, \Phi, H(1, \cdot), \Omega_0)\\ &= \deg(F, \mathrm{id}_{\Omega_0}, f, \Omega_0) = \deg(F - f, \Omega_0, 0).\end{aligned}$$

For the fourth inequality, we have used Proposition 11.32. This proves the uniqueness, and also the compatibility with the non-oriented case follows from the corresponding property of the Brouwer degree.

To prove the existence, we define the degree in the above situation by

$$\deg(F, \Phi, \varphi, \Omega) := \deg(F - f, \Omega_0, 0), \tag{12.4}$$

where f is an $(\partial\Omega_0, Y)$-admissible simplifier for (F, Φ, φ) on $\overline{\Omega_0}$. We already know that such an f exists, and we have to show that this definition is independent of the particular choice of Ω_0 and $f_0 := f$. Thus, let Ω_1 and f_1 be different such choices. We have to show that

$$\deg(F - f_0, \Omega_0, 0) = \deg(F - f_1, \Omega_1, 0). \tag{12.5}$$

We assume first $\Omega_0 = \Omega_1$. In this case, we apply the uniqueness assertion of Theorem 11.36 with $M := \overline{\Omega_0}$ and $M_0 := \partial\Omega_0$. We find that there is a continuous $h_0\colon [0,1] \times M \to Y$ with $h_0(0,\cdot) = f_0$, $h_0(1,\cdot) = f_1$, and $\operatorname{coin}_{M_0}(F, h_0(t,\cdot)) = \varnothing$ for all $t \in [0,1]$. It follows that the closed and thus compact set $(F - h_0)^{-1}(0) \subseteq [0,1] \times M$ is even contained in $\Omega_0 = \Omega_1$. Hence, (12.5) follows in case $\Omega_0 = \Omega_1$ by the homotopy invariance of the Brouwer degree.

In case $\Omega_0 \neq \Omega_1$, we put $\Omega_2 := \Omega_0 \cap \Omega_1$. For $i = 0, 1$, we apply the existence assertion of Theorem 11.36 with $M := \overline{\Omega_i}$ and $M_0 := M \setminus \Omega_2$ to find that there is an (M_0, Y)-admissible simplifier g_i for (F, Φ, φ) on Ω_i. Applying the special case of the uniqueness which we have just shown, and using the excision property of the Brouwer degree, we obtain that

$$\deg(F - f_i, \Omega_i, 0) = \deg(F - g_i, \Omega_i, 0) = \deg(F - g_i, \Omega_2, 0)$$

for $i = 0, 1$. Now we note that g_0 and g_1 are both $(\partial\Omega_2, Y)$-admissible simplifiers for (F, Φ, φ) on $\overline{\Omega_2}$. Applying once more the special case of the uniqueness which we have just shown, we obtain

$$\deg(F - g_0, \Omega_2, 0) = \deg(F - g_1, \Omega_2, 0),$$

and (12.5) follows by combining the above equalities.

It is clear from the definition that the degree (12.4) satisfies the excision property. If $f = \varphi \circ \Phi$ is single-valued then it is continuous by Propositions 2.80 and 2.94. Proposition 11.31 implies that f is a simplifier, and the homotopy $H(t,z) = \varphi(z)$ shows that it is a (M_0, Y)-admissible simplifier. Hence, we have

$$\deg(F, \Phi, \varphi, \Omega) = \deg(F - (\varphi \circ \Phi), \Omega, 0).$$

This implies in particular the normalization property, and applying the normalization property with $\varphi \circ \Phi$ in place of φ, we obtain the remaining equality in (12.3).

We show now that deg also satisfies the homotopy invariance in the last function. Thus, let $h\colon [0,1] \times \Omega \to Y$ be an oriented homotopy with compact

$$K := \{(t,x) \in [0,1] \times \Omega : F(x) \in h(t, \Phi(x))\}.$$

By Corollary 9.3, there is an open set $\Omega_0 \subseteq \Omega$ with $[0,1] \times K \subseteq \Omega_0$ and compact $\overline{\Omega_0} \subseteq \Omega$. We find by our definition of the degree a $(\partial\Omega_0, Y)$-admissible simplifier f for $(F, \Phi, h(1,\cdot))$ on $\overline{\Omega_0}$. In particular, there is a continuous function $H\colon [0,1] \times \Phi(M) \to Y$ with $H(0,\cdot) = h(1,\cdot)$ and $\operatorname{coin}_{M_0}(F, \Phi, H(t,\cdot)) = \varnothing$ for all $t \in [0,1]$ such that f_1 is a simplifier for $(F, \Phi, H(1,\cdot))$ on $\overline{\Omega_0}$. We consider the concatenated homotopy

$$H_0(t,x) := \begin{cases} h(2t, x) & \text{if } 2t \le 1, \\ H(2t-1, x) & \text{if } 2t \ge 1, \end{cases}$$

which is continuous by the glueing lemma (Lemma 2.93). Then H_0 proves that f_1 is an $(\partial\Omega_0, Y)$-admissible simplifier f for $(F, \Phi, h(1, \cdot))$ on $\overline{\Omega}_0$. By the definition of the degree, we have thus shown that

$$\deg(F, \Phi, h(i, \cdot), \Omega) = \deg(F - f, \Omega_0, 0)$$

for $i = 0, 1$, and so the homotopy invariance in the last function is established.

We prove now the equivalence invariance (with $\mathcal{T}$ replaced by $\mathcal{T}_0$). We can assume that

$$(F, \Phi, \varphi, \Omega, Y, \Gamma) \sqsubseteq (F, \widetilde{\Phi}, \widetilde{\varphi}, \Omega, Y, \widetilde{\Gamma}),$$

since the general case follows by induction. By our hypothesis, (F, Φ, φ) and $(F, \widetilde{\Phi}, \widetilde{\varphi})$ are in standard form, say $\Phi = p^{-1}$ and $\widetilde{\Phi} = \widetilde{p}^{-1}$. We obtain from Proposition 11.19 that there is a continuous map $J\colon \Gamma \to \widetilde{\Gamma}$ with

$$\widetilde{p} \circ J = p \quad \text{and} \quad \widetilde{\varphi} \circ J = \varphi.$$

Let $\Omega_0 \subseteq X$ be an open neighborhood of the compact set $\operatorname{coin}(F, \widetilde{\Phi}, \widetilde{\varphi})$ with compact $\overline{\Omega}_0 \subseteq \Omega$. By (11.6), we also have $\operatorname{coin}(F, \Phi, \varphi) \subseteq \Omega_0$. Let f be a $(\partial\Omega_0, Y)$-admissible simplifier for $(F, \widetilde{\Phi}, \widetilde{\varphi})$ on $\overline{\Omega}_0$. This means that there is $H \in C([0, 1] \times \widetilde{\Phi}(\overline{\Omega}_0), Y)$ with $H(0, \cdot) = \widetilde{\varphi}$ and $\operatorname{coin}_{\partial\Omega_0}(F, \widetilde{\Phi}, H(t, \cdot)) = \varnothing$ for all $t \in [0, 1]$ such that $H(1, \cdot) = f \circ \widetilde{p}$. Then the continuous function $H_0\colon [0, 1] \times \Phi(\overline{\Omega}_0) \to Y$, $H_0(t, z) := H(t, J(z))$, has the property that $H_0(0, \cdot) = \varphi$, and by (11.3), we have $H_0(t, \Phi(x)) \subseteq H(t, \widetilde{\Phi}(x))$ and thus $\operatorname{coin}_{\partial\Omega_0}(F, \Phi, H_0(t, \cdot)) = \varnothing$ for all $t \in [0, 1]$. Finally, $H_0(1, \cdot) = f \circ \widetilde{p} \circ J = f \circ p$. Consequently, H_0 proves that f is a $(\partial\Omega_0, Y)$-admissible simplifier for (F, Φ, φ) on $\overline{\Omega}_0$. We obtain by the definition of the degree that

$$\deg(F, \Phi, \varphi, \Omega) = \deg(F - f, \Omega_0, 0) = \deg(F, \widetilde{\Phi}, \widetilde{\varphi}, \Omega),$$

and the equivalence invariance for the class $\mathcal{T}_0$ is established.

Finally, we pass to the general case. Given $(F, \Phi, \varphi, \Omega) \in \mathcal{T}_{\mathrm{Br}}(X, Y)$, we let $(F, \widetilde{\Phi}, \widetilde{\varphi})$ denote the corresponding standard form. Theorem 11.23 implies that $(F, \widetilde{\Phi}, \widetilde{\varphi}) \in \mathcal{T}_{\mathrm{Br}}(X, Y)$ is in standard form and

$$(F, \widetilde{\Phi}, \widetilde{\varphi}) \precsim (F, \Phi, \varphi).$$

The uniqueness of the degree thus follows from the weak equivalence invariance and from the uniqueness of the degree for triples from $\mathcal{T}_0$. To define the degree, we put

$$\deg(F, \Phi, \varphi, \Omega) := \deg(F, \widetilde{\Phi}, \widetilde{\varphi}, \Omega), \tag{12.6}$$

recalling that we have already proved that the latter is well-defined. Since

$$\operatorname{coin}(F, \Phi, \varphi) = \operatorname{coin}(F, \widetilde{\Phi}, \widetilde{\varphi}) \tag{12.7}$$

by Proposition 11.8, we obtain the excision property from the excision property for $\mathcal{T}_0$ and from Proposition 11.25. Similarly, we obtain the compatibility of the oriented and non-oriented cases and the (single-valued) normalization properties from the corresponding property for $\mathcal{T}_0$. To see the equivalence invariance, we note that Theorem 11.28 implies that if two triples are $\mathcal{T}$-equivalent then their standard forms are $\mathcal{T}_0$-equivalent. Hence, also the equivalence invariance follows from the corresponding property for $\mathcal{T}_0$. It remains to prove the homotopy invariance in the last function. Recall that the definition of the standard form means that the standard form of $(F, \Phi, h(t,\cdot), \Omega, Y, \Gamma)$ has the form $(F, \widetilde{\Phi}, \widetilde{h}(t,\cdot), \Omega, Y, \tilde{\Gamma})$ with $\tilde{\Gamma} := \operatorname{graph}(\Phi)$, $\widetilde{\Phi}(x,z) := x$, and $\widetilde{h}(t,(x,z)) := h(t,z)$. In particular, $\widetilde{h}$ is continuous. By (12.7), we have

$$
\begin{aligned}
&\{(t,x) \in [0,1] \times \Omega : F(x) \in \widetilde{h}(t, \widetilde{\Phi}(x))\} \\
&\quad = \bigcup_{t\in[0,1]} (\{t\} \times \operatorname{coin}(F, \widetilde{\Phi}, \widetilde{h}(t,\cdot))) = \bigcup_{t\in[0,1]} (\{t\} \times \operatorname{coin}(F, \Phi, h(t,\cdot))) \\
&\quad = \{(t,x) \in [0,1] \times \Omega : F(x) \in h(t, \Phi(x))\}.
\end{aligned}
$$

Hence, if this set is compact, the homotopy invariance of the degree for triples in standard form implies

$$
\deg(F, \widetilde{\Phi}, \widetilde{h}(0,\cdot), \Omega) = \deg(F, \widetilde{\Phi}, \widetilde{h}(1,\cdot), \Omega).
$$

By the definition (12.6) of the degree, this equality means (12.2). □

Remark 12.5. The above uniqueness proof holds even if one fixes F (and its orientation) and requires the properties of Definition 12.2 only for that fixed F (and its restrictions).

Remark 12.6. In view of Remark 9.41, we emphasize that Theorem 12.4 and the subsequent properties of the Brouwer triple degree hold also if X and Y are oriented and if for the map F only the correspondingly induced orientations are considered.

In fact, the existence (and further properties) of such a degree are clear, since it is a special case of the degree we consider above. Only whether the uniqueness is guaranteed by its properties may be unclear, but concerning uniqueness, we have the even stronger result of Remark 12.5.

Remark 12.7. One could also define a C^r Brouwer triple degree ($1 \le r \le \infty$) where only maps F of class C^r are considered (one has to assume that X is of class C^r in this case).

In fact, the existence (and further properties) of such a degree are special cases of the setting with continuous F, and the uniqueness of such a degree follows from Remark 12.5.

Of course, the degree is actually just the restriction of the degree which we had considered above. However, for maps $F \in C^1(\Omega, Y) = \mathcal{F}_0(\Omega, Y)$ we have a different notion of orientation (Definition 8.25) which is compatible with the orientation which we used above (Definition 9.16) in the sense the orientations can be transformed into each other by means of (9.2) (and Proposition 9.18). However, the notions are not the same, so we must make precise:

When we speak about the C^1 *Brouwer triple degree*, we mean the orientation in the sense of Definition 8.25. In this case, in the normalization property, we also mean the corresponding C^1 Brouwer degree with the orientation in the sense of Definition 8.25 (cf. Remark 9.73).

From the compatibility of both notions of orientations (Proposition 9.18), it follows that there is exactly one C^1 Brouwer triple degree which satisfies the same properties as the Brouwer triple degree.

The subsequent proof of the additivity is typical for all proofs of properties of the Brouwer triple degree: Essentially, one has to repeat the above uniqueness proof carefully, since it instructs how to actually "calculate" the degree (by means of the ordinary Brouwer degree).

Theorem 12.8. *The Brouwer triple degree has the following properties for each* $(F, \Phi, \varphi, \Omega) \in \mathcal{T}_{\mathrm{Br}}(X, Y)$.

($\mathrm{H}_{\mathcal{T}_{\mathrm{Br}}}$) **(Additivity).** *If* $\Omega = \Omega_1 \cup \Omega_2$ *with disjoint open subsets* $\Omega_i \subseteq \Omega$ *then*

$$\deg(F, \Phi, \varphi, \Omega) = \deg(F, \Phi, \varphi, \Omega_1) + \deg(F, \Phi, \varphi, \Omega_2).$$

($\mathrm{I}_{\mathcal{T}_{\mathrm{Br}}}$) **(Existence).** *If* $\deg(F, \Phi, \varphi, \Omega) \neq 0$ *then* $\operatorname{coin}_\Omega(F, \Phi, \varphi) \neq \varnothing$.

($\mathrm{J}_{\mathcal{T}_{\mathrm{Br}}}$) **(Excision-Additivity).** *If* $\Omega_i \subseteq \Omega$ $(i \in I)$ *is a family of pairwise disjoint open sets with* $\operatorname{coin}_\Omega(F, \Phi, \varphi) \subseteq \bigcup_{i \in I} \Omega_i$ *and* $\operatorname{coin}_{\Omega_i}(F, \Phi, \varphi)$ *is compact for all* $i \in I$, *then*

$$\deg(F, \Phi, \varphi, \Omega) = \sum_{i \in I} \deg(F, \Phi, \varphi, \Omega_i),$$

where in the sum at most a finite number of summands is nonzero.

Proof. To prove the additivity, we recall that the (weak) equivalence invariance and Theorem 11.23 imply that the degree of a triple is that of its standard form. By

Proposition 11.25 the standard form of $(F, \Phi, \varphi, \Omega_i)$ $(i = 1, 2)$ is the restriction of the standard form of $(F, \Phi, \varphi, \Omega)$. We have to show the additivity for these standard forms. Hence, without loss of generality, we may assume that $(F, \Phi, \varphi, \Omega)$ is in standard form, say $\Phi = p^{-1}$. By Proposition 9.1, there is an open neighborhood $\Omega_0 \subseteq \Omega$ of $K := \operatorname{coin}(F, \Phi, \varphi)$ with compact $M := \overline{\Omega_0} \subseteq \Omega$. Then $M_0 := \partial\Omega_0 = \overline{\Omega_0} \setminus \Omega_0 \subseteq \Omega$ is compact, and $\operatorname{coin}_{M_0}(F, \Phi, \varphi) = \varnothing$. By Theorem 11.36, there is an (M_0, Y)-admissible simplifier f for (F, Φ, φ) on M. Hence, there is a continuous function $H: [0, 1] \times \Phi(M) \to Y$ with $H(0, \cdot) = \varphi$ and $\operatorname{coin}_{M_0}(F, \Phi, H(t, \cdot)) = \varnothing$ for all $t \in [0, 1]$. With the convention of Remark 12.3, Proposition 11.6 implies that $K_0 := \operatorname{coin}_{[0,1]\times M}(F, \Phi, H)$ is closed in $[0, 1]\times M$ and thus compact. By construction of H, we have $K_0 \subseteq [0, 1]\times\Omega_0$, and by Proposition 2.29 also the closed subsets $K_i := K_0 \setminus ([0, 1] \times \Omega_{3-i})(i = 1, 2)$ are compact. Using the excision property, the homotopy invariance in the last function, Proposition 11.32, the (weak) equivalence invariance, and the normalization property, we calculate

$$\begin{aligned}\deg(F, \Phi, \varphi, \Omega) &= \deg(F, \Phi, \varphi, \Omega_0) = \deg(F, \Phi, H(1, \cdot), \Omega_0)\\ &= \deg(F, \mathrm{id}_{\Omega_0}, f, \Omega_0) = \deg(F - f, \Omega_0, 0).\end{aligned}$$

Carrying out the same calculation with $\Omega_i \cap \Omega_0$ in place of Ω (recall the compactness of K_i), we obtain also

$$\deg(F, \Phi, \varphi, \Omega_i) = \deg(F - f, \Omega_i, 0) \quad \text{for } i = 1, 2.$$

Now the additivity follows from the additivity of the Brouwer degree.

The existence and excision-additivity properties follow from the excision and additivity by the same arguments that we had used for the Brouwer degree. □

Now we come to the deeper properties of the Brouwer triple degree.

The Brouwer triple degree would not be very useful if it would not satisfy a much stronger homotopy invariance property than that from Definition 12.2.

Roughly speaking, the idea of the proof of the generalized homotopy invariance which we prove now is to find a simplifier simultaneously for every time t. This is possible by applying Theorem 11.36 to an auxiliary triple $(G, \widetilde{H}, h)$. The crucial point is that the simplifier f one obtains in this way is then actually continuous in both variables (t, x).

Theorem 12.9. *The Brouwer triple degree has the following property:*

($K_{\mathcal{T}_{\mathrm{Br}}}$) **(Generalized Homotopy Invariance).** *Let (G, H, h, W, Y, Γ) be a generalized proper acyclic* homotopy triple for $(G_t, H_t, h_t, W_t, Y, \Gamma_t)$ $(t \in [0, 1])$ where $W \subseteq [0, 1] \times X$ is open, and $G \in C(W, Y)$ a generalized* *(oriented) homotopy. Then* $(G_t, H_t, h_t, W_t) \in \mathcal{T}_{\mathrm{Br}}(X, Y)$ *for all* $t \in [0,1]$, *and*

$$\deg(G_t, H_t, h_t, W_t) \text{ is independent of } t \in [0,1].$$

Proof. Proposition 11.41 implies the assertion $(G_t, H_t, h_t, W_t) \in \mathcal{T}_{\mathrm{Br}}(X, Y)$. Moreover, with $\widetilde{H}(t,x) := \{t\} \times H(t,x)$ as in Proposition 11.41, we have that $(G, \widetilde{H}, h)$ is a proper acyclic* homotopy triple. We can assume in view of Proposition 11.42 and the (weak) equivalence invariance without loss of generality that $(G, \widetilde{H}, h)$ is in standard form.

By Proposition 9.1, there is an open neighborhood $U \subseteq W$ of the set $K := \operatorname{coin}(G, H, h)$ with compact $M := \overline{U} \subseteq W$. Then $M_0 := \partial U = \overline{U} \setminus U \subseteq W$ is compact, and $\operatorname{coin}_{M_0}(G, H, h) = \varnothing$. By Theorem 11.36, there is an (M_0, Y)-admissible simplifier f for (G, H, h) on M. Hence, there is a continuous function $\hat{h}\colon [0,1] \times H(M) \to Y$ with $\hat{h}(0, \cdot) = h$ and $\operatorname{coin}_{M_0}(G, H, \hat{h}(s, \cdot)) = \varnothing$ for all $s \in [0,1]$ such that f is a simplifier for $(G, H, \hat{h}(1, \cdot))$. With the convention of Remark 12.3, Proposition 11.6 implies that $K_0 := \operatorname{coin}_{[0,1]\times M}(G, H, \hat{h})$ is closed in $[0,1] \times M$ and thus compact. By construction of H, we have $K_0 \subseteq [0,1] \times U$. Using the excision property, the homotopy invariance in the last function, Proposition 11.32, the (weak) equivalence invariance and the normalization property, we calculate with $U_t := \{x : (t,x) \in U\}$ that

$$\begin{aligned}\deg(G_t, H_t, h_t, W_t) &= \deg(G_t, H_t, \hat{h}(0, t, \cdot), U_t) = \deg(G_t, H_t, \hat{h}(1, t, \cdot), U_t)\\ &= \deg(G_t, \mathrm{id}_{U_t}, f(t, \cdot), U_t) = \deg(G(t, \cdot) - f(t, \cdot), U_t, 0).\end{aligned}$$

Note also that by Proposition 11.32 the set $K_1 := M \cap (G - f)^{-1}(0) = \operatorname{coin}_M(G, H, \hat{h}(1, \cdot))$ is disjoint from M and thus contained in U. Since K_1 is a closed subset of M and thus compact, we obtain by the generalized homotopy invariance of the Brouwer degree that the last degree is independent of $t \in [0,1]$. □

The generalized homotopy invariance can even be extended to a bordism invariance for abstract homotopy triples.

Theorem 12.10. *The Brouwer triple degree has the following property:*

($L_{\mathcal{T}_{\mathrm{Br}}}$) **(Bordism Invariance for C^1 Manifolds).** *Let* W *be a* C^1 *manifold over* $\mathbb{R} \times E_X$ *with boundary* ∂W. *Let* (G, H, h, W, Y, Γ) *be a proper acyclic* function triple such that* $C := \operatorname{coin}(G, H, h)$ *is compact.* $\Omega_0, \Omega_1 \subseteq \partial W$ *be open in* ∂W *and disjoint with* $C \cap \partial W \subseteq \Omega_0 \cup \Omega_1$. *Then we have for* $i = 0, 1$ *with* $G_i := G|_{\Omega_i}$, $H_i := H|_{\Omega_i}$, $h_i := h|_{H(\Omega_i)}$ *that* $(G_i, H_i, h_i, \Omega_i) \in \mathcal{T}_{\mathrm{Br}}(\Omega_i, Y)$, *and*

$$\deg_{(\Omega_0, Y)}(G_0, H_0, h_0, \Omega_0) = \deg_{(\Omega_1, Y)}(G_1, H_1, h_1, \Omega_1).$$

It is admissible that $\Omega_i = \varnothing$ *in which case the corresponding degree in this formula is considered as* 0. *The orientation of* G_i *in the oriented case is described below.*

In the oriented case, it is assumed that W and Y be oriented, and ∂W be oriented as the boundary as described in Definition 9.62. The orientation of G_0 is then assumed to be that induced by the orientations of Y and $\Omega_0 \subseteq \partial W$ in the sense of Proposition 9.34, while the orientation of G_1 is assumed to be the opposite of the orientation induced by the orientations of Y and $\Omega_1 \subseteq \partial W$.

Proof. By Proposition 9.1, we find for $i = 0, 1$ that there is an open in Ω_i neighborhood $U_i \subseteq \Omega_i$ of the compact set $C \cap \Omega_i$ such that $W_i := \overline{U}_i$ is a compact subset of Ω_i. Putting $\Gamma_i := H_i(\Omega_i)$, we obtain that (G, H, h, W, Y, Γ) is an abstract homotopy triple for the family $(G_i, H_i, h_i, W_i, Y, \Gamma_i)$ $(i \in \{0, 1\})$. By the excision and restriction property, we are to show that

$$\deg_{(U_0,Y)}(G_0, H_0, h_0, U_0) = \deg_{(U_1,Y)}(G_1, H_1, h_1, U_1). \tag{12.8}$$

Now we pass to the standard form of (G, H, h). According to Proposition 11.46, this corresponds to passing to the standard forms of (G_i, H_i, h_i, U_i), and so the (weak) equivalence invariance of the degree implies that none of the above degrees changes. Hence, without loss of generality, we can assume that (G, H, h) is in standard form.

By Proposition 9.1, there is an open neighborhood $U \subseteq W$ of $C \cup W_0 \cup W_1$ with compact $M := \overline{U} \subseteq W$. Then $M_0 := \partial U = \overline{U} \setminus U \subseteq W$ is compact, and $\operatorname{coin}_{M_0}(G, H, h) = \varnothing$. By Theorem 11.36, there is an (M_0, Y)-admissible simplifier f for (G, H, h) on M. Hence, there is a continuous function $\hat{h}\colon [0, 1] \times H(M) \to Y$ with $\hat{h}(0, \cdot) = h$ and $\operatorname{coin}_{M_0}(G, H, \hat{h}(t, \cdot)) = \varnothing$ for all $t \in [0, 1]$ such that f is a simplifier for $(G, H, \hat{h}(1, \cdot))$. With the convention of Remark 12.3, Proposition 11.6 implies that $C_0 := \operatorname{coin}_{[0,1]\times M}(G, H, \hat{h})$ is closed in $[0, 1] \times M$ and thus compact. By construction of H, we have $C_0 \subseteq [0, 1] \times U$. Using the homotopy invariance in the last function, Proposition 11.32, the (weak) equivalence invariance, and the normalization property, we calculate for $i = 0, 1$ with $f_i := f|_{U_i}$ that

$$\begin{aligned}\deg_{(U_i,Y)}(G_i, H_i, h_i, U_i) &= \deg_{(U_i,Y)}(G_i, H_i, \hat{h}(1, \cdot), U_i)\\ &= \deg_{(U_i,Y)}(G_i, \mathrm{id}_{U_i}, f_i, U_i) = \deg_{(U_i,Y)}(G_i - f_i, U_i, 0).\end{aligned}$$

Note now that $G_i - f_i = (G - f)|_{U_i}$ where $G - f \in C(M, Y)$. Moreover, by Proposition 11.32 the set $M \cap (G - f)^{-1}(0) = \operatorname{coin}_M(G, H, \hat{h}(1, \cdot))$ is compact and disjoint from M_0 and thus contained in U. Applying the bordism invariance of the Brouwer degree, we find that the last degree is actually independent of $i \in \{0, 1\}$ which implies (12.8). □

Proposition 12.11. *The Brouwer triple degree has the following properties for each* $(F, \Phi, \varphi, \Omega) \in \mathcal{T}_{\mathrm{Br}}(X, Y)$.

($\mathrm{M}_{\mathcal{T}_{\mathrm{Br}}}$) **(Topological-Isomorphic Invariance).** *Let J_1 be a homeomorphism of an open subset of a Banach manifold X_0 onto Ω, and J_2 an isomorphism of Y onto a real topological vector space Y_0. Then*

$$\deg_{(X,Y)}(F, \Phi, \varphi, \Omega) = \deg_{(X_0,Y_0)}(J_2 \circ F \circ J_1, \Phi \circ J_1, J_2 \circ \varphi, J_1^{-1}(\Omega)).$$

In the oriented case, J_1 and J_2 are assumed to be oriented, and the orientation of $J_2 \circ F \circ J_1$ is the composite orientation of Corollary 9.30.

($\mathrm{N}_{\mathcal{T}_{\mathrm{Br}}}$) **(Diffeomorphic-Isomorphic Invariance).** *Let J_1 be a diffeomorphism of an open subset of a Banach manifold X_0 onto Ω and J_2 an isomorphism of Y onto a real normed vector space Y_0. Then*

$$\deg_{(X,Y)}(F, \Phi, \varphi, \Omega) = \deg_{(X_0,Y_0)}(J_2 \circ F \circ J_1, \Phi \circ J_1, J_2 \circ \varphi, J_1^{-1}(\Omega)).$$

In the oriented case, the orientation of $J_2 \circ F \circ J_1$ is understood in the sense of Corollary 9.31.

($\mathrm{O}_{\mathcal{T}_{\mathrm{Br}}}$) **(Restriction).** *Let $X_0 \subseteq X$ be open. Then*

$$\deg_{(X_0,Y)} = \deg_{(X,Y)}|_{\mathcal{T}_{\mathrm{Br}}(X_0,Y)}. \tag{12.9}$$

Proof. For ($\mathrm{O}_{\mathcal{T}_{\mathrm{Br}}}$), we note that the right-hand side of (12.9) is a map which has all properties required in Definition 12.2 when X is replaced by X_0: Concerning the normalization property, this follows from the restriction property of the Brouwer degree. The uniqueness of the Brouwer triple degree thus implies (12.9).

Now we prove ($\mathrm{M}_{\mathcal{T}_{\mathrm{Br}}}$). To this end, we consider J_1, J_2 and $\Omega \subseteq X$ as fixed, and we define a map d for $(F, \Phi, \varphi, \Omega_0) \in \mathcal{T}_{\mathrm{Br}}(\Omega, Y)$ by

$$d(F, \Phi, \varphi, \Omega_0) := \deg_{(X_0,Y_0)}(J_2 \circ F \circ J_1, \Phi \circ J_1, J_2 \circ \varphi, J_1^{-1}(\Omega_0)).$$

A straightforward calculation shows that d satisfies all properties of Definition 12.2 (when (X, Ω) is replaced by (Ω, Ω_0)). This is nontrivial only concerning the normalization property. For that property, we note that

$$(J_2 \circ F \circ J_1, \mathrm{id}_{\Omega_0} \circ J_1, J_2 \circ \varphi) \approx (J_2 \circ F \circ J_1, J_1^{-1} \circ \mathrm{id}_{\Omega_0} \circ J_1, J_2 \circ \varphi \circ J_1),$$

and so we find by the (weak) equivalence invariance and normalization properties of $\deg_{(X_0,Y_0)}$ that

$$\begin{aligned} d(F, \mathrm{id}_{\Omega_0}, \varphi) &= \deg_{(X_0,Y_0)}(J_2 \circ F \circ J_1, \mathrm{id}_{J_1^{-1}(\Omega_0)}, J_2 \circ \varphi \circ J_1, J_1^{-1}(\Omega_0)) \\ &= \deg_{(X_0,Y_0)}(J_2 \circ F \circ J_1 - J_2 \circ \varphi \circ J_1, J_1^{-1}(\Omega_0), 0) \\ &= \deg_{(X_0,Y_0)}(J_2 \circ (F - \varphi) \circ J_1, J_1^{-1}(\Omega_0), 0) \\ &= \deg_{(\Omega,Y)}(F - \varphi, \Omega_0, 0). \end{aligned}$$

Here, we have used that J_2 is linear and for the last equality the homeomorphic invariance of the Brouwer degree.

By the uniqueness of the Brouwer triple degree, we conclude finally that $d = \deg_{(\Omega,Y)}$. Using the restriction property of the Brouwer triple degree which we had already proved, we obtain that

$$d(F, \Phi, \varphi, \Omega) = \deg_{(X_0,Y_0)}(F, \Phi, \varphi, \Omega),$$

and so the topological-isomorphic invariance ($\mathrm{M}_{\mathcal{T}_{\mathrm{Br}}}$) of the Brouwer triple degree is established. The diffeomorphic-isomorphic invariance ($\mathrm{N}_{\mathcal{T}_{\mathrm{Br}}}$) is only the special case of ($\mathrm{M}_{\mathcal{T}_{\mathrm{Br}}}$) when the diffeomorphism J_1 and the isomorphism J_2 carry the natural orientations. □

In order to extend the Brouwer triple degree to an infinite-dimensional setting, we will make essential use of the following property.

Theorem 12.12. *The Brouwer triple degree has the following property for each* $(F, \Phi, \varphi, \Omega) \in \mathcal{T}_{\mathrm{Br}}(X, Y)$.

($\mathrm{P}_{\mathcal{T}_{\mathrm{Br}}}$) **(Reduction)**. *Suppose that there is an open neighborhood* $\Omega_0 \subseteq \Omega$ *of* $\mathrm{coin}_\Omega(F, \Phi, \varphi)$ *with* $F \in C^1(\Omega_0, Y)$ *and a linear subspace* $Y_0 \subseteq Y$ *with*

(a) $\varphi(\Phi(\Omega_0)) \subseteq Y_0$.
(b) Y_0 *is transversal to* F *on* Ω_0.

Then $X_0 := \Omega_0 \cap F^{-1}(Y_0)$ *is a* C^1 *submanifold of* X *of the same dimension as* Y_0, *and if this dimension is positive, we have with* $F_0 := F|_{X_0} \in C^1(X_0, Y_0)$ *that*

$$\deg_{(X,Y)}(F, \Phi, \varphi, \Omega) = \deg_{(X_0,Y_0)}(F_0, \Phi, \varphi, X_0).$$

If F *is oriented, the orientation of the restriction* F_0 *is the inherited orientation in the sense of Definition* 9.19.

Remark 12.13. To be formally correct, we should actually write on the right-hand side $\deg_{(X_0,Y_0)}(F_0, \Phi_0, \varphi_0, X_0)$ where Φ_0 and φ_0 denote corresponding restrictions of Φ and φ, respectively. However, we use here the same slightly sloppy notation that we had already used several times. Only for F_0 we made an exception to remind of the notation which we used for the inherited orientation.

Proof. Theorem 8.55 implies that X_0 is a manifold of the same dimension as Y_0. Using Theorem 11.23 and the (weak) equivalence invariance of the degree, we can assume that (F, Φ, φ) is in standard form. By Proposition 9.1, there is an open neighborhood $U \subseteq X$ of $K := \operatorname{coin}_\Omega(F, \Phi, \varphi)$ with compact $M := \overline{U} \subseteq \Omega$. Then also $M_0 := \partial U = M \setminus U$ and $M_1 := M_0 \cap X_0$ are compact. By Theorem 11.36, there is an (M_0, Y_0)-admissible simplifier f for (F, Φ, φ) on M. Hence, there is a homotopy $h\colon [0,1] \times \Phi(M) \to Y_0$ with $\operatorname{coin}_{M_0}(F, \Phi, h(t, \cdot)) = \varnothing$ for all $t \in [0,1]$ such that $h(0, \cdot) = \varphi$, and f is a simplifier for $(F, \Phi, h(1, \cdot))$. The excision property, homotopy invariance, Proposition 11.32, the (weak) equivalence invariance, and the normalization property now imply

$$\begin{aligned}\deg_{(X,Y)}(F, \Phi, \varphi, \Omega) &= \deg_{(X,Y)}(F, \Phi, \varphi, U) = \deg_{(X,Y)}(F, \Phi, h(1, \cdot), U)\\ &= \deg_{(X,Y)}(F, \operatorname{id}_U, f, U) = \deg_{(X,Y)}(F - f, U, 0).\end{aligned}$$

Since h and f assume only values from Y_0, we obtain analogously that

$$\begin{aligned}\deg_{(X_0,Y_0)}(F_0, \Phi, \varphi, X_0) &= \deg_{(X_0,Y_0)}(F_0, \Phi, h(1, \cdot), U \cap X_0)\\ &= \deg_{(X_0,Y_0)}(F_0, \operatorname{id}_{U \cap X_0}, f, U \cap X_0)\\ &= \deg_{(X_0,Y_0)}(F_0 - f, U \cap X_0, 0).\end{aligned}$$

Since f assumes only values from Y_0 and Y_0 is transversal to F on U, the equality of the above expressions now follows from the C^0 reduction property of the Brouwer degree (Theorem 10.1). □

Remark 12.14. Theorem 12.12 is the only result which used the full strength of Theorem 11.36: For all of our other applications of Theorem 11.36, the special case $Y_0 = Y$ in Theorem 11.36 was sufficient, but now we had to use that F need not necessarily assume its values in Y_0.

For completeness, we show now also that the Brouwer triple degree inherits the behaviour under Cartesian products from that of the Brouwer degree:

Proposition 12.15. *The Brouwer triple degree has the following property:*

($Q_{\mathcal{T}_{\mathrm{Br}}}$) **(Cartesian Product).** *For $i = 1, 2$, let X_i be a manifold without boundary of class C^1 over the real vector space E_{X_i}, and let $Y_i = E_{Y_i}$ be a real* *vector space with* $0 < \dim E_{X_i} = \dim E_{Y_i} < \infty$. *For* $(F_i, \Phi_i, \varphi_i, \Omega_i) \in \mathcal{T}_{\mathrm{Br}}(X_i, Y_i)$, *we put* $X := X_1 \times X_2$, $\Omega := \Omega_1 \times \Omega_2$, $Y := Y_1 \times Y_2$, $F := F_1 \otimes F_2$, $\Phi := \Phi_1 \otimes \Phi_2$, *and* $\varphi := \varphi_1 \otimes \varphi_2$. *Then* $(F, \Phi, \varphi, \Omega) \in \mathcal{T}_{\mathrm{Br}}(X, Y)$, *and*

$$\begin{aligned}&\deg_{(X,Y)}(F, \Phi, \varphi, \Omega)\\ &\quad = \deg_{(X_1,Y_1)}(F_1, \Phi_1, \varphi_1, \Omega_1) \deg_{(X_2,Y_2)}(F_2, \Phi_2, \varphi_2, \Omega_2).\end{aligned}$$

In the oriented case, F is equipped with the product orientation.

Proof. It is evident that the standard form corresponding to (F, Φ, φ) is the product of the corresponding standard forms of (F_i, Φ_i, φ_i) $(i = 1,2)$ in the same sense as above. Hence, using Theorem 11.23 and the (weak) equivalence invariance of the degree, we can assume that (F, Φ, φ) and (F_i, Φ_i, φ_i) $(i = 1,2)$ are in standard form.

By Proposition 9.1, we find for $i = 1,2$ open neighborhoods $U_i \subseteq X$ of $K_i := \mathrm{coin}_{\Omega_i}(F_i, \Phi_i, \varphi_i)$ with compact $\overline{U}_i \subseteq \Omega_i$. By Theorem 11.36, there are $(\partial U_i, Y_i)$-admissible simplifiers f_i for (F_i, Φ_i, φ_i) on $\overline{U}_i$. Hence, there are homotopies $h_i\colon [0,1] \times \Phi(M_i) \to Y_i$ with $\mathrm{coin}_{M_i}(F_i, \Phi_i, h_i(t, \cdot)) = \varnothing$ for all $t \in [0,1]$ such that $h_i(0, \cdot) = \varphi$ and f_i is a simplifier for $(F_i, \Phi_i, h_i(1, \cdot))$. The excision property, homotopy invariance, Proposition 11.32, the (weak) equivalence invariance, and the normalization property now imply on the one hand

$$\begin{aligned}\deg_{(X_i,Y_i)}(F_i, \Phi_i, \varphi_i, \Omega_i) &= \deg_{(X_i,Y_i)}(F_i, \Phi_i, \varphi_i, U_i)\\ &= \deg_{(X_i,Y_i)}(F_i, \Phi_i, h_i(1, \cdot), U_i)\\ &= \deg_{(X_i,Y_i)}(F_i, \mathrm{id}_{U_i}, f_i, U_i) = \deg_{(X_i,Y_i)}(F_i - f_i, U_i, 0)\end{aligned}$$

for $i = 1, 2$, and on the other hand, we obtain with $U := U_1 \times U_2$, $f := f_1 \otimes f_2$, and $h(t, z_1, z_1) := (h_1(t, z_1), h_2(t, z_2))$ analogously

$$\begin{aligned}\deg_{(X,Y)}(F, \Phi, \varphi, \Omega) &= \deg_{(X,Y)}(F, \Phi, \varphi, U) = \deg_{(X,Y)}(F, \Phi, h(1, \cdot), U)\\ &= \deg_{(X,Y)}(F, \mathrm{id}_U, f, U) = \deg_{(X,Y)}(F - f, U, 0).\end{aligned}$$

Hence, the assertion follows from the Cartesian product property of the Brouwer degree. □

12.2 The Triple Variant of the Benevieri–Furi Degree

Our aim in this section is to extend the Brouwer triple degree of Section 12.1 to the infinite-dimensional situation. Although the Brouwer degree is topological-isomorphic invariant, it is not completely obvious how the reduction to a finite-dimensional situation should be done, and in fact this problem was essentially only solved in [142]. The latter was our motivation to choose the reduction property as the key property for the definition of the Benevieri–Furi degree. Now we can actually use the more or less same idea also for function triples, and we will see by the reduction property of the Brouwer triple degree that this leads to a well-defined degree.

Throughout this section, let X be a C^1 Banach manifold without boundary over a real Banach space $E = E_X$, and let $Y = E_Y$ be a Banach space. We assume throughout the non-degeneracy hypothesis (10.8). Recall that with AC, the latter means just $E_X \neq \{0\}$ and $E_Y \neq \{0\}$.

Definition 12.16. We write $(F, \Phi, \varphi, \Omega) \in \mathcal{T}_{\mathrm{gen}}(X, Y)$ if $\Omega \subseteq X$ is open, $F\colon \Omega \to Y$, and if there is a Hausdorff space Γ with $\Phi\colon \Omega \multimap \Gamma$, $\varphi\colon \Gamma \to Y$, and if there is an open neighborhood $\Omega_0 \subseteq \Omega$ of

$$\operatorname{coin}(F, \Phi, \varphi) = \{x \in \Omega : F(x) \in \varphi(\Phi(x))\}$$

with the following properties:

(a) $F|_{\Omega_0} \in \mathcal{F}_0(\Omega_0, Y)$.

(b) $\Phi|_{\Omega_0}$ is acyclic*.

(c) $\varphi|_{\Phi(\Omega_0)}$ is continuous.

If additionally $\operatorname{coin}(F, \Phi, \varphi)$ is compact, we write $(F, \Phi, \varphi, \Omega) \in \mathcal{T}_{\mathrm{prop}}(X, Y)$.

If $(F, \Phi, \varphi, \Omega) \in \mathcal{T}_{\mathrm{prop}}(X, Y)$ and additionally $\varphi(\Phi(\Omega_0))$ is contained in a finite-dimensional linear subspace of Y, we write $(F, \Phi, \varphi, \Omega) \in \mathcal{T}_{\mathrm{BF}}(X, Y)$. The set $\Omega_0 \subseteq \Omega$ is called an *admissible neighborhood.*

The degree will assume its values in $\mathbb{Z}_2$. If we want to obtain a degree which assumes its values in $\mathbb{Z}$, we have to assume that $F|_{\Omega_0}$ is oriented on Ω_0 (for some admissible neighborhood), and we call this setting the oriented case.

Note that we require all hypotheses only in a neighborhood of $\operatorname{coin}(F, \Phi, \varphi)$. In applications, one usually knows more and thus can use the following more convenient test.

Proposition 12.17. *Let* $\Omega \subseteq X$ *be open,* $F \in \mathcal{F}_0(\Omega, Y)$, Γ *be a Hausdorff space,* $\Phi\colon \Omega \multimap \Gamma$ *be acyclic**, *and* $\varphi \in C(\Gamma, Y)$. *Then* $(F, \Phi, \varphi, \Omega) \in \mathcal{T}_{\mathrm{gen}}(X, Y)$ *is a closed function triple, and the following statements are equivalent:*

(a) $(F, \Phi, \varphi, \Omega)$ *belongs to* $\mathcal{T}_{\mathrm{prop}}(X, Y)$.

(b) $\operatorname{coin}(F, \Phi, \varphi)$ *is relatively compact in* Ω.

Proof. Since $F \in C(\Omega, Y)$ and Y is Hausdorff, $\operatorname{graph}(F)$ is closed by Corollary 2.117. Hence, the assertion follows from Proposition 11.6. □

Proposition 12.18. *Let* $(F, \Phi, \varphi, \Omega)$ *belong to* $\mathcal{T}_{\mathrm{gen}}(X, Y)$ $(\mathcal{T}_{\mathrm{prop}}(X, Y))$*, and* Ω_0 *be admissible. Then* (F, Φ, φ) *is a function triple, and the restriction of this triple to* Ω_0 *is a (proper) acyclic* function triple.*

Proof. $\mathrm{graph}(F|_{\Omega_0})$ is closed in $\Omega_0 \times Y$ by Corollary 2.117. □

Hence, after passing to appropriate subsets, we actually only deal with proper acyclic* function triples.

Definition 12.19. The *Benevieri-Furi triple degree* is an operator $\deg = \deg_{(X,Y)}$ which associates to each $(F, \Phi, \varphi, \Omega) \in \mathcal{T}_{\mathrm{BF}}(X, Y)$ a number from $\mathbb{Z}_2$ (or from $\mathbb{Z}$ in the oriented case) such that the following holds:

($\mathrm{A}_{\mathcal{T}_{\mathrm{BF}}}$) **(Reduction)**. Let $\Omega_0 \subseteq \Omega$ be an admissible open neighborhood of $\mathrm{coin}_\Omega(F, \Phi, \varphi)$, that is $F \in \mathcal{F}_0(\Omega_0, Y)$ (being oriented on Ω_0 in the oriented case), $\Phi|_{\Omega_0}$ is acyclic*, and $\varphi|_{\Phi(\Omega_0)}$ is continuous and assumes its values in a finite-dimensional subspace $Y_0 \subseteq Y$. If additionally Y_0 is transversal to F on Ω_0 then $X_0 := \Omega_0 \cap F^{-1}(Y_0)$ is a C^1 submanifold of X of the same dimension as Y_0, and if this dimension is positive, we have with $F_0 := F|_{X_0} \in \mathcal{F}_0(X_0, Y_0)$ that

$$\deg_{(X,Y)}(F, \Phi, \varphi, \Omega) = \deg_{(X_0,Y_0)}(F_0, \Phi, \varphi, X_0), \tag{12.10}$$

where the right-hand side denotes the C^1 Brouwer triple degree (cf. Remark 12.7). In case $X_0 = \varnothing$, the right-hand side is defined as zero. In the oriented case, the orientation of F_0 is the inherited orientation in the sense of Definition 8.65.

Theorem 12.20. *For fixed* X *and* Y *there is exactly one Benevieri–Furi triple degree. Moreover, this degree automatically has the following properties for any* $(F, \Phi, \varphi, \Omega) \in \mathcal{T}_{\mathrm{BF}}(X, Y)$*:*

($\mathrm{B}_{\mathcal{T}_{\mathrm{BF}}}$) **(Equivalence Invariance)**. *With* $\mathcal{T} = \mathcal{T}_{\mathrm{BF}}(X, Y)$*, we have*

$$(F, \Phi, \varphi) \sim_{\mathcal{T}} (F, \widetilde{\Phi}, \widetilde{\varphi}) \implies \deg(F, \Phi, \varphi, \Omega) = \deg(F, \widetilde{\Phi}, \widetilde{\varphi}, \Omega).$$

($\mathrm{C}_{\mathcal{T}_{\mathrm{BF}}}$) **(Compatibility with the Brouwer Triple Degree)**. *If* $\dim E_X = \dim E_Y < \infty$ *then* $\deg$ *is the* C^1 *Brouwer triple degree (cf. Remark* 12.7*).*

($\mathrm{D}_{\mathcal{T}_{\mathrm{BF}}}$) **(Compatibility with the Non–Oriented Case)**. *The degrees for the oriented and non-oriented case are the same modulo* 2 *(if the oriented case applies).*

($E_{\mathcal{T}_{\mathrm{BF}}}$) **(Excision).** *If* $\Omega_0 \subseteq \Omega$ *is open and contains* $\operatorname{coin}(F, \Phi, \varphi)$ *then*

$$\deg(F, \Phi, \varphi, \Omega_0) = \deg(F, \Phi, \varphi, \Omega).$$

Proof. The uniqueness follows from (12.10), if we can prove that for each $(F, \Phi, \varphi, \Omega) \in \mathcal{T}_{\mathrm{BF}}(X, Y)$ there actually *is* some Ω_0 and Y_0 as in the formulation of the reduction property, i.e. that there is an admissible neighborhood Ω_0 such that $\varphi(\Omega_0)$ is contained in a finite-dimensional subspace Y_0 which is transversal to F on Ω_0. However, the latter follows immediately from Proposition 8.70 in view of Remark 8.71.

For the existence, we choose such Ω_0 and Y_0 and define the degree by (12.10). We have to show that this is well-defined, that is, independent of the particular choice of Ω_0 and Y_0. Thus, let Ω_1 and Y_1 be possibly different choices. We find by Proposition 8.70 and Remark 8.71 a finite-dimensional subspace $Y_2 \subseteq Y$ containing $Y_0 + Y_1$ which is transversal to F on some open neighborhood $\Omega_2 \subseteq \Omega_0 \cap \Omega_1$ of $\operatorname{coin}_\Omega(F, \Phi, \varphi)$. Then in particular $Y_0, Y_1 \subseteq Y_2$. For $i = 0, 1, 2$, we put $X_i := \Omega_i \cap F^{-1}(Y_i)$, and let $F_i := F|_{X_i} \in \mathcal{F}_0(X_i, Y_i)$ with the inherited orientation. For $i = 0, 1$, we put $\hat{X}_i := \Omega_2 \cap X_i$ and calculate with the reduction property of the Brouwer triple degree (Theorem 12.12) together with the restriction and excision properties for the Brouwer triple degree

$$\deg_{(X_2,Y_2)}(F_2, \Phi, \varphi, X_2) = \deg_{(\hat{X}_i,Y_i)}(F_i, \Phi, \varphi, \hat{X}_i)$$
$$\deg_{(X_i,Y_i)}(F_i, \Phi, \varphi, \Omega_2 \cap X_i) = \deg_{(X_i,Y_i)}(F_i, \Phi, \varphi, X_i) \quad \text{for } i = 0, 1.$$

Since the left-hand side is independent of $i \in \{0, 1\}$, so must be the right-hand side, and so the Benevieri–Furi triple degree is well-defined.

The excision property follows from the very definition, since in the reduction property also any smaller admissible neighborhood Ω_0 can be chosen (and must give the same value for the Benevieri–Furi triple degree, since we know that it is well-defined). This definition via the reduction property also implies the compatibility with the non-oriented case by the corresponding property of the Brouwer triple degree. The compatibility with the Brouwer triple degree follows from the choice $Y_0 = Y$ in the reduction property. For the proof of the equivalence invariance, it suffices by induction to consider the case

$$(F, \Phi, \varphi) \sqsubseteq (F, \widetilde{\Phi}, \widetilde{\varphi}).$$

In this case, we choose the sets Ω_0 and Y_0 for the definition of the degree according to the triple $(F, \widetilde{\Phi}, \widetilde{\varphi})$. It follows that the same sets can be used in the definition of the degree for (F, Φ, φ), and then the equalities of the corresponding degrees follows from (12.10) and the equivalence invariance of the C^1 Brouwer triple degree (recall Proposition 11.16). □

Theorem 12.21. *The Benevieri–Furi triple degree has the subsequent properties for every* $(F, \Phi, \varphi, \Omega) \in \mathcal{T}_{\mathrm{BF}}(X, Y)$.

($F_{\mathcal{T}_{\mathrm{BF}}}$) **(Homotopy Invariance in the Last Function).** *Let* $\Omega \subseteq X$ *be open,* $F \in \mathcal{F}_0(\Omega, Y)$ *(oriented on* Ω*),* Γ *a Hausdorff space, and* $\Phi\colon \Omega \multimap \Gamma$ *acyclic*. If* $Y_0 \subseteq Y$ *is a finite-dimensional subspace and* $h \in C([0,1] \times \Phi(\Omega), Y_0)$ *is such that*

$$\operatorname{coin}_{[0,1]\times\Omega}(F, \Phi, h) = \{(t, x) \in [0,1] \times \Omega : F(x) \in h(t, \Phi(x))\}$$

is compact then

$$\deg(F, \Phi, h(0, \cdot), \Omega) = \deg(F, \Phi, h(1, \cdot), \Omega).$$

($G_{\mathcal{T}_{\mathrm{BF}}}$) **(Weak Equivalence Invariance).** *If* $(F, \widetilde{\Phi}, \widetilde{\varphi})$ *is an acyclic* function triple then*

$$(F, \widetilde{\Phi}, \widetilde{\varphi}) \precsim (F, \Phi, \varphi) \implies \deg(F, \widetilde{\Phi}, \widetilde{\varphi}, \Omega) = \deg(F, \Phi, \varphi, \Omega).$$

($H_{\mathcal{T}_{\mathrm{BF}}}$) **(Normalization).** *If* $\Phi = \mathrm{id}_\Omega$ *then*

$$\deg(F, \mathrm{id}_\Omega, \varphi, \Omega) = \deg(F, \varphi, \Omega),$$

where the right-hand side denotes the Benevieri–Furi coincidence degree.

($I_{\mathcal{T}_{\mathrm{BF}}}$) **(Single-Valued Normalization).** *If* $\varphi \circ \Phi$ *is single-valued then*

$$\deg(F, \Phi, \varphi, \Omega) = \deg(F, \mathrm{id}_\Omega, \varphi \circ \Phi, \Omega) = \deg(F, \varphi \circ \Phi, \Omega), \quad (12.11)$$

where the right-hand side denotes the Benevieri–Furi coincidence degree.

Proof. The weak equivalence invariance is in view of (11.8), a special case of the equivalence invariance, which has been proved in Theorem 12.20. Concerning ($F_{\mathcal{T}_{\mathrm{BF}}}$), we note that

$$K := \bigcup_{t\in[0,1]} \operatorname{coin}_\Omega(F, \Phi, h(t, \cdot)))$$

is compact by Corollary 2.101. Proposition 8.70 implies in view of Remark 8.71 that there is an open neighborhood $\Omega_0 \subseteq \Omega$ of K and a finite-dimensional subspace $Y_1 \subseteq Y$ containing Y_0 which is transversal to F on Ω_0. By the reduction property, we obtain with $X_0 := \Omega_0 \cap F^{-1}(Y_1)$ and $F_0 := F|_{X_0} \in \mathcal{F}_0(X_0, Y_1)$ that

$$\deg(F, \Phi, h(i, \cdot), \Omega) = \deg_{(X_0,Y_1)}(F_0, \Phi, h(i, \cdot), X_0)$$

for $i = 0, 1$. Now the homotopy invariance in the last function of the C^1 Brouwer triple degree implies that the right-hand side is independent of $i \in \{0, 1\}$, and ($F_{\mathcal{T}_{BF}}$) is established. Concerning the single-valued normalization property, we recall that Proposition 8.70 implies that there is an admissible neighborhood $\Omega_0 \subseteq \Omega$ and a finite-dimensional subspace $Y_0 \subseteq Y$ which is transversal to F on Ω_0 and contains $\varphi(\Phi(\Omega_0))$. The reduction property of the Benevieri–Furi triple degree implies for $X_0 := \Omega_0 \cap F^{-1}(Y_0)$ and $F_0 := F|_{X_0} \in \mathcal{F}_0(X_0, Y_0)$ that

$$\deg(F, \Phi, \varphi, \Omega) = \deg_{(X_0,Y_0)}(F, \Phi, \varphi, X_0),$$

while the reduction property of the Benevieri–Furi coincidence degree implies

$$\deg(F, \varphi \circ \Phi, \Omega) = \deg_{(X_0,Y_0)}(F - (\varphi \circ \Phi)|_{X_0}, X_0, 0).$$

By the single valued normalization property of the C^1 Brouwer triple degree, these expressions are the same. The normalization property is a special case and implies the remaining equality in (12.11). □

We have defined the Benevieri–Furi triple degree analogously to the definition of the Benevieri–Furi coincidence degree and proved in particular that it has the properties ($E_{\mathcal{T}_{BF}}$)–($H_{\mathcal{T}_{BF}}$).

Perhaps, it might have been a more natural approach to define the degree by the properties ($E_{\mathcal{T}_{BF}}$)–($H_{\mathcal{T}_{BF}}$): This would be an analogous definition to the Brouwer triple degree, but only replacing the Brouwer degree by the Benevieri–Furi coincidence degree in the definition.

The following theorem states that actually both approaches lead to the same degree under a mild additional hypothesis concerning separation axioms for the spaces occurring in the considered function triples. It is unknown to the author whether without this additional hypothesis the degree is uniquely determined by the properties ($E_{\mathcal{T}_{BF}}$)–($H_{\mathcal{T}_{BF}}$). This is the reason why Definition 12.19 was chosen.

For a moment, let $\mathcal{T}_0$ be the subclass of all function triples $(F, \Phi, \varphi, \Omega) \in \mathcal{T}_{BF}(X, Y)$ with the additional property that there is an admissible neighborhood Ω_0 with the property that $[0,1] \times \overline{\Omega_0} \times \Phi(\Omega_0)$ is T_5. If $(F, \Phi, \varphi, \Omega)$ is in standard form, we require only that $\overline{\Omega_0}$ and $[0,1] \times \Phi(\Omega_0)$ are T_5.

Theorem 12.22. *Let* deg *be any operator which associates to $(F, \Phi, \varphi, \Omega) \in \mathcal{T}_0$ (see above) a number from $\mathbb{Z}_2$ (or $\mathbb{Z}$ in the oriented case) with the properties* ($E_{\mathcal{T}_{BF}}$)–($H_{\mathcal{T}_{BF}}$)*. Then* deg *is the restriction of the Benevieri–Furi triple degree to $\mathcal{T}_0$.*

Proof. Let $A := (F, \Phi, \varphi, \Omega) \in \mathcal{T}_0$. We have to show that $\deg(A)$ is the Benevieri–Furi triple degree of A. Passing to the standard form if necessary (using the weak equivalence invariance and Theorem 11.23), we can assume that A is in standard form. For the standard form there is an admissible neighborhood Ω_0 such that $\overline{\Omega_0}$ and $[0, 1] \times \Phi(\Omega_0)$ are T_5. Shrinking Ω_0 if necessary, we can assume that there is a finite-dimensional subspace $Y_0 \subseteq Y$ which is transversal to F on Ω_0 and which contains $\varphi(\Phi(\Omega_0))$. Recall that $X_0 := \Omega_0 \cap F^{-1}(Y_0)$ is a finite-dimensional submanifold by Theorem 8.55. In particular, Proposition 9.1 implies that there is an open in X_0 neighborhood $U \subseteq X_0$ of $K := \operatorname{coin}_\Omega(F, \Phi, \varphi)$ whose closure $\overline{U}$ (in X_0) is compact. In particular, also $M_0 := \partial_{X_0} U = \overline{U} \setminus U$ is compact. Since $\overline{U} \subseteq \Omega_0$ and U is open in X_0, it follows that the sets U and $B := (X_0 \setminus \overline{U}) \cup \partial\Omega_0$ are separated in X and thus separated in the T_5 space $\overline{\Omega_0}$. Applying Proposition 2.35(c), we find disjoint open in $\overline{\Omega_0}$ sets $\Omega_1, W \subseteq \overline{\Omega_0}$ with $U \subseteq \Omega_1$ and $B \subseteq W$. It follows that $\Omega_1 \subseteq \Omega_0$ is open in X, and its closure $M := \overline{\Omega_1}$ in X is contained in Ω_0 and satisfies $\overline{\Omega_1} \cap X_0 \subseteq \overline{U}$. In particular, $\partial_X \Omega_1 \cap X_0 = M_0$ and $M \subseteq \Omega_0$. Since Ω_0 and $[0, 1] \times \Phi(\Omega_0)$ are T_5, it follows that the spaces M and $[0, 1] \times \Phi(M)$ are T_4, and so Y_0 is an AE for these spaces by Corollary 2.67. Thus, we can apply Remark 11.39 to find that there is an (M_0, Y_0)-admissible simplifier f for A on M. This means that there is $H \in C([0, 1] \times \Phi(M), Y_0)$ with $\operatorname{coin}_{M_0}(F, \Phi, H(t, \cdot)) = \varnothing$ for all $t \in [0, 1]$ and $H(0, \cdot) = \varphi|_{\Phi(M)}$ such that f is a simplifier for $(F, \Phi, H(1, \cdot))$. We put $C := \operatorname{coin}_{[0,1] \times M}(F, \Phi, H)$ (recall the convention of Remark 12.3). Since H assumes its values in Y_0, we have $C \subseteq [0, 1] \times (M \cap F^{-1}(Y_0))$. Since $M \cap F^{-1}(Y_0) \subseteq X_0$ and $M \cap X_0 = \overline{U}$, we have even $C \subseteq [0, 1] \times \overline{U}$. Hence, C is compact. Moreover, since $\operatorname{coin}_{[0,1] \times M_0}(F, \Phi, H) = \varnothing$, we have $C \subseteq [0, 1] \times U \subseteq [0, 1] \times \Omega_1$. Thus, using the excision property, applying the homotopy invariance in the last function with $h := H|_{[0,1] \times \Omega_1}$, using the weak equivalence invariance, and the normalization property of the degree, we calculate

$$\begin{aligned}\deg(F, \Phi, \varphi, \Omega) = \deg(F, \Phi, \varphi, \Omega_1) &= \deg(F, \Phi, h(1, \cdot), \Omega_1)\\ &= \deg(F, \mathrm{id}_{\Omega_1}, f, \Omega) = \deg(F, f, \Omega),\end{aligned}$$

where the last "deg" denotes the Benevieri–Furi coincidence degree. Repeating the same calculation for the Benevieri–Furi triple degree, we obtain that the given map deg assumes at A the same value as the Benevieri–Furi triple degree. □

Theorem 12.23. *The Benevieri–Furi triple degree has the following property:*

($J_{\mathcal{T}_{BF}}$) **(Generalized Homotopy Invariance).** *Let (G, H, h, W, Y, Γ) be a generalized proper acyclic* homotopy triple for $(G_t, H_t, h_t, W_t, Y, \Gamma_t)$ ($t \in$*

$[0,1]$) *where* $W \subseteq [0,1] \times X$ *is open,* $G\colon W \to Y$ *a generalized (oriented) Fredholm homotopy of index* 0*, and* $h(H(W)) \subseteq Y_0$ *for some finite-dimensional subspace* $Y_0 \subseteq Y$*. Then* $(G_t, H_t, h_t, W_t) \in \mathcal{T}_{\mathrm{BF}}(X,Y)$ *for all* $t \in [0,1]$*, and*

$$\deg(G_t, H_t, h_t, W_t) \text{ is independent of } t \in [0,1].$$

Proof. Proposition 11.41 implies the assertion $(G_t, H_t, h_t, W_t) \in \mathcal{T}_{\mathrm{BF}}(X,Y)$. Moreover, with $\widetilde{H}(t,x) := \{t\} \times H(t,x)$ as in Proposition 11.41, we have that $(G, \widetilde{H}, h)$ is a proper acyclic* homotopy triple. We can assume in view of Proposition 11.42 and the (weak) equivalence invariance without loss of generality that $(G, \widetilde{H}, h)$ is in standard form. Shrinking W and increasing Y_0 if necessary, we can assume by Proposition 8.70 (and Remark 8.71) without loss of generality that Y_0 is transversal to G on W. It suffices to show by Proposition 2.19 that

$$d(t) := \deg(H(t,\cdot), h(t,\cdot), W_t)$$

is locally constant on $[0,1]$. Thus, let $t_0 \in [0,1]$, and we have to show that d is constant in some neighborhood of t_0.

Putting $X_t := G_t^{-1}(Y_0)$ and $Z_t := \{t\} \times X_t$, we find by Corollary 8.63 that $Z = \bigcup_{t\in[0,1]} Z_t$ is a partial C^1 manifold. In particular, Z is a finite-dimensional C^0 manifold, and so we find by Proposition 9.1 that there is some open neighborhood $V \subseteq Z$ of $\operatorname{coin}_W(G,H,h)$ whose closure C in Z is compact. By the reduction property, we have

$$d(t) = \deg_{(X_t,Y_0)}(G_t, H_t, h_t, X_t),$$

where the right-hand side denotes the Brouwer triple degree. We apply Theorem 11.36 with the compact sets $M_0 := C \setminus V$ and $M := C$ and thus find that there is an (M_0, Y_0)-admissible simplifier f for (G,H,h) on M. Hence, there is $\hat{H} \in C([0,1] \times H(M), Y_0)$ with $\operatorname{coin}_{M_0}(G,H,h,\hat{H}(s,\cdot)) = \varnothing$ for all $s \in [0,1]$, $\hat{H}(0,\cdot) = h|_{H(M)}$ such that f is a simplifier for $(G,H,\hat{H}(1,\cdot))$. With the convention of Remark 12.3, we find by Proposition 11.6 that

$$K_t := \operatorname{coin}_{[0,1]\times(X_t\cap C)}(G_t, H_t, \hat{H})$$

is a closed hence compact subset of $[0,1] \times C$. From $\operatorname{coin}_{M_0}(G,H,h,\hat{H}(s,\cdot)) = \varnothing$, we obtain that $K_t \subseteq [0,1] \times (X_t \cap V)$. Using the excision property, the homotopy invariance, the (weak) equivalence invariance and the normalization property of the Brouwer triple degree, we obtain

$$\begin{aligned} d(t) &= \deg_{(X_t,Y_0)}(G_t, H_t, \hat{H}(1,\cdot), V \cap X_t) \\ &= \deg_{(X_t,Y_0)}(G_t, \mathrm{id}_{V\cap X_t}, f(t,\cdot), V \cap X_t) \\ &= \deg_{(X_t,Y_0)}(G(t,\cdot) - f(t,\cdot), V \cap X_t, 0). \end{aligned}$$

Note also that $f \in C(M, Y_0)$ and that $\operatorname{coin}_M(G, f)$ is closed and thus compact and disjoint from M_0. Hence, $\operatorname{coin}_V(G, f)$ is compact.

Now repeating the argument which we had used for the homotopy invariance of the Benevieri–Furi coincidence degree (Theorem 10.9) (using (G, f) in place of (H, h) in that proof), we find that the last expression is independent of t in some neighborhood of t_0.

Under the mild additional assumption that the compact set $\operatorname{coin}_V(G, f)$ has a T_4 neighborhood $N \subseteq X$, we need not repeat the *proof* of Theorem 10.9, but we can use a trick to *apply* Theorem 10.9 instead: Using Corollary 2.67, we can extend f to some $f \in C(N, Y_0)$. Let $U_0 \subseteq X$ be an open neighborhood of $\operatorname{coin}_V(G, f)$ with $U_0 \subseteq N$. By definition of the inherited topology there is an open set $U_1 \subseteq X$ with $V = U_1 \cap Z$. Then $U := U_0 \cap U_1$ is open in X with $U \cap Z \subseteq V$. Since $f(N) \subseteq Y_0$, we have $\operatorname{coin}_U(G, f) \subseteq U \cap G^{-1}(Y_0) = U \cap Z \subseteq V$, and thus actually $\operatorname{coin}_U(G, f) = \operatorname{coin}_V(G, f)$ is compact. The (generalized) homotopy invariance of the Benevieri–Furi coincidence degree implies that

$$\deg_{(X,Y)}(G(t, \cdot), f(t, \cdot), U)$$

is independent of t, and by the reduction property of the Benevieri–Furi coincidence degree, this is just $d(t)$. □

Proposition 12.24. *The Benevieri–Furi triple degree has the following properties for each* $(F, \Phi, \varphi) \in \mathcal{T}_{\mathrm{BF}}(X, Y)$.

($\mathrm{K}_{\mathcal{T}_{\mathrm{BF}}}$) **(Additivity).** *If* $\Omega = \Omega_1 \cup \Omega_2$ *with disjoint open subsets* $\Omega_i \subseteq \Omega$ *then*

$$\deg(F, \Phi, \varphi, \Omega) = \deg(F, \Phi, \varphi, \Omega_1) + \deg(F, \Phi, \varphi, \Omega_2).$$

($\mathrm{L}_{\mathcal{T}_{\mathrm{BF}}}$) **(Existence).** *If* $\deg(F, \Phi, \varphi, \Omega) \neq 0$ *then* $\operatorname{coin}_\Omega(F, \Phi, \varphi) \neq \emptyset$.

($\mathrm{M}_{\mathcal{T}_{\mathrm{BF}}}$) **(Excision-Additivity).** *If* $\Omega_i \subseteq \Omega$ $(i \in I)$ *is a family of pairwise disjoint open sets with* $\operatorname{coin}_\Omega(F, \Phi, \varphi) \subseteq \bigcup_{i \in I} \Omega_i$, *and if* $\operatorname{coin}_{\Omega_i}(F, \Phi, \varphi)$ *is compact for all* $i \in I$, *then*

$$\deg(F, \Phi, \varphi, \Omega) = \sum_{i \in I} \deg(F, \Phi, \varphi, \Omega_i),$$

where in the sum at most a finite number of summands is nonzero.

Proof. For the proof of the additivity, let $Y_0 \subseteq Y$ be a finite-dimensional subspace which is transversal to F an admissible open neighborhood $\Omega_0 \subseteq \Omega$ of $\operatorname{coin}_\Omega(F, \Phi, \varphi)$ (Proposition 8.70 and Remark 8.71). Then Y_0 is also transversal to F on $\Omega_{i,0} := \Omega_0 \cap \Omega_{i,0}$ $(i = 1,2)$. Putting $X_0 := \Omega_0 \cap F^{-1}(Y_0)$, and $X_i := \Omega_{i,0} \cap F^{-1}(Y_0)$ $(i = 1,2)$, we obtain by the reduction property that

$$\begin{aligned}\deg(F,\Phi,\varphi,\Omega) &= \deg_{(X_0,Y_0)}(F,\Phi,\varphi,X_0),\\ \deg(F,\Phi,\varphi,\Omega_i) &= \deg_{(X_i,Y_0)}(F,\Phi,\varphi,X_i)\\ &= \deg_{(X_0,Y_0)}(F,\Phi,\varphi,X_i) \quad \text{for } i = 1,2,\end{aligned}$$

where we used the restriction property of the Brouwer triple degree for the last equality. Now the additivity of the Benevieri–Furi triple degree follows from the additivity of the Brouwer triple degree.

The excision and additivity properties imply the existence and excision-additivity properties by the same arguments that we had already used for the Brouwer degree. □

Proposition 12.25. *The Benevieri–Furi triple degree has the following properties for each* $(F,\Phi,\varphi) \in \mathcal{T}_{\mathrm{BF}}(X,Y)$.

($N_{\mathcal{T}_{\mathrm{BF}}}$) **(Diffeomorphic-Isomorphic Invariance).** *Let* J_1 *be a diffeomorphism of an open subset of a Banach manifold* X_0 *onto* Ω *and* J_2 *an isomorphism of* Y *onto a real normed vector space* Y_0. *Then*

$$\deg_{(X,Y)}(F,\Phi,\varphi,\Omega) = \deg_{(X_0,Y_0)}(J_2 \circ F \circ J_1, \Phi \circ J_1, J_2 \circ \varphi, J_1^{-1}(\Omega)).$$

In the oriented case, the orientation of $J_2 \circ F \circ J_1$ *is understood in the sense of Proposition* 8.38.

($O_{\mathcal{T}_{\mathrm{BF}}}$) **(Restriction).** *Let* $X_0 \subseteq X$ *be open. Then*

$$\deg_{(X_0,Y)} = \deg_{(X,Y)}|_{\mathcal{T}_{\mathrm{BF}}(X_0,Y)}.$$

Proof. Let $Y_1 \subseteq Y$ be a finite-dimensional subspace which is transversal to F on an admissible open neighborhood $\Omega_0 \subseteq \Omega$ of $\operatorname{coin}_\Omega(F,\Phi,\varphi)$ (Proposition 8.70 and Remark 8.71). By the reduction property, we have with $X_1 := \Omega_0 \cap F^{-1}(Y_0)$ that

$$\deg_{(X,Y)}(F,\Phi,\varphi,\Omega) = \deg_{(X_1,Y_1)}(F,\Phi,\varphi,X_1).$$

By the chain rule, we have also that $Y_2 := J_2(Y_1) \subseteq Y_0$ is transversal to $F_1 := J_2 \circ F \circ J_1$ on $\Omega_1 := J_1^{-1}(\Omega_0)$, and so the reduction property for $\deg_{(X_0,Y_0)}$ implies similarly with $X_2 := \Omega_1 \cap F_1^{-1}(Y_2)$ that

$$\deg_{(X_0,Y_0)}(F_1, \Phi \circ J_1, J_2 \circ \varphi, J_1^{-1}(\Omega)) = \deg_{(X_2,Y_2)}(F_1, \Phi \circ J_1, J_2 \circ \varphi, X_2).$$

Now the diffeomorphic-isomorphic invariance of the Brouwer triple degree implies that these degrees are the same. Hence, ($N_{\mathcal{T}_{BF}}$) is established.

The restriction property is the special case $J_2 = \mathrm{id}_Y$ and $J_1 = \mathrm{id}_\Omega \colon \Omega \to X$. □

Proposition 12.26. *The Benevieri–Furi triple degree has the following property:*

($P_{\mathcal{T}_{BF}}$) **(Cartesian Product).** *For $i = 1, 2$, let X_i be a manifold without boundary of class C^1 over the real Banach space E_{X_i}, and let $Y_i = E_{Y_i}$ be a real Banach space. For $(F_i, \Phi_i, \varphi_i, \Omega_i) \in \mathcal{T}_{BF}(X_i, Y_i)$, we put $X := X_1 \times X_2$, $\Omega := \Omega_1 \times \Omega_2$, $Y := Y_1 \times Y_2$, $F := F_1 \otimes F_2$, $\Phi := \Phi_1 \otimes \Phi_2$, and $\varphi := \varphi_1 \otimes \varphi_2$. Then $(F, \Phi, \varphi, \Omega) \in \mathcal{T}_{BF}(X, Y)$ and*

$$\begin{aligned} &\deg_{(X,Y)}(F, \Phi, \varphi, \Omega) \\ &\quad = \deg_{(X_1,Y_1)}(F_1, \Phi_1, \varphi_1, \Omega_1) \deg_{(X_2,Y_2)}(F_2, \Phi_2, \varphi_2, \Omega_2). \end{aligned}$$

In the oriented case, F is equipped with the product orientation.

Proof. Putting $K := \mathrm{coin}_\Omega(F, \Phi, \varphi)$ and $K_i := \mathrm{coin}_{\Omega_i}(F_i, \Phi_i, \varphi_i)$ $(i = 1, 2)$, we have $K = K_1 \times K_2$, and so K is compact by Theorem 2.63. Hence, $(F, \Phi, \varphi, \Omega) \in \mathcal{T}_{BF}(X, Y)$.

For $i = 1, 2$, we apply Proposition 8.70 and Remark 8.71: There are finite-dimensional subspaces $Y_{i,0} \subseteq Y_i$ which are transversal to F_i on admissible open neighborhoods $\Omega_{i,0} \subseteq \Omega_i$ of K_i. Then $Y_0 := Y_{1,0} \times Y_{2,0}$ is a finite-dimensional subspace of Y which is transversal to F on $\Omega_0 := \Omega_{1,0} \times \Omega_{2,0}$. Applying the reduction property, we thus obtain the assertion from the Cartesian product property of the Brouwer triple degree. □

Chapter 13

The Degree for Compact Fredholm Triples

The aim of this chapter is to extend the Benevieri–Furi triple degree theory to function triples (F, Φ, φ) which need not satisfy the artificial requirement that the range of φ be contained in a finite-dimensional subspace.

For the classical Brouwer degree (in Banach spaces) it is well-known that a corresponding extension of degree theory exists for maps of the form $F = \mathrm{id} - \varphi$ where φ is continuous and locally compact: This is the famous Leray–Schauder degree [97]. In our terminology, this is just a degree for the particular function triples $(\mathrm{id}_\Omega, \mathrm{id}_\Omega, \varphi)$ with locally compact φ. We will develop in Section 13.1 even a degree theory for function triples of the form (F, Φ, φ) where F is (oriented) Fredholm, Φ is acyclic*, and φ is continuous and locally compact.

We will actually reduce this degree theory analogously to the original approach of Leray and Schauder to the finite-dimensional situation. This may appear now rather straightforward, but this is only so simple, because we already solved the main difficulty related with this approach: Namely, we already established the existence of the Benevieri–Furi degree for function triples so that, roughly speaking, we have completely settled the finite-dimensional situation, already, even within an infinite-dimensional framework. In the original approach of Leray and Schauder this difficulty can be solved much easier by using the reduction property of the fixed point index in subspaces (recall the remarks after Theorem 10.1). However, such a simple property is not available in our situation, and we really had to use the Brouwer triple degree on *manifolds* in order to develop the Benevieri–Furi triple degree, even if we should only be interested in Banach *spaces*.

13.1 The Leray–Schauder Triple Degree

Throughout this section, let X be a C^1 Banach manifold without boundary over a real Banach space $E = E_X$, and let $Y = E_Y$ be a real Banach space. We assume throughout the non-degeneracy hypothesis (10.8), that is, $E_X \neq \{0\}$ and $E_Y \neq \{0\}$ (if we assume AC).

Definition 13.1. We write $(F, \Phi, \varphi, \Omega) \in \mathcal{T}_{\mathrm{LS}}(X, Y)$ if the following holds.

$\Omega \subseteq X$ is open, $F: \Omega \to Y$, and there is a Hausdorff space Γ with $\Phi: \Omega \multimap \Gamma$ and $\varphi: \Gamma \to Y$. The set

$$\mathrm{coin}(F, \Phi, \varphi) = \{x \in \Omega : F(x) \in \varphi(\Phi(x))\}$$

is compact and has an open neighborhood $\Omega_0 \subseteq \Omega$ such that the following holds:

(a) $F|_{\Omega_0} \in \mathcal{F}_0(\Omega_0, Y)$.

(b) $\Phi|_{\Omega_0}$ is acyclic*.

(c) $\varphi|_{\Phi(\Omega_0)}$ has a closed graph in $\Phi(\Omega_0) \times Y$.

(d) $\varphi \circ \Phi|_{\Omega_0}$ is locally compact.

The set $\Omega_0 \subseteq \Omega$ is called an *admissible neighborhood.*

For the oriented version of the degree, we assume in addition that $F|_{\Omega_0}$ is oriented on Ω_0.

Remark 13.2. The requirement that $\varphi|_{\Phi(\Omega_0)}$ has a closed graph is equivalent to $\varphi|_{\Phi(\Omega_0)} \in C(\Phi(\Omega_0), Y)$ by Corollaries 2.117 and 2.124.

In particular, $\mathcal{T}_{\mathrm{LS}}(X, Y) \subseteq \mathcal{T}_{\mathrm{prop}}(X, Y)$.

Proposition 13.3. *Let $\Omega \subseteq X$ be open, $F \in \mathcal{F}_0(\Omega, Y)$, Γ be a Hausdorff space, $\Phi: \Omega \multimap \Gamma$ be acyclic*, and $\varphi: \Gamma \to Y$ have a closed graph in $\Gamma \times Y$, and $\varphi \circ \Phi$ be locally compact. Then (F, Φ, φ) is a closed function triple, and the following statements are equivalent:*

(a) $(F, \Phi, \varphi, \Omega) \in \mathcal{T}_{\mathrm{LS}}(X, Y)$.

(b) $(F, \Phi, \varphi, \Omega) \in \mathcal{T}_{\mathrm{prop}}(X, Y)$.

(c) $\mathrm{coin}(F, \Phi, \varphi)$ *is relatively compact in* Ω.

Proof. In view of Remark 13.2 the assertion follows from Proposition 12.17. □

Definition 13.4. The *Leray-Schauder triple degree* is an operator $\deg = \deg_{(X,Y)}$ which associates to each $(F, \Phi, \varphi, \Omega) \in \mathcal{T}_{\mathrm{LS}}(X, Y)$ a number from $\mathbb{Z}_2$ (or from $\mathbb{Z}$ in the oriented case) such that the following properties are satisfied:

($\mathrm{A}_{\mathcal{T}_{\mathrm{LS}}}$) **(Homotopy Invariance in the Last Function).** Let $\Omega \subseteq X$ be open, $F \in \mathcal{F}_0(\Omega, Y)$ (oriented on Ω), Γ a Hausdorff space, and $\Phi: \Omega \multimap \Gamma$ acyclic*. If $h \in C([0,1] \times \Phi(\Omega), Y)$ is compact and such that

$$\mathrm{coin}_{[0,1]\times\Omega}(F, \Phi, h) = \bigcup_{t\in[0,1]} (\{t\} \times \mathrm{coin}_{\Omega}(F, \Phi, h(t, \cdot)))$$

is compact then

$$\deg(F, \Phi, h(0, \cdot), \Omega) = \deg(F, \Phi, h(1, \cdot), \Omega).$$

($B_{\mathcal{T}_{LS}}$) **(Excision).** If $\Omega_0 \subseteq \Omega$ is open and contains $\operatorname{coin}(F, \Phi, \varphi)$ then

$$\deg(F, \Phi, \varphi, \Omega_0) = \deg(F, \Phi, \varphi, \Omega).$$

($C_{\mathcal{T}_{LS}}$) **(Reduction).** Let $\Omega_0 \subseteq \Omega$ be an open neighborhood of $\operatorname{coin}_\Omega(F, \Phi, \varphi)$ such that $F \in \mathcal{F}_0(\Omega_0, Y)$ (being oriented on Ω_0 in the oriented case), $\Phi|_{\Omega_0}$ is acyclic*, and $\varphi|_{\Phi(\Omega_0)}$ is continuous and assumes its values in a finite-dimensional subspace $Y_0 \subseteq Y$. If additionally Y_0 is transversal to F on Ω_0 then $X_0 := \Omega_0 \cap F^{-1}(Y_0)$ is a C^1 submanifold of X of the same dimension as Y_0, and if this dimension is positive, we have with $F_0 := F|_{X_0} \in \mathcal{F}_0(X_0, Y_0)$ that

$$\deg_{(X,Y)}(F, \Phi, \varphi, \Omega) = \deg_{(X_0,Y_0)}(F_0, \Phi, \varphi, X_0),$$

where the right-hand side denotes the C^1 Brouwer triple degree (cf. Remark 12.7). In case $X_0 = \varnothing$, the right-hand side is defined as zero. In the oriented case, the orientation of F_0 is the inherited orientation in the sense of Definition 8.65.

We will show in a moment:

Theorem 13.5. *For each fixed X and Y there is exactly one Leray–Schauder triple degree. Moreover, in case $\Phi = \mathrm{id}_\Omega$ the degree $\deg(F, \mathrm{id}_\Omega, \varphi, \Omega)$ is uniquely determined by the restriction of the above properties to function triples of the form $(F, \mathrm{id}_\Omega, \varphi, \Omega)$. In addition, the Leray–Schauder degree has the following properties for each $(F, \Phi, \varphi, \Omega) \in \mathcal{T}_{LS}$:*

($D_{\mathcal{T}_{LS}}$) **(Generalized Homotopy Invariance).** *Let (G, H, h, W, Y, Γ) be a generalized proper acyclic* homotopy triple for $(G_t, H_t, h_t, W_t, Y, \Gamma_t)$ ($t \in [0, 1]$) where $W \subseteq [0, 1] \times X$ is open, $G \colon W \to Y$ a generalized (oriented) homotopy, and $h \circ \widetilde{H}$ is locally compact with $\widetilde{H}(t, x) := (t, H(t, x))$. Then $(G_t, H_t, h_t, W_t) \in \mathcal{T}_{LS}(X, Y)$ for all $t \in [0, 1]$, and*

$$\deg(G_t, H_t, h_t, W_t) \text{ is independent of } t \in [0, 1].$$

($E_{\mathcal{T}_{LS}}$) **(Compatibility with the Brouwer Triple Degree).** *Let $\dim E_X = \dim E_Y > 0$ be finite. Then $\deg$ is the C^1 Brouwer triple degree (cf. Remark 12.7).*

($\mathrm{F}_{\mathcal{T}_{\mathrm{LS}}}$) **(Compatibility with the Benevieri–Furi Triple Degree).** *If there is a neighborhood* $\Omega_0 \subseteq \Omega$ *of* $\operatorname{coin}_\Omega(F, \Phi, \varphi)$ *such that* $\varphi(\Phi(\Omega_0))$ *is contained in a finite-dimensional subspace of* Y *then* $\deg(F, \Phi, \varphi, \Omega)$ *is the Benevieri–Furi triple degree of* $(F, \Phi, \varphi, \Omega)$.

($\mathrm{G}_{\mathcal{T}_{\mathrm{LS}}}$) **(Compatibility with the Benevieri–Furi Coincidence Degree).** *If there is a neighborhood* $\Omega_0 \subseteq \Omega$ *of* $\operatorname{coin}_\Omega(F, \Phi, \varphi)$ *such that* $\varphi(\Phi(\Omega_0))$ *is contained in a finite-dimensional subspace of* Y *and if* $\varphi \circ \Phi$ *is single-valued then*

$$\deg(F, \Phi, \varphi, \Omega) = \deg(F, \varphi \circ \Phi, \Omega),$$

where the degree on the right-hand side is the Benevieri–Furi coincidence degree.

($\mathrm{H}_{\mathcal{T}_{\mathrm{LS}}}$) **(Compatibility with the Non-Oriented Case).** *The degrees for the oriented and non-oriented case are the same modulo 2 (if the oriented case applies).*

Later, we will show that the Leray–Schauder triple degree satisfies actually much more properties like the additivity.

We prepare the proof of Theorem 13.5 by the following observations.

Proposition 13.6. *The Leray–Schauder triple degree has automatically the properties* ($\mathrm{E}_{\mathcal{T}_{\mathrm{LS}}}$)–($\mathrm{G}_{\mathcal{T}_{\mathrm{LS}}}$). *An analogous assertion holds for the restriction to function triples of the form* $(F, \mathrm{id}_\Omega, \varphi, \Omega)$.

Proof. Since the Leray–Schauder triple degree (resp. its restrictions to function triples of the form $(F, \mathrm{id}_\Omega, \varphi, \Omega)$) satisfies the same reduction property as the Benevieri–Furi triple degree (resp. as the Benevieri–Furi coincidence degree for (F, φ, Ω)), and since this reduction property characterizes the Benevieri–Furi triple (resp. coincidence) degree, we obtain the compatibility with the Benevieri–Furi triple (coincidence) degree. The other compatibilities follow from the compatibility of the Benevieri–Furi triple degree with the Benevieri–Furi coincidence degree and the Brouwer triple degree. □

Proposition 13.7. *Let* $(F, \Phi, \varphi, \Omega) \in \mathcal{T}_{\mathrm{LS}}(X, Y)$. *Then* (F, Φ, φ) *is a function triple, and there is an admissible neighborhood* Ω_0 *with the additional property that* $\varphi(\Phi(\Omega_0))$ *is relatively compact in* Y.

Proof. By Remark 13.2, the first assertion follows from Proposition 12.18. The second assertion follows from Proposition 2.125. □

The key property for the Leray–Schauder approach is that continuous maps with a compact range may be arbitrarily good approximated by continuous maps with values in a finite-dimensional space. This is usually achieved by means of so-called Schauder projections. Although it is not hard to construct these projections directly, we show that they come out as a trivial special case of our generalization of Dugundji's extension theorem (Theorem 3.86). By that theorem, we can even consider a closed ε-net instead of just a finite ε-net in the following which may be useful in some occasions.

Proposition 13.8 (Schauder Projection). (AC). *Let $K \subseteq Y$, and $M \subseteq Y$ be a closed ε-net for K. Then for each $\delta > \varepsilon$ there is $\rho \in C(K \cup M, \operatorname{conv} M)$ satisfying $\rho(x) = x$ for each $x \in M$ and*

$$\|\rho(x) - x\| < \delta \quad \textit{for each } x \in K.$$

Proof. We apply Theorem 3.86 with $X := K \cup M$, $A := M$, $f := \mathrm{id}_A$, and $L(x) := \delta/\varepsilon$. Hence, we find some $\rho := F \in C(K \cup M, \operatorname{conv} M)$ satisfying $F|_M = f = \mathrm{id}_M$ and (3.30). The latter means for $x \in K$ in view of $\operatorname{dist}(x, M) < \varepsilon$ that

$$\rho(x) \in \operatorname{conv}\{y \in Y : d(x, y) \le L(x) \operatorname{dist}(x, M)\} \subseteq \operatorname{conv} B_\delta(x) = B_\delta(x).$$

For the last equality, we have used Proposition 3.3. □

Remark 13.9. Proposition 13.8 holds also if Y is an arbitrary normed space, not necessarily a Banach space. Moreover, it is not necessary that M is closed: It suffices that $K \cap M$ is closed in K.

In view of Remark 3.89, we do not need AC if $K \setminus M$ is separable: AC_ω is sufficient in this case.

Also the following Corollary 13.10 holds if Y is a normed space and not necessarily a Banach space.

Corollary 13.10. *For each topological space Γ, each $\varphi \in C(\Gamma, Y)$ for which $\varphi(\Gamma)$ is relatively compact, and each $\varepsilon > 0$ there is a finite-dimensional subspace $Y_0 \subseteq Y$ and $\varphi_0 \in C(\Gamma, Y_0)$ such that $\varphi_0(\Gamma) \subseteq \operatorname{conv} \varphi(\Gamma)$ and $\|\varphi(z) - \varphi_0(z)\| < \varepsilon$ for all $z \in \Gamma$.*

Proof. We apply Proposition 13.8 with $K := \varphi(\Gamma)$ and $\delta := \varepsilon$: By Proposition 3.26 and (3.1), we have $\chi_K(K) = 0$ and thus find a finite $\varepsilon/2$-net $M \subseteq K$ for K. With ρ as in Proposition 13.8, we obtain that $\varphi_0 := \rho \circ \varphi$ has the required properties. □

Let $(F, \Phi, \varphi, \Omega) \in \mathcal{T}_{\mathrm{LS}}(X, Y)$. By Proposition 13.7, there is an admissible neighborhood $\Omega_0 \subseteq \Omega$ such that $C := \varphi(\Phi(\Omega_0))$ is relatively compact. By Corollary 3.62, it follows that $\overline{\operatorname{conv}} C$ is compact. In view of Corollary 8.74, there is a neighborhood $N \subseteq \Omega_0$ of $\operatorname{coin}_\Omega(F, \Phi, \varphi)$ such that $F|_N$ is proper. In particular, $K := N \cap F^{-1}(\overline{\operatorname{conv}} C)$ is compact. Let $U \subseteq X$ be open with $\operatorname{coin}_\Omega(F, \Phi, \varphi) \subseteq U \subseteq N$. Then $K \setminus U$ is compact with $\operatorname{coin}_{K\setminus U}(F, \Phi, \varphi) = \varnothing$. Proposition 3.15 implies that

$$\varepsilon := \min_{x \in K\setminus U} \operatorname{dist}(F(x), \varphi(\Phi(x))) > 0.$$

By Corollary 13.10, there is a finite-dimensional subspace $Y_0 \subseteq Y$ and a function $\varphi_0 \in C(\Phi(N), Y_0 \cap \operatorname{conv} C)$ satisfying

$$\sup_{z \in \Phi(K\setminus U)} \|\varphi(z) - \varphi_0(z)\| < \varepsilon. \tag{13.1}$$

Proposition 13.11. *In the situation described above, we have* $(F, \Phi, \varphi_0, U) \in \mathcal{T}_{\mathrm{BF}}(X, Y) \subseteq \mathcal{T}_{\mathrm{LS}}(X, Y)$ *and*

$$\operatorname{coin}_N(F, \Phi, \varphi_0) \subseteq U. \tag{13.2}$$

Moreover, the Leray–Schauder triple degree must automatically satisfy

$$\deg(F, \Phi, \varphi, \Omega) = \deg(F, \Phi, \varphi_0, U). \tag{13.3}$$

Proof. We define $h \in C([0,1] \times \Phi(N), \operatorname{conv} C)$ by $h(t,z) := t\varphi_0(z) + (1-t)\varphi(z)$. The set

$$C_h := \operatorname{coin}_{[0,1]\times N}(F, \Phi, h) = \{(t,x) \in [0,1] \times N : F(x) \in h(t, \Phi(x))\}$$

is by Proposition 11.6 a closed subset of $[0,1] \times K$. Since $[0,1] \times K$ is compact (Theorem 2.63), the set C_h is compact by Proposition 2.29. For each $(t,x) \in [0,1] \times (K \setminus U)$, we have by (13.1) that

$$\delta := \sup_{z \in \Phi(U)} \|h(t,z) - \varphi(z)\| < \varepsilon \le \operatorname{dist}(F(x), \varphi(\Phi(x))),$$

and so $\operatorname{dist}(F(x), h(t, \Phi(x))) \ge \varepsilon - \delta > 0$. Hence, $C_h \subseteq [0,1] \times U$. This implies (13.2). Using the homotopy invariance in the last function, we obtain also that $(F, \Phi, h(i, \cdot), U) \in \mathcal{T}_{\mathrm{LS}}(X, Y)$ for $i = 0, 1$ and

$$\deg(F, \Phi, h(0, \cdot), U) = \deg(F, \Phi, h(1, \cdot), U) = \deg(F, \Phi, \varphi_0, U).$$

Since the excision property implies

$$\deg(F, \Phi, \varphi, \Omega) = \deg(F, \Phi, \varphi, U) = \deg(F, \Phi, h(0, \cdot), U),$$

we also obtain (13.3). □

Proof of Theorem 13.5. We understand the right-hand side of (13.3) as the Benevieri–Furi triple degree (or in case $\Phi = \mathrm{id}_\Omega$ as the Benevieri–Furi coincidence degree of (F, φ_0)). In view of Proposition 13.6, the uniqueness assertion follows.

Concerning the existence, we use (13.3) to define the value $\deg(F, \Phi, \varphi, \Omega)$. Let us show first that this is well-defined, that is, independent of the particular choice of Ω_0, U, N and φ_0.

We show first that this definition is independent of the particular choice of φ_0. Thus, let φ_1 be a possibly different choice. We define $h \in C([0,1] \times \Phi(N), \operatorname{conv} C)$ by $h(t,z) := t\varphi_1(z) + (1-t)\varphi_0(z)$. The set

$$C_h := \operatorname{coin}_{[0,1]\times N}(F, \Phi, h) = \{(t,x) \in [0,1] \times N : F(x) \in h(t, \Phi(x))\}$$

is by Proposition 11.6 a closed subset of $[0,1] \times K$. Since $[0,1] \times K$ is compact (Theorem 2.63), the set C_h is compact by Proposition 2.29. For each $(t,x) \in [0,1] \times (K \setminus U)$, we have by (13.1) that

$$\delta := \sup_{z \in \Phi(U)} \|h(t,z) - \varphi(z)\| < \varepsilon \leq \operatorname{dist}(F(x), \varphi(\Phi(x))),$$

and so $\operatorname{dist}(F(x), h(t, \Phi(x))) \geq \varepsilon - \delta > 0$. Hence, $C_h \subseteq [0,1] \times U$. Using the homotopy invariance in the last function of the Benevieri–Furi triple degree, we obtain that $(F, \Phi, h(i, \cdot), U) \in \mathcal{T}_{\mathrm{BF}}(X,Y)$ for $i = 0, 1$ and

$$\deg(F, \Phi, h(0, \cdot), U) = \deg(F, \Phi, h(1, \cdot), U).$$

In particular, $(F, \Phi, \varphi_i, U) \in \mathcal{T}_{\mathrm{BF}}(X,Y)$ for $i = 0, 1$, and

$$\deg(F, \Phi, \varphi_0, U) = \deg(F, \Phi, \varphi_1, U).$$

We show now that the definition is also independent of the particular choice of Ω_0, N, and U as well. Thus, let $\widetilde{\Omega}_0$, $\widetilde{N}$, and $\widetilde{U}$ be possibly different choices (with a corresponding set $\widetilde{K}$). We put $C_0 := \varphi(\Phi(\Omega_0 \cup \widetilde{\Omega}_0))$, $N_0 := N \cup \widetilde{N}$, $K_0 := N_0 \cap F^{-1}(C_0)$, and $U_0 := U \cap \widetilde{U}$. By what we have shown above, we are still free to choose the corresponding functions φ_0 and $\widetilde{\varphi}_0$, since for each choice, we will obtain the same degree. We note that

$$\varepsilon_0 := \min_{x \in K_0 \setminus U_0} \operatorname{dist}(F(x), \varphi(\Phi(x))) > 0.$$

By Corollary 13.10, there are finite-dimensional subspaces $Y_0, \widetilde{Y}_0 \subseteq Y$ and functions $\varphi_0 \in C(\Phi(N), Y_0 \cap \operatorname{conv} C)$ and $\widetilde{\varphi}_0 \in C(\Phi(N), \widetilde{Y}_0 \cap \operatorname{conv} C)$ satisfying

$$\sup_{z \in \Phi(K \setminus U_0)} \|\varphi(z) - \varphi_0(z)\| < \varepsilon_0 \quad \text{and} \quad \sup_{z \in \Phi(\widetilde{K} \setminus U_0)} \|\varphi(z) - \widetilde{\varphi}_0(z)\| < \varepsilon_0.$$

Then (restrictions of) $\varphi_0, \widetilde{\varphi}_0$ are admissible for Proposition 13.11, and moreover, they are also admissible for the definition of the degree (by means of (13.3)) with the choices $\Omega_0 \cup \widetilde{\Omega}_0$, N_0, and U_0. By Proposition 13.11, we have

$$\operatorname{coin}_N(F, \Phi, \varphi_0) \cup \operatorname{coin}_{\widetilde{N}}(F, \Phi, \widetilde{\varphi}_0) \subseteq U_0,$$

and thus by the excision property

$$\deg(F, \Phi, \varphi_0, U) = \deg(F, \Phi, \varphi_0, U_0), \tag{13.4}$$

$$\deg(F, \Phi, \widetilde{\varphi}_0, \widetilde{U}) = \deg(F, \Phi, \widetilde{\varphi}_0, U_0). \tag{13.5}$$

Since φ_0 and $\widetilde{\varphi}_0$ are admissible for the definition of the degree with the choices $\Omega_0 \cup \widetilde{\Omega}_0$, N_0, and U_0, the right-hand sides of (13.4) and (13.5) are the same by what we have shown above.

Thus, we can indeed use (13.3) to define the degree. We have to show that this degree has all properties required in Definition 13.4 and Theorem 13.5. The excision property is clear from the very definition by means of (13.3) (since that definition is independent of the choice of Ω_0). The compatibility with the non-oriented case follows also from that definition: The formula (13.3) holds in the oriented and in the non-oriented case, and so the values of the degree are the same modulo 2, since this is the case for the Benevieri–Furi triple degree.

Also the reduction property is obvious from that definition: If $(F, \Phi, \varphi, \Omega)$ is as in that property, we can choose $\varphi_0 := \varphi$, and thus have that

$$\deg(F, \Phi, \varphi, \Omega) = \deg(F, \Phi, \varphi, U)$$

is the Benevieri–Furi triple degree. By the reduction property of the Benevieri–Furi triple degree (and by the excision and restriction property of the Brouwer triple degree), we obtain the reduction property of the Leray–Schauder triple degree.

It remains to prove the generalized homotopy invariance, since the homotopy invariance in the last function is a special case. Thus, let (G, H, h, W, Y, Γ) be a generalized proper acyclic* homotopy triple for the family $(G_t, H_t, h_t, W_t, Y, \Gamma_t)$ $(t \in [0, 1])$ where $W \subseteq [0, 1] \times X$ is open, $G \colon W \to Y$ a generalized (oriented) homotopy, and $h \circ \widetilde{H}$ is locally compact with $\widetilde{H}(t, x) := (t, H(t, x))$.

By Proposition 2.125, there is an open neighborhood $V \subseteq W$ of the compact set $K_0 := \operatorname{coin}_W(G, \widetilde{H}, h)$ such that $C := h(\widetilde{H}(V))$ is relatively compact. By Corollary 3.62, it follows that $\overline{\operatorname{conv}}\, C$ is compact. In view of Theorem 8.73, there is a neighborhood $N \subseteq V$ of K_0 such that $G|_N$ is proper. Let $U \subseteq [0, 1] \times X$ be open with $K_0 \subseteq U \subseteq N$. Putting $K := N \cap G^{-1}(C)$, we find that $K \setminus U$ is compact with $\operatorname{coin}_{K \setminus U}(G, \widetilde{H}, h) = \varnothing$. Proposition 3.15 implies that

$$\varepsilon := \min_{(t,x) \in K \setminus U} \operatorname{dist}(G(t, x), h(t, H(t, x))) > 0.$$

By Corollary 13.10, there is a finite-dimensional subspace $Y_0 \subseteq Y$ and a function $\widetilde{h} \in C(\Gamma, Y_0 \cap \operatorname{conv} C)$ satisfying

$$\sup_{(t,z)\in\widetilde{H}(K\setminus U)} \|h(t,z) - \widetilde{h}(t,z)\| < \varepsilon.$$

We put now $U_t := \{x : (t,x) \in U\}$ and $\widetilde{h}_t := \widetilde{h}(t,\,\cdot)$. By the definition (13.3) of the degree, we have

$$\deg(G_t, H_t, h_t, W_t) = \deg(G_t, H_t, \widetilde{h}_t, U_t). \tag{13.6}$$

Moreover, (13.2) implies that the compact set $\operatorname{coin}_N(G, \widetilde{H}, \widetilde{h})$ is contained in U. Hence, $\operatorname{coin}_U(G, \widetilde{H}, \widetilde{h})$ is actually compact, and so $(G, H, \widetilde{h}, U, Y, \Gamma)$ is a proper acyclic* homotopy triple for $(G_t, H_t, \widetilde{h}_t, U_t)$ $(t \in [0,1])$. The generalized homotopy invariance of the Benevieri–Furi triple degree (Theorem 12.23) thus implies that the right-hand side of (13.6) is independent of $t \in [0,1]$. □

Definition 13.12. We write $(F, \varphi, \Omega) \in \mathcal{P}_{\mathrm{LS}}(X, Y)$ if $\Omega \subseteq X$ is open, $F, \varphi \colon \Omega \to Y$, $\operatorname{coin}(F, \varphi)$ is compact, and if there is an open neighborhood $\Omega_0 \subseteq \Omega$ of $\operatorname{coin}(F, \varphi)$ such that

(a) $F|_{\Omega_0} \in \mathcal{F}_0(\Omega_0, Y)$.

(b) $\varphi|_{\Omega_0} \in C(\Omega_0, Y)$ is locally compact.

In the oriented case, we assume that $F|_{\Omega_0}$ is oriented on Ω_0.

Definition 13.13. The *Leray-Schauder coincidence degree* $\deg = \deg_{(X,Y)}$ is the map which associates to each $(F, \varphi, \Omega) \in \mathcal{P}_{\mathrm{LS}}(X, Y)$ a number from $\mathbb{Z}_2$ (or $\mathbb{Z}$ in the oriented case) such that the following properties are satisfied for each $(F, \varphi, \Omega) \in \mathcal{P}_{\mathrm{LS}}(X, Y)$.

($A_{\mathcal{P}_{\mathrm{LS}}}$) **(Homotopy Invariance)**. If $\Omega \subseteq X$ is open, $F \in \mathcal{F}_0(\Omega, Y)$ (oriented on Ω) and $h\colon [0,1] \times \Omega \to Y$ is continuous and locally compact and such that

$$\operatorname{coin}_{[0,1]\times\Omega}(F, h) = \{(t,x) \in [0,1] \times \Omega : F(x) = h(t,x)\}$$

is compact, then

$$\deg(F, h(0,\,\cdot), \Omega) = \deg(F, h(1,\,\cdot), \Omega).$$

($B_{\mathcal{P}_{\mathrm{LS}}}$) **(Excision)**. If $\Omega_0 \subseteq \Omega$ is open and contains $\operatorname{coin}(F, \Phi, \varphi)$ then

$$\deg(F, \varphi, \Omega_0) = \deg(F, \varphi, \Omega).$$

($C_{\mathcal{P}_{\mathrm{LS}}}$) **(Reduction)**. Let $\Omega_0 \subseteq \Omega$ be an open neighborhood of $\operatorname{coin}(F,\varphi)$ with $F|_{\Omega_0} \in \mathcal{F}_0(\Omega_0, Y)$, and $Y_0 \neq \{0\}$ a finite-dimensional subspace of Y with $\varphi|_{\Omega_0} \in C(\Omega_0, Y_0)$ and such that Y_0 is transversal to F on Ω_0. Then $X_0 := \Omega_0 \cap F^{-1}(Y_0)$ is empty or a submanifold of X of the same dimension as Y_0, and the map $G := (F - \varphi)|_{X_0} \in C(X_0, Y_0)$ satisfies $(G, X_0, 0) \in \mathcal{B}^0(X_0, Y_0)$ and

$$\deg_{(X,Y)}(F, \varphi, \Omega) = \deg_{(X_0,Y_0)}(G, X_0, 0),$$

where the right-hand side denotes the C^0 Brouwer degree; in case $X_0 = \varnothing$, we define the right-hand side as 0. The orientation of G for the oriented case is as described after Definition 10.7.

Theorem 13.14. *For fixed X and Y there is exactly one Leray–Schauder coincidence degree. Moreover, the Leray–Schauder triple degree has the following property for each $(F, \Phi, \varphi, \Omega) \in \mathcal{T}_{\mathrm{LS}}(X, Y)$:*

($I_{\mathcal{T}_{\mathrm{LS}}}$) **(Normalization).** *If $\Phi = \mathrm{id}_\Omega$ then*

$$\deg(F, \mathrm{id}_\Omega, \varphi, \Omega) = \deg(F, \varphi, \Omega),$$

where the right-hand side denotes the Leray–Schauder coincidence degree.

Proof. To see the existence of the Leray–Schauder coincidence degree, we can simply define it by

$$\deg(F, \varphi, \Omega) := \deg(F, \mathrm{id}_\Omega, \Omega)$$

and observe that all required properties follow from the corresponding properties of the Leray–Schauder triple degree. The normalization property of the Leray–Schauder triple degree is clear from that definition.

The additional remark in Theorem 13.5 (claiming that the uniqueness is obtained by restricting the corresponding properties to triples of the form $(F, \mathrm{id}_\Omega, \Omega)$) shows the uniqueness of the Leray–Schauder coincidence degree: In fact, the properties of Definition 13.13 correspond exactly to the properties of Definition 13.4 for this restriction. □

Remark 13.15. The classical Leray–Schauder degree of a map $\mathrm{id}_\Omega - \varphi$ with $X = Y$ and locally compact φ is a special case of the Leray–Schauder coincidence degree with $F = \mathrm{id}_\Omega$, equipped with the natural orientation as a diffeomorphism.

This follows from Proposition 13.11 applied with $F = \mathrm{id}_\Omega$ and $\Phi = \mathrm{id}_\Omega$ and the definition of the classical Leray–Schauder degree (which we will not repeat in this monograph in detail, since it is in fact just defined as in the mentioned special case).

The notion "Leray–Schauder coincidence degree" is somewhat a misnomer: In fact, this is the degree which was developed in [19] and thus should perhaps be called Benevieri–Furi coincidence degree. However, we reserved the latter name already for the special case of maps with values in finite dimensional subspaces. Since the only difference to the Benevieri–Furi coincidence degree is the usage of the Leray–Schauder approximation procedure in the above approach by means of Proposition 13.11, we have selected the names of Leray and Schauder for the more general (locally compact) setting.

Theorem 13.16. *The Leray–Schauder triple degree has the following properties for each* $(F, \Phi, \varphi, \Omega) \in \mathcal{T}_{\mathrm{LS}}(X, Y)$*:*

($\mathrm{J}_{\mathcal{T}_{\mathrm{LS}}}$) **(Weak Equivalence Invariance).** *If* $(F, \widetilde{\Phi}, \widetilde{\varphi}) \in \mathcal{T}_{\mathrm{LS}}(X, Y)$ *then*

$$(F, \widetilde{\Phi}, \widetilde{\varphi}) \precsim (F, \Phi, \varphi) \implies \deg(F, \widetilde{\Phi}, \widetilde{\varphi}, \Omega) = \deg(F, \Phi, \varphi, \Omega).$$

($\mathrm{K}_{\mathcal{T}_{\mathrm{LS}}}$) **(Equivalence Invariance).** *With* $\mathcal{T} = \mathcal{T}_{\mathrm{LS}}(X, Y)$*, we have*

$$(F, \Phi, \varphi) \sim_{\mathcal{T}} (F, \widetilde{\Phi}, \widetilde{\varphi}) \implies \deg(F, \Phi, \varphi, \Omega) = \deg(F, \widetilde{\Phi}, \widetilde{\varphi}, \Omega).$$

($\mathrm{L}_{\mathcal{T}_{\mathrm{LS}}}$) **(Single-Valued Normalization).** *If* $\varphi \circ \Phi$ *is single-valued then*

$$\deg(F, \Phi, \varphi, \Omega) = \deg(F, \mathrm{id}_{\Omega}, \varphi \circ \Phi, \Omega) = \deg(F, \varphi \circ \Phi, \Omega),$$

where the right-hand side denotes the Leray–Schauder coincidence degree.

Proof. To prove the equivalence invariance, an induction shows that it suffices to prove that if $(F, \Phi, \varphi, \Omega), (F, \widetilde{\Phi}, \widetilde{\varphi}, \Omega) \in \mathcal{T}$ satisfy $(F, \widetilde{\Phi}, \widetilde{\varphi}) \sqsubseteq (F, \Phi, \varphi)$ then they have the same Leray–Schauder triple degree. Thus, let $J \colon \widetilde{\Phi}(\Omega) \to \Phi(\Omega)$ be continuous with

$$J \circ \widetilde{\Phi} \subseteq \Phi \quad \text{and} \quad \varphi \circ J = \widetilde{\varphi}.$$

Let $\Omega_0 \subseteq \Omega$, $C := \varphi(\Phi(\Omega_0))$, $N \subseteq \Omega_0$, $K := N \cap F^{-1}(\overline{\mathrm{conv}}\, C)$, $U \subseteq N$, ε, and φ_0 be as described before Proposition 13.11, and so (13.3) holds, also when we understand the right-hand side as the Benevieri–Furi triple degree. We put $\widetilde{C} := \widetilde{\varphi}(\widetilde{\Phi}(\Omega_0))$ and $\widetilde{K} := N \cap F^{-1}(\overline{\mathrm{conv}}\, \widetilde{C})$. Then $\widetilde{C} \subseteq K$, $\widetilde{K} \subseteq K$, and

$$\min_{x \in \widetilde{K} \setminus U} \operatorname{dist}(F(x), \widetilde{\varphi}(\widetilde{\Phi}(x))) \geq \varepsilon.$$

Hence, $\widetilde{\varphi}_0 := \varphi_0 \circ J$ can be used in Proposition 13.11 with $(F, \widetilde{\Phi}, \widetilde{\varphi}, \Omega)$, and so

$$\deg(F, \widetilde{\Phi}, \widetilde{\varphi}, \Omega) = \deg(F, \widetilde{\Phi}, \widetilde{\varphi}_0, U),$$

when we understand the right-hand side as the Benevieri–Furi triple degree. Comparing this formula with (13.3) and using the equivalence invariance of the Benevieri–Furi triple degree, we obtain

$$\deg(F, \widetilde{\Phi}, \widetilde{\varphi}, \Omega) = \deg(F, \Phi, \varphi, \Omega),$$

which we had to prove. The weak equivalence invariance is a special case of the equivalence invariance.

Note that Theorem 11.23 implies that for any $(F, \Phi, \varphi, \Omega) \in \mathcal{T}$ also the corresponding standard form $(F, \widetilde{\Phi}, \widetilde{\varphi}, \Omega)$ belongs to $\mathcal{T}$. Moreover, if Φ is upper semi-continuous on Ω_0, φ is continuous on $\Phi(\Omega_0)$, and $\varphi \circ \Phi$ is single-valued then $\varphi \circ \Phi$ is continuous on Ω_0 in view of Proposition 2.94. Hence, also $(F, \mathrm{id}_\Omega, \varphi \circ \Phi, \Omega)$ belongs to $\mathcal{T}$. By Proposition 11.29, we obtain $(F, \Phi, \varphi) \sim_{\mathcal{T}} (F, \mathrm{id}_\Omega, \varphi)$, and so the equivalence invariance implies the first equality in the single-valued normalization property. The second equality follows from the normalization property. □

Theorem 13.17. *The Leray–Schauder triple degree satisfies the following properties for each* $(F, \Phi, \varphi, \Omega) \in \mathcal{T}_{\mathrm{LS}}(X, Y)$*:*

($\mathrm{M}_{\mathcal{T}_{\mathrm{LS}}}$) **(Additivity).** *If* $\Omega = \Omega_1 \cup \Omega_2$ *with disjoint open subsets* $\Omega_1, \Omega_2 \subseteq \Omega$ *then*

$$\deg(F, \Phi, \varphi, \Omega) = \deg(F, \Phi, \varphi, \Omega_1) + \deg(F, \Phi, \varphi, \Omega_2).$$

($\mathrm{N}_{\mathcal{T}_{\mathrm{LS}}}$) **(Existence).** *If* $\deg(F, \Phi, \varphi, \Omega) \neq 0$ *then* $\mathrm{coin}_\Omega(F, \Phi, \varphi) \neq \varnothing$.

($\mathrm{O}_{\mathcal{T}_{\mathrm{LS}}}$) **(Excision-Additivity).** *If* $\Omega_i \subseteq \Omega$ $(i \in I)$ *is a family of pairwise disjoint open sets with* $\mathrm{coin}_\Omega(F, \Phi, \varphi) \subseteq \bigcup_{i \in I} \Omega_i$ *and* $\mathrm{coin}_{\Omega_i}(F, \Phi, \varphi)$ *is compact for all* $i \in I$ *then*

$$\deg(F, \Phi, \varphi, \Omega) = \sum_{i \in I} \deg(F, \Phi, \varphi, \Omega_i),$$

where in the sum at most a finite number of summands is nonzero.

($\mathrm{P}_{\mathcal{T}_{\mathrm{LS}}}$) **(Diffeomorphic-Isomorphic Invariance).** *Let* J_1 *be a diffeomorphism of an open subset of a Banach manifold* X_0 *onto* Ω*, and* J_2 *an isomorphism of* Y *onto a real normed vector space* Y_0*. Then*

$$\deg_{(X,Y)}(F, \Phi, \varphi, \Omega) = \deg_{(X_0,Y_0)}(J_2 \circ F \circ J_1, \Phi \circ J_1, J_2 \circ \varphi, J_1^{-1}(\Omega)).$$

In the oriented case, the orientation of $J_2 \circ F \circ J_1$ *is understood in the sense of Proposition* 8.38.

($Q_{\mathcal{T}_{LS}}$) **(Restriction).** *Let $X_0 \subseteq X$ be open. Then*

$$\deg_{(X_0,Y)} = \deg_{(X,Y)}|_{\mathcal{T}_{LS}(X_0,Y)}.$$

Proof. Let Ω_0, $C := \varphi(\Phi(\Omega_0))$, $N \subseteq \Omega_0$, ε, $K := N \cap F^{-1}(\overline{\operatorname{conv}}\, C)$, U, and φ_0 be as described for Proposition 13.11. Then we have in particular (13.3), also when we understand the right-hand side as the Benevieri–Furi triple degree.

For the additivity, we put $N_i := N \cap \Omega_i$ $(i = 1, 2)$ and note that $K_i := N_i \cap F^{-1}(Y_0) = K \setminus \Omega_{3-i}$ $(i = 1, 2)$ is a closed and thus compact subset of K (Proposition 2.29). Hence, using Proposition 13.11, we obtain with $U_i := U \cap \Omega_i$ that

$$\deg(F, \Phi, \varphi, \Omega_i) = \deg(F, \Phi, \varphi_0, U_i) \quad \text{for } i = 1, 2.$$

Combining this with (13.3) and the additivity of the Benevieri–Furi triple degree, we obtain ($M_{\mathcal{T}_{LS}}$).

For ($P_{\mathcal{T}_{LS}}$), we find by Proposition 3.56 numbers $0 < c_1 \le c_2 < \infty$ such that

$$c_1\|y\| \le \|J_2 y\| \le c_2\|y\| \quad \text{for all } y \in Y.$$

In view of Corollary 13.10, we can assume that the function φ even satisfies

$$\sup_{z \in \Phi(K \setminus U)} \|\varphi(z) - \varphi_0(z)\| < c_2^{-1} c_1 \varepsilon.$$

Putting $\widetilde{\Omega} := J_1^{-1}(\Omega)$, $\widetilde{F} := J_2 \circ F \circ J_1$, $\widetilde{\Phi} := \Phi \circ J_1$, $\widetilde{\varphi} := J_2 \circ \varphi$, $\widetilde{\varphi}_0 := J_2 \circ \varphi_0$, $\widetilde{K} := J_1^{-1}(K)$, $\widetilde{U} := J_1^{-1}(U)$, and

$$\widetilde{\varepsilon} := \min_{x \in \widetilde{K} \setminus \widetilde{U}} \operatorname{dist}(\widetilde{F}(x), \widetilde{\varphi}(\widetilde{\Phi}(x))),$$

we have $\widetilde{\varepsilon} \ge c_1 \varepsilon$ and thus

$$\sup_{z \in \widetilde{\Phi}(\widetilde{K} \setminus \widetilde{U})} \|\widetilde{\varphi}(z) - \widetilde{\varphi}_0(z)\| < \widetilde{\varepsilon}.$$

By Proposition 13.11, we obtain

$$\deg_{(X_0,Y_0)}(\widetilde{F}, \widetilde{\Phi}, \widetilde{\varphi}, \widetilde{\Omega}) = \deg_{(X_0,Y_0)}(\widetilde{F}, \widetilde{\Phi}, \widetilde{\varphi}_0, \widetilde{U}_i),$$

where we understand the right-hand side as the Benevieri–Furi triple degree. Comparing this formula with (13.3) and using the diffeomorphic-isomorphic invariance of the Benevieri–Furi triple degree, we obtain ($P_{\mathcal{T}_{LS}}$).

The restriction property ($Q_{\mathcal{T}_{LS}}$) is the special case $J_2 = \mathrm{id}_Y$ and $J_1 = \mathrm{id}_\Omega \colon \Omega \to X$ of the diffeomorphic-isomorphic invariance. The excision and additivity properties imply the existence and excision-additivity properties by the same arguments that we had already used for the Brouwer degree. □

Proposition 13.18. *The Leray–Schauder triple degree satisfies the following property:*

($R_{\mathcal{T}_{LS}}$) **(Cartesian Product).** *For $i = 1, 2$, let X_i be a manifold without boundary of class C^1 over the real Banach space E_{X_i}, and let $Y_i = E_{Y_i}$ be a real Banach space. For $(F_i, \Phi_i, \varphi_i, \Omega_i) \in \mathcal{T}_{LS}(X_i, Y_i)$, we put $X := X_1 \times X_2$, $\Omega := \Omega_1 \times \Omega_2$, $Y := Y_1 \times Y_2$, $F := F_1 \otimes F_2$, $\Phi := \Phi_1 \otimes \Phi_2$, and $\varphi := \varphi_1 \otimes \varphi_2$. Then $(F, \Phi, \varphi, \Omega) \in \mathcal{T}_{LS}(X, Y)$ and*

$$\deg_{(X,Y)}(F, \Phi, \varphi, \Omega) \\ = \deg_{(X_1,Y_1)}(F_1, \Phi_1, \varphi_1, \Omega_1) \deg_{(X_2,Y_2)}(F_2, \Phi_2, \varphi_2, \Omega_2).$$

In the oriented case, F is equipped with the product orientation.

Proof. For $i = 1, 2$, let $\Omega_{0,i}$, $C_i := \varphi(\Phi(\Omega_{0,i}))$, $N_i \subseteq \Omega_{0,i}$, ε_i, $K_i := N_i \cap F_i^{-1}(\overline{\operatorname{conv}} C_i)$, U_i, and $\varphi_{0,i}$ be as described in front of Proposition 13.11 corresponding to $(F_i, \Phi_i, \varphi_i, \Omega_i)$. Then we have in particular

$$\deg_{(X_i,Y_i)}(F_i, \Phi_i, \varphi_i, \Omega_i) = \deg_{(X_i,Y_i)}(F_i, \Phi_i, \varphi_{0,i}, U_i),$$

where we understand the degree on the right-hand side as the Benevieri–Furi triple degree. Putting $\Omega_0 := \Omega_{0,1} \times \Omega_{0,2}$, $C := C_1 \times C_2$, $N := N_1 \times N_2$, $K := K_1 \times K_2$, $U := U_1 \times U_2$, $\varphi_0 := \varphi_{0,1} \otimes \varphi_{0,2}$, and $\varepsilon := \varepsilon_1 + \varepsilon_2$, we see that all properties as described in front of Proposition 13.11 are satisfied, and so (13.3) holds, where we understand the degree on the right-hand side as the Benevieri–Furi triple degree. The assertion thus follows from the Cartesian product property of the Benevieri–Furi triple degree. □

13.2 The Leray–Schauder Coincidence Degree and Classical Applications

Recall that we defined the Leray–Schauder coincidence degree already in Definition 13.13. For many applications this Leray–Schauder coincidence degree is already sufficient.

Therefore, although it is actually only a special case of the Leray–Schauder triple degree, we formulate in this section all relevant properties. We emphasize once more that this special case actually goes back to P. Benevieri and M. Furi [19] (although not in this general form; for instance, the homotopy invariance was proved in [19] only for $H \in C^1$) so that perhaps the name Benevieri–Furi coincidence degree might be more appropriate (if we wouldn't have reserved this name already). Note that in [19] actually not only a coincidence degree for a couple

(F, φ) but even a degree for the map $F - \varphi$ is developed (according to personal communication, this was the reason for the stronger hypothesis $H \in C^1$ in [19]).

Theorem 13.19. *The Leray–Schauder coincidence degree satisfies besides of the properties* $(A_{\mathcal{P}_{LS}})$–$(C_{\mathcal{P}_{LS}})$ *of Definition* 13.13 *the following properties for each* $(F, \varphi, \Omega) \in \mathcal{P}_{LS}(X, Y)$.

$(D_{\mathcal{P}_{LS}})$ **(Generalized Homotopy Invariance).** *Let* $W \subseteq [0,1] \times X$ *be open, and* $H: W \to Y$ *be an (oriented) generalized Fredholm homotopy of index* 0. *Let* $h \in C(W, Y)$ *be locally compact and such that* $\operatorname{coin}_W(H, h)$ *is compact. Put* $W_t := \{x : (t, x) \in W\}$. *Then* $(H(t,\cdot), h(t,\cdot), W_t) \in \mathcal{P}_{LS}(X, Y)$, *and*

$$\deg(H(t,\cdot), h(t,\cdot), W_t) \text{ is independent of } t \in [0,1].$$

$(E_{\mathcal{P}_{LS}})$ **(Compatibility with the Brouwer Degree).** *If* $0 < \dim E_X = \dim E_Y < \infty$ *then*

$$\deg(F, \varphi, \Omega) = \deg(F - \varphi, \Omega, 0),$$

where the right-hand side denotes that C^0 *Brouwer degree. In the oriented case, the orientation of* $F - \varphi$ *is as described after Theorem* 10.9.

$(F_{\mathcal{P}_{LS}})$ **(Compatibility with the Benevieri–Furi Coincidence Degree).** *If there is a neighborhood* $\Omega_0 \subseteq \Omega$ *of* $\operatorname{coin}_\Omega(F, \varphi)$ *such that* $\varphi(\Omega_0)$ *is contained in a finite-dimensional subspace of* Y *then* $\deg(F, \varphi, \Omega)$ *is the Benevieri–Furi coincidence degree of* (F, φ, Ω).

$(G_{\mathcal{P}_{LS}})$ **(Compatibility with the Non-Oriented Case).** *The degrees for the oriented and non-oriented case are the same modulo* 2 *(if the oriented case applies).*

$(H_{\mathcal{P}_{LS}})$ **(Additivity).** *If* $\Omega = \Omega_1 \cup \Omega_2$ *with disjoint open subsets* $\Omega_1, \Omega_2 \subseteq \Omega$ *then*

$$\deg(F, \varphi, \Omega) = \deg(F, \varphi, \Omega_1) + \deg(F, \varphi, \Omega_2).$$

$(I_{\mathcal{P}_{LS}})$ **(Existence).** *If* $\deg(F, \varphi, \Omega) \neq 0$ *then* $\operatorname{coin}_\Omega(F, \varphi) \neq \varnothing$.

$(J_{\mathcal{P}_{LS}})$ **(Excision-Additivity).** *If* $\Omega_i \subseteq \Omega$ $(i \in I)$ *is a family of pairwise disjoint open sets with* $\operatorname{coin}_\Omega(F, \varphi) \subseteq \bigcup_{i \in I} \Omega_i$ *and* $\operatorname{coin}_{\Omega_i}(F, \varphi)$ *is compact for all* $i \in I$, *then*

$$\deg(F, \varphi, \Omega) = \sum_{i \in I} \deg(F, \varphi, \Omega_i),$$

where in the sum at most a finite number of summands is nonzero.

($K_{\mathcal{P}_{LS}}$) **(Diffeomorphic-Isomorphic Invariance).** *Let J_1 be a diffeomorphism of an open subset of a Banach manifold X_0 onto Ω, and J_2 an isomorphism of Y onto a real normed vector space Y_0. Then*

$$\deg_{(X,Y)}(F,\varphi,\Omega) = \deg_{(X_0,Y_0)}(J_2 \circ F \circ J_1, J_2 \circ \varphi \circ J_1, J_1^{-1}(\Omega)).$$

In the oriented case, the orientation of $J_2 \circ F \circ J_1$ is understood in the sense of Proposition 8.38.

($L_{\mathcal{P}_{LS}}$) **(Restriction).** *Let $X_0 \subseteq X$ be open. Then*

$$\deg_{(X_0,Y)} = \deg_{(X,Y)}|_{\mathcal{P}_{LS}(X_0,Y)}.$$

($M_{\mathcal{P}_{LS}}$) **(Cartesian Product).** *For $i = 1, 2$, let X_i be a manifold without boundary of class C^1 over the real Banach space E_{X_i}, and let $Y_i = E_{Y_i}$ be a real Banach space. For $(F_i, \varphi_i, \Omega_i) \in \mathcal{P}_{LS}(X_i, Y_i)$, we put $X := X_1 \times X_2$, $\Omega := \Omega_1 \times \Omega_2$, $Y := Y_1 \times Y_2$, $F := F_1 \otimes F_2$, and $\varphi := \varphi_1 \otimes \varphi_2$. Then $(F, \Phi, \varphi, \Omega) \in \mathcal{P}_{LS}(X, Y)$ and*

$$\deg_{(X,Y)}(F,\varphi,\Omega) = \deg_{(X_1,Y_1)}(F_1,\varphi_1,\Omega_1)\deg_{(X_2,Y_2)}(F_2,\varphi_2,\Omega_2).$$

In the oriented case, F is equipped with the product orientation.

Proof. In view of Theorem 13.14, all properties follow from the corresponding properties of the Leray–Schauder triple degree with $\Phi = \mathrm{id}_\Omega$. □

Some general remark on our approach is appropriate: We have obtained the Leray–Schauder coincidence degree now as a trivial special case of the Leray–Schauder triple degree. Of course, one could prove it also without referring to any degree for function triples by just arguing analogously to our proofs in case $\Phi = \mathrm{id}_\Omega$.

However, it seems that it is not possible that we could have argued the other way, that is: If we would have established *first* the Leray–Schauder coincidence degree (without using function triples), and then would attempt to use that degree to define the Leray–Schauder triple degree, we would not have succeeded; at least no such way is known to the author: All which could have been used is the special case of the Benevieri–Furi coincidence degree in the way that we have used it in Section 13.1, that is, a previous extension of the Benevieri–Furi coincidence degree to a Leray–Schauder coincidence degree would have been in vain.

13.3 Classical Applications of the Leray–Schauder Degree

As already remarked after Definition 13.13, the "classical" Leray–Schauder degree is the special case $X = Y = E$ and $F = \mathrm{id}_\Omega$ (with the natural orientation). This special case is of course well-known and was in fact the first degree theory systematically developed.

In this section, we present some classical famous applications of this special case. So this section is mainly meant for the reader who is unfamiliar with the classical Leray–Schauder degree, although perhaps also the advanced reader might find some slight generalizations of well-known results.

For the rest of this section, we assume that $X = Y = E$ is a real Banach space, and we work only with the Leray–Schauder coincidence degree of the form $\deg(\mathrm{id}_\Omega, \varphi, \Omega)$ with locally compact φ. It is convenient to think of this as a degree for the map $F := \mathrm{id}_\Omega - \varphi$, and in fact, this is how the classical Leray–Schauder degree is usually defined in literature. However, since we consider this degree only in this section, we will not introduce a special notation for it but just consider it as a special case of the Leray–Schauder coincidence degree.

We show that essentially all applications of the Brouwer degree which we had given in Section 9.3 have "infinite-dimensional analogues" in Banach spaces for (locally) compact maps φ. The proofs consist just in applying analogous arguments, using the Leray–Schauder degree instead of the Brouwer degree.

Theorem 13.20 (Schauder Fixed Point). *Let M be a nonempty closed convex subset of a Banach space E, and $\varphi \in C(M, M)$ be compact. Then φ has a fixed point.*

Proof. By Theorem 4.36, M is an CE_M for the class of T_4 spaces. Since φ is compact, we can thus extend φ to some compact $\varphi \in C(E, M)$. In particular $C := \overline{\varphi(E)}$ is a compact subset of M. Note that $m(t, x) := tx$ is continuous, and so $m([0, 1] \times C)$ is compact. It follows that the map $h \in C([0, 1] \times E, E)$, $h(t, x) := t\varphi(x) = m(t, \varphi(x))$ is compact. Moreover, $\mathrm{coin}(\mathrm{id}_E, h)$ is a closed subset of the compact set $[0, 1] \times m([0, 1] \times C)$ and thus compact. The homotopy invariance of the Leray–Schauder coincidence degree implies that

$$\deg(\mathrm{id}_E, h(1, \cdot), E) = \deg(\mathrm{id}_E, h(0, \cdot), E) = \deg(\mathrm{id}_E, E, 0) = 1,$$

where the latter follows by the reduction property of the Leray–Schauder coincidence degree and the normalization of the Brouwer degree. The existence property implies that $\mathrm{coin}(\mathrm{id}_E, h(1, \cdot))$ contains some point x. Then $x = \varphi(x) \in M$ implies that x is a fixed point of the original map φ. □

The above proof was rather analogous to Theorem 9.76: Essentially, the only difference is that we had to replace the Brouwer degree by the Leray–Schauder (coincidence) degree. The same holds for the following result:

Theorem 13.21 (Leray-Schauder Alternative). *Let Ω be an open subset of a Banach space E. Let $\varphi \in C(\Omega, E)$ be locally compact. Then at least one of the following holds:*

(a) *φ has a fixed point.*

(b) *For each $x_0 \in \Omega$ the set*

$$C := \bigcup_{\lambda>1} \{x \in \Omega : \varphi(x) - x_0 = \lambda(x - x_0)\}$$

fails to be relatively compact in Ω.

Proof. Assume by contradiction that both properties fail, that is, φ has no fixed points, and there is $x_0 \in \Omega$ such that the corresponding set $\overline{C}$ is a compact subset of Ω. Since φ has no fixed points, it follows for the locally compact homotopy $h(t,x) := x_0 + t(\varphi(x) - x_0)$ that $\operatorname{coin}_\Omega(\operatorname{id}_\Omega, h(t, \cdot)) \subseteq \overline{C}$ for each $t \in [0,1]$. Hence,

$$K := \operatorname{coin}_{[0,1]\times\Omega}(\operatorname{id}_\Omega, h) \subseteq [0,1] \times \overline{C}.$$

Since K is closed in $[0,1] \times \Omega$ and thus closed in the compact set $[0,1] \times \overline{C}$, it follows that K is compact. The (generalized) homotopy invariance of the Leray–Schauder coincidence degree implies that

$$\deg(\operatorname{id}_\Omega, h(1, \cdot), \Omega) = \deg(\operatorname{id}_\Omega, h(0, \cdot), \Omega) = \deg(\operatorname{id}_\Omega, x_0, \Omega) = 1,$$

where the latter follows by the reduction property of the Leray–Schauder coincidence degree and the normalization of the Brouwer degree. The existence property implies that $\operatorname{coin}(\operatorname{id}_\Omega, h(1, \cdot)) \neq \varnothing$ which means that φ has fixed point. □

It seems that it is not so well-known that Theorem 13.21 holds for *locally* compact maps: In most text books, one can only find the more restrictive hypothesis that φ be compact or at least compact on bounded subsets of Ω. Also, it seems not to be so well-known that we need not assume for Theorem 13.21 that φ has a continuous extension to $\overline{\Omega}$.

Corollary 13.22 (Rothe's Fixed Point Theorem). *Let Ω be a nonempty open convex subset of a Banach space E, and $\varphi \in C(\overline{\Omega}, E)$ be compact. If $\varphi(\partial\Omega) \subseteq \overline{\Omega}$ then φ has a fixed point.*

Proof. The same argument as in Corollary 9.78 shows that the result is a special case of Theorem 13.21. □

Of course, using the full power of the Leray–Schauder coincidence degree, we could also formulate results similar to Theorem 13.21 for the case $F \neq \mathrm{id}_\Omega$. More general, using the Leray–Schauder triple degree, we could formulate existence results for the coincidence inclusion

$$F(x) \in \varphi(\Phi(x))$$

for the case that $\varphi \circ \Phi$ is locally compact. However, we will do this only later (Section 14.5), since we first want to discuss how to relax the hypothesis that $\varphi \circ \Phi$ be locally compact.

Mainly for completeness, we also discuss a variant of Borsuk's Theorem 9.84 for the Leray–Schauder degree and some of its consequences.

This theorem cannot be obtained by just imitating the proof of the finite-dimensional case, since an induction on the dimension of the space was involved. Instead, we reduce the result to the finite-dimensional setting and apply the finite-dimensional result.

Theorem 13.23 (Borsuk). *Let E be a Banach space, and $\Omega \subseteq E$ be open with $-\Omega = \Omega$. Let $\varphi \in C(\Omega, E)$ be locally compact and odd, that is $\varphi(-x) = -\varphi(x)$ for all $x \in \Omega$. If $\operatorname{fix}\varphi := \{x \in \Omega : \varphi(x) = x\}$ is compact then*

$$\deg(\mathrm{id}_\Omega, \varphi, \Omega) \text{ is } \begin{cases} \text{odd} & \text{if } 0 \in \Omega, \\ \text{even} & \text{if } 0 \notin \Omega. \end{cases} \tag{13.7}$$

Proof. By Proposition 2.125, there is an open neighborhood $\Omega_0 \subseteq \Omega$ of $\operatorname{fix}\varphi$ such that $\varphi|_{\Omega_0}$ is compact. Replacing Ω by $\Omega_0 \cap (-\Omega_0)$ if necessary, we can assume in view of the excision property without loss of generality that $\Omega = \Omega_0$. By Corollary 3.62, we may thus assume that $C := \overline{\operatorname{conv}}\,\varphi(\Omega)$ is compact. By Corollary 2.48, there is an open neighborhood $U \subseteq E$ of $\operatorname{fix}\varphi$ (or of $\{0\} \cup \operatorname{fix}\varphi$ in case $0 \in \Omega$) with $N := \overline{U} \subseteq \Omega$. Replacing U by $U \cap (-U)$ if necessary, we can assume without loss of generality that $-U = U$ and $-N = N$. Then also $K := N \cap \mathrm{id}_\Omega^{-1}(C) = \Omega \cap C$ satisfies $-K = K$. Proposition 3.15 implies that

$$\varepsilon := \min_{x \in K \setminus U} \operatorname{dist}(x, \varphi(x)) > 0.$$

By Corollary 13.10, there is a finite-dimensional subspace $Y_0 \subseteq E$ and a function $\varphi_0 \in C(N, Y_0 \cap \operatorname{conv} C)$ satisfying (13.1). Replacing φ_0 by $\hat{\varphi}_0(x) := \frac{1}{2}(\varphi(x) -$

$\varphi_0(-x))$ if necessary, we can assume that φ_0 is odd. By Proposition 13.11, we obtain

$$\deg(\mathrm{id}_\Omega, \varphi, \Omega) = \deg(\mathrm{id}_U, \varphi_0, U).$$

The right-hand side can also be interpreted as the Benevieri–Furi coincidence degree. By the reduction property of the latter, we obtain if view of $U_0 := \mathrm{id}_U^{-1}(Y_0) = Y_0 \cap U$ that this degree is the same as the Brouwer degree

$$\deg_{(U_0,Y_0)}(\mathrm{id}_{U_0} - \varphi_0, U_0, 0) = \deg_{(Y_0,Y_0)}(\mathrm{id}_{U_0} - \varphi_0, U_0, 0),$$

where we used the restriction property of the Brouwer degree. Note that $-U_0 = U_0$ and that $\mathrm{id}_{U_0} - \varphi_0$ is an odd function. Since $0 \in U_0$ if and only if $0 \in \Omega$, the assertion thus follows from Borsuk's Theorem 9.84 for the Brouwer degree. □

In infinite dimensions, it is more convenient to formulate the analogue of Corollary 9.86 in terms of the map $\varphi = \mathrm{id}_\Omega - F$ instead of F so that the name "fixed point theorem" is really justified (recall Remark 9.88).

Similarly as for the Leray–Schauder alternative, we point out that it is perhaps not so well-known that the following variant of Borsuk's fixed point theorem holds also for *locally* compact maps, unbounded Ω, and if the map φ does not necessarily possess a continuous extension to $\overline{\Omega}$.

Theorem 13.24 (Borsuk Fixed Point). *Let E be a Banach space, and $\Omega \subseteq E$ be open with $-\Omega = \Omega$. Let $\varphi \in C(\Omega, E)$ be locally compact and such that*

$$C := \{x \in \Omega : \|x + \varphi(-x)\|(x - \varphi(x)) = -\|x - \varphi(x)\|(x + \varphi(-x))\}$$

is compact; this holds in particular if $\varphi \in C(\overline{\Omega}, E)$ is compact and satisfies

$$\|x + \varphi(-x)\|(x - \varphi(x)) \neq -\|x - \varphi(x)\|(x + \varphi(-x)) \quad \text{for all } x \in \partial\Omega. \tag{13.8}$$

Then (13.7) *holds, and in case $0 \in \Omega$, the map φ has a fixed point in Ω.*

Proof. We put $F := \mathrm{id}_\Omega - \varphi$. In case (13.8), we have

$$C = \{x \in \overline{\Omega} : \|F(-x)\|F(x) = \|F(x)\|F(-x)\},$$

and by the continuity of F and of the norm, the set on the right-hand side is closed in E. Hence, (13.8) implies that C is closed in E. Moreover, $x \in C$ implies for $x \neq \varphi(x)$ that

$$x = \frac{\|x + \varphi(-x)\|\varphi(x) + \|x - \varphi(x)\|(-\varphi(-x))}{\|x + \varphi(-x)\| + \|x - \varphi(x)\|} \in \operatorname{conv}\{\varphi(x), -\varphi(-x)\}.$$

Thus, C is contained in the set $\overline{\operatorname{conv}}(\varphi(\Omega) \cup (-\varphi(\Omega)))$ which is compact if φ is compact (Corollary 3.62). Consequently, in case (13.8), we have that C is a closed subset of a compact set and thus compact.

To prove the main assertion, we define $H: [0,1] \times \Omega \to E$ by

$$H(t,x) := F(x) - tF(-x).$$

Then

$$h(t,x) := x - \frac{H(t,x)}{t+1} = \frac{\varphi(x) - t\varphi(-x)}{1+t}$$

is a locally compact homotopy. Moreover, we have $x = h(t,x)$ if and only if $F(x) = tF(-x)$. In case $F(x) \neq 0$, this implies that $F(x)$ is a positive multiple of $F(-x)$, and so $x \in C$. In case $F(x) = 0$, we have trivially $x \in C$. Hence, $\operatorname{coin}_{[0,1]\times\Omega}(\operatorname{id}_\Omega, h)$ is a closed subset of the compact set $[0,1]\times C$ and thus compact. The homotopy invariance of the Leray–Schauder coincidence degree thus implies

$$\deg(\operatorname{id}_\Omega, \varphi, \Omega) = \deg(\operatorname{id}_\Omega, h(0,\,\cdot\,), \Omega) = \deg(h(1,\,\cdot\,), \Omega, 0).$$

Since $h(1,\,\cdot\,)$ is an odd map, Borsuk's Theorem 13.23 implies that the latter degree satisfies (13.7). In case $0 \in \Omega$, the existence property thus implies that φ has a fixed point in Ω. □

We have intentionally used the notation of Corollary 9.86 in the proof of Theorem 13.24 to make clear that the proof of the main assertion is practically identical: One only has to use the Leray–Schauder coincidence degree $\deg(\operatorname{id}_\Omega, \varphi, \Omega)$ instead of the Brouwer degree $\deg(F, \Omega, 0)$ and apply Theorem 13.23. The only real difference in the proof is that we could not directly put $h(t,x) = x - H(t,x)$ but also had to divide the last term by $t+1$ to obtain a locally compact homotopy.

In the same manner, one can generalize Theorem 9.94.

Theorem 13.25 (Invariance of Domain). *Let E be a finite-dimensional normed space, and $\Omega \subseteq E$ be open. If $F \in C(\Omega, E)$ is locally one-to-one and $\varphi := \operatorname{id}_\Omega - F$ is locally compact then $F(\Omega)$ is open in E.*

Proof. The proof is analogous to the proof of Theorem 9.94, just replacing the Brouwer degree $\deg(F, \Omega_0, 0)$ by the Leray–Schauder degree $\deg(\operatorname{id}_\Omega, \varphi, \Omega_0)$. The only difference is that we cannot work directly with the homotopy $h(t,x) = x - H(t,x)$, but we have to use instead the homotopy $h(t,x) := x - H(t,x)/(1+t)$, or alternatively $h(t,x) := x - H(t, x/(1+t))$, to obtain a (locally) compact map. □

We close this section by noting that it is also possible to prove a more general version of the Borsuk and Invariance of Domain Theorems for the Leray–Schauder coincidence degree when $F \neq \mathrm{id}$, see e.g. [14], [32], but we restricted ourselves in this section to the more classical examples.

Chapter 14

The Degree for Noncompact Fredholm Triples

Since Darbo's celebrated fixed point theorem [36], many attempts have been made to transfer topological results from the "compact setting" to a "condensing setting". Condensing variants of the classical Leray-Schauder degree have been developed by R. D. Nussbaum [116], [118] and B. N. Sadovskiĭ [129], see also e.g. [2], [38], [83]. Variants of these results in the multivalued case can be found in the earlier mentioned papers [53], [55], [56], [58], [59], [144], [148]. The condensing version of the Benevieri–Furi coincidence degree was developed in [15]. The earlier mentioned papers [121], [146] with a multivalued variant (similar to our triple degree) contain also corresponding results for the condensing case.

Since it turned out that one can often estimate measures of noncompactness only on countable subsets (see [76], [109] or [138, §11 and §12]), it appears important to have also corresponding "countable condensing" variants of the results. Early corresponding results involving fixed points of single-valued maps can be found in [35], [80], [108], [133], the corresponding variant for the classical degree and a more general homotopical theory can be found in [136] and [139], respectively. More general degree theories in the multivalued case can be treated with [137].

Simultaneously, it has also become clear that condensing maps are usually related with convexity assumptions. This makes it very hard to employ these methods on nonconvex sets, for instance for maps on spheres, which occur naturally if one deals with positively homogeneous operators. Especially R. D. Nussbaum has considered maps in such cases [117], [119].

It turned out that the concept of a so-called fundamental set can be transferred also to the nonconvex setting [6] (see also [5]), and that this concept leads at least in the convex setting to results for countably condensing maps. This led the author in [141] to an axiomatic approach which describes a general scheme how to transfer a degree theory for "compact" function triples to a degree theory for noncompact function triples. We can obtain now even stronger results than by just applying the results from [141] to the Leray–Schauder triple degree, since the Leray–Schauder triple degree has stronger properties than those required in [141]. In particular, we can avoid to work with closures of $\Omega \subseteq X$ which was one of the main hypotheses about the given degree in [141]. We point out that in view of

Remark 2.86, we obtain thus a much more natural theory.

Although the details are somewhat involved, it is rather easy to describe the rough idea of our approach: The idea is to call a set $K \subseteq Y$ fundamental for a function triple (F, Φ, φ) if the restriction of $\varphi \circ \Phi$ to $F^{-1}(K)$ is compact and maps *into* K, and if "everything which is important concerning the degree" occurs on $F^{-1}(K)$. Then one can prove that there is a degree theory for all triples which possess a fundamental set. Afterwards, one can verify that, for instance, "countably condensing triples" possess a convex fundamental set and thus one has a corresponding degree theory for such triples.

The plan of this chapter is as follows.

In Section 14.1, we will develop the corresponding degree theory for triples which possess a fundamental set. The notion of a fundamental set is introduced axiomatically by describing the properties which we need for developing the degree. Unfortunately, the notion of a fundamental set is somewhat involved, and so the axiomatic description may appear hard to verify. However, this is actually not the case: In Section 14.2, we will give simpler (sufficient) criteria for fundamental sets. Using these criteria, we will give in Section 14.3 a sufficient criterion for convex fundamental sets. We will see that the corresponding degree theory becomes even nicer for these triples. In Section 14.4, we show that this degree theory applies for "countably condensing triples". In Section 14.5, we transfer some of the classical applications of the Brouwer and Leray–Schauder degree (of Sections 9.6 and 13.3) to the setting of "countably condensing function triples". In the final Section 14.6, we give an example demonstrating how the theory developed in this monograph can be applied to a wide class of boundary value problems.

The advantage of our approach is not only that we obtain a clearer presentation by separating the *usage* of fundamental sets in degree theory and how to *verify* that such sets exist. There is also the mathematical advantage that the results in Section 14.1 and 14.2 are not related to convexity and thus might also be used in certain "nonconvex" situations, for instance to obtain results for fixed points of triples on spheres. However, the latter is currently a topic of future research, although special cases of the so-called "pushing condition" which occurs naturally in our approach were already considered [60], [61].

14.1 The Degree for Fredholm Triples with Fundamental Sets

Throughout, we assume that X is a C^1 Banach manifold without boundary over a real Banach space E_X, and that $Y = E_Y$ is a real Banach space such that the non-degeneracy hypothesis (10.8) holds, that is $E_X \neq \{0\}$ and $E_Y \neq \{0\}$ (if we assume AC).

Actually, we will define two slightly different variants of the degree for fundamental sets. In order to treat both variants simultaneously, we introduce the following notation.

Definition 14.1. For a topological space Γ and $K \subseteq Z \subseteq Y$, we denote by $\mathrm{Comp}_Z(\Gamma, K)$ the set of all $\varphi \in C(\Gamma, K)$ which are compact into Z, that is:

(a) $\varphi\colon \Gamma \to Y$ is continuous.

(b) $\varphi(\Gamma) \subseteq K$, and

(c) $\varphi(\Gamma)$ is contained in a compact subset of Z.

By $\mathrm{Comp}(\Gamma, K)$ we mean either $\mathrm{Comp}_K(\Gamma, K)$ or $\mathrm{Comp}_Y(\Gamma, K)$, the choice being made once and for all for the rest of this section.

Depending on how we have chosen $\mathrm{Comp}(\Gamma, K)$, we will obtain two degree theories under slightly different assumptions. If K is closed in Y (which is the most important case), we have

$$\mathrm{Comp}_Y(\Gamma, K) = \mathrm{Comp}_K(\Gamma, K),$$

so that in this important case the two degree theories actually coincide.

The idea for the definition of the degree for a function triple (F, Φ, φ) is to fix an appropriate set $K \subseteq Y$, and to replace φ by a function $\varphi_0 \in \mathrm{Comp}(\Gamma, K)$ which coincides with φ on $\Phi(F^{-1}(K))$. We intend to call K fundamental when this idea leads to a well-defined degree.

The choice of φ_0 is reasonable as the following lemma shows:

Lemma 14.2. *Let $(F, \Phi, \varphi, \Omega, Y, \Gamma)$ be some function triple, and $K \subseteq Y$ be a set containing* $\mathrm{fix}_\Omega(F, \Phi, \varphi)$. *If $\Omega_0 \subseteq \Omega$ contains* $\mathrm{coin}_\Omega(F, \Phi, \varphi)$, *and if $\varphi_0\colon \Omega_0 \to K$ satisfies $\varphi_0(z) = \varphi(z)$ for all $z \in \Phi(\Omega_0 \cap F^{-1}(K))$, then*

$$\mathrm{coin}_{\Omega_0}(F, \Phi, \varphi_0) = \mathrm{coin}_\Omega(F, \Phi, \varphi). \tag{14.1}$$

Proof. Since $\varphi_0(\Omega_0) \subseteq K$, we have

$$\mathrm{coin}_{\Omega_0}(F, \Phi, \varphi_0) = \mathrm{coin}_{\Omega_0 \cap F^{-1}(K)}(F, \Phi, \varphi_0) = \mathrm{coin}_{\Omega_0 \cap F^{-1}(K)}(F, \Phi, \varphi).$$

Since $\mathrm{fix}_\Omega(F, \Phi, \varphi) \subseteq K$, we thus obtain

$$\mathrm{coin}_{\Omega_0}(F, \Phi, \varphi_0) = \mathrm{coin}_{\Omega_0}(F, \Phi, \varphi).$$

The last set is $C := \mathrm{coin}_\Omega(F, \Phi, \varphi)$, because $C \subseteq \Omega_0$. □

Before we can give the definition of fundamental sets, we have to introduce an auxiliary definition. We denote by deg the Leray–Schauder triple degree.

Definition 14.3. Let $(F, \Phi, \varphi, \Omega) \in \mathcal{T}_{\text{prop}}(X, Y)$. We call $K \subseteq Y$ a *retraction candidate* if there is an open neighborhood $\Omega_0 \subseteq \Omega$ of $\text{coin}_\Omega(F, \Phi, \varphi)$ such that the following holds.

(i) $\text{fix}_\Omega(F, \Phi, \varphi) \subseteq K$.

(ii) There is $\varphi_0 \in \text{Comp}(\Phi(\Omega_0), K)$ satisfying $\varphi_0(z) = \varphi(z)$ for all $z \in \Phi(\Omega_0 \cap F^{-1}(K))$.

(iii) For each open neighborhood $\Omega_1 \subseteq \Omega_0$ of $\text{coin}_\Omega(F, \Phi, \varphi)$ and each $\varphi_0, \varphi_1 \in \text{Comp}(\Phi(\Omega_1), K)$ satisfying $\varphi_0(z) = \varphi_1(z) = \varphi(z)$ for all $z \in \Phi(\Omega_1 \cap F^{-1}(K))$, we have

$$\deg(F, \Phi, \varphi_0, \Omega_1) = \deg(F, \Phi, \varphi_1, \Omega_1).$$

Note that (ii) implies in particular that $(\varphi \circ \Phi)(\Omega_1 \cap F^{-1}(K))$ is a subset of K and relatively compact in Y or K (depending on the choice of $\text{Comp}(\Gamma, K)$, recall Definition 14.1).

We will see in Theorem 14.19 that this necessary condition together with (i) is already sufficient that K is a retraction candidate, provided that K is an CNE_Y or CNE_K, respectively (if $\Phi(\Omega)$ satisfies certain separation properties). Recall in particular that K is an CNE_K if it is an ANR (Corollary 4.38) or a neighborhood retract of Y (Theorem 4.36). This connection with neighborhood retracts motivated our choice of the terminology "retraction candidate".

Now we come to the definition of fundamental sets. This definition depends on the choice of a family $\mathcal{A}$ of subsets $K \subseteq Y$. In connection with condensing maps, we will use the family $\mathcal{A}$ of all closed convex subsets of Y. This is in a sense the simplest case and will be discussed later in more detail.

Definition 14.4. Let $\mathcal{A}$ denote a fixed family of subsets $K \subseteq Y$, and $(F, \Phi, \varphi, \Omega) \in \mathcal{T}_{\text{prop}}(X, Y)$. We call K *$\mathcal{A}$-fundamental* for $(F, \Phi, \varphi, \Omega)$ if K is a retraction candidate and if either $K = \emptyset$ or if $K \in \mathcal{A}$ and for each retraction candidate $K_0 \in \mathcal{A}$ with $K \cap K_0 \neq \emptyset$ and $K \neq K_0$ at least one of the sets $K_1 := K \cap K_0$ or $K_1 := K \cup K_0$ has the following property.

There is an open neighborhood $\Omega_0 \subseteq \Omega$ of $\text{coin}_\Omega(F, \Phi, \varphi)$ such that the following holds:

(i) There is $\varphi_1 \in \text{Comp}(\Phi(\Omega_0), K_1)$ satisfying $\varphi_1(z) = \varphi(z)$ for all $z \in \Phi(\Omega_0 \cap F^{-1}(K_1))$.

(ii) For each open neighborhood $\Omega_1 \subseteq \Omega_0$ of $\operatorname{coin}_\Omega(F, \Phi, \varphi)$, each $\varphi_0 \in \operatorname{Comp}(\Phi(\Omega_1), K)$ satisfying $\varphi_0(z) = \varphi(z)$ for all $z \in \Phi(\Omega_1 \cap F^{-1}(K))$, and each $\varphi_1 \in \operatorname{Comp}(\Phi(\Omega_1), K_1)$ satisfying $\varphi_1(z) = \varphi(z)$ for all $z \in \Phi(\Omega_1 \cap F^{-1}(K_1))$, we have

$$\deg(F, \Phi, \varphi_0, \Omega_1) = \deg(F, \Phi, \varphi_1, \Omega_1).$$

Remark 14.5. The term "fundamental set" is in literature often defined rather differently: Often, this term is reserved for what we call (almost) convex-fundamental in Section 14.3 and which turns out to be a special case of the above definition if $\mathcal{A}$ denotes the family of closed convex subsets of Y.

The above definition is much more general and describes axiomatically the core property needed to develop a degree theory for function triples which possess a fundamental set. For this reason, the above definition appears to the author rather natural for the term "fundamental". There is no reason to tie this notion with convexity or closedness. In fact, any not too large family of neighborhood retracts of Y (which need neither be closed nor convex) is a good choice for $\mathcal{A}$, as we will see.

There is no monotonicity of Definition 14.4 with respect to the family $\mathcal{A}$: If the family $\mathcal{A}$ becomes larger, the number of retraction candidates in $\mathcal{A}$ increases, but simultaneously the condition (ii) of Definition 14.4 becomes more restrictive if that number increases. An extreme example for this is the following.

Example 14.6. Let $\mathcal{A} = \{K\}$ contain only one nonempty set $K \subseteq Y$. In that case, the set K is $\mathcal{A}$-fundamental if and only if it is a retraction candidate.

Definition 14.7. We denote by $\mathcal{T}_{\text{fund}}(X, Y, \mathcal{A})$ the class of all $(F, \Phi, \varphi, \Omega) \in \mathcal{T}_{\text{prop}}(X, Y)$ which possess an $\mathcal{A}$-fundamental set. The *$\mathcal{A}$-fundamental Fredholm triple degree* is an operator $\operatorname{DEG} = \operatorname{DEG}_{(X,Y,\mathcal{A})}$ which associates to each $(F, \Phi, \varphi, \Omega) \in \mathcal{T}_{\text{fund}}(X, Y, \mathcal{A})$ a number from $\mathbb{Z}_2$ (or $\mathbb{Z}$ in the oriented case) such that the following property holds for every $(F, \Phi, \varphi, \Omega) \in \mathcal{T}_{\text{fund}}(X, Y, \mathcal{A})$:

($\mathrm{A}_{\mathcal{T}_{\text{fund}}}$) **(Permanence)**. If K is $\mathcal{A}$-fundamental for $(F, \Phi, \varphi, \Omega)$, $\Omega_0 \subseteq \Omega$ is an open neighborhood of $\operatorname{coin}_\Omega(F, \Phi, \varphi)$, and $\varphi_0 \in \operatorname{Comp}(\Phi(\Omega_0), K)$ satisfies $\varphi_0(z) = \varphi(z)$ for all $z \in \Phi(\Omega_0 \cap F^{-1}(K))$ then

$$\operatorname{DEG}(F, \Phi, \varphi, \Omega) = \deg(F, \Phi, \varphi_0, \Omega_0),$$

where the right-hand side denotes the Leray–Schauder triple degree.

We point out that the $\mathcal{A}$-fundamental Fredholm triple degree need not satisfy the strong form of the equivalence invariance as the other triple degrees which we considered. Nevertheless, the property ($D_{\mathcal{T}_{\text{fund}}}$) in the subsequent Theorem 14.8 shows that we may suppress Γ in the notation, that is, the degree depends only on the subset $\Phi(\Omega) \subseteq \Gamma$ (and its topology) and not on the whole space Γ.

We also point out that this is the first time that we use a different symbol DEG for the degree. We do this to make clear that this degree does not necessarily possess the strong compatibility properties which we had previously. In particular, it need not be an extension of the Leray–Schauder degree. These difficulties are caused by the fact that the degree will depend on the choice of the family $\mathcal{A}$, in general. In case $Y \in \mathcal{A}$, we have the compatibility with the previous degrees, as we will see.

We will see that DEG exists and is unique, and that it has properties which are rather analogous to the properties which we know from the other degree theories, although we need some additional hypotheses in some cases.

Before we attack this program, let us give a bit motivation about the meaning of the degree and a small historical comparison. This is best be done in the particular case $F = \mathrm{id}_\Omega$ (with the natural orientation) and $\Phi = \mathrm{id}_\Omega$. In this case, if K is $\mathcal{A}$-fundamental, we must necessarily have $\varphi(K \cap \Omega_0) \subseteq K$ for some open neighborhood $\Omega_0 \subseteq \Omega$ of the fixed point set of φ. Moreover, the degree actually depends only on the restriction $\varphi|_K$. The degree is thus a count for the number of fixed points of $\varphi \in C(\Omega_K, K)$ where $\Omega_K \subseteq K$ is open in K. Hence, it makes sense to call the degree in this point the fixed point index on K.

It will follow from the results of Section 14.2 (Theorem 14.19) that K is $\{K\}$-fundamental for $(\mathrm{id}_\Omega, \mathrm{id}_\Omega, \varphi, \Omega)$ (recall Example 14.6) if K is an ANR and $\varphi \in \mathrm{Comp}(\Omega_0 \cap K, K)$ (and if we made the choice $\mathrm{Comp}(\Gamma, K) = \mathrm{Comp}_K(\Gamma, K)$).

Hence, our degree actually includes the fixed point index for compact maps on ANR spaces which was originally introduced by A. Granas [75] and generalized by R. D. Nussbaum for noncompact maps [115], [116].

However, if we consider more maps F and Φ than only restrictions of given fixed maps, it is obviously not true that the degree depends only on the restriction of the function triple $(F, \Phi, \varphi, \Omega)$ to $\Omega_0 \cap F^{-1}(K)$. Thus, in this more general situation, it is not possible to speak about a corresponding "coincidence point index" theory. Therefore, we call DEG a degree, although it is actually a generalization of what historically is called the *fixed point index*. (Note that for the case $X = Y$, $\mathcal{A} = \{Y\}$, $F = \Phi = \mathrm{id}_\Omega$, the above degree is actually the fixed point index of φ which we discussed after Theorem 10.1.)

Theorem 14.8. *For each $(X, Y, \mathcal{A})$ there is exactly one $\mathcal{A}$-fundamental Fredholm triple degree. This degree has automatically the following properties for each $(F, \Phi, \varphi, \Omega) \in \mathcal{T}_{\text{fund}}(X, Y, \mathcal{A})$:*

$(\mathrm{B}_{\mathcal{T}_{\text{fund}}})$ **(Excision).** *If $\Omega_0 \subseteq \Omega$ is open and contains $\operatorname{coin}(F, \Phi, \varphi)$ then*

$$\operatorname{DEG}(F, \Phi, \varphi, \Omega_0) = \operatorname{DEG}(F, \Phi, \varphi, \Omega).$$

$(\mathrm{C}_{\mathcal{T}_{\text{fund}}})$ **(Existence).** *If $\operatorname{DEG}(F, \Phi, \varphi, \Omega) \neq 0$ then $\operatorname{coin}_\Omega(F, \Phi, \varphi) \neq \varnothing$.*

$(\mathrm{D}_{\mathcal{T}_{\text{fund}}})$ **(Restriction in the Last Function).** *We denote the function triple more verbosely as $(F, \Phi, \varphi, \Omega, Y, \Gamma)$. This function triple has the same $\mathcal{A}$-fundamental sets as $(F, \Phi, \varphi|_{\Phi(\Omega)}, \Omega, Y, \Phi(\Omega))$, and*

$$\operatorname{DEG}(F, \Phi, \varphi|_{\Phi(\Omega)}, \Omega) = \operatorname{DEG}(F, \Phi, \varphi, \Omega).$$

$(\mathrm{E}_{\mathcal{T}_{\text{fund}}})$ **(Compatibility with the Non-Oriented Case).** *The degrees for the oriented and non-oriented case are the same modulo 2 (if the oriented case applies).*

$(\mathrm{F}_{\mathcal{T}_{\text{fund}}})$ **(Weak Compatibility with the Leray–Schauder Triple Degree).** *If $(F, \Phi, \varphi, \Omega) \in \mathcal{T}_{\text{LS}}(X, Y)$ and if there is an $\mathcal{A}$-fundamental set K which contains $\varphi(\Phi(\Omega_0))$ for some neighborhood $\Omega_0 \subseteq \Omega$ of $\operatorname{coin}_\Omega(F, \Phi, \varphi)$ then*

$$\operatorname{DEG}(F, \Phi, \varphi, \Omega) = \deg(F, \Phi, \varphi, \Omega).$$

Proof. Let K be $\mathcal{A}$-fundamental for $(F, \Phi, \varphi, \Omega)$, and $\Omega_0 \subseteq \Omega$ be as in Definition 14.3. For any open neighborhood $\Omega_1 \subseteq \Omega_0$ of $C := \operatorname{coin}_\Omega(F, \Phi, \varphi) = \operatorname{coin}_{\Omega_0}(F, \Phi, \varphi)$, we find some $\varphi_0 \in \operatorname{Comp}(\Phi(\Omega_1), K)$ with $\varphi_0(z) = \varphi(z)$ for each $z \in \Phi(\Omega_0 \cap F^{-1}(K))$, namely the restriction of the function of Definition 14.3(ii) to Ω_1. The uniqueness assertion follows now, since the permanence property implies

$$\operatorname{DEG}(F, \Phi, \varphi, \Omega) = \deg(F, \Phi, \varphi_0, \Omega_1). \tag{14.2}$$

For the existence assertion, we use (14.2) to define the left-hand side. We must show that this is well-defined, that is, independent of the particular choice of K, Ω_0, Ω_1, and φ_0.

We show first the independence of the choice of $(\Omega_0, \Omega_1, \varphi_0)$. Let $(\hat{\Omega}_0, \hat{\Omega}_1, \varphi_1)$ be possibly different choices. We put $\Omega_2 := \Omega_1 \cap \hat{\Omega}_1$. Definition 14.3 implies (with Ω_2 replacing Ω_1) that

$$\deg(F, \Phi, \varphi_0, \Omega_2) = \deg(F, \Phi, \varphi_1, \Omega_2). \tag{14.3}$$

We recall that (14.1) implies

$$\operatorname{coin}_{\Omega_1}(F,\Phi,\varphi_0) = C. \tag{14.4}$$

Hence, the excision property of the Leray–Schauder triple degree implies

$$\deg(F,\Phi,\varphi_0,\Omega_1) = \deg(F,\Phi,\varphi_0,\Omega_2).$$

Since an analogous calculation shows

$$\deg(F,\Phi,\varphi_1,\hat{\Omega}_1) = \deg(F,\Phi,\varphi_1,\Omega_2),$$

we obtain in view of (14.3) that

$$\deg(F,\Phi,\varphi_0,\Omega_1) = \deg(F,\Phi,\varphi_1,\hat{\Omega}_1).$$

This shows the independence of (14.2) of the particular choice of $(\Omega_0,\Omega_1,\varphi_0)$.

As a side result, we note that (14.4) and the existence property of the Leray–Schauder coincidence degree imply that the right-hand side of (14.2) vanishes in case $C = \varnothing$; in particular, the existence property follows.

Now we show the independence of the choice of K. Thus, let $K_0 \neq K$ be a different choice. Assume first that $K \cap K_0 = \varnothing$. Since $F(C) \subseteq K$ and $F(C) \subseteq K_0$, we must have $C = \varnothing$. In this case, we have already seen that the right-hand side of (14.2) vanishes; hence, in this case, we obtain certainly the (same!) value 0 by our definition (14.2). Thus, it remains to consider the case $K \cap K_0 \neq \varnothing$. Then one of $K_1 = K \cap K_0$ or $K_1 = K \cup K_0$ has the properties required in Definition 14.4. Since we have already seen the independence of our definition of the particular choice of Ω_0, namely that we can replace it by any smaller open neighborhood of $\operatorname{coin}_{\Omega}(F,\Phi,\varphi)$, we can assume without loss of generality that the same set Ω_0 can be used in Definition 14.4 as for K in Definition 14.3 and, moreover, that the same set Ω_0 can be used for both definitions if the roles of K and K_0 are exchanged (we replace Ω_0 by the intersection of the corresponding four sets if necessary).

Assume now first that $K_1 = K \cap K_0$ has the property required in Definition 14.4. Then there is $\varphi_1 \in \operatorname{Comp}(\Phi(\Omega_1), K_1)$ with $\varphi_1(z) = \varphi(z)$ for all $z \in \Phi(\Omega_1 \cap F^{-1}(K_1))$. Then $\varphi_1(z) = \varphi(z)$ for all $z \in \Phi(\Omega_1 \cap F^{-1}(K))$ and for all $z \in \Phi(\Omega_1 \cap F^{-1}(K_0))$, and so we can choose the function $\varphi_0 := \varphi_1$ in the definition (14.2) of the degree for K and for K_0. Hence, we obtain the same value for the degree when we choose K_0 instead of K.

Assume now that $K_1 = K \cup K_0$ has the property required in Definition 14.4. Then there is $\varphi_2 \in \operatorname{Comp}(\Phi(\Omega_1), K_1)$ with $\varphi_2(z) = \varphi(z)$ for all $z \in \Phi\big(\Omega_2 \cap F^{-1}(K_1)\big)$. By Definition 14.3, there are $\varphi_0 \in \operatorname{Comp}(\Phi(\Omega_1), K)$ with $\varphi_0(z) =$

$\varphi(z)$ for all $z \in \Phi(\Omega_1 \cap F^{-1}(K))$ and $\varphi_1 \in \mathrm{Comp}(\Phi(\Omega_1), K_0)$ with $\varphi_1(z) = \varphi(z)$ for all $z \in \Phi(\Omega_1 \cap F^{-1}(K_0))$. By Definition 14.4, we obtain

$$\deg(F, \Phi, \varphi_0, \Omega_1) = \deg(F, \Phi, \varphi_2, \Omega_1). \tag{14.5}$$

Now we apply Definition 14.4 with the roles of K and K_0 exchanged: For the case that $\hat{K}_1 = K_0 \cap K_1$ has the property required in this definition, we have already shown that we obtain the same values for the right-hand side of (14.2). In the opposite case, $K_1 = K_0 \cup K_1$ must have the property required in this definition, and so we obtain analogously to (14.5) that

$$\deg(F, \Phi, \varphi_1, \Omega_1) = \deg(F, \Phi, \varphi_2, \Omega_1).$$

Together with (14.5), we conclude that

$$\deg(F, \Phi, \varphi_0, \Omega_1) = \deg(F, \Phi, \varphi_1, \Omega_1),$$

and so the definition (14.2) is indeed independent of the choice of the $\mathcal{A}$-fundamental set K.

The independence of the choice of Ω_0 in our definition implies that the thus defined degree satisfies the excision and permanence properties. Concerning ($\mathrm{D}_{\mathcal{T}_{\mathrm{fund}}}$), it suffices to note that neither Definition 14.3 nor Definition 14.4 nor anything from the above considerations required to consider any point or subset from $\Gamma \setminus \Phi(\Omega)$ or of the restriction of φ to that set. The compatibility of the degree with the non-oriented case follows immediately from the definition (14.2) and the corresponding property of the Leray–Schauder triple degree. For the weak compatibility with the Leray–Schauder triple degree, we note that, in view of Proposition 13.7, we can assume that $\varphi|_{\Phi(\Omega_0)} \in \mathrm{Comp}(\Phi(\Omega_0), K)$, and so we can choose $\varphi_0 = \varphi$ in the definition (14.2). □

Corollary 14.9. *The $\mathcal{A}$-fundamental Fredholm triple degree has the following property:*

($\mathrm{G}_{\mathcal{T}_{\mathrm{fund}}}$) **(Compatibility with the Leray–Schauder Triple Degree).** *Suppose that $Y \in \mathcal{A}$. If $(F, \Phi, \varphi, \Omega) \in \mathcal{T}_{\mathrm{LS}}(X, Y)$ then Y is $\mathcal{A}$-fundamental for $(F, \Phi, \varphi, \Omega)$, and*

$$\mathrm{DEG}(F, \Phi, \varphi, \Omega) = \deg(F, \Phi, \varphi, \Omega).$$

Proof. We verify first that $K = Y$ is a retraction candidate. Indeed, let Ω_0 be as in Proposition 13.7. Since $\Phi(\Omega_0 \cap F^{-1}(K)) = \Phi(\Omega_0)$, we can in Definition 14.3(ii) choose $\varphi_0 := \varphi|_{\Phi(\Omega_0)}$, and the property (iii) of Definition 14.3 is trivial, since

$\varphi_0 = \varphi_1$. These considerations imply also that $K = Y$ satisfies the two properties of Definition 14.4 with $K_1 = K \cup K_0 = Y$. Hence, Y is $\mathcal{A}$-fundamental. The assertion now follows from the weak compatibility with the Leray–Schauder triple degree. □

The following example shows that even in simple finite-dimensional situations it can happen that DEG differs from deg, so that the hypothesis $Y \in \mathcal{A}$ in Corollary 14.9 is really essential.

Example 14.10. Let $\Omega = X = Y = \mathbb{R}^2$, $F = \Phi = \mathrm{id}_\Omega$, and $\varphi = 2\,\mathrm{id}_\Omega$. Then $\mathrm{coin}(F, \Phi, \varphi) = \{0\}$, and the normalization property of the Leray–Schauder triple degree implies

$$\deg(F, \Phi, \varphi, \Omega) = \deg(F - \varphi, \Omega, 0) = \deg(-\mathrm{id}_\Omega, \Omega, 0) = \operatorname{sgn}\det(-\mathrm{id}_{\mathbb{R}^2}) = 1.$$

On the other hand, we choose now $\mathcal{A} := \{K\}$ with $K := \mathbb{R} \times \{0\}$. In order to show that K is $\mathcal{A}$-fundamental for (F, Φ, φ), it suffices in view of Example 14.6 to show that K is a retraction candidate. We verify the properties of Definition 14.3 even for any bounded open neighborhood $\Omega_0 \subseteq \Omega$ of 0. The property (ii) follows with the choice $\varphi_0(x_1, x_2) = (2x_1, 0)$: We have $\varphi_0 \in \mathrm{Comp}(\Omega_0, K)$ by the Heine–Borel theorem (Proposition 3.59), since Ω_0 is bounded. It remains to verify property (iii). Thus, let $\Omega_1 \subseteq \Omega_0$ be an open neighborhood of 0. Now let $\varphi_0, \varphi_1 \in \mathrm{Comp}(\Omega_1, K)$ satisfy $\varphi_i(x_1, 0) = \varphi(x_1, 0) = (2x_1, 0)$ $(i = 0, 1)$ for all $(x_1, x_2) = (x_1, 0) \in \Omega_1 \cap K$. Note that $Y_0 := K$ is a linear subspace of Y which is transversal to F, and so the reduction and normalization property implies with $X_0 := \Omega_1 \cap F^{-1}(Y_0) = \Omega_1 \cap K$ that

$$\deg(F, \Phi, \varphi_i, \Omega_1) = \deg_{(X_0,Y_0)}(F, \Phi, \varphi, X_0) = \deg_{(X_0,Y_0)}(\mathrm{id}_{X_0}, \varphi, X_0)$$

for $i = 0, 1$, where the last degree denotes the Benevieri–Furi coincidence degree. Putting $\widetilde{\Omega}_1 := \{x_1 : (x_1, 0) \in \Omega_1\}$, we thus find by the diffeomorphic-isomorphic invariance and the compatibility with the C^1 Brouwer degree that

$$\deg(F, \Phi, \varphi_i, \Omega_1) = \deg_{(\mathbb{R},\mathbb{R})}(\mathrm{id}_{\widetilde{\Omega}_1} - 2\,\mathrm{id}_{\widetilde{\Omega}_1}, \widetilde{\Omega}_1) = \deg_{(\mathbb{R},\mathbb{R})}(-\mathrm{id}_{\widetilde{\Omega}_1}, \widetilde{\Omega}_1) = -1$$

for $i = 0, 1$. Hence, K is indeed a retraction candidate and thus $\mathcal{A}$-fundamental. The above calculation shows also in view of the permanence property that

$$\mathrm{DEG}_{(X,Y,\mathcal{A})}(F, \Phi, \varphi, \Omega) = -1 \neq 1 = \deg(F, \Phi, \varphi, \Omega).$$

On the other hand, we have:

Example 14.11. If we choose $\mathcal{A}_0 = \{K, Y\}$ in Example 14.10 then Corollary 14.9 implies that

$$\mathrm{DEG}_{(X,Y,\mathcal{A}_0)}(F, \Phi, \varphi, \Omega) = \deg(F, \Phi, \varphi, \Omega) = 1.$$

As we have seen in Example 14.10, the set K is a retraction candidate. However, K cannot be $\mathcal{A}$-fundamental, since the calculation in Example 14.10 would imply that the permanence property is violated for K.

The above example shows in particular also that the permanence property does not hold for retraction candidates, in general.

This explains why our definition of $\mathcal{A}$-fundamental sets is so cumbersome: We had indeed to require some additional properties, since the properties of retraction candidates alone would not be sufficient to define a degree by means of the permanence property.

The above examples demonstrate also that the value of the degree depends on the choice of the set $\mathcal{A}$, in general. Fortunately, there is an important situation in which we know that we obtain the same value.

Proposition 14.12. *The $\mathcal{A}$-fundamental Fredholm triple degree has the following property:*

($\mathrm{H}_{\mathcal{T}_{\mathrm{fund}}}$) **(Weak Compatibility with Respect to $\mathcal{A}$).** *Let $\mathcal{A}$, $\mathcal{A}_0$ be two families of subsets of Y, and let $(F, \Phi, \varphi, \Omega) \in \mathcal{T}_{\mathrm{prop}}$ be such that there is a set K which is $\mathcal{A}$-fundamental and simultaneously $\mathcal{A}_0$-fundamental. Then $(F, \Phi, \varphi, \Omega) \in \mathcal{T}_{\mathrm{fund}}(X, Y, \mathcal{A}) \cap \mathcal{T}_{\mathrm{fund}}(X, Y, \mathcal{A}_0)$, and*

$$\mathrm{DEG}_{(X_0,Y,\mathcal{A})}(F, \Phi, \varphi, \Omega) = \mathrm{DEG}_{(X,Y,\mathcal{A}_0)}(F, \Phi, \varphi, \Omega).$$

Proof. Since K is a retraction candidate, there is an open neighborhood $\Omega_0 \subseteq \Omega$ of $\mathrm{coin}_\Omega(F, \Phi, \varphi)$ and $\varphi_0 \in \mathrm{Comp}(\Phi(\Omega_0), K)$ satisfying $\varphi_0(z) = \varphi(z)$ for all $z \in \Phi(\Omega_0 \cap F^{-1}(K))$. The permanence property thus implies

$$\mathrm{DEG}_{(X,Y,\mathcal{A})}(F, \Phi, \varphi, \Omega) = \deg(F, \Phi, \varphi_0, \Omega_0).$$

On the other hand, the permanence property of $\mathrm{DEG}_{(X,Y,\mathcal{A}_0)}$ implies

$$\mathrm{DEG}_{(X,Y,\mathcal{A}_0)}(F, \Phi, \varphi, \Omega) = \deg(F, \Phi, \varphi_0, \Omega_0).$$

A comparison of these formulas implies the assertion. □

Remark 14.13. Examples 14.10 and 14.11 show that it would not be sufficient to assume only

$$(F, \Phi, \varphi, \Omega) \in \mathcal{T}_{\mathrm{fund}}(X, Y, \mathcal{A}) \cap \mathcal{T}_{\mathrm{fund}}(X, Y, \mathcal{A}_0)$$

in Proposition 14.12: We really need that the *same* set is $\mathcal{A}$-fundamental and $\mathcal{A}_0$-fundamental.

Now we come to a first formulation of the generalized homotopy invariance property for the $\mathcal{A}$-fundamental Fredholm triple degree. A formulation which is easier to apply will be given in Theorem 14.23.

Theorem 14.14. *The $\mathcal{A}$-fundamental Fredholm triple degree* DEG *has the following property:*

($\mathrm{I}_{\mathcal{T}_{\mathrm{fund}}}$) **(Generalized Homotopy Invariance I).** *Let (G, H, h, W, Y, Γ) be a generalized proper acyclic* homotopy triple for $(G_t, H_t, h_t, W_t, Y, \Gamma_t)$ $(t \in [0,1])$ where $W \subseteq [0,1] \times X$ is open, and $G\colon W \to Y$ is a generalized (oriented) Fredholm homotopy of index 0. We put $\widetilde{H}(t,x) := \{t\} \times H(t,x)$. Suppose that there is a set $K \subseteq Y$ such that the following holds:*

(a) *K is $\mathcal{A}$-fundamental for (G_t, H_t, h_t, W_t) for every $t \in [0,1]$.*
(b) *There is an open neighborhood $U \subseteq W$ of* $\operatorname{coin}(G, \widetilde{H}, h)$ *and $\widetilde{h} \in$* $\operatorname{Comp}(\widetilde{H}(U), K)$ *such that $\widetilde{h}(t,z) = h(t,z)$ for all $(t,z) \in \widetilde{H}(U \cap G^{-1}(K))$.*

Then $(G_t, H_t, h_t, W_t) \in \mathcal{T}_{\mathrm{fund}}(X, Y, \mathcal{A})$ for all $t \in [0,1]$, and

$$\mathrm{DEG}(G_t, H_t, h_t, W_t) \text{ is independent of } t \in [0,1].$$

Proof. The assertion $(G_t, H_t, h_t, W_t) \in \mathcal{T}_{\mathrm{fund}}(X, Y, \mathcal{A})$ follows in view of Proposition 11.41 and since K is $\mathcal{A}$-fundamental for (G_t, H_t, h_t, W_t).

By hypothesis, there are an open neighborhood $U \subseteq W$ of the set $C := \operatorname{coin}_W(G, \widetilde{H}, h)$ and $\widetilde{h} \in \operatorname{Comp}(\widetilde{H}(U), K)$ satisfying $\widetilde{h}(t,z) = h(t,z)$ for all $(t,z) \in \widetilde{H}(U \cap G^{-1}(K))$. Since K is a retraction candidate for (G_t, H_t, h_t, W_t) for every $t \in [0,1]$, we have $\operatorname{fix}_W(G, \widetilde{H}, h) \subseteq K$. Hence, Lemma 14.2 implies

$$\operatorname{coin}_U(G, \widetilde{H}, \widetilde{h}) = C.$$

We obtain with $U_t := \{x : (t,x) \in U\}$ and $\widetilde{h}_t := \widetilde{h}(t, \cdot) \in \operatorname{Comp}(U_t, K)$ that

$$\operatorname{coin}_{U_t}(G_t, H_t, \widetilde{h}_t) = \operatorname{coin}_{W_t}(G_t, H_t, h_t).$$

We conclude that $(G, H, \widetilde{h})$ is a proper generalized acyclic* homotopy triple for the family $(G_t, H_t, \widetilde{h}_t)$. Since K is $\mathcal{A}$-fundamental for every (G_t, H_t, h_t), the permanence property implies

$$\mathrm{DEG}(G_t, H_t, h_t, W_t) = \deg(G_t, H_t, \widetilde{h}_t, U_t).$$

The generalized homotopy invariance of the Leray–Schauder triple degree shows that the latter value is independent of $t \in [0, 1]$. □

For the proof of the additivity properties of the $\mathcal{A}$-fundamental Fredholm triple degree, we have to require also that the *same* set K is $\mathcal{A}$-fundamental for $(F, \Phi, \varphi, \Omega)$ as for the restrictions $(F, \Phi, \varphi, \Omega_i)$:

Theorem 14.15. *The $\mathcal{A}$-fundamental Fredholm triple degree* DEG *has the following properties for every $(F, \Phi, \varphi, \Omega) \in \mathcal{T}_{\mathrm{fund}}(X, Y, \mathcal{A})$:*

($\mathrm{J}_{\mathcal{T}_{\mathrm{fund}}}$) **(Weak Additivity).** *Let $\Omega = \Omega_1 \cup \Omega_2$ with disjoint open subsets $\Omega_1, \Omega_2 \subseteq \Omega$. If there is a set $K \subseteq Y$ which is $\mathcal{A}$-fundamental for $(F, \Phi, \varphi, \Omega)$ and for $(F, \Phi, \varphi, \Omega_i)$ $(i = 1, 2)$ then*

$$\mathrm{DEG}(F, \Phi, \varphi, \Omega) = \mathrm{DEG}(F, \Phi, \varphi, \Omega_1) + \mathrm{DEG}(F, \Phi, \varphi, \Omega_2).$$

($\mathrm{K}_{\mathcal{T}_{\mathrm{fund}}}$) **(Weak Excision-Additivity).** *Let $\Omega_i \subseteq \Omega$ $(i \in I)$ be a family of pairwise disjoint open sets with $\mathrm{coin}_\Omega(F, \Phi, \varphi) \subseteq \bigcup_{i \in I} \Omega_i$ and such that $\mathrm{coin}_{\Omega_i}(F, \Phi, \varphi)$ is compact for all $i \in I$. If there is a set $K \subseteq Y$ which is $\mathcal{A}$-fundamental for $(F, \Phi, \varphi, \Omega)$ and for all $(F, \Phi, \varphi, \Omega_i)$ $(i \in I)$ then*

$$\mathrm{DEG}(F, \Phi, \varphi, \Omega) = \sum_{i \in I} \mathrm{DEG}(F, \Phi, \varphi, \Omega_i),$$

where in the sum at most a finite number of summands is nonzero.

Proof. It suffices to prove ($\mathrm{K}_{\mathcal{T}_{\mathrm{fund}}}$), since ($\mathrm{J}_{\mathcal{T}_{\mathrm{fund}}}$) is a special case. There is an open set $\Omega_0 \subseteq \Omega$ with $\mathrm{coin}_\Omega(F, \Phi, \varphi) \subseteq \Omega$ and a function $\varphi_0 \in \mathrm{Comp}(\Omega_0, K)$ with $\varphi_0(z) = \varphi(z)$ for all $z \in \Phi(\Omega_0 \cap F^{-1}(K))$. By Lemma 14.2, we have

$$\mathrm{coin}_{\Omega_0}(F, \Phi, \varphi_0) = \mathrm{coin}_\Omega(F, \Phi, \varphi).$$

The latter implies with $\Omega_{0,i} := \Omega_0 \cap \Omega_i$ that

$$\mathrm{coin}_{\Omega_{0,i}}(F, \Phi, \varphi_0) = \mathrm{coin}_{\Omega_i}(F, \Phi, \varphi) \quad \text{for all } i \in I.$$

By the permanence property, we have

$$\mathrm{DEG}(F, \Phi, \varphi, \Omega) = \deg(F, \Phi, \varphi_0, \Omega_0),$$

where the right-hand side denotes the Leray–Schauder triple degree. Since K is $\mathcal{A}$-fundamental for every $(F, \Phi, \varphi, \Omega_i)$, the permanence property implies also

$$\mathrm{DEG}(F, \Phi, \varphi, \Omega_i) = \deg(F, \Phi, \varphi_0, \Omega_{0,i}).$$

A comparison of the above formulas implies the assertion by the excision-additivity property of the Leray–Schauder triple degree. □

For the diffeomorphic-isomorphic invariance, we must be somewhat careful with isomorphisms of Y, since the degree also depends on the family $\mathcal{A}$. However, we have no problem with isomorphisms which map the corresponding families onto each other. More precisely, the following holds:

Theorem 14.16. *The $\mathcal{A}$-fundamental Fredholm triple degree has the following property for every $(F, \Phi, \varphi, \Omega) \in \mathcal{T}_{\mathrm{fund}}(X, Y, \mathcal{A})$:*

($\mathrm{L}_{\mathcal{T}_{\mathrm{fund}}}$) **(Diffeomorphic-Isomorphic Invariance).** *Let J_1 be a diffeomorphism of an open subset of a Banach manifold X_0 onto Ω and J_2 an isomorphism of Y onto a real normed vector space Y_0, and let $\mathcal{A}_0 = \{J_2(K) : K \in \mathcal{A}\}$. Then*

$$\begin{aligned}&\mathrm{DEG}_{(X,Y,\mathcal{A})}(F, \Phi, \varphi, \Omega)\\&\quad = \mathrm{DEG}_{(X_0,Y_0,\mathcal{A}_0)}(J_2 \circ F \circ J_1, \Phi \circ J_1, J_2 \circ \varphi, J_1^{-1}(\Omega)).\end{aligned}$$

In the oriented case, the orientation of $J_2 \circ F \circ J_1$ is understood in the sense of Proposition 8.38.

($\mathrm{M}_{\mathcal{T}_{\mathrm{fund}}}$) **(Restriction).** *Let $X_0 \subseteq X$ be open. Then*

$$\mathrm{DEG}_{(X_0,Y,\mathcal{A})} = \mathrm{DEG}_{(X,Y,\mathcal{A})}|_{\mathcal{T}_{\mathrm{fund}}(X_0,Y,\mathcal{A})}.$$

Proof. The diffeomorphic-isomorphic invariance of the Leray–Schauder triple degree implies that K is a retraction candidate for $A := (F, \Phi, \varphi, \Omega)$ if and only if $J_2(K)$ is a retraction candidate for $\widetilde{A} := (J_2 \circ F \circ J_1, \Phi \circ J_1, J_2 \circ \varphi, J_1^{-1}(\Omega))$. In view of the choice of $\mathcal{A}_0$, it follows similarly that K is $\mathcal{A}$-fundamental for A if and only if $J_2(K)$ is $\mathcal{A}_0$-fundamental for $\widetilde{A}$. Let K be $\mathcal{A}$-fundamental for A. There is an open set $\Omega_0 \subseteq \Omega$ with $\mathrm{coin}_\Omega(F, \Phi, \varphi) \subseteq \Omega$ and a function $\varphi_0 \in \mathrm{Comp}(\Omega_0, K)$ with $\varphi_0(z) = \varphi(z)$. Then Ω_0 and φ_0 are as in the permanence property for A, and $J_1^{-1}(\Omega_0)$ and $J_2 \circ \varphi_0$ are as in the permanence property for $\widetilde{A}$, and so we have

$$\begin{aligned}\mathrm{DEG}_{(X,Y,\mathcal{A})}(A) &= \deg_{(X,Y)}(F, \Phi, J_2 \circ \varphi_0, \Omega_0),\\ \mathrm{DEG}_{(X_0,Y_0,\mathcal{A}_0)}(\widetilde{A}) &= \deg_{(X_0,Y_0)}(J_2 \circ F \circ J_1, \Phi \circ J_1, J_2 \circ \varphi_0, J_1^{-1}(\Omega_0)).\end{aligned}$$

Hence, the diffeomorphic-isomorphic invariance of the Leray–Schauder triple degree implies ($L_{\mathcal{T}_{\rm fund}}$).

The restriction property is the special case $J_2 = \mathrm{id}_Y$ and $J_1 = \mathrm{id}_\Omega : \Omega \to X$ of ($L_{\mathcal{T}_{\rm fund}}$). □

Taking care about the associated family $\mathcal{A}$ for each space, we can also formulate a corresponding Cartesian product property:

Theorem 14.17. *The $\mathcal{A}$-fundamental Fredholm triple degree has the following property:*

($N_{\mathcal{T}_{\rm fund}}$) **(Cartesian Product).** *For $i = 1, 2$, let X_i be a manifold without boundary of class C^1 over the real Banach space E_{X_i}, and let $Y_i = E_{Y_i}$ be a real Banach space. Let $\mathcal{A}_i$ be a family of subsets of Y_i, and $\mathcal{A}$ a family of subsets of $Y := Y_1 \times Y_2$. For $(F_i, \Phi_i, \varphi_i, \Omega_i) \in \mathcal{T}_{\rm prop}(X_i, Y_i)$, we put $X := X_1 \times X_2$, $\Omega := \Omega_1 \times \Omega_2$, $F := F_1 \otimes F_2$, $\Phi := \Phi_1 \otimes \Phi_2$, and $\varphi := \varphi_1 \otimes \varphi_2$. Suppose that, for $i = 1, 2$, K_i is $\mathcal{A}_i$-fundamental for $(F_i, \Phi_i, \varphi_i, \Omega_i)$, and that $K := K_1 \times K_2$ is $\mathcal{A}$-fundamental for $(F, \Phi, \varphi, \Omega)$. Then*

$$\begin{aligned}&\mathrm{DEG}_{(X,Y,\mathcal{A})}(F, \Phi, \varphi, \Omega)\\&\quad= \mathrm{DEG}_{(X_1,Y_1,\mathcal{A}_1)}(F_1, \Phi_1, \varphi_1, \Omega_1)\,\mathrm{DEG}_{(X_2,Y_2,\mathcal{A}_2)}(F_2, \Phi_2, \varphi_2, \Omega_2).\end{aligned}$$

In the oriented case, F is equipped with the product orientation.

Proof. For $i = 1, 2$, there are an open neighborhood $\Omega_{0,i} \subseteq \Omega_i$ of the set $\mathrm{coin}_{\Omega_i}(F_i, \Phi_i, \varphi_i)$ and $\varphi_{0,i} \in \mathrm{Comp}(\Phi_i(\Omega_{0,i}), K_i)$ with $\varphi_{0,i}(z) = \varphi(z)$ for all $z \in \Phi_i(\Omega_{0,i} \cap F_i^{-1}(K_i))$. We put $\Omega_0 := \Omega_{0,1} \times \Omega_{0,2}$. Then the function $\varphi_0 := \varphi_{0,1} \otimes \varphi_{0,2}$ belongs to $\mathrm{Comp}(\Phi(\Omega_0), K)$ and satisfies $\varphi_0(z) = \varphi(z)$ for all $z \in \Phi(\Omega_0 \cap F^{-1}(K))$. By the permanence property of the three degrees, we obtain

$$\begin{aligned}\mathrm{DEG}_{(X,Y,\mathcal{A})}(F, \Phi, \varphi, \Omega) &= \deg(F, \Phi, \varphi_0, \Omega_0),\\ \mathrm{DEG}_{(X_i,Y_i,\mathcal{A}_i)}(F_i, \Phi_i, \varphi_i, \Omega_i) &= \deg(F_i, \Phi_i, \varphi_{0,i}, \Omega_{0,i})\end{aligned}$$

for $i = 1, 2$. Hence, the assertion follows from the Cartesian product property of the Leray–Schauder triple degree. □

Also for the equivalence invariance, we have to require that we can choose the *same* $\mathcal{A}$-fundamental sets:

Theorem 14.18. *The $\mathcal{A}$-fundamental Fredholm triple degree has the following properties.*

($O_{\mathcal{T}_{\text{fund}}}$) **(Very Weak Equivalence Invariance).** *Suppose that* $(F, \widetilde{\Phi}, \widetilde{\varphi}, \Omega) \approx (F, \Phi, \varphi, \Omega)$ *both belong to* $\mathcal{T}_{\text{prop}}(X, Y)$*. Then both function triples have the same retraction candidates and the same* $\mathcal{A}$*-fundamental sets, and if they have some* $\mathcal{A}$*-fundamental set then*

$$\text{DEG}(F, \widetilde{\Phi}, \widetilde{\varphi}, \Omega) = \text{DEG}(F, \Phi, \varphi, \Omega).$$

($P_{\mathcal{T}_{\text{fund}}}$) **(Mild Equivalence Invariance).** *For* $K \subseteq Y$*, let* $\mathcal{T}_K$ *denote the subclass of triples from* $\mathcal{T}_{\text{fund}}(X, Y, \mathcal{A})$ *which have* K *as an* $\mathcal{A}$*-fundamental set. If* $(F, \Phi, \varphi) \sim_{\mathcal{T}_K} (F, \widetilde{\Phi}, \widetilde{\varphi})$ *then*

$$\text{DEG}(F, \Phi, \varphi, \Omega) = \text{DEG}(F, \widetilde{\Phi}, \widetilde{\varphi}, \Omega).$$

Proof. We show first ($P_{\mathcal{T}_{\text{fund}}}$). By induction, we can assume that $(F, \Phi, \varphi) \sqsubseteq (F, \widetilde{\Phi}, \widetilde{\varphi})$ where both triples belong to $\mathcal{T}_K$. Let $\Omega_0 \subseteq \Omega$ be an open neighborhood of $\text{coin}_\Omega(F, \widetilde{\Phi}, \widetilde{\varphi})$, and let $\widetilde{\varphi}_0 \in \text{Comp}(\widetilde{\Phi}(\Omega_0), K)$ satisfy $\widetilde{\varphi}_0(z) = \widetilde{\varphi}(z)$ for all $z \in \widetilde{\Phi}(\Omega_0 \cap F^{-1}(K))$. The permanence property implies

$$\text{DEG}(F, \widetilde{\Phi}, \widetilde{\varphi}, \Omega) = \text{DEG}(F, \widetilde{\Phi}, \widetilde{\varphi}_0, \Omega_0). \tag{14.6}$$

Let J denote the function of Definition 11.7, that is, $J \circ \Phi \subseteq \widetilde{\Phi}$ and $\varphi = \widetilde{\varphi} \circ J$. Then in particular $J(\Phi(\Omega_0)) \subseteq \widetilde{\Phi}(\Omega_0)$. Then the function $\varphi_0 := \widetilde{\varphi}_0 \circ J|_{\Phi(\Omega_0)}$ belongs to $\text{Comp}(\Phi(\Omega_0), K)$ and satisfies for all $z \in \Phi(\Omega_0 \cap F^{-1}(K))$ by (11.3) that $J(z) \in \widetilde{\Phi}(\Omega_0 \cap F^{-1}(K))$. Hence, $\varphi_0(z) = \widetilde{\varphi}(J(z)) = \varphi(z)$ holds in view of (11.3). Note also that (11.6) implies $\text{coin}_\Omega(F, \Phi, \varphi) \subseteq \Omega_0$. The permanence property thus implies

$$\text{DEG}(F, \Phi, \varphi, \Omega) = \text{DEG}(F, \Phi, \varphi_0, \Omega_0).$$

Hence, the assertion follows with (14.6) and the equivalence invariance of the Leray–Schauder triple degree.

To prove ($O_{\mathcal{T}_{\text{fund}}}$), let $(F, \Phi, \varphi, \Omega) \approx (F, \widetilde{\Phi}, \widetilde{\varphi}, \Omega)$ both belong to $\mathcal{T}_{\text{prop}}(X, Y)$. We show first that both function triples have the same retraction candidates. Since $\approx$ is symmetric (recall Proposition 11.10), it suffices to show that each retraction candidate $K \subseteq Y$ for $(F, \Phi, \varphi, \Omega)$ is also a retraction candidate for $(F, \widetilde{\Phi}, \widetilde{\varphi}, \Omega)$.

Let J be a homeomorphism of $\Phi(\Omega)$ onto $\widetilde{\Phi}(\Omega)$ satisfying $J \circ \Phi = \widetilde{\Phi}$ and $\varphi = \widetilde{\varphi} \circ J$. By (11.8), we have

$$C := \text{coin}_\Omega(F, \Phi, \varphi, \Omega) = \text{coin}_\Omega(F, \widetilde{\Phi}, \widetilde{\varphi}, \Omega)$$

and $\text{fix}_\Omega(F, \widetilde{\Phi}, \widetilde{\varphi}, \Omega) \subseteq K$. In particular, property (i) of Definition 14.3 holds for $(F, \widetilde{\Phi}, \widetilde{\varphi}, \Omega)$.

Let Ω_0 be as in Definition 14.3 for $(F, \Phi, \varphi, \Omega)$. If $\Omega_1 \subseteq \Omega_0$ is an open neighborhood of C and $\varphi_0 \in \mathrm{Comp}(\Phi(\Omega_1), K)$ satisfies $\varphi_0(z) = \varphi(z)$ for all $z \in \Phi(\Omega_1 \cap F^{-1}(K))$, then $\widetilde{\varphi}_0 := \varphi \circ J^{-1}$ belongs to $\mathrm{Comp}(\widetilde{\Phi}(\Omega_1), K)$ and satisfies $\widetilde{\varphi}_0(z) = \widetilde{\varphi}(z)$ for all $z \in \widetilde{\Phi}(\Omega_1 \cap F^{-1}(K))$. Hence, also property (ii) of Definition 14.3 holds for $(F, \widetilde{\Phi}, \widetilde{\varphi}, \Omega)$.

To see that also property (iii) of Definition 14.3 holds for $(F, \widetilde{\Phi}, \widetilde{\varphi}, \Omega)$, let $\widetilde{\varphi}_0, \widetilde{\varphi}_1 \in \mathrm{Comp}(\widetilde{\Phi}(\Omega_1), K)$ satisfy $\widetilde{\varphi}_0(z) = \widetilde{\varphi}_1(z) = \widetilde{\varphi}(z)$ for all $z \in \widetilde{\Phi}(\Omega_1 \cap F^{-1}(K))$. For $i = 0, 1$, we put $\varphi_i := \widetilde{\varphi}_i \circ J|_{\Phi(\Omega_1)}$, and find that $\varphi_i \in \mathrm{Comp}(\Phi(\Omega_1), K)$ satisfy $\varphi_i(z) = \widetilde{\varphi}(J(z)) = \varphi(z)$ for all $z \in \Phi(\Omega_1 \cap F^{-1}(K))$. We obtain that

$$\deg(F, \Phi, \varphi_0, \Omega_1) = \deg(F, \Phi, \varphi_1, \Omega),$$

and by the (weak) equivalence invariance of the Leray–Schauder degree, we obtain also that

$$\deg(F, \Phi, \varphi_i, \Omega_1) = \deg(F, \widetilde{\Phi}, \widetilde{\varphi}_i, \Omega_1)$$

for $i = 0, 1$. Combining these equalities, we find

$$\deg(F, \widetilde{\Phi}, \widetilde{\varphi}_0, \Omega_1) = \deg(F, \widetilde{\Phi}, \widetilde{\varphi}_1, \Omega).$$

Hence, K is a retraction candidate for $(F, \widetilde{\Phi}, \widetilde{\varphi}, \Omega)$.

Now that we know that (F, Φ, φ) and $(F, \widetilde{\Phi}, \widetilde{\varphi})$ have exactly the same retraction candidates, we obtain by a completely analogous reasoning that they have also the same $\mathcal{A}$-fundamental sets. In particular, if such a set K exists, we have $(F, \Phi, \varphi) \sim_{\mathcal{T}_K} (F, \widetilde{\Phi}, \widetilde{\varphi})$, and so $(\mathrm{P}_{\mathcal{T}_{\mathrm{fund}}})$ applies. □

14.2 Homotopic Tests for Fundamental Sets

The aim of this section is to provide sufficient conditions for the properties in Definition 14.3 and 14.4 which are simpler to verify. Moreover, we want to provide the same for the hypotheses for the generalized homotopy invariance (Theorem 14.14).

Concerning retraction candidates, this is surprisingly simple.

We use throughout the notations of Section 14.1. We recall that we consider two cases in Definition 14.1: Either $\mathrm{Comp}(\Gamma, K) = \mathrm{Comp}_K(\Gamma, K)$ or $\mathrm{Comp}(\Gamma, K) = \mathrm{Comp}_Y(\Gamma, K)$. In the former case, we say that a set K is an CNE if it is a CNE_K. In the latter case, we say that K is an CNE if it is a CNE_Y. For the case that K is closed, the two cases become identical (recall Proposition 4.35).

Recall that when we consider the case $\mathrm{Comp}(\Gamma, K) = \mathrm{Comp}_K(\Gamma, K)$, then any ANR (LCNR) K is an CNE by Corollary 4.38 (Theorem 4.36).

In particular, the following result implies that in this case any neighborhood retract of Y is a retraction candidate under the necessary hypotheses mentioned after Definition 14.3.

As already announced after Definition 14.3, this is our motivation for the terminology "retraction candidate". This result is also the main reason, why the class CNE_Z is so important for us.

Theorem 14.19. *Let* $(F, \Phi, \varphi, \Omega) \in \mathcal{T}_{\mathrm{prop}}(X, Y)$, *and* $K \subseteq Y$. *Suppose that there is an open neighborhood* $\Omega_0 \subseteq \Omega$ *of* $C := \mathrm{coin}_\Omega(F, \Phi, \varphi)$ *such that the following holds.*

(a) $\mathrm{fix}_\Omega(F, \Phi, \varphi) \subseteq K$.

(b) $K = \varnothing$, *or for every open neighborhood* $\Omega_1 \subseteq \Omega_0$ *of* C *there is a neighborhood* Z_0 *of* $\Phi(C)$ *in* $\Phi(\Omega_1)$ *such that* K *is an* CNE *for* $[0, 1] \times Z_0$.

(c) $K = \varnothing$, *or there is a neighborhood* Z *of* $\Phi(C)$ *in* $\Phi(\Omega_0)$ *such that* K *is an* CNE *for* Z, *and the restriction of* φ *to*

$$\Gamma_0 := Z \cap \overline{\Phi(\Omega_0 \cap F^{-1}(K))}$$

belongs to $\mathrm{Comp}(\Gamma_0, K)$.

Then K *is a retraction candidate.*

We postpone the proof for a moment.

Remark 14.20. The reader may wonder why we formulated Theorem 14.19 so general and not only for the apparently most natural case $Z_0 := \Phi(\Omega_1)$ and $Z := \Phi(\Omega_0)$.

The reason is that, typically, K is only a CNE for a class of T_4 spaces. Now the condition that K is an CNE for $[0, 1] \times \Phi(\Omega_1)$ can in view of Proposition 2.35 usually only be satisfied if $[0, 1] \times \Phi(\Omega_0)$ is T_5.

In contrast, by choosing a possibly different set Z_0, we are in this case in the setting of locally $[0, 1]$-normal triples (recall Definition 11.47) and thus can treat the case that $[0, 1] \times \Phi(\Omega_0)$ is only T_4.

The proof of Theorem 14.19 uses the following two lemmas which we will also use in a more general context.

Lemma 14.21. *Let $(F, \Phi, \varphi, \Omega, Y, \Gamma)$ be a function triple, $K \subseteq Y$, and $\Omega_0 \subseteq \Omega$ (not necessarily open) contain $C := \operatorname{coin}_\Omega(F, \Phi, \varphi)$. Suppose that there is a neighborhood Z of $\Phi(C)$ in $\Phi(\Omega_0)$ such that the following holds:*

(a) $\operatorname{fix}_\Omega(F, \Phi, \varphi) \subseteq K$.

(b) $K = \varnothing$, *or K is an* CNE *for Z.*

(c) *The restriction of φ to*

$$\Gamma_0 := Z \cap \overline{\Phi(\Omega_0 \cap F^{-1}(K))}$$

belongs to $\operatorname{Comp}(\Gamma_0, K)$.

(d) $\Phi|_U$ *is upper semicontinuous for some neighborhood $U \subseteq \Omega$ of C.*

Then there is an open in Ω_0 set $\Omega_1 \subseteq \Omega_0$ with $C \subseteq \Omega_1$ and a function $\varphi_0 \in \operatorname{Comp}(\Phi(\Omega_1), K)$ with $\varphi_0(z) = \varphi(z)$ for each $z \in \Phi(\Omega_1 \cap F^{-1}(K))$.

Proof. We first treat the degenerate case $K = \varnothing$. In this case, we must have $\operatorname{fix}_\Omega(F, \Phi, \varphi) = \varnothing$ and thus also $C = \varnothing$. Hence, we can choose $\Omega_1 = \varnothing$.

Now assume that $K \neq \varnothing$, that is, K is an CNE for Z. Note that Proposition 2.10 implies that Γ_0 is closed in Z. Hence, there is a neighborhood Γ_1 of Γ_0 in Z and an extension $\varphi_0 \in \operatorname{Comp}(\Gamma_1, K)$ of $\varphi|_{\Gamma_0}$. We note that $\operatorname{fix}_\Omega(F, \Phi, \varphi) \subseteq K$ implies $\Phi(C) \subseteq \Gamma_0$.

We can assume that U is open in Ω. Then the set $\Omega_2 := U \cap \Omega_0$ is open in Ω_0, and Proposition 2.90 implies that $\Psi := \Phi|_{\Omega_2} \colon \Omega_2 \multimap Z_0 := \Phi(\Omega_0)$ is upper semicontinuous. Since $\Gamma_1 \subseteq Z$ is a neighborhood of $\Phi(C)$, and $Z \subseteq Z_0$ is a neighborhood of $\Phi(C)$, we find that there is an open set $O \subseteq Z_0$ with $\Phi(C) \subseteq O \subseteq \Gamma_1$. Hence, Proposition 2.92 implies that $\Omega_1 := \Psi^-(O) = \Omega_2 \cap \Phi^-(O)$ is open in Ω_0 with $C \subseteq \Omega_1$. It follows that the restriction of φ_0 to $\Phi(\Omega_1) \subseteq O \subseteq \Gamma_1$ has the required property. □

Lemma 14.22. *Let $(F, \Phi, \varphi, \Omega, Y, \Gamma)$ be a function triple, $K \subseteq Y$, and $\Omega_0 \subseteq \Omega$ (not necessarily open) contain $C := \operatorname{coin}_\Omega(F, \Phi, \varphi)$.*

Let $\varphi_0, \varphi_1 \in \operatorname{Comp}(\Phi(\Omega_0), K)$ satisfy $\varphi_0(z) = \varphi_1(z) = \varphi(z)$ for all $z \in \Phi(\Omega_0 \cap F^{-1}(K))$. Suppose also that there is a neighborhood Z of $\Phi(C)$ in $\Phi(\Omega_0)$ such that the following holds:

(a) $\operatorname{fix}_\Omega(F, \Phi, \varphi) \subseteq K$.

(b) $K = \varnothing$, *or K is an* CNE *for* $[0, 1] \times Z$

(c) $\Phi|_U$ *is upper semicontinuous for some neighborhood $U \subseteq \Omega$ of C.*

Then there is an open in Ω_0 *set* $\Omega_1 \subseteq \Omega_0$ *with* $C \subseteq \Omega_1$ *and a function* $h \in \mathrm{Comp}([0,1]\times\Phi(\Omega_1), K)$ *with* $h(i,\cdot) = \varphi_i|_{\Phi(\Omega_1)}$ *for* $i = 0, 1$ *such that* $h(t,z) = \varphi(z)$ *for all* $(t,z) \in [0,1] \times \Phi(\Omega_1 \cap F^{-1}(K))$.

Proof. In the degenerate case $K = \varnothing$, we must have $C = \varnothing$ and thus can choose $\Omega_1 = \varnothing$. Thus, assume that $K \neq \varnothing$. Then K is an CNE for $[0,1] \times Z$. Let Γ_0 denote the closure of $M := \Phi(\Omega_0 \cap F^{-1}(K))$ in Z. Since Y is Hausdorff and $\varphi_0|_M = \varphi_1|_M = \varphi|_M$, we obtain by Lemma 2.55 that $\varphi_0|_{\Gamma_0} = \varphi_1|_{\Gamma_0} = \varphi|_{\Gamma_0}$. By the both-sided homotopy extension theorem (Theorem 4.45), there is a neighborhood Γ_1 of Γ_0 in Z and $h \in \mathrm{Comp}([0,1] \times \Gamma_1, K)$ with $h(i,\cdot)|_{\Gamma_1} = \varphi_i|_{\Gamma_1}$ for $i = 0, 1$. Since $\mathrm{fix}_\Omega(F, \Phi, \varphi) \subseteq K$, we have $\Phi(C) \subseteq \Gamma_0$.

We can assume that $U \subseteq \Omega$ is open. Then the set $\Omega_2 := U \cap \Omega_0$ is open in Ω_0, and Proposition 2.90 implies that $\Psi := \Phi|_{\Omega_2} \colon \Omega_2 \multimap Z_0 := \Phi(\Omega_0)$ is upper semicontinuous. Since $\Gamma_1 \subseteq Z$ is a neighborhood of $\Phi(C)$, and $Z \subseteq Z_0$ is a neighborhood of $\Phi(C)$, we find that there is an open set $O \subseteq Z_0$ with $\Phi(C) \subseteq O \subseteq \Gamma_1$. Hence, Proposition 2.92 implies that $\Omega_1 := \Psi^-(O) = \Omega_2 \cap \Phi^-(O)$ is open in Ω_0 with $C \subseteq \Omega_1$. It follows that the restriction of h to $[0,1] \times \Phi(\Omega_1) \subseteq [0,1] \times O \subseteq [0,1] \times \Gamma_1$ has the required property. □

Proof of Theorem 14.19. We can assume $K \neq \varnothing$, since otherwise we have $C = \varnothing$ and can choose $\Omega_0 = \varnothing$. By Lemma 14.21, there is an open neighborhood $\widetilde{\Omega}_0 \subseteq \Omega_0$ of C and $\varphi_0 \in \mathrm{Comp}(\Phi(\widetilde{\Omega}_0), K)$ with $\varphi_0(z) = \varphi(z)$ for each $z \in \Phi(\widetilde{\Omega}_0 \cap F^{-1}(K))$. We show that this is the set required for Definition 14.3. We just verified the property (ii) of that definition, and property (i) holds by hypothesis. It remains to verify property (iii) of Definition 14.3.

Thus, let $\Omega_1 \subseteq \widetilde{\Omega}_0$ be an open neighborhood of C, and suppose that $\varphi_0, \varphi_1 \in \mathrm{Comp}(\Phi(\Omega_1), K)$ satisfy $\varphi_0(z) = \varphi_1(z) = \varphi(z)$ for all $z \in \Phi(\Omega_1 \cap F^{-1}(K))$. By Lemma 14.2, we have $\mathrm{coin}_{\Omega_1}(F, \Phi, \varphi_i) = C$ for $i = 0, 1$. Lemma 14.22 implies that there is an open neighborhood $\Omega_2 \subseteq \Omega_1$ of C and a function $h \in \mathrm{Comp}([0,1]\times\Phi(\Omega_2), K)$ with $h(i,\cdot) = \varphi_i|_{\Phi(\Omega_2)}$ for $i = 0, 1$ such that $h(t,z) = \varphi(z)$ for all $(t,z) \in [0,1]\times\Phi(\Omega_2 \cap F^{-1}(K))$. In view of Lemma 14.2, we obtain

$$\mathrm{coin}_{\Omega_2}(F, \Phi, h(t,\cdot)) = C = \mathrm{coin}_{\Omega_1}(F, \Phi, \varphi_i)$$

for all $t \in [0,1]$ and $i = 0, 1$. In particular, the set $\mathrm{coin}_{[0,1]\times\Omega_2}(F, \Phi, h) = [0,1]\times C$ is compact by Theorem 2.63. Hence, the excision and homotopy invariance implies

$$\begin{aligned}\deg_{\Omega_1}(F, \Phi, \varphi_0, \Omega_1) &= \deg_{\Omega_2}(F, \Phi, h(0,\cdot), \Omega_2)\\ &= \deg_{\Omega_2}(F, \Phi, h(1,\cdot), \Omega_2) = \deg_{\Omega_1}(F, \Phi, \varphi_1, \Omega_1),\end{aligned}$$

and so Definition 14.3(iii) holds. □

As a side result, Lemma 14.21 implies the announced version of the generalized homotopy invariance which is easier to apply.

Theorem 14.23. *The $\mathcal{A}$-fundamental Fredholm triple degree* DEG *has the following property:*

($Q_{\mathcal{T}_{\mathrm{fund}}}$) **(Generalized Homotopy Invariance II).** *Let (G, H, h, W, Y, Γ) be a generalized proper acyclic* homotopy triple for $(G_t, H_t, h_t, W_t, Y, \Gamma_t)$ ($t \in [0,1]$) where $W \subseteq [0,1] \times X$ is open, and $G\colon W \to Y$ is a generalized (oriented) Fredholm homotopy of index* 0. *We put $\widetilde{H}(t,x) := \{t\} \times H(t,x)$. Suppose that there is an open neighborhood $U \subseteq W$ of $C := \mathrm{coin}_W(G, \widetilde{H}, h)$, a set $K \subseteq Y$, and a neighborhood Z of the compact set $\widetilde{H}(C)$ in $\widetilde{H}(U)$ such that the following holds:*

(a) *K is $\mathcal{A}$-fundamental for (G_t, H_t, h_t, W_t) for every $t \in [0,1]$.*
(b) *$K = \varnothing$, or K is an* CNE *for Z.*
(c) *The restriction of h to*

$$\Gamma_0 := Z \cap \overline{\widetilde{H}(U \cap G^{-1}(K))}$$

belongs to $\mathrm{Comp}(\Gamma_0, K)$.

Then $(G_t, H_t, h_t, W_t) \in \mathcal{T}_{\mathrm{fund}}(X, Y, \mathcal{A})$ for all $t \in [0,1]$, and

$$\mathrm{DEG}(G_t, H_t, h_t, W_t) \text{ is independent of } t \in [0,1].$$

Note that the compactness of $\widetilde{H}(C)$ is automatic by Proposition 2.100.

Proof. We apply Lemma 14.21 with the function triple $(G, \widetilde{H}, h, U, Y, \Gamma)$, noting that $\widetilde{H}$ is upper semicontinuous by Proposition 11.41. With the set $\widetilde{U} := \Omega_1$ and the function $\widetilde{h} := \varphi_0$ of Lemma 14.21, we find that the hypotheses of Theorem 14.14 are satisfied. □

Lemma 14.21 also implies that the property (i) of Definition 14.3 can be verified rather easily:

Theorem 14.24. *Let $\mathcal{A}$ be a family of subsets of Y, $(F, \Phi, \varphi, \Omega) \in \mathcal{T}_{\mathrm{prop}}(X, Y)$, and $K, K_0 \in \mathcal{A}$ be two retraction candidates such that the following holds with either $K_1 = K \cap K_0$ or $K_1 = K \cup K_0$.*

For each open neighborhood $\Omega_0 \subseteq \Omega$ of $C := \mathrm{coin}_\Omega(F, \Phi, \varphi)$ there is a neighborhood Z of $\Phi(C)$ in $\Phi(\Omega_0)$ such that K_1 is a CNE *for Z.*

Then the property (i) *of Definition* 14.4 *holds with that K_1 and some open neighborhood $\Omega_0 \subseteq \Omega$ of C.*

Proof. Since K, K_0 are both retraction candidates for $(F, \Phi, \varphi, \Omega)$, there are open neighborhoods $\Omega_1, \Omega_2 \subseteq \Omega$ of C and $\varphi_1 \in \mathrm{Comp}(\Phi(\Omega_1), K)$, $\varphi_2 \in \mathrm{Comp}(\Phi(\Omega_2), K_0)$ satisfying $\varphi_1(z) = \varphi(z)$ for all $z \in \Phi(\Omega_1 \cap F^{-1}(K))$ and $\varphi_2(z) = \varphi(z)$ for all $z \in \Phi(\Omega_2 \cap F^{-1}(K_0))$. We put $\Omega_0 := \Omega_1 \cap \Omega_2$. Note that $\varphi(z) = \varphi_1(z)$ for all $z \in M_1 := \Phi(\Omega_0 \cap F^{-1}(K))$ and $\varphi(z) = \varphi_2(z)$ for all $z \in M_2 := \Phi(\Omega_0 \cap F^{-1}(K_0))$. For $i = 1, 2$, let Γ_i denote the closure of M_i in $Z_0 := \Phi(\Omega_0)$. By Lemma 2.55, we have $\varphi|_{\Gamma_i} = \varphi_i|_{\Gamma_i}$.

We put now $M_0 := \Phi(\Omega_0 \cap F^{-1}(K_1))$, and let Γ_0 denote the closure of M_0 in Z_0. In case $K_1 = K \cap K_0$, we have $M_0 \subseteq M_1 \cap M_2$ and thus $\Gamma_0 \subseteq \Gamma_1 \cap \Gamma_2$. In case $K_2 = K \cup K_0$, we have $M_0 = M_1 \cup M_2$ and thus $\Gamma_0 = \Gamma_1 \cup \Gamma_2$. In both cases, we obtain that $\varphi|_{\Gamma_0} \in \mathrm{Comp}(\Gamma_0, K_1)$. Note that, by hypothesis, there is a neighborhood Z of $\Phi(C)$ in Z_0 such that K_1 is a CNE for Z. By Proposition 2.10, we have

$$\Gamma_0 = Z_0 \cap \overline{\Phi(\Omega_0 \cap F^{-1}(K_1))}.$$

Hence, applying Lemma 14.21 with K_1 in place of K and with $\Gamma_0 \cap Z$ in place of Γ_0, we find that there is an open neighborhood $\widetilde{\Omega}_0 \subseteq \Omega_0$ of C and $\varphi_2 \in \mathrm{Comp}(\Phi(\widetilde{\Omega}_0), K_1)$ satisfying $\varphi_2(z) = \varphi(z)$ for all $z \in \Phi(\widetilde{\Omega}_0 \cap F^{-1}(K_1))$. □

Theorem 14.24 suggests some natural choices for the family $\mathcal{A}$. Indeed, natural such choices for $\mathcal{A}$ are:

(a) the family of all closed convex subsets of Y (recall Theorem 4.36).

(b) the family of all convex subsets of Y in case $\mathrm{Comp}(\Gamma, K) = \mathrm{Comp}_K(\Gamma, K)$ (recall Theorem 4.36).

(c) the family of all finite unions of convex compact subsets of Y (recall Proposition 4.39).

(d) the family of all locally finite unions of convex closed subsets of Y.

Of course, several other choices are possible, as well. The last family corresponds to the degree theory (actually fixed point index theory) developed by R. D. Nussbaum [116].

The difference between the family of convex subsets of Y and the family of closed convex subsets of Y corresponds to the difference of the classes CNE_K and CNE_Y and is one of the reasons why the choice $\mathrm{Comp}(\Gamma, K) = \mathrm{Comp}_K(\Gamma, K)$ might be interesting. In contrast, the choice $\mathrm{Comp}(\Gamma, K) = \mathrm{Comp}_Y(\Gamma, K)$ has the advantage that the condition (c) of Theorem 14.19 can be verified easier (for instance, by using measures of noncompactness). So none of the two cases is clearly superior over the other, and this is the reason, why we consider both of them.

As Example 14.11 shows, the property (ii) of Definition 14.4 must be of a more restrictive nature: In that example, K, $K_0 := Y$, $K \cap K_0 = K$ and $K \cup K_0 = Y$ are all CNE for all subsets of Γ, and even $\varphi \in \mathrm{Comp}(\Phi(\Omega_0), Y)$ and $\varphi \in \mathrm{Comp}(\Phi(\Omega_0) \cap F^{-1}(K), K)$, but K is not $\mathcal{A}$-fundamental for the family $\mathcal{A} = \{K, Y\}$ so that property (ii) of Definition 14.4 must be violated nevertheless for $K_1 = K$ and $K_1 = Y$.

We give now a sufficient homotopic test for that property.

Definition 14.25. Let $(F, \Phi, \varphi, \Omega) \in \mathcal{T}_{\mathrm{prop}}(X, Y)$, and $K_1 \subseteq K_2 \subseteq Y$. We call K_1 a *local* $(F, \Phi, \varphi, \Omega)$*-deformation retract of* K_2, if there are an open neighborhood $\Omega_0 \subseteq \Omega$ of $\mathrm{coin}_\Omega(F, \Phi, \varphi)$ and $H \in C([0,1] \times K_2, K_2)$ with the following properties:

(a) $H(0, \cdot) = \mathrm{id}_{K_2}$, $H(\{1\} \times K_2) \subseteq K_1$.

(b) $H(t, y) = y$ for all $y \in (\varphi \circ \Phi)(\Omega_0 \cap F^{-1}(K_1))$ and all $t \in [0, 1]$.

(c) For every $t \in [0, 1]$, we have

$$\mathrm{fix}_{\Omega_0 \cap F^{-1}(K_2)}(F, \Phi, H(t, \varphi(\cdot))) \subseteq K_1.$$

In case $\mathrm{Comp} = \mathrm{Comp}_Y$, we require in addition that H has an extension to $H \in C([0,1] \times \overline{K}_2, Y)$.

Theorem 14.26. *Suppose that* K_1 *is a retraction candidate for* $(F, \Phi, \varphi, \Omega) \in \mathcal{T}_{\mathrm{prop}}(X, Y)$*. Assume in addition that* K_1 *is a local* $(F, \Phi, \varphi, \Omega)$*-deformation retract of* K_2*. Let* Ω_0 *be as in Definition* 14.25*. Then for each open neighborhood* $\Omega_1 \subseteq \Omega_0$ *of* $\mathrm{coin}_\Omega(F, \Phi, \varphi)$ *and each* $\varphi_i \in \mathrm{Comp}(\Phi(\Omega_1), K_i)$ *satisfying* $\varphi_i(z) = \varphi(z)$ *for all* $z \in \Phi(\Omega_1 \cap F^{-1}(K_i))$ *for* $i = 1, 2$*, we have*

$$\deg(F, \Phi, \varphi_1, \Omega_1) = \deg(F, \Phi, \varphi_2, \Omega_1). \tag{14.7}$$

Proof. With H as in Definition 14.25, we define $h, h_0 \in C([0,1] \times \Phi(\Omega_1), K_2)$ by $h(t, z) := H(t, \varphi_2(z))$ and $h_0(t, z) := H(t, \varphi(z))$. By Lemma 14.2, we have

$$\mathrm{coin}_{\Omega_1}(F, \Phi, h(t, \cdot)) = \mathrm{coin}_{\Omega_1}(F, \Phi, h_0(t, \cdot)).$$

Note that by hypothesis $\mathrm{fix}_{\Omega_1}(F, \Phi, h_0(t, \cdot)) \subseteq K_1$ and $H(t, y) = y$ for all $y \in (\varphi \circ \Phi)(\Omega_0 \cap F^{-1}(K_1))$. This implies

$$\begin{aligned} \mathrm{coin}_{\Omega_1}(F, \Phi, h_0(t, \cdot)) &= \mathrm{coin}_{\Omega_1 \cap F^{-1}(K_1)}(F, \Phi, h_0(t, \cdot)) \\ &= \mathrm{coin}_{\Omega_1 \cap F^{-1}(K_1)}(F, \Phi, \varphi). \end{aligned}$$

Recall that K_1 is a retraction candidate and thus $\operatorname{fix}_{\Omega_0}(F, \Phi, \varphi) \subseteq K_1$. Since Ω_1 contains $C := \operatorname{coin}_{\Omega_0}(F, \Phi, \varphi)$, we obtain

$$\operatorname{coin}_{\Omega_1 \cap F^{-1}(K_1)}(F, \Phi, \varphi) = \operatorname{coin}_{\Omega_1}(F, \Phi, \varphi) = C.$$

Summarizing, we have shown with the convention of Remark 12.3 that

$$\operatorname{coin}_{[0,1]\times\Omega_1}(F, \Phi, h) = [0, 1] \times C$$

is compact in view of Theorem 2.63.

Moreover, the set $C_0 := \overline{\varphi_2(\Omega_1)}$ is compact and contained in K_2 or, in case $\operatorname{Comp} = \operatorname{Comp}_Y$, in $\overline{K}_2$. In both cases, $H([0, 1] \times C_0) \subseteq Y$ is compact and contains $h_0([0, 1] \times \Phi(\Omega_2))$. Hence, h_0 is compact into Y, and so the homotopy invariance of the Leray–Schauder triple degree in the last function implies that

$$\deg(F, \Phi, \varphi_2, \Omega_1) = \deg(F, \Phi, h(0, \cdot), \Omega_1) = \deg(F, \Phi, h(1, \cdot), \Omega_1).$$

The function $\varphi_0 := h(1, \cdot)$ belongs to $\operatorname{Comp}(\Phi(\Omega_1), K_1)$ and satisfies $\varphi_0(z) = \varphi_2(z) = \varphi(z)$ for all $z \in \Phi(\Omega_1 \cap F^{-1}(K_1))$. Since K_1 is a retraction candidate, this implies

$$\deg(F, \Phi, \varphi_0, \Omega_1) = \deg(F, \Phi, \varphi_1, \Omega_1).$$

Combining the above equations, we obtain (14.7). □

Corollary 14.27. *Let $\mathcal{A}$ denote a fixed family of subsets $K \subseteq Y$. Let K be a retraction candidate for $(F, \Phi, \varphi, \Omega) \in \mathcal{T}_{\mathrm{prop}}(X, Y)$. Assume that either $K = \emptyset$ or that $K \in \mathcal{A}$ and for each retraction candidate $K_0 \in \mathcal{A}$ with $K \cap K_0 \neq \emptyset$ and $K \neq K_0$ there is an open neighborhood $\Omega_0 \subseteq \Omega$ of $C := \operatorname{coin}_\Omega(F, \Phi, \varphi)$ such that at least one of the following holds:*

(a) *For each open neighborhood $\Omega_1 \subseteq \Omega_0$ of C there is a neighborhood Z of $\Phi(C)$ in $\Phi(\Omega_1)$ such that $K \cup K_0$ is an* CNE *for Z. Moreover, K is a local $(F, \Phi, \varphi, \Omega)$-deformation retract of $K \cup K_0$.*

(b) *For each open neighborhood $\Omega_1 \subseteq \Omega_0$ of C there is a neighborhood Z of $\Phi(C)$ in $\Phi(\Omega_1)$ such that $K \cap K_0$ is an* CNE *for Z. Moreover, $K \cap K_0$ is a retraction candidate, and $K \cap K_0$ is a local $(F, \Phi, \varphi, \Omega)$-deformation retract of K.*

Then K is $\mathcal{A}$-fundamental for $(F, \Phi, \varphi, \Omega)$.

Proof. We obtain property (i) of Definition 14.4 by Theorem 14.24, and property (ii) by Theorem 14.26 with $K_1 = K$ and $K_2 = K \cup K_0$ or with $K_1 = K \cap K_0$ and $K_2 = K$, respectively. □

In particular, we have shown the following result:

Theorem 14.28. *Let $\mathcal{A}$ denote a family of subsets $K \subseteq Y$. Let $(F, \Phi, \varphi, \Omega) \in \mathcal{T}_{\mathrm{prop}}(X, Y)$, and $K \subseteq Y$. Suppose that there is an open neighborhood $\Omega_0 \subseteq \Omega$ of $C := \mathrm{coin}_\Omega(F, \Phi, \varphi)$ such that the following holds.*

(a) $\mathrm{fix}_\Omega(F, \Phi, \varphi) \subseteq K$.

(b) $K = \varnothing$ *or* $K \in \mathcal{A}$.

(c) *For every* $K_0 \in \mathcal{A}$ *with* $K \cap K_0 \neq \varnothing$ *and every open neighborhood* $\Omega_1 \subseteq \Omega_0$ *of* C *there is a neighborhood* Z_0 *of* $\Phi(C)$ *in* $\Phi(\Omega_1)$ *such that* $K \cup K_0$ *is an* CNE *for* $[0, 1] \times Z_0$.

(d) *For every* $K_0 \in \mathcal{A}$ *with* $K \cap K_0 \neq \varnothing$ *the set* K *is a local* $(F, \Phi, \varphi, \Omega)$*-deformation retract of* $K \cup K_0$.

(e) $K = \varnothing$, *or there is a neighborhood* Z *of* $\Phi(C)$ *in* $\Phi(\Omega_0)$ *such that* K *is an* CNE *for* Z, *and the restriction of* φ *to*

$$\Gamma_0 := Z \cap \overline{\Phi(\Omega_0 \cap F^{-1}(K))}$$

belongs to $\mathrm{Comp}(\Gamma_0, K)$.

Then K *is* $\mathcal{A}$*-fundamental for* $(F, \Phi, \varphi, \Omega)$.

Proof. Theorem 14.19 implies that K is a retraction candidate, and so the assertion follows from Corollary 14.27(a). □

Note that the condition $\varphi|_{\Gamma_0} \in \mathrm{Comp}(\Gamma_0, K)$ of Theorem 14.28 is satisfied if:

(a) K is compact.

(b) $\varphi(\Phi(\Omega_0 \cap F^{-1}(K))) \subseteq K$.

In the case $F = \Phi = \mathrm{id}_\Omega$, the latter hypothesis is sometimes called *pushing condition* in literature [60], [61]. It is sometimes satisfied in situations where the special case of degree theory for condensing operators (which we develop in the next sections) does not apply. We refer to [4] and [3] for some applications of this condition.

14.3 The Degree for Fredholm Triples with Convex-fundamental Sets

In this section, we consider a particular case of Sections 14.1 and 14.2: Throughout this section, we let $\mathcal{A}$ denote the family of all closed convex subsets $K \subseteq Y$.

Since any $K \in \mathcal{A}$ is closed in Y, we have $\mathrm{Comp}_K(\Gamma, K) = \mathrm{Comp}_Y(\Gamma, K)$, and K is an CNE_K if and only if it is a CNE_Y (recall Proposition 4.35). In particular, we need not distinguish these cases as in the previous sections.

However, instead of requiring the existence of a fundamental set, we will require the existence of a convex-fundamental set in the sense of the following definition.

Definition 14.29. Let $(F, \Phi, \varphi, \Omega) \in \mathcal{T}_{\mathrm{gen}}(X, Y)$, and let $\Omega_0 \subseteq \Omega$ be an open neighborhood of $\mathrm{coin}_\Omega(F, \Phi, \varphi)$.

A set $K \subseteq Y$ is *almost convex-fundamental* for $(F, \Phi, \varphi, \Omega)$ on Ω_0 if the following holds:

(a) K is convex and closed in Y.

(b) $\varphi(\Phi(\Omega_0 \cap F^{-1}(K))) \subseteq K$.

(c) For all $x \in \Omega_0$ we have the implication

$$F(x) \in \mathrm{conv}(K \cup \varphi(\Phi(x))) \implies F(x) \in K. \tag{14.8}$$

If additionally K is compact, we call K *convex-fundamental* for $(F, \Phi, \varphi, \Omega)$.

We will see that "countably condensing triples" possess a convex-fundamental set and that, under minor additional assumptions, each convex-fundamental set is $\mathcal{A}$-fundamental, and so the degree theory of the previous sections can be used. Moreover, we will see that for triples with convex-fundamental sets, this theory has even stronger properties than in the previous sections.

Remark 14.30. Historically, almost convex-fundamental sets were called "fundamental" in [88], and they were one of the first approaches to degree theory for noncompact maps.

Proposition 14.31. *Let K be almost convex-fundamental for $(F, \Phi, \varphi, \Omega)$. Then* $\mathrm{fix}_\Omega(F, \Phi, \varphi) \subseteq K$.

Proof. Since $\mathrm{coin}_\Omega(F, \Phi, \varphi) \subseteq \Omega_0$, we have $\mathrm{fix}_\Omega(F, \Phi, \varphi) = \mathrm{fix}_{\Omega_0}(F, \Phi, \varphi)$, and so the assertion follows from (14.8). □

Proposition 14.32. *Every closed convex set $K \subseteq Y$ containing $(\varphi \circ \Phi)(\Omega_0)$ is almost convex-fundamental for $(F, \Phi, \varphi, \Omega)$ on Ω_0.*

Proof. The assertion follows immediately from Definition 14.29. □

The following criterion gives a convenient test for the existence of convex-fundamental sets (even for families of triples). This criterion even can decide whether there exists a convex-fundamental set containing a given prescribed set $M \subseteq Y$.

We will see in Section 14.4 that this criterion implies in particular that "countably condensing triples" have a convex-fundamental set containing a prescribed set $M \subseteq Y$.

In view of the subsequent homotopy invariance, we formulate the criterion even for families of function triples.

Theorem 14.33. *We consider a family of function triples $(F_t, \Phi_t, \varphi_t, \Omega_t) \in \mathcal{T}_{\mathrm{gen}}(X, Y)$ $(t \in I)$ and a family $\Omega_{0,t} \subseteq \Omega_t$ $(t \in I)$ of open neighborhoods of* $\mathrm{coin}_{\Omega_t}(F_t, \Phi_t, \varphi_t, \Omega_t)$. *Let $\mathcal{A}_0$ denote the family of all subsets $K \subseteq Y$ which are almost convex-fundamental for $(F_t, \Phi_t, \varphi_t, \Omega_t)$ on $\Omega_{0,t}$ for every $t \in [0, 1]$. Then the following holds:*

(a) *For each $M \subseteq Y$ there is a smallest set K satisfying $M \subseteq K \in \mathcal{A}_0$. This set satisfies*

$$K = \overline{\mathrm{conv}}(M \cup \bigcup_{t \in I} \varphi_t(\Phi_t(\Omega_{0,t} \cap F_t^{-1}(K)))). \tag{14.9}$$

This set is compact if and only if there is a set containing M which is convex-fundamental for $(F_t, \Phi_t, \varphi_t, \Omega_t)$ on $\Omega_{0,t}$ for every $t \in [0, 1]$.

(b) *Let $\Psi\colon \mathcal{A} \multimap Y$ satisfy*

$$K \subseteq K_0 \implies \varphi_t(\Phi_t(\Omega_{0,t} \cap F_t^{-1}(K))) \subseteq \Psi(K) \subseteq \Psi(K_0)$$

for all $K, K_0 \in \mathcal{A}$ and all $t \in I$. Then there is a smallest set K satisfying $\Psi(K) \subseteq K \in \mathcal{A}_0$. This set satisfies

$$K = \overline{\mathrm{conv}}(\Psi(K)). \tag{14.10}$$

This set is compact if and only if there is a set K_0 satisfying $\Psi(K_0) \subseteq K_0$ such that K_0 is convex-fundamental for $(F_t, \Phi_t, \varphi_t, \Omega_t)$ on $\Omega_{0,t}$ for every $t \in [0, 1]$.

Proof. To see (b), let $\mathcal{K}$ denote the family of all $K \in \mathcal{A}_0$ which satisfy $\Psi(K) \subseteq K$. We note that Proposition 14.32 implies $\mathcal{K} \neq \varnothing$. Hence, $K_0 := \bigcap \mathcal{K}$ exists. Since intersections of closed convex sets are closed and convex (recall Proposition 2.51), we have $K_0 \in \mathcal{A}$.

Now we put $K_1 := \overline{\mathrm{conv}}(\Psi(K_0))$. Each $K \in \mathcal{K}$ is closed and convex, and we have $\Psi(K_0) \subseteq \Psi(K) \subseteq K$. We obtain that $K_1 \subseteq K$ for all $K \in \mathcal{K}$, and so $K_1 \subseteq K_0$.

We claim that $K_1 \in \mathcal{K}$. To see this, we note first that $K_1 \subseteq K_0$ implies $\Psi(K_1) \subseteq \Psi(K_0) \subseteq K_1$. It remains to show that $K_1 \in \mathcal{A}_0$. For each $t \in I$, we have in view of $K_1 \subseteq K_0$ that

$$\varphi_t(\Phi_t(\Omega_{0,t} \cap F_t^{-1}(K_1))) \subseteq \Psi(K_1) \subseteq \Psi(K_0) \subseteq K_1,$$

and if $x \in \Omega_{0,t}$ satisfies $F_t(x) \in \operatorname{conv}(K_1 \cup \varphi_t(\Phi_t(x)))$ then we have for each $K \in \mathcal{K}$ in view of $K_1 \subseteq K_0 \subseteq K$ that $F_t(x) \in \operatorname{conv}(K \cup \varphi_t(\Phi_t(x)))$. Hence, we obtain from (14.8) that $F_t(x) \in K$ for each $K \in \mathcal{A}$ which implies $F_t(x) \in K_0$. It follows that $x \in F_t^{-1}(K_0)$ and thus

$$F_t(x) \in \operatorname{conv}(K_1 \cup \varphi_t(\Phi_t(\Omega_{0,t} \cap F_t^{-1}(K_0)))) \subseteq \operatorname{conv}(K_1 \cup \Psi(K_0)) \subseteq K_1.$$

Hence, we have indeed $K_1 \in \mathcal{A}_0$ and thus $K_1 \in \mathcal{K}$.

It follows that $K_0 \subseteq K_1 \in \mathcal{K}$, and since $K_1 \subseteq K_0$, we have actually $K_0 = K_1 \in \mathcal{K}$.

Hence, we have shown $K_0 \in \mathcal{K}$ and $K_0 = \overline{\operatorname{conv}}(\Psi(K_0))$. This implies the first assertions of (b). For the last assertion of (b), it suffices to note that, since $K_0 = \bigcap \mathcal{K}$ is closed and $K_0 \in \mathcal{K}$, Proposition 2.29 implies that $\mathcal{K}$ contains a compact set if and only if K_0 is compact.

For the proof of (a), we put for $K \in \mathcal{A}$

$$\Psi_0(K) := \bigcup_{t \in I} \varphi_t(\Phi_t(\Omega_{0,t} \cap F_t^{-1}(K)))$$

and note that $K \in \mathcal{A}_0$ implies $\Psi_0(K) \subseteq K$. Hence, for $K \in \mathcal{A}_0$, we have $M \subseteq K$ if and only if $\Psi_0(K) \cup M \subseteq K$. Thus, we obtain (a) by applying (b) with $\Psi(K) := \Psi_0(K) \cup M$. □

If additionally $(F, \Phi, \varphi, \Omega)$ is locally $[0, 1]$-normal (recall Definition 11.47) and belongs to $\mathcal{T}_{\mathrm{prop}}(X, Y)$, we introduce a new notation:

Definition 14.34. We write $(F, \Phi, \varphi, \Omega) \in \mathcal{T}_{\mathrm{conv}}(X, Y)$ if $(F, \Phi, \varphi, \Omega) \in \mathcal{T}_{\mathrm{prop}}(X, Y)$ is locally $[0, 1]$-normal and has a convex-fundamental set.

Theorem 14.35. *Every $(F, \Phi, \varphi, \Omega) \in \mathcal{T}_{\mathrm{conv}}(X, Y)$ belongs to $\mathcal{T}_{\mathrm{fund}}(X, Y, \mathcal{A})$.*

If K is convex-fundamental then K is $\mathcal{A}$-fundamental, and we have for every open set $\Omega_0 \subseteq \Omega$ with compact $\operatorname{coin}_{\Omega_0}(F, \Phi, \varphi)$ that $(F, \Phi, \varphi, \Omega_0) \in \mathcal{T}_{\mathrm{conv}}(X, Y)$ and that K is convex-fundamental and $\mathcal{A}$-fundamental for $(F, \Phi, \varphi, \Omega_0)$.

Proof. We show first that K is $\mathcal{A}$-fundamental for $(F, \Phi, \varphi, \Omega)$. To this end, we apply Theorem 14.28. Let Ω_0 be as in Definition 14.29. Note that we have indeed

$\mathrm{fix}_\Omega(F, \Phi, \varphi) \subseteq K$ by Proposition 14.31. If $K_0 \in \mathcal{A}$ satisfies $K \cap K_0 \neq \emptyset$ then $K_1 := K \cup K_0$ is an CNE_{K_1} (even an CE_{K_1}) for the class of T_4 spaces by Proposition 4.39. Since (F, Φ, φ) is a locally $[0,1]$-normal triple and since $\Phi(C)$ is compact by Proposition 2.100, we thus find that the only hypothesis of Theorem 14.28 which remains to be checked is that K is a local $(F, \Phi, \varphi, \Omega)$-deformation retract of $K \cup K_0$.

To see this, we recall that by Corollary 4.26, there is a retraction $\rho \in C(K \cup K_0, K)$ onto K. We claim that a map $H \in C([0,1] \times (K \cup K_0), K \cup K_0)$ with the properties of Definition 14.25 is given by

$$H(t, y) := (1 - t)y + t\rho(y).$$

Indeed, $H(0, \cdot) = \mathrm{id}_{K \cup K_0}$ and $H(\{1\} \times (K \cup K_0)) \subseteq \rho(K \cup K_0) = K$. Moreover, for all $y \in (\varphi \circ \Phi)(\Omega_0 \cap F^{-1}(K)) \subseteq K$, we have $\rho(y) = y$ and thus $H(t, y) = y$ $(t \in [0,1])$. It remains to show that

$$\mathrm{fix}_{\Omega_0 \cap F^{-1}(K \cup K_0)}(F, \Phi, H(t, \varphi(\cdot))) \subseteq K \tag{14.11}$$

for all $t \in [0,1]$. Thus, assume that $t \in [0,1]$ and $x \in \Omega_0 \cap F^{-1}(K \cup K_0)$ satisfy

$$F(x) \in H(t, \varphi(\Phi(x))) \subseteq \mathrm{conv}(K \cup \varphi(\Phi(x))).$$

By (14.8), we have $F(x) \in K$, and so (14.11) is shown. Hence, K is indeed $\mathcal{A}$-fundamental for $(F, \Phi, \varphi, \Omega)$.

For the last assertion, we note that if $\Omega_1 \subseteq \Omega$ is open and such that $\mathrm{coin}_{\Omega_1}(F, \Phi, \varphi)$ is compact then Definition 14.29 immediately implies that K is also convex-fundamental for $(F, \Phi, \varphi, \Omega_1)$. Similarly, Definition 11.47 implies that $(F, \Phi, \varphi, \Omega_1)$ is locally $[0,1]$-normal. Hence, $(F, \Phi, \varphi, \Omega_1) \in \mathcal{T}_{\mathrm{conv}}(X, Y)$, and by what we had shown, it follows that K is $\mathcal{A}$-fundamental for $(F, \Phi, \varphi, \Omega_1)$. □

In view of Theorem 14.35 the following definition makes sense:

Definition 14.36. The *convex-fundamental Fredholm triple degree* $\mathrm{Deg} = \mathrm{Deg}_{(X,Y)}$ is the operator which associates to each $(F, \Phi, \varphi, \Omega) \in \mathcal{T}_{\mathrm{conv}}(X, Y)$ the number from $\mathbb{Z}_2$ (or $\mathbb{Z}$ in the oriented case) which is the $\mathcal{A}$-fundamental Fredholm degree of that function triple for the class $\mathcal{A}$ of closed convex subsets of Y. More briefly:

$$\mathrm{Deg}_{(X,Y)} := \mathrm{DEG}_{(X,Y,\mathcal{A})}|_{\mathcal{T}_{\mathrm{conv}}(X,Y)}.$$

We use the new symbol Deg to indicate the particular choice of $\mathcal{A}$ and that we require the existence of a convex-fundamental set instead of only the existence of an $\mathcal{A}$-fundamental set. Under this slightly more restrictive hypothesis, we obtain more natural properties for the degree, as we will show now: Most of the additional hypotheses which we had to require for the corresponding properties in Section 14.1 can be dropped. Moreover, we can now also formulate a single-valued normalization property under reasonable assumptions.

Theorem 14.37. *The convex-fundamental Fredholm degree has the following properties for every $(F, \Phi, \varphi, \Omega) \in \mathcal{T}_{\mathrm{conv}}(X, Y)$:*

($\mathrm{A}_{\mathcal{T}_{\mathrm{conv}}}$) **(Permanence).** *If the set K is convex-fundamental or $\mathcal{A}$-fundamental for $(F, \Phi, \varphi, \Omega)$, $\Omega_0 \subseteq \Omega$ is an open neighborhood of $\mathrm{coin}_\Omega(F, \Phi, \varphi)$, and $\varphi_0 \in \mathrm{Comp}(\Phi(\Omega_0), K)$ satisfies $\varphi_0(z) = \varphi(z)$ for all $z \in \Phi(\Omega_0 \cap F^{-1}(K))$ then*

$$\mathrm{Deg}(F, \Phi, \varphi, \Omega) = \deg(F, \Phi, \varphi_0, \Omega_0).$$

($\mathrm{B}_{\mathcal{T}_{\mathrm{conv}}}$) **(Strong Permanence).** *If the set K is almost convex-fundamental for $(F, \Phi, \varphi, \Omega)$ and for $(F, \Phi, \varphi_0, \Omega) \in \mathcal{T}_{\mathrm{conv}}(X, Y)$, and if there is an open set $\Omega_0 \subseteq \Omega$ with*

$$\mathrm{coin}_\Omega(F, \Phi, \varphi) \cup \mathrm{coin}_\Omega(F, \Phi, \varphi_0) \subseteq \Omega_0$$

with $\varphi_0(z) = \varphi(z)$ for all $z \in \Phi(\Omega_0 \cap F^{-1}(K))$ then

$$\mathrm{coin}_\Omega(F, \Phi, \varphi) = \mathrm{coin}_\Omega(F, \Phi, \varphi_0) \tag{14.12}$$

and

$$\mathrm{Deg}(F, \Phi, \varphi, \Omega) = \mathrm{Deg}(F, \Phi, \varphi_0, \Omega).$$

($\mathrm{C}_{\mathcal{T}_{\mathrm{conv}}}$) **(Excision).** *If $\Omega_0 \subseteq \Omega$ is open and contains $\mathrm{coin}(F, \Phi, \varphi)$ then*

$$\mathrm{Deg}(F, \Phi, \varphi, \Omega_0) = \mathrm{Deg}(F, \Phi, \varphi, \Omega).$$

($\mathrm{D}_{\mathcal{T}_{\mathrm{conv}}}$) **(Generalized Homotopy Invariance).** *Let (G, H, h, W, Y, Γ) be a generalized proper acyclic* homotopy triple for $(G_t, H_t, h_t, W_t, Y, \Gamma_t)$ ($t \in [0, 1]$) where $W \subseteq [0, 1] \times X$ is open, and $G\colon W \to Y$ is a generalized (oriented) Fredholm homotopy of index 0. Suppose that the following holds with $\widetilde{H}(t, x) := \{t\} \times H(t, x)$ and an open neighborhood $U \subseteq W$ of $C := \mathrm{coin}_W(G, \widetilde{H}, h)$.*

(a) *There is $K \subseteq Y$ which is convex-fundamental for (G_t, H_t, h_t, W_t) on $U_t := \{x : (t, x) \in U\}$ for every $t \in [0, 1]$.*

(b) *The compact set* $\widetilde{H}(C)$ *has a* T_4 *neighborhood in* $\widetilde{H}(U)$.

(c) (G_t, H_t, h_t) *is locally* $[0,1]$*-normal for every* $t \in [0,1]$.

Then $(G_t, H_t, h_t, W_t) \in \mathcal{T}_{\text{conv}}(X,Y)$ *for all* $t \in [0,1]$, *and*

$$\text{Deg}(G_t, H_t, h_t, W_t) \text{ is independent of } t \in [0,1].$$

($\text{E}_{\mathcal{T}_{\text{conv}}}$) **(Existence).** *If* $\text{Deg}(F, \Phi, \varphi, \Omega) \neq 0$ *then* $\text{coin}_\Omega(F, \Phi, \varphi) \neq \emptyset$.

($\text{F}_{\mathcal{T}_{\text{conv}}}$) **(Equivalence Invariance).** *Let* $\mathcal{T} := \mathcal{T}_{\text{conv}}(X,Y)$. *If* $(F, \Phi, \varphi) \sim_{\mathcal{T}} (F, \widetilde{\Phi}, \widetilde{\varphi})$ *then*

$$\text{Deg}(F, \Phi, \varphi, \Omega) = \text{Deg}(F, \widetilde{\Phi}, \widetilde{\varphi}, \Omega).$$

Moreover, if

$$(F, \Phi, \varphi, \Omega) \sqsubseteq (F, \widetilde{\Phi}, \widetilde{\varphi}, \Omega) \tag{14.13}$$

and K *is (almost) convex-fundamental for* $(F, \widetilde{\Phi}, \widetilde{\varphi}, \Omega)$ *then* K *is (almost) convex-fundamental for* $(F, \Phi, \varphi, \Omega)$.

($\text{G}_{\mathcal{T}_{\text{conv}}}$) **(Single-Valued Normalization).** *If* $\varphi \circ \Phi$ *is single-valued then the triples* $(F, \Phi, \varphi, \Omega)$ *and* $(F, \text{id}_\Omega, \varphi \circ \Phi, \Omega)$ *have the same (almost) convex-fundamental sets. If* $(F, \Phi, \varphi, \Omega) \in \mathcal{T}_{\text{conv}}(X,Y)$ *and if* $(F, \text{id}_\Omega, \varphi \circ \Phi, \Omega)$ *and the standard form of* $(F, \Phi, \varphi, \Omega)$ *are locally* $[0,1]$*-normal then we have* $(F, \text{id}_\Omega, \varphi \circ \Phi, \Omega) \in \mathcal{T}_{\text{conv}}(X,Y)$ *and*

$$\text{Deg}(F, \Phi, \varphi, \Omega) = \text{Deg}(F, \text{id}_\Omega, \varphi \circ \Phi, \Omega).$$

($\text{H}_{\mathcal{T}_{\text{conv}}}$) **(Compatibility with the Non-Oriented Case).** *The degrees for the oriented and non-oriented case are the same modulo 2 (if the oriented case applies).*

($\text{I}_{\mathcal{T}_{\text{conv}}}$) **(Compatibility with the Leray–Schauder Triple Degree).**

If $(F, \Phi, \varphi, \Omega) \in \mathcal{T}_{\text{LS}}(X,Y)$ *then* $(F, \Phi, \varphi, \Omega) \in \mathcal{T}_{\text{fund}}(X, Y, \mathcal{A})$, *and*

$$\text{DEG}(F, \Phi, \varphi, \Omega) = \deg(F, \Phi, \varphi, \Omega). \tag{14.14}$$

If $(F, \Phi, \varphi, \Omega) \in \mathcal{T}_{\text{LS}}(X,Y)$ *is locally* $[0,1]$*-normal then* $(F, \Phi, \varphi, \Omega) \in \mathcal{T}_{\text{conv}}(X,Y)$ *and*

$$\text{Deg}(F, \Phi, \varphi, \Omega) = \deg(F, \Phi, \varphi, \Omega). \tag{14.15}$$

($J_{\mathcal{T}_{\mathrm{conv}}}$) **(Additivity).** *Let $\Omega = \Omega_1 \cup \Omega_2$ with disjoint open subsets $\Omega_1, \Omega_2 \subseteq \Omega$. Then $(F, \Phi, \varphi, \Omega_i) \in \mathcal{T}_{\mathrm{conv}}(X, Y)$ for $i = 1, 2$, and*

$$\mathrm{Deg}(F, \Phi, \varphi, \Omega) = \mathrm{Deg}(F, \Phi, \varphi, \Omega_1) + \mathrm{Deg}(F, \Phi, \varphi, \Omega_2).$$

($K_{\mathcal{T}_{\mathrm{conv}}}$) **(Excision-Additivity).** *Let $\Omega_i \subseteq \Omega$ ($i \in I$) be a family of pairwise disjoint open sets with $\mathrm{coin}_\Omega(F, \Phi, \varphi) \subseteq \bigcup_{i \in I} \Omega_i$ and such that $\mathrm{coin}_{\Omega_i}(F, \Phi, \varphi)$ is compact for all $i \in I$. Then $(F, \Phi, \varphi, \Omega_i) \in \mathcal{T}_{\mathrm{conv}}(X, Y)$ for all $i \in I$, and*

$$\mathrm{Deg}(F, \Phi, \varphi, \Omega) = \sum_{i \in I} \mathrm{Deg}(F, \Phi, \varphi, \Omega_i),$$

where in the sum at most a finite number of summands is nonzero.

($L_{\mathcal{T}_{\mathrm{conv}}}$) **(Diffeomorphic-Isomorphic Invariance).** *Let J_1 be a diffeomorphism of an open subset of a Banach manifold X_0 onto Ω and J_2 an isomorphism of Y onto a real normed vector space Y_0. Then*

$$\mathrm{Deg}_{(X,Y)}(F, \Phi, \varphi, \Omega) = \mathrm{Deg}_{(X_0,Y_0)}(J_2 \circ F \circ J_1, \Phi \circ J_1, J_2 \circ \varphi, J_1^{-1}(\Omega)).$$

In the oriented case, the orientation of $J_2 \circ F \circ J_1$ is understood in the sense of Proposition 8.38.

($M_{\mathcal{T}_{\mathrm{conv}}}$) **(Restriction).** *Let $X_0 \subseteq X$ be open. Then*

$$\mathrm{Deg}_{(X_0,Y)} = \mathrm{Deg}_{(X,Y)}|_{\mathcal{T}_{\mathrm{conv}}(X_0,Y)}.$$

($N_{\mathcal{T}_{\mathrm{conv}}}$) **(Cartesian Product).** *For $i = 1, 2$, let X_i be a manifold without boundary of class C^1 over the real Banach space E_{X_i}, and let $Y_i = E_{Y_i}$ be a real Banach space. For $(F_i, \Phi_i, \varphi_i, \Omega_i) \in \mathcal{T}_{\mathrm{conv}}(X_i, Y_i)$, we put $X := X_1 \times X_2$, $\Omega := \Omega_1 \times \Omega_2$, $F := F_1 \otimes F_2$, $\Phi := \Phi_1 \otimes \Phi_2$, and $\varphi := \varphi_1 \otimes \varphi_2$. If $(F, \Phi, \varphi, \Omega)$ is locally $[0, 1]$-normal then $(F, \Phi, \varphi, \Omega) \in \mathcal{T}_{\mathrm{conv}}(X, Y)$, and*

$$\begin{aligned} &\mathrm{Deg}_{(X,Y)}(F, \Phi, \varphi, \Omega) \\ &\quad = \mathrm{Deg}_{(X_1,Y_1)}(F_1, \Phi_1, \varphi_1, \Omega_1)\, \mathrm{Deg}_{(X_2,Y_2)}(F_2, \Phi_2, \varphi_2, \Omega_2). \end{aligned}$$

In the oriented case, F is equipped with the product orientation.

Remark 14.38. In the generalized homotopy invariance ($D_{\mathcal{T}_{\mathrm{conv}}}$) the compactness of $\widetilde{H}(C)$ is automatic by Proposition 2.100. Moreover, the hypothesis about the T_4 neighborhood is trivially satisfied if $(G, \widetilde{H}, h)$ is locally normal (recall Definition 11.47).

Remark 14.39. In the single-valued normalization property ($\mathrm{G}_{\mathcal{T}_{\mathrm{conv}}}$), the hypotheses that $(F, \mathrm{id}_\Omega, \varphi \circ \Phi, \Omega)$ and the standard form of $(F, \Phi, \varphi, \Omega)$ are locally $[0,1]$-normal are satisfied if $[0,1] \times \Omega \times \Phi(\Omega)$ is a T_5 space.

Proof. The permanence property ($\mathrm{A}_{\mathcal{T}_{\mathrm{conv}}}$) follows from the definitions and Theorem 14.35. For ($\mathrm{B}_{\mathcal{T}_{\mathrm{conv}}}$), we note first that (14.8) implies

$$\mathrm{fix}_\Omega(F, \Phi, \varphi) \cup \mathrm{fix}_\Omega(F, \Phi, \varphi_0) \subseteq K,$$

and so (14.12) follows from

$$\begin{aligned}\mathrm{coin}_\Omega(F, \Phi, \varphi) &= \mathrm{coin}_{\Omega_0 \cap F^{-K}}(F, \Phi, \varphi) \\ &= \mathrm{coin}_{\Omega_0 \cap F^{-K}}(F, \Phi, \varphi_0) = \mathrm{coin}_\Omega(F, \Phi, \varphi_0).\end{aligned}$$

Next, we observe that Theorem 14.33 implies that K contains a set K_0 which is simultaneously convex-fundamental for $(F, \Phi, \varphi, \Omega)$ and for $(F, \Phi, \varphi_0, \Omega)$. In particular, $K_0 \subseteq K$ is $\mathcal{A}$-fundamental and thus a retraction candidate for $(F, \Phi, \varphi, \Omega)$ and $(F, \Phi, \varphi_0, \Omega)$. We thus find an open neighborhood $\Omega_1 \subseteq \Omega_0$ of (14.12) and $\varphi_1 \in \mathrm{Comp}(\Phi(\Omega_1), K_0)$ satisfying $\varphi_1(z) = \varphi(z) = \varphi_0(z)$ for all $z \in \Phi(\Omega_1 \cap F^{-1}(K_0))$. The permanence property implies

$$\mathrm{Deg}(F, \Phi, \varphi, \Omega) = \deg(F, \Phi, \varphi_1, \Omega_1) = \mathrm{Deg}(F, \Phi, \varphi_0, \Omega),$$

and so the strong permanence property is proved.

The properties ($\mathrm{C}_{\mathcal{T}_{\mathrm{conv}}}$), ($\mathrm{E}_{\mathcal{T}_{\mathrm{conv}}}$), ($\mathrm{H}_{\mathcal{T}_{\mathrm{conv}}}$), and ($\mathrm{M}_{\mathcal{T}_{\mathrm{conv}}}$) are special cases of the corresponding properties of the $\mathcal{A}$-fundamental Fredholm triple degree.

For the proof of the generalized homotopy invariance ($\mathrm{D}_{\mathcal{T}_{\mathrm{conv}}}$), we verify all hypotheses of Theorem 14.23. By Theorem 14.35, we find that K is $\mathcal{A}$-fundamental for (G_t, H_t, h_t, W_t) for every $t \in [0,1]$. Recall that $K = \varnothing$ or that K is a CE_K (or equivalently CE_Y, recall Proposition 4.35) for the class of T_4 spaces by Theorem 4.36. In particular, if Z denotes a T_4 neighborhood of $\widetilde{H}(C)$ in $\widetilde{H}(U)$, we obtain that K is a CNE_K (and CNE_Y) for Z. Moreover, since K is $\mathcal{A}$-fundamental for (G_t, H_t, h_t, W_t) on U_t, we have in particular

$$h_t(\{t\} \times H_t(U_t \cap G_t^{-1}(K))) \subseteq K$$

for all $t \in [0,1]$. It follows that

$$h(\widetilde{H}(U \cap G^{-1}(K))) \subseteq K.$$

Since K is closed, we obtain in view of Proposition 2.85 with

$$\Gamma_0 := Z \cap \overline{\widetilde{H}(U \cap G^{-1}(K))}$$

that

$$h(\Gamma_0) \subseteq h(\overline{\widetilde{H}(U \cap G^{-1}(K))}) \subseteq \overline{K} = K.$$

Hence, the restriction of h to Γ_0 belongs to $\mathrm{Comp}_Y(\Gamma_0, K) = \mathrm{Comp}_K(\Gamma_0, K)$. We thus verified all hypotheses of Theorem 14.23, and so

$$\mathrm{Deg}(G_t, H_t, h_t, W_t) = \mathrm{DEG}(G_t, H_t, h_t, W_t)$$

is independent of $t \in [0, 1]$.

Concerning the equivalence invariance, we show first the second assertion of ($\mathrm{F}_{\mathcal{T}_{\mathrm{conv}}}$). Thus, assume that (14.13) holds, and that K is (almost) convex-fundamental for the function triple $(F, \widetilde{\Phi}, \widetilde{\varphi}, \Omega)$ on an open neighborhood $\Omega_0 \subseteq \Omega$ of $\mathrm{coin}_\Omega(F, \widetilde{\Phi}, \widetilde{\varphi}, \Omega)$. Then (11.6) implies that Ω_0 is a neighborhood of $\mathrm{coin}_\Omega(F, \Phi, \varphi, \Omega)$. We claim that K is (almost) convex-fundamental for the function triple $(F, \Phi, \varphi, \Omega)$ on Ω_0. To see this, we note that in view of (11.5), we have

$$\varphi(\Phi(\Omega_0 \cap F^{-1}(K))) \subseteq \widetilde{\varphi}(\widetilde{\Phi}(\Omega_0 \cap F^{-1}(K))) \subseteq K,$$

and that for each $x \in \Omega_0$ the relation

$$F(x) \in \mathrm{conv}(K \cup \varphi(\Phi(x)))$$

implies

$$F(x) \in \mathrm{conv}(K \cup \widetilde{\varphi}(\widetilde{\Phi}(x)))$$

and thus $F(x) \in K$. Hence, K is indeed (almost) convex-fundamental for $(F, \Phi, \varphi, \Omega)$ on Ω_0.

In particular, if (14.13) holds where both function triples belong to $\mathcal{T}$ and if K is convex-fundamental for $(F, \widetilde{\Phi}, \widetilde{\varphi}, \Omega) \in \mathcal{T}$ then K is simultaneously $\mathcal{A}$-fundamental for $(F, \widetilde{\Phi}, \widetilde{\varphi}, \Omega)$ and for $(F, \Phi, \varphi, \Omega)$ by Theorem 14.35, and so Theorem 14.18($\mathrm{P}_{\mathcal{T}_{\mathrm{fund}}}$) implies

$$\mathrm{Deg}(F, \Phi, \varphi, \Omega) = \mathrm{Deg}(F, \widetilde{\Phi}, \widetilde{\varphi}, \Omega).$$

This shows the equivalence invariance in the special case (14.13), and the general case follows by a trivial induction.

For the proof of ($\mathrm{G}_{\mathcal{T}_{\mathrm{conv}}}$), we note first that

$$C := \mathrm{coin}_\Omega(F, \Phi, \varphi) = \mathrm{coin}_\Omega(F, \mathrm{id}_\Omega, \varphi \circ \Phi).$$

Moreover, Propositions 2.80 and 2.94 imply that $\varphi \circ \Phi|_{\Omega_0}$ is continuous for some open neighborhood $\Omega_0 \subseteq \Omega$ of C, and so $(F, \mathrm{id}_\Omega, \varphi \circ \Phi) \in \mathcal{T}_{\mathrm{prop}}(X, Y)$. Since

$$(\varphi \circ \Phi)(\mathrm{id}_\Omega(\Omega_0 \cap F^{-1}(K))) = \varphi(\Phi(\Omega_0 \cap F^{-1}(K)))$$

and

$$\operatorname{conv}(K \cup (\varphi \circ \Phi)(\mathrm{id}_\Omega(x))) = \operatorname{conv}(K \cup \varphi(\Phi(x))),$$

it follows that K is (almost) convex-fundamental for $(F, \Phi, \varphi, \Omega)$ on Ω_0 if and only if K is (almost) convex-fundamental for $(F, \mathrm{id}_\Omega, \varphi \circ \Phi, \Omega)$.

In particular, if $(F, \Phi, \varphi, \Omega) \in \mathcal{T} := \mathcal{T}_{\mathrm{conv}}(X, Y)$ and $(F, \mathrm{id}_\Omega, \varphi \circ \Phi, \Omega)$ is locally $[0, 1]$-normal, it follows that $(F, \mathrm{id}_\Omega, \varphi \circ \Phi, \Omega) \in \mathcal{T}$. Let $(F, \widetilde{\Phi}, \widetilde{\varphi}, \Omega)$ denote the standard form of $(F, \Phi, \varphi, \Omega)$, and let K be convex-fundamental for $(F, \Phi, \varphi, \Omega)$. Note that Proposition 11.29 implies

$$(F, \Phi, \varphi) \gtrsim (F, \widetilde{\Phi}, \widetilde{\varphi}) \lesssim (F, \mathrm{id}_\Omega, \varphi \circ \Phi).$$

By the second assertion of the equivalence invariance, we obtain that K is convex-fundamental for $(F, \widetilde{\Phi}, \widetilde{\varphi}, \Omega)$, and so $(F, \widetilde{\Phi}, \widetilde{\varphi}, \Omega)$ belongs to $\mathcal{T}$ if it is locally $[0, 1]$-normal. Hence, the equivalence invariance of Deg implies the assertion.

The first part of $(\mathrm{I}_{\mathcal{T}_{\mathrm{conv}}})$ follows from Corollary 14.9 since $K = Y$ is convex and closed and belongs to $Y \in \mathcal{A}$. For the second part, we note that Proposition 13.7 implies for every $(F, \Phi, \varphi, \Omega) \in \mathcal{T}_{\mathrm{LS}}(X, Y)$ that there is an open neighborhood $\Omega_0 \subseteq \Omega$ of $\operatorname{coin}_\Omega(F, \Phi, \varphi)$ such that $C := \varphi(\Phi(\Omega_0))$ is relatively compact in Y. Corollary 3.62 implies that $K := \overline{\operatorname{conv}} C$ is compact, and so Proposition 14.32 implies that K is convex-fundamental for $(F, \Phi, \varphi, \Omega)$ on Ω_0. Hence, if additionally $(F, \Phi, \varphi, \Omega)$ is locally $[0, 1]$-normal, it follows that $(F, \Phi, \varphi, \Omega) \in \mathcal{T}_{\mathrm{conv}}(X, Y)$, and (14.15) follows from the definition of Deg and (14.14).

For the additivity and excision-additivity, we note that if K is convex-fundamental for $(F, \Phi, \varphi, \Omega)$ then Theorem 14.35 implies that K is simultaneously convex-fundamental and thus $\mathcal{A}$-fundamental for $(F, \Phi, \varphi, \Omega_i)$ for all $i = 1, 2$ or $i \in I$, respectively. Hence, the additivity and excision-additivity follows from Theorem 14.15.

The diffeomorphic-isomorphic invariance property follows from the corresponding property of the $\mathcal{A}$-fundamental Fredholm triple degree, since any isomorphism J_2 of Y onto Y_0 automatically has the property that $K \subseteq Y$ is closed and convex if and only if $J_2(K)$ is closed and convex.

For the Cartesian product property, we first show that if K_i is convex-fundamental for $(F_i, \Phi_i, \varphi_i, \Omega_i)$ on $\Omega_{0,i} \subseteq \Omega_i$ for $i = 1, 2$ then $K = K_1 \times K_2$ is convex-fundamental for $(F, \Phi, \varphi, \Omega)$ on $\Omega_0 := \Omega_{0,1} \times \Omega_{0,2}$. Indeed, K is convex and compact by Theorem 2.63, and

$$\begin{aligned}\varphi(\Phi(\Omega_0 \cap F^{-1}(K))) &= \varphi_1(\Phi_1(\Omega_{0,1} \cap F_1^{-1}(K))) \times \varphi_2(\Phi_2(\Omega_{0,2} \cap F_2^{-1}(K))) \\ &\subseteq K_1 \times K_2 = K.\end{aligned}$$

Moreover, if $x \in \Omega_0$, that is $x = (x_1, x_2)$ with $x_i \in \Omega_{0,i}$ for $i = 1, 2$, we have

$$\begin{aligned}\operatorname{conv}(K \cup \varphi(\Phi(x))) &\subseteq \operatorname{conv}((K_1 \cup \varphi_1(\Phi_1(x_1))) \times (K_2 \cup \varphi_2(\Phi_2(x_2))))\\ &= \operatorname{conv}(K_1 \cup \varphi_1(\Phi_1(x_1))) \times \operatorname{conv}(K_2 \cup \varphi_2(\Phi_2(x_2))),\end{aligned}$$

and $F(x) = (F_1(x_1), F_2(x_2))$, and so (14.8) holds. Since $(F, \Phi, \varphi, \Omega)$ is locally $[0, 1]$-normal, it follows that $(F, \Phi, \varphi, \Omega) \in \mathcal{T}_{\mathrm{conv}}(X, Y)$. By Theorem 14.35, we conclude that K is $\mathcal{A}$-fundamental for $(F, \Phi, \varphi, \Omega)$ where $\mathcal{A}$ denotes the family of all closed convex subsets of Y. Let $\mathcal{A}_0$ denote the family of all sets of the form $K_1 \times K_2$ where $K_i \subseteq Y_i$ is closed and convex for $i = 1, 2$. Then $\mathcal{A}_0 \subseteq \mathcal{A}$. Since $K \in \mathcal{A}_0$ is $\mathcal{A}$-fundamental and $\mathcal{A}_0 \subseteq \mathcal{A}$, we obtain immediately from Definition 14.4 that K is $\mathcal{A}_0$-fundamental. Using Theorem 14.17, we find that

$$\begin{aligned}&\mathrm{DEG}_{(X,Y,\mathcal{A}_0)}(F, \Phi, \varphi, \Omega)\\ &\quad = \mathrm{Deg}_{(X_1,Y_1)}(F_1, \Phi_1, \varphi_1, \Omega_1)\,\mathrm{Deg}_{(X_2,Y_2)}(F_2, \Phi_2, \varphi_2, \Omega_2).\end{aligned}$$

Moreover, using Proposition 14.12, we find

$$\mathrm{DEG}_{(X,Y,\mathcal{A}_0)}(F, \Phi, \varphi, \Omega) = \mathrm{DEG}_{(X,Y,\mathcal{A})}(F, \Phi, \varphi, \Omega) = \mathrm{Deg}_{(X,Y)}(F, \Phi, \varphi, \Omega),$$

and so the Cartesian product property is proved. □

14.4 Countably Condensing Triples

In this section, we discuss how to verify that a function triple $(F, \Phi, \varphi, \Omega)$ belongs to $\mathcal{T}_{\mathrm{conv}}(X, Y)$ (recall Definition 14.34). More general, in view of the generalized homotopy invariance of the degree, we discuss how to verify that there is a set $K \subseteq Y$ which is convex-fundamental for a family of function triples $(F_t, \Phi_t, \varphi_t, \Omega_t)$ (recall Definition 14.29).

We use throughout the notations of Section 14.3.

The starting point of our considerations is the criterion formulated in Theorem 14.33: By that criterion, it suffices in particular to verify that any set $K \subseteq Y$ satisfying the set inequality (14.9) or (14.10) is compact.

Since we want to obtain conditions on countable sets, we use the idea to replace K by a certain countable subset $\hat{K}$ which "almost" (up to some closures) satisfies the set inequalities (14.9) or (14.10) and such that the compactness of K follows from the compactness of certain auxiliary sets associated with $\hat{K}$.

We keep the above description of the idea so rough, because there are two "dual" approaches to achieve this aim. These approaches lead to rather similar results which, however, vary in some details. So we will actually develop both approaches.

Originally, the author has developed both approaches in [141], basing them on two "dual" lemmas from [137]. We present now a new lemma which is even slightly simpler to formulate than the mentioned "dual" lemmas and which contains both of them as special cases.

Thus, both approach can actually be based on this lemma. Roughly speaking, this lemma states that if we have "almost" the set equality $F(M) = \overline{\operatorname{conv}}\,\Psi(M)$, namely (14.16), then there exist "arbitrarily large" countable subsets $C \subseteq M$ which also satisfy "almost and up to closures" an analogous equality (14.17), (14.18). In some cases, this becomes even really a set equality (14.19) for closures.

Lemma 14.40. *Let Z be a locally convex space, M be a set, and $F, \Psi\colon M \multimap Z$ be such that $F(x)$ and $\Psi(x)$ are separable for every $x \in M$. Assume also that $\overline{\operatorname{conv}}\,\Psi(M)$ is metrizable. Let I be an (at most) countable index set, and $V_i \subset Z$ $(i \in I)$ be such that*

$$\bigcup_{i\in I}(V_i \cap \operatorname{conv}\Psi(M)) \subseteq F(M) \subseteq \overline{\operatorname{conv}}\,\Psi(M). \tag{14.16}$$

Then for each countable $C_0 \subseteq M$ there is a countable $C \subseteq M$ with $C_0 \subseteq C$,

$$\overline{F(C)} \subseteq \overline{\operatorname{conv}}\,\Psi(C), \tag{14.17}$$

$$\overline{V_i \cap \operatorname{conv}\Psi(C)} \subseteq \overline{V_i \cap F(C)} \subseteq \overline{V_i \cap \overline{\operatorname{conv}}\,\Psi(C)} \quad \textit{for all } i \in I. \tag{14.18}$$

For those $i \in I$ for which $(\overline{\operatorname{conv}}\,\Psi(M)) \setminus V_i$ is closed in Z, (14.18) *is even equivalent to*

$$\overline{V_i \cap F(C)} = \overline{V_i \cap \operatorname{conv}\Psi(C)}. \tag{14.19}$$

Proof. We show first that for each countable $C_n \subseteq M$ there is a countable $C_{n+1} \subseteq M$ satisfying

$$C_n \subseteq C_{n+1}, \tag{14.20}$$

$$F(C_n) \subseteq \overline{\operatorname{conv}}\,\Psi(C_{n+1}), \tag{14.21}$$

$$V_i \cap \operatorname{conv}\Psi(C_n) \subseteq \overline{V_i \cap F(C_{n+1})} \quad \text{for all } i \in I. \tag{14.22}$$

Indeed, the second inclusion of (14.16) implies $F(C_n) \subseteq \overline{\operatorname{conv}}\,\Psi(M)$. Since $F(C_n)$ is a countable union of separable sets and thus separable, Proposition 3.55 implies that there is a countable $P_n \subseteq \Psi(M)$ with $F(C_n) \subseteq \overline{\operatorname{conv}}\,P_n$. There is a countable $A_n \subseteq M$ with $P_n \subseteq \Psi(A_n)$. Hence, $F(C_n) \subseteq \overline{\operatorname{conv}}\,\Psi(A_n)$.

Since $\Psi(C_n)$ is a countable union of separable sets and thus separable, Proposition 3.55 implies that $V_{n,i} := V_i \cap \operatorname{conv}\Psi(C_n) \subseteq \overline{\operatorname{conv}}\,\Psi(C_n)$ are separable.

Since I is countable, it follows that there exist countable $Q_n \subseteq \bigcup_{i \in I} V_{n,i}$ with $V_{n,i} \subseteq \overline{V_{n,i} \cap Q_n}$ for all $i \in I$. Note that the first inclusion of (14.16) implies $Q_n \subseteq F(M)$. Consequently, there are countable $B_n \subseteq M$ with $Q_n \subseteq F(B_n)$. We thus have

$$V_i \cap \operatorname{conv} \Psi(C_n) = V_{n,i} \subseteq \overline{V_{n,i} \cap F(B_n)} \subseteq \overline{V_i \cap F(B_n)}.$$

It follows that $C_{n+1} := A_n \cup B_n \cup C_n$ satisfies (14.20)–(14.22).

Hence, for each countable $C_0 \subseteq M$, we can define by induction a sequence of countable sets $C_1, C_2, \ldots \subseteq M$ satisfying (14.20)–(14.22) for all $n = 0, 1, \ldots$ We claim that $C = \bigcup_{n=1}^{\infty} C_n$ has the required property. Indeed, C is countable as a union of countably many countable sets. By (14.20), we have $C_0 \subseteq C$. From (14.21), we obtain $F(C_n) \subseteq \overline{\operatorname{conv}}\, \Psi(C)$ for all n, and so $F(C) \subseteq \overline{\operatorname{conv}}\, \Psi(C)$ follows. This implies (14.17) and the second inclusion of (14.18). Finally, (14.22) implies with $K_n := \operatorname{conv} \Psi(C_n)$ for each fixed $i \in I$ that

$$V_i \cap K_n \subseteq \overline{V_i \cap F(C)}$$

holds for all n. In particular, $K := \bigcup_{n=1}^{\infty} K_n$ satisfies

$$V_i \cap K \subseteq \overline{V_i \cap F(C)}.$$

Note that K is convex by Proposition 2.51, because (14.20) implies $K_n \subseteq K_{n+1}$ for all n. In view of $\Psi(C) \subseteq K$, we have $\operatorname{conv} \Psi(C) \subseteq K$ and thus obtain the first inclusion of (14.18).

For the last assertion, we apply Proposition 2.11 with $Y := V_i$, $M := \operatorname{conv} \Psi(C)$, and $N := \overline{\operatorname{conv}}\, \Psi(M)$. We thus find with Proposition 3.54 that actually

$$\overline{V_i \cap \overline{\operatorname{conv}}\, \Psi(C)} = \overline{V_i \cap \operatorname{conv} \Psi(C)}$$

if $N \setminus V_i$ is closed. □

Now one approach consists in a rather straightforward application of the above Lemma 14.40 with $F = \mathrm{id}_M$ and $M = K$ where K is as in Theorem 14.33. For simplicity, we formulate this approach only in the particular situation of Theorem 14.33(a), although a similar result could also be formulated for Theorem 14.33(b).

Theorem 14.41. *We consider a family of function triples* $(F_t, \Phi_t, \varphi_t, \Omega_t) \in \mathcal{T}_{\mathrm{gen}}(X, Y)$ $(t \in I)$ *and a family* $\Omega_{0,t} \subseteq \Omega_t$ $(t \in I)$ *of open neighborhoods of* $\operatorname{coin}_{\Omega_t}(F_t, \Phi_t, \varphi_t, \Omega_t)$. *Let* $M \subseteq Y$ *be separable, and let* $K \subseteq Y$ *be the corresponding set of Theorem* 14.33(a). *We define* $\Psi\colon K \multimap Y$ *by*

$$\Psi(x) := \bigcup_{t \in I} \varphi_t(\Phi_t(\Omega_{0,t} \cap F_t^{-1}(x))). \tag{14.23}$$

Suppose that $\Psi(x)$ *is separable for every* $x \in K$. *Let* $V_n \subseteq Y$ $(n \in N)$ *where* N *is (at most) countable. Suppose that for any countable* $C \subseteq K$ *the inclusions*

$$\overline{C} \subseteq \overline{\operatorname{conv}}(M \cup \Psi(C)), \tag{14.24}$$

$$\overline{V_n \cap \operatorname{conv}(M \cup \Psi(C))} \subseteq \overline{V_n \cap C} \subseteq \overline{V_n \cap \overline{\operatorname{conv}}(M \cup \Psi(C))} \quad \text{for all } n \in N, \tag{14.25}$$

and for those $n \in N$ *for which* $(\overline{\operatorname{conv}}(M \cup \Psi(K))) \setminus V_n$ *is closed in* Y *also*

$$\overline{V_n \cap C} = \overline{V_n \cap \operatorname{conv}(M \cup \Psi(C))} \tag{14.26}$$

imply that C *is relatively compact in* Y. *Then* $K \supseteq M$ *is convex-fundamental for* $(F_t, \Phi_t, \varphi_t, \Omega_t)$ *on* $\Omega_{0,t}$ *for every* $t \in I$.

Proof. We have to show that K is compact. Assume that this is not the case. Since K is closed and thus complete by Lemma 3.8, Proposition 3.26 implies that $\beta(K) > 0$. Hence, there is a countable $C_0 \subseteq K$ with $\beta(C_0) > 0$. We apply Lemma 14.40 with $M = K$, $F = \mathrm{id}_M$, and the multivalued function $\widetilde{\Psi}(x) := M \cup \Psi(x)$. Note that $\widetilde{\Psi}(x)$ is separable, since M and $\Psi(x)$ are assumed to be separable. Moreover, (14.9) means $K = \overline{\operatorname{conv}}\, \widetilde{\Psi}(K)$, that is $\mathrm{id}_M(M) = \overline{\operatorname{conv}}\, \widetilde{\Psi}(M)$. Hence (14.16) is satisfied. Lemma 14.40 thus shows that there is a countable $C \subseteq M$ with $C_0 \subseteq C$ and satisfying (14.24) and (14.25) or even (14.26) respectively. Since $C_0 \subseteq C$, we have $\beta(C) \geq \beta(C_0) > 0$, and so C is not relatively compact by Proposition 3.26 which contradicts the hypotheses. □

Remark 14.42. For $(F_t, \Phi_t, \varphi_t, \Omega_t)$ $(t \in I)$ and $\Omega_{0,t} \subseteq \Omega_t$, we put $\Omega := \{(t,x) : x \in \Omega_{0,t}\}$, $F(t,x) := F_t(x)$, $\widetilde{\Phi}(t,x) := (t, \Phi_t(x))$, and $\varphi(t,z) := \varphi_t(z)$. Then the function (14.23) is just a restriction of

$$\Psi := \varphi \circ \widetilde{\Phi} \circ F|_{\Omega_0}^{-1}.$$

For the particular case that we consider only one function triple $(F, \Phi, \varphi, \Omega) = (F_t, \Phi_t, \varphi_t, \Omega_t)$ and $\Omega_0 = \Omega_{0,t} \subseteq \Omega$, the function (14.23) is just a restriction of

$$\Psi := \varphi \circ \Phi \circ F|_{\Omega_0}^{-1}.$$

Thus, roughly speaking, Ψ is just the composition $\varphi \circ \Phi \circ F^{-1}$. Hence, Ψ is actually a function which is very naturally associated with the family of function triples.

As a special case of Theorem 14.41, we obtain that "countably condensing triples" and even "countably condensing triple families" (that is, families for which the associated map Ψ is countably 1-condensing) possess a (common) convex-fundamental set. Moreover, it can even be arranged that this set contains a given prescribed set M. More precisely, the following holds.

Theorem 14.43. *We consider a family of function triples* $(F_t, \Phi_t, \varphi_t, \Omega_t) \in \mathcal{T}_{\mathrm{gen}}(X, Y)$ $(t \in I)$ *and a family* $\Omega_{0,t} \subseteq \Omega_t$ $(t \in I)$ *of open neighborhoods of* $\operatorname{coin}_{\Omega_t}(F_t, \Phi_t, \varphi_t, \Omega_t)$. *Let* $M \subseteq Y$ *be separable, and let* $K \subseteq Y$ *be the corresponding set of Theorem* 14.33(a). *Assume that* (14.23) *is seperable for every* $x \in K$ *and at least one of the following holds.*

(a) *There is a regular monotone measure of noncompactness* γ *on* Y *satisfying* $\gamma(M \cup C) = \gamma(C)$ *for each countable* $C \subseteq K$ *and such that* (14.23) *is* $(1, \frac{\gamma^c}{\gamma^c})$*-condensing on* K.

(b) M *is separable and the function* (14.23) *has the property that for each countable* $C \subseteq K$ *and each countable* $C_0 \subseteq \Psi(C) \subseteq K$ *for which* C *is not relatively compact there is a monotone measure of noncompactness* γ *on* Y *satisfying* $\gamma(M \cup C_0) < \gamma(C)$.

Then $K \supseteq M$ *is convex-fundamental for* $(F_t, \Phi_t, \varphi_t, \Omega_t)$ *on* $\Omega_{0,t}$ *for every* $t \in I$.

Proof. We assume first (b) and verify that the hypotheses of Theorem 14.41 are satisfied. Thus, let $C \subseteq K$ be countable and satisfy (14.24). We are to show that C is relatively compact. Assume by contradiction that this is not the case. Since $\overline{C}$ is separable, we obtain from (14.24) by Proposition 3.55 that there is a countable $A \subseteq M \cup \Psi(C)$ with $\overline{C} \subseteq \overline{\operatorname{conv}}\, A$. In particular, there is a countable $C_0 \subseteq \Psi(C)$ with $\overline{C} \subseteq \overline{\operatorname{conv}}(M \cup C_0)$. Note that $\Psi(C) \subseteq K$ holds automatically in view of $C \subseteq K$ by (14.9). Now if γ is chosen corresponding to C and C_0 as in (b), we can calculate

$$\gamma(C) = \gamma(\overline{C}) \leq \gamma(\overline{\operatorname{conv}}(M \cup C_0)) = \gamma(M \cup C_0),$$

which is a contradiction. Now we show that (a) implies (b). Indeed, if γ is as in (a) then we have in particular $\gamma(M) = \gamma(M \cup \varnothing) = \gamma(\varnothing) = 0$, and so M is relatively compact and thus separable by Corollary 3.27. Moreover, if $C \subseteq K$ is not relatively compact then $\gamma(C) > 0$, and so we have for every countable $C_0 \subseteq \Phi(C) \subseteq K$ that $\gamma(M \cup C_0) = \gamma(C_0) < \gamma(C)$. □

Remark 14.44. In Theorem 14.43, the main difference between the two hypotheses is that in (b) the function γ may depend on C and C_0 while this is not the case for (a).

Now we discuss the mentioned "dual" approach. This approach is a bit more subtle than just applying Lemma 14.40 with $F = \mathrm{id}_K$. Instead, we apply

Lemma 14.40 with the function F from the given function triple (F, Φ, φ) and with $M := F^{-1}(K)$.

We formulate the result immediately for families of function triples. It is more convenient to formulate the result first for a rather general case with an auxiliary map Ψ:

Theorem 14.45. *For a family $(F_t, \Phi_t, \varphi_t, \Omega_t) \in \mathcal{T}_{\mathrm{gen}}(X, Y)$ $(t \in I)$, let $\Omega_{0,t} \subseteq \Omega_t$ $(t \in I)$ be a family of open neighborhoods of $\mathrm{coin}_{\Omega_t}(F_t, \Phi_t, \varphi_t)$. For $C \subseteq \widetilde{\Omega} := \bigcup_{t\in I} \Omega_{0,t}$, we put*

$$F(C) := \bigcup_{t\in I} F_t(C \cap \Omega_{0,t}).$$

Let $\Psi\colon \widetilde{\Omega} \multimap Y$ be such that

$$\varphi_t(\Phi_t(x)) \subseteq \Psi(x) \text{ and } \Psi(x) \text{ is separable for all } x \in \Omega_{0,t}$$

for every $t \in I$. Assume that for every countable subset $C \subseteq \widetilde{\Omega}$ the inclusions

$$\overline{F(\widetilde{\Omega}) \cap \operatorname{conv} \Psi(C)} \subseteq \overline{F(C)} \subseteq \overline{\operatorname{conv}}\, \Psi(C) \tag{14.27}$$

and if $(\overline{\operatorname{conv}}\, \Psi(\widetilde{\Omega})) \setminus F(\widetilde{\Omega})$ is closed in Y together with the equality

$$\overline{F(C)} = \overline{F(\widetilde{\Omega}) \cap \operatorname{conv} \Psi(C)} \tag{14.28}$$

imply that $\Psi(C)$ is relatively compact in Y.

Then there is a set $K \subseteq Y$ which is convex-fundamental for $(F_t, \Phi_t, \varphi_t, \Omega_t)$ on $\Omega_{0,t}$ and satisfies $\Psi(\Omega_{0,t} \cap F_t^{-1}(K)) \subseteq K$ for all $t \in I$.

Proof. For $K \subseteq Y$, we put

$$M_K := \bigcup_{t\in I} \{x \in \Omega_{0,t} : F_t(x) \in K\}.$$

Applying Theorem 14.33 with

$$\hat{\Psi}(K) := \bigcup_{t\in I} \Psi(\Omega_{0,t} \cap F_t^{-1}(K)) = \Psi(M_K),$$

we find that there is a smallest set $K \subseteq Y$ which is almost convex-fundamental for $(F_t, \Phi_t, \varphi_t, \Omega_t)$ on $\Omega_{0,t}$ for every $t \in I$ and which satisfies $\hat{\Psi}(K) \subseteq K$. Moreover, this set K satisfies

$$K = \overline{\operatorname{conv}}\, \hat{\Psi}(K) = \overline{\operatorname{conv}}\, \Psi(M_K).$$

We are to show that K is compact. Putting $M := M_K$ and $V_0 := F(\widetilde{\Omega})$, we have

$$V_0 \cap \overline{\operatorname{conv}}\, \Psi(M) = F(\widetilde{\Omega}) \cap K = F(M) \subseteq K = \overline{\operatorname{conv}}\, \Psi(M).$$

Suppose that $K = \overline{\operatorname{conv}}\, \Psi(M)$ fails to be compact. Since K is complete (recall Lemma 3.8), we obtain by Proposition 3.26 that $\chi_Y(K) > 0$. Theorem 3.61 implies $\chi_Y(\Psi(M)) > 0$. In view of (3.1), we have $\beta(\Psi(M)) > 0$, and so there is a countable $A \subseteq \Psi(M)$ with $\beta(A) > 0$. Let $C_0 \subseteq M$ be countable with $A \subseteq \Psi(C_0)$. Applying Lemma 14.40, we obtain in view of $V_0 = F(\widetilde{\Omega}) \supseteq F(C)$ a countable set $C \subseteq M$ satisfying $C_0 \subseteq C$ and (14.27). Moreover, if

$$Y_0 := (\overline{\operatorname{conv}}\, \Psi(\widetilde{\Omega})) \setminus F(\widetilde{\Omega})$$

is closed in Y then also $Y_0 \cap K = (\overline{\operatorname{conv}}\, \Psi(M)) \setminus V_0$ is closed in Y, and so (14.19) implies in view of $V_0 \supseteq F(C)$ that (14.28) holds. Since $C \supseteq C_0$, we have $\beta(\Psi(C)) \geq \beta(\Psi(C_0)) \geq \beta(A) > 0$. Hence, $\Psi(C)$ is not relatively compact in Y, contradicting our hypothesis. □

For practically all applications, it is sufficient to deal with the following simpler special case (usually with $M = \varnothing$):

Theorem 14.46. *For a family $(F_t, \Phi_t, \varphi_t, \Omega_t) \in \mathcal{T}_{\mathrm{gen}}(X, Y)$ $(t \in I)$, let $\Omega_{0,t} \subseteq \Omega_t$ $(t \in I)$ be a family of open neighborhoods of $\operatorname{coin}_{\Omega_t}(F_t, \Phi_t, \varphi_t)$. For $C \subseteq \widetilde{\Omega} := \bigcup_{t \in I} \Omega_{0,t}$, we put*

$$F(C) := \bigcup_{t \in I} F_t(C \cap \Omega_{0,t}),$$

and we assume that $\Psi\colon \widetilde{\Omega} \multimap Y$,

$$\Psi(x) := \bigcup_{\substack{t \in I \\ x \in \Omega_{0,t}}} \varphi_t(\Phi_t(x)) \text{ is separable for all } x \in \widetilde{\Omega}. \tag{14.29}$$

Let $M \subseteq Y$ be separable and such that for every countable subset $C \subseteq \widetilde{\Omega}$ the inclusions

$$\overline{F(\widetilde{\Omega}) \cap \operatorname{conv}(M \cup \Psi(C))} \subseteq \overline{F(C)} \subseteq \overline{\operatorname{conv}}(M \cup \Psi(C)) \tag{14.30}$$

and if $(\overline{\operatorname{conv}}(M \cup \Psi(\widetilde{\Omega}))) \setminus F(\widetilde{\Omega})$ is closed in Y together with the equality

$$\overline{F(C)} = \overline{F(\widetilde{\Omega}) \cap \operatorname{conv} \Psi(C)}$$

imply that $\Psi(C)$ is relatively compact in Y.

Then there is a set $K \supseteq M$ which is convex-fundamental for $(F_t, \Phi_t, \varphi_t, \Omega_t)$ on $\Omega_{0,t}$ for every $t \in I$.

Proof. The assertion follows by applying Theorem 14.45 with the multivalued function $\widetilde{\Psi}(x) := M \cup \Psi(x)$. □

Before we formulate Theorems 14.45 or 14.46 in terms of measures of noncompactness, we note that the condition (14.29) is in all situations of interest satisfied automatically:

Proposition 14.47. *Let $(F, \Phi, \varphi, W, Y, \Gamma)$ be a generalized homotopy triple for the family $(F_t, \Phi_t, \varphi_t, \Omega_{0,t}, Y, \Gamma_t)$ $(t \in I)$. Then* (14.29) *holds if one of the following is true:*

(a) *I is compact and $W = I \times \Omega_{0,0}$.*

(b) *I is separable and metrizable, and Γ is metrizable.*

Proof. In case $W = I \times \Omega_{0,0}$, we have $\Omega_{0,t} = \Omega_{0,0}$ for all $t \in I$, and so $\Psi(x) = \varphi(\Phi(I \times \{x\}))$. Hence, if I is compact then two applications of Proposition 2.100 imply that $\Psi(x)$ is a compact subset of Y and thus separable by Corollary 3.27.

If I is a separable metric space, then also the subset $I_x := \{t \in I : x \in \Omega_{0,t}\}$ is separable by Corollary 3.17. With $\Gamma_x := \Phi(I_x \times \{x\})$, the map $\Phi\colon I_x \times \{x\} \multimap \Gamma_x$ is upper semicontinuous by Proposition 2.90 and thus upper semicontinuous in the uniform sense by Proposition 3.19 if Γ is a metric space. Since $\Phi(t, x)$ is compact and thus separable by Corollary 3.27, we obtain by Corollary 3.21 that Γ_x is separable. Applying Corollary 3.21 once more, we obtain that $\varphi(\Gamma_x) = \Psi(x)$ is separable. □

Theorem 14.48. *Let $(F, \Phi, \varphi, W, Y, \Gamma)$ be a generalized homotopy triple for a family $(F_t, \Phi_t, \varphi_t, \Omega_{0,t}, Y, \Gamma_t)$ $(t \in [0,1])$. Suppose $W = [0.1] \times \Omega_{0,0}$ or that Γ is metrizable. For $C \subseteq \widetilde{\Omega} := \bigcup_{t\in[0,1]} \Omega_{0,t}$, we put*

$$F(C) := \bigcup_{t\in[0,1]} F_t(C \cap \Omega_{0,t}),$$

and we define $\Psi\colon \widetilde{\Omega} \to Y$ by

$$\Psi(x) := \bigcup_{\substack{t\in[0,1]\\ x\in\Omega_{0,t}}} \varphi_t(\Phi_t(x)).$$

Let $M \subseteq Y$ be separable, and suppose that for each countable $C \subseteq \widetilde{\Omega}$ with noncompact $\overline{\Psi(C)}$ and each countable $C_0 \subseteq \Psi(C)$ there is a monotone measure of noncompactness γ on Y satisfying

$$\gamma(M \cup C_0) < \gamma(F(C)). \tag{14.31}$$

Then there is a set $K \supseteq M$ *which is convex-fundamental for* $(F_t, \Phi_t, \varphi_t, \Omega_t)$ *on* $\Omega_{0,t}$ *for every* $t \in [0, 1]$.

Proof. We apply Theorem 14.46. Since (14.29) holds by Proposition 14.47, it suffices to verify that for each countable $C \subseteq \widetilde{\Omega}_{0,0}$ with (14.30) the set $\overline{\Psi(C)}$ is compact. Assume by contradiction that this is not the case. Since $\Psi(C)$ is a countable union of separable sets and thus separable, there is a countable dense subset $C_0 \subseteq \Psi(C)$. Using the last inclusion of (14.30), we find for every monotone measure of noncompactness γ that

$$\begin{aligned}\gamma(F(C)) &= \gamma(\overline{F(C)}) \le \gamma(\overline{\operatorname{conv}}(M \cup \Psi(C))) \\ &= \gamma(M \cup \Psi(C)) \le \gamma(\overline{M \cup C_0}) = \gamma(M \cup C_0),\end{aligned}$$

contradicting our hypothesis. □

Roughly speaking, the condition (14.31) means that "Ψ is more compact than F is proper".

14.5 Classical Applications in the General Framework

We apply now Theorem 14.46 to use the homotopy invariance in the last function to obtain generalizations of some of the results of Section 9.6 to a richer class of operators.

Throughout this section, we use the notations of Section 14.3.

Theorem 14.49 (Continuation Principle). *Let* $\Omega \subseteq X$ *be open and* $F \in \mathcal{F}_0(\Omega, Y)$, *oriented or non-oriented. Let* Γ *be a Hausdorff space such that* $[0, 1] \times \Gamma$ *is* T_5. *Let* $\Phi\colon \Omega \multimap \Gamma$ *be acyclic**, *and* $\varphi \in C(\Gamma, Y)$. *Suppose that there is* $y_0 \in Y$ *such that the following holds:*

(a) *The set*

$$A := \bigcup_{0 \le t \le 1} \{x \in \Omega : F(x) - y_0 \in t(\varphi(\Phi(x)) - y_0)\} \tag{14.32}$$

has at least one of the following properties:

(1) *A is relatively compact in Ω.*

(2) *$F(A)$ fails to be relatively compact in Y.*

(b) *There is an open neighborhood* $\Omega_0 \subseteq \Omega$ *of* A *such that for each countable* $C \subseteq \Omega_0$ *the inclusions*

$$\overline{F(\Omega_0) \cap \operatorname{conv}(\{y_0\} \cup \varphi(\Phi(C)))} \subseteq \overline{F(C)} \subseteq \overline{\operatorname{conv}}(\{y_0\} \cup \varphi(\Phi(C)))$$

and if $(\overline{\operatorname{conv}}(\{y_0\} \cup \varphi(\Phi(C)))) \setminus F(\Omega_0)$ *is closed in* Y *together with the equality*

$$\overline{F(C)} = \overline{F(\Omega_0) \cap \operatorname{conv}(\{y_0\} \cup \varphi(\Phi(C)))} \tag{14.33}$$

imply that $\varphi(\Phi(C))$ *is relatively compact.*

Then A *is relatively compact in* Ω, $(F, \Phi, \varphi, \Omega) \in \mathcal{T}_{\mathrm{conv}}(X, Y)$, *and*

$$\operatorname{Deg}(F, \Phi, \varphi, \Omega) = \deg(F, \Omega, y_0),$$

where the right-hand side denotes the Benevieri–Furi degree. In particular, if that degree is nonzero, we have $\operatorname{coin}(F, \Phi, \varphi) \neq \varnothing$.

Proof. We apply the generalized homotopy invariance with $W := [0,1] \times \Omega$, $U := [0,1] \times \Omega_0$, $G(t,x) := F(x)$, $H(t,x) := \Phi(x)$, $\widetilde{H}(t,x) := \{t\} \times \Phi(x)$, and $h(t,z) := t\varphi(z) + (1-t)y_0$. Then (G, H, h, W, Y, Γ) is an acyclic* generalized homotopy triple for the family (G_t, H_t, h_t, W_t).

Using Theorem 14.46 with $M = \{y_0\}$, we find that our hypotheses imply that there is a set $K \subseteq Y$ which is convex-fundamental for $(G_t, H_t, h_t, W_t) = (F, \Phi, t\varphi + (1-t)y_0, \Omega)$ on $U_t = \Omega_0$: The hypotheses of Theorem 14.46 correspond to our hypotheses, because

$$\operatorname{conv} \Psi(C) = \operatorname{conv}(\{y_0\} \cup \varphi(\Phi(C)))$$

for every $C \subseteq \Omega_0$ by the definition of h.

Since K contains $\operatorname{fix}_{W_t}(G_t, H_t, h_t)$ for every $t \in [0,1]$ by Proposition 14.31, the definition of A implies $F(A) \subseteq K$. Since K is compact, we obtain that $F(A)$ is relatively compact, and so the hypothesis implies that A is relatively compact in Ω. Note that the definition of A implies $\operatorname{coin}_W(F, \widetilde{H}, h) \subseteq [0,1] \times A$. Hence, if $K_0 \subseteq \Omega$ denotes a compact set containing A then Proposition 11.41 implies that $\operatorname{coin}_W(F, \widetilde{H}, h)$ is a closed and thus compact subset of $P := [0,1] \times K_0$ (Proposition 2.29 and Theorem 2.63). Hence, the homotopy triple (G, H, h, W, Y, Γ) is proper. Finally, since we assume that $[0,1] \times \Gamma$ is T_5, we obtain that (G_t, H_t, h_t, U_t) is locally $[0,1]$-normal.

The generalized homotopy invariance thus implies with $\varphi_0(x) \equiv y_0$ that

$$\operatorname{Deg}(F, \Phi, \varphi, \Omega) = \operatorname{Deg}(F, \Phi, \varphi_0, \Omega).$$

By the compatibility with the Leray–Schauder triple degree, the last degree can also be interpreted as the Leray–Schauder triple degree. In view of the single-valued normalization property of the latter, this is the Benevieri–Furi coincidence degree $\deg(F, \varphi_0, \Omega)$, and by the compatibility with the Benevieri–Furi degree, the latter is $\deg(F, \Omega, y_0)$ □

Corollary 14.50 (Continuation Principle). *Let $\Omega \subseteq X$ be open and $F \in \mathcal{F}_0(\Omega, Y)$, oriented or non-oriented. Let Γ be a Hausdorff space such that $[0,1] \times \Gamma$ is T_5. Let $\Phi\colon \overline{\Omega} \multimap \Gamma$ be acyclic*, and $\varphi \in C(\Gamma, Y)$. Suppose that there is $y_0 \in Y$ such that the following holds:*

(a) *The map Φ has an extension to an upper semicontinuous map $\Phi\colon \overline{\Omega} \multimap \Gamma$ with closed values, F has an extension to a map $F \in C(\overline{\Omega}, Y)$, and*

$$F(x) - y_0 \neq t(\varphi(\Phi(x)) - y_0) \quad \text{for all } (t,x) \in [0,1] \times \partial\Omega.$$

(b) *For each countable $C \subseteq \Omega$ with noncompact $\overline{C}$ and each countable $C_0 \subseteq \varphi(\Phi(C))$ there is a regular monotone measure of noncompactness γ on Y with*

$$\gamma(\{y_0\} \cup C_0) < \gamma(F(C)).$$

Then $(F, \Phi, \varphi, \Omega) \in \mathcal{T}_{\mathrm{conv}}(X, Y)$, and

$$\operatorname{Deg}(F, \Phi, \varphi, \Omega) = \deg(F, \Omega, y_0),$$

where the right-hand side denotes the Benevieri–Furi degree. In particular, if that degree is nonzero, we have $\operatorname{coin}(F, \Phi, \varphi) \neq \emptyset$.

Proof. We verify the hypotheses of Theorem 14.49. Concerning property (a), we note that the hypothesis implies that the set (14.32) is the same as

$$A = \bigcup_{0 \le t \le 1} \{x \in \overline{\Omega} : F(x) - y_0 \in t(\varphi(\Phi(x)) - y_0)\}.$$

Since

$$\{(t,x) \in [0,1] \times \overline{\Omega} : F(x) - y_0 \in t(\varphi(\Phi(x)) - y_0)\}$$

is closed in $[0,1] \times \overline{\Omega}$ by Corollary 2.115 and $[0,1]$ is compact, it follows from Corollary 2.112 that A is closed in $\overline{\Omega}$. In view of $A \subseteq \Omega$, it follows that A is closed in Ω. Hence, if A fails to be relatively compact in Ω then A fails to be compact. Since A is closed in X, we obtain $\beta(A) > 0$. There is a countable subset $C \subseteq A$ with $\beta(C) > 0$, and so $\overline{C}$ fails to be compact. By our hypothesis (b), we obtain $\gamma(F(C)) > 0$ for some regular set function γ, and so $F(C) \subseteq F(A)$ fails to be relatively compact in Y. Hence, hypothesis (a) of Theorem 14.49 holds.

To verify hypothesis (b), we argue analogously to the proof of Theorem 14.48 to show that if $C \subseteq \Omega$ is countable and satisfies

$$\overline{F(C)} \subseteq \overline{\operatorname{conv}}(\{y_0\} \cup \varphi(\Phi(C))),$$

then $M := \varphi(\Phi(C))$ is relatively compact. Indeed, since M is a countable union of separable sets and thus separable, there is a countable dense $C_0 \subseteq M$. Then we have for every monotone measure of noncompactness γ that

$$\begin{aligned}\gamma(F(C)) &= \gamma(\overline{F(C)}) \leq \gamma(\overline{\operatorname{conv}}(\{y_0\} \cup M)) \\ &= \gamma(\{y_0\} \cup M) \leq \gamma(\{y_0\} \cup \overline{C}_0) = \gamma(\{y_0\} \cup C_0).\end{aligned}$$

By hypothesis, this implies that $\overline{C} \subseteq \overline{\Omega}$ is compact, and so $\varphi(\Phi(\overline{C}))$ is compact by Proposition 2.100. Hence, M is relatively compact in Y. □

Theorem 14.49 can be considered as an abstract form of the Leray–Schauder alternative. In fact, for the particular case $F = \mathrm{id}_\Omega$, we obtain as a special case a non-compact Leray–Schauder alternative:

Theorem 14.51 (Leray–Schauder Alternative). *Let $X = Y$, $\Omega \subseteq X$ be open, and Γ be a Hausdorff space such that $[0,1] \times \Gamma$ is a T_5. Let $\Phi\colon \Omega \multimap \Gamma$ be acyclic*, and $\varphi \in C(\Gamma, X)$. Suppose that there is $x_0 \in \Omega$ such that the following holds:*

(a) *The set*

$$A := \bigcup_{\lambda > 1} \{x \in \Omega : \lambda(x - x_0) \in \varphi(\Phi(x)) - x_0\}$$

is either relatively compact in Ω or fails to be relatively compact in X. Alternatively, assume that Φ has an upper semicontinuous extension $\Phi\colon \overline{\Omega} \multimap \Gamma$ with closed values and

$$\lambda(x - x_0) \notin \varphi(\Phi(x)) - x_0 \quad \textit{for all } \lambda \geq 1. \tag{14.34}$$

(b) *There is an open neighborhood $\Omega_0 \subseteq \Omega$ of A with $x_0 \in A$ such that for each countable $C \subseteq \Omega_0$ the equality*

$$\overline{C} = \overline{\Omega_0 \cap \operatorname{conv}(\{x_0\} \cup \varphi(\Phi(C)))} \tag{14.35}$$

implies that $\varphi(\Phi(C))$ is relatively compact.

Then $\varphi \circ \Phi$ has a fixed point in Ω.

Proof. We apply Theorem 14.49 with $F = \mathrm{id}_\Omega$ and $y_0 = x_0$, observing that the normalization property for the diffeomorphism F of the Benevieri–Furi degree implies $\deg(\mathrm{id}_\Omega, \Omega, y_0) = 1$ when we consider either the non-oriented case or the natural orientation of F.

Assume by contradiction that $\varphi \circ \Phi$ has no fixed point in Ω. Then

$$A \cup \{x_0\} = \bigcup_{0 \leq t \leq 1} \{x \in \Omega : F(x) - y_0 \in t(\varphi(\Phi(x)) - y_0)\},$$

is the set (14.32) of Theorem 14.49. Hence, if $A \cup \{x_0\}$ fails to be relatively compact in Ω then also A fails to be relatively compact in Ω, and so A fails to be relatively compact in X, by hypothesis. Thus also $F(A \cup \{x_0\}) = A \cup \{x_0\}$ fails to be relatively compact in X. Under the assumption (14.34), we can show as in the proof of Corollary 14.50 that $A \cup \{x_0\}$ is closed in Ω, and so also in this case this set fails to be relatively compact in X if it fails to be relatively compact in Ω. In both cases, we have proved hypothesis (a) of Theorem 14.49. Moreover, since $A \cup \{x_0\} \subseteq \Omega_0$, it follows that Ω_0 is a neighborhood of the set (14.32).

Concerning hypothesis (b) of Theorem 14.49, it thus suffices to note that the equality (14.35) is just (14.33) in our special case. We are allowed to consider (14.33), because the set

$$(\overline{\operatorname{conv}}(\{x_0\} \cup \varphi(\Phi(C)))) \setminus F(\Omega_0)$$

is closed, since $F(\Omega_0) = \Omega_0$ is open. Summarizing, all hypotheses of Theorem 14.49 are satisfied, and so $\varphi \circ \Phi$ has a fixed point in Ω which is a contradiction to our assumption. □

Although Theorem 14.51 can be generalized also for the case that Φ is acyclic (and not only acyclic*), see [137], we point out that no such generalization is known for Theorem 14.49 (unless F is a homeomorphism).

In the author's opinion, Theorem 14.51 demonstrates the power and beauty of our approach. In the single-valued case, special cases of Theorem 14.51 and of the subsequent consequence are due to H. Mönch [108].

Corollary 14.52 (Multivalued Mönch–Rothe Fixed Point Theorem). *Let $X = Y$, $\Omega \subseteq X$ be open and convex, and Γ be a Hausdorff space such that $[0,1] \times \Gamma$ is T_5. Let $\Phi\colon \overline{\Omega} \multimap \Gamma$ be acyclic*, and $\varphi \in C(\Gamma, X)$. Suppose that $\varphi(\Phi(\partial\Omega)) \subseteq \overline{\Omega}$ and that there is $x_0 \in \Omega$ and an open neighborhood $\Omega_0 \subseteq \Omega$ of*

$$\bigcup_{\lambda>1} \{x \in \Omega : \lambda(x - x_0) \in \varphi(\Phi(x)) - x_0\}$$

with $x_0 \in \Omega_0$ such that for each countable $C \subseteq \Omega_0$ the equality (14.35) *implies that $\varphi(\Phi(C))$ is relatively compact.*

Then $\varphi \circ \Phi$ has a fixed point in $\overline{\Omega}$.

Proof. We can assume that there is no fixed point on $\partial\Omega$. Then the hypothesis (14.34) holds, since if there were $x \in \partial\Omega$, $\lambda \geq 1$, and $z \in \Phi(x)$ with $\lambda(x - x_0) \in \varphi(z) - x_0$ then $\lambda > 1$ by our assumption, and we have with $t := 1 - \lambda^{-1} \in (0,1]$ in view of $\varphi(z) \in \overline{\Omega}$ that $x = (1-t)\varphi(z) + tx_0 \in \Omega$ by Lemma 4.41 which is a contradiction. Thus, the result follows from Theorem 14.51. □

Remark 14.53. The compactness hypothesis in Theorem 14.51 and Corollary 14.52 is in particular satisfied if one of the following holds:

(a) There is a regular monotone measure of noncompactness γ on X with $\gamma(\{x_0\} \cup C) = \gamma(C)$ for all countable $C \subseteq X$ and such that $\varphi \circ \Phi$ is $(1, {}^{\gamma^c}_{\gamma^c})$-condensing on Ω_0.

(b) For each countable $C \subseteq \Omega_0$ and each countable $C_0 \subseteq \varphi(\Phi(C))$ with noncompact $\overline{C}_0$ there is a monotone measure of noncompactness γ on X with $\gamma(\{x_0\} \cup C_0) < \gamma(C)$.

Indeed, if $C \subseteq \Omega_0$ is countable and such that $M := \varphi(\Phi(C))$ is not relatively compact then we observe that M is a countable union of compact hence separable sets, and thus we find a countable dense $C_0 \subseteq M$, in particular $\overline{C}_0 = \overline{M}$ is noncompact. It follows that (14.35) must fail since otherwise we would have with the corresponding γ from the hypothesis that

$$\begin{aligned}\gamma(C) &\le \gamma(\overline{\operatorname{conv}}(\{x_0\} \cup M)) = \gamma(\{x_0\} \cup M) \\ &\le \gamma(\{x_0\} \cup \overline{C}_0) = \gamma(\{x_0\} \cup C_0),\end{aligned}$$

contradicting the hypothesis.

Corollary 14.54 (Multivalued Countable Darbo–Rothe Fixed Point Theorem). *Let $X = Y$, $\Omega \subseteq X$ be open and convex, and Γ be a Hausdorff space such that $[0,1] \times \Gamma$ is T_5. Let $\Phi\colon \overline{\Omega} \multimap \Gamma$ be acyclic*, and $\varphi \in C(\Gamma, X)$. Suppose that $\varphi(\Phi(\partial\Omega)) \subseteq \overline{\Omega}$ and that there is $x_0 \in \Omega$ and a regular monotone measure of noncompactness γ on X satisfying $\gamma(\{x_0\} \cup C) = \gamma(C)$ for all countable $C \subseteq X$. If $\varphi \circ \Phi$ is $(1, {}^{\gamma^c}_{\gamma^c})$-condensing then it has a fixed point in $\overline{\Omega}$.*

Proof. In view of Remark 14.53, this is a special case of $\Omega_0 = \Omega$ of Corollary 14.52 □

The original theorem of Darbo dealt only with single-valued $(q, {}^{\alpha}_{\alpha})$-bounded maps f with $q < 1$ on bounded sets and used a rather different approach by iterating sets under the map $M \mapsto \overline{\operatorname{conv}}\, f(M)$. Daher [35] pointed out that the assertion of Darbo's fixed point theorem holds also if the hypotheses concerning measures of noncompactness are satisfied only on countable sets, and this was sharpened (and applied) by Mönch [108], [109].

14.6 A Sample Application for Boundary Value Problems

Without going into much detail, we sketch in this section how degree theory for function triples can be applied for boundary value problems in Banach spaces. Let X be a Banach space, A be the generator of a C_0 semigroup in X, $f\colon [0,T] \times X \multimap X$, and $F, G\colon X \to X$. We consider the differential inclusion

$$x'(t) \in Ax(t) + f(t, x(t)) \qquad (0 \le t \le T) \tag{14.36}$$

together with the boundary condition

$$F(x(0)) = G(x(T)). \tag{14.37}$$

We equip $C_X := C([0,T], X)$ with the max-norm

$$\|x\|_{C_X} := \max_{t\in[0,T]} \|x(t)\|.$$

There are various Aronszajn type results available for (14.36) which state that under some natural conditions on A and f the map $\Phi\colon X \multimap C_X$, which associates to every $x_0 \in X$ the set of all solutions of (14.36) satisfying the initial condition $x(0) = x_0$, is upper semicontinuous and has the property that $\Phi(x_0)$ is a nonempty R_δ for every $x_0 \in X$, see e.g. [83, Corollary 5.2.1 and Theorem 5.3.1]. We will assume this property in the following. Then Φ is acyclic*.

We assume that $G \in C(X, X)$ and define $\varphi, \varphi_0 \in C(C_X, X)$ by $\varphi_0(x) := x(T)$ and $\varphi := G \circ \varphi_0$. Then $\varphi \circ \Phi$ is the map which associates to every x_0 the set of all values $G(x(T))$ where $x \in C_X$ is a solution of (14.36) satisfying $x(0) = x_0$. It follows that (14.36), (14.37) can be rewritten in the form

$$F(x_0) \in \varphi(\Phi(x_0)).$$

We point out that the multivaluedness of this problem need not come from the multivaluedness of f in (14.36). Rather, it comes from the fact that the solution of the initial value problem corresponding to (14.36) need not be unique: It is well-known that this can occur for (14.36) also in the single-valued case if $f(t,\cdot)$ fails to be locally Lipschitz. A simple example for the latter is the differential equation $x'(t) = \sqrt{|x(t)|}$ (in $X = \mathbb{R}$) where the problem corresponding to the initial value $x(0) = 0$ has for each $\tau \in [0,T]$ the solution

$$x(t) = \begin{cases} \frac{1}{4}(t-\tau)^2 & \text{if } t \ge \tau, \\ 0 & \text{if } t \le \tau. \end{cases}$$

Note that the set $\Phi(0)$ of all these solutions is an R_δ, because it is homeomorphic to $[0,T]$ so that this is not in contradiction with the mentioned Aronszajn type results.

Theorem 14.55. *Under the above hypotheses on Φ and G, let $\Omega \subseteq X$ be open and $F \in \mathcal{F}_0(\Omega, X)$ such that the following holds:*

(a) *F has an extension $F \in C(\overline{\Omega}, X)$, and for every $\lambda \in [0,1]$ the equation* (14.36) *has no solution x with $x(0) \in \partial\Omega$ and $F(x(0)) = \lambda G(x(T))$.*

(b) *For every countable $C \subseteq \Omega$ with noncompact $\overline{C}$ and every countable $C_0 \subseteq \varphi(\Phi(C))$ there is a regular monotone measure of noncompactness γ with*

$$\gamma(C_0) < \gamma(F(C)).$$

Then the Benevieri–Furi degree $\deg(F,\Omega,0)$ *exists, and if it is nonzero then the boundary value problem* (14.36), (14.37) *has a solution with* $x(0) \in \Omega$.

Proof. $\Phi|_{\overline{\Omega}}$ is acyclic*, by hypothesis. Hence, the assertion follows from the continuation principle (Corollary 14.50), applied with $Y = X$ and $y_0 = 0$. □

Condition (b) of Theorem 14.55 looks technical but is actually satisfied in most applications if Ω is bounded: If f satisfies natural compactness hypotheses then one can prove estimates of the type

$$\gamma_0(C_0) \le k\gamma_1(C)$$

(with $\gamma_1, \gamma_0 \in \{\chi_X, \alpha, \beta\}$) for each countable $C \subseteq \Omega$ and each countable $C_0 \subseteq \varphi_0(\Phi(C))$, see e.g. [83, Theorem 6.3.1]. Hence, if G is $(k_0, {}^{\gamma_1^c}_{\gamma_0^c})$-bounded, one can obtain estimates of the type

$$\gamma_2(C_0) \le L\gamma_1(C)$$

for each countable $C \subseteq \Omega$ and each countable $C_0 \subseteq \varphi(\Phi(C))$ (with $L = k_0 k$). In particular, if F satisfies

$$\gamma_2(F(C)) > L\gamma_1(C)$$

for every countable $C \subseteq \Omega$ with $\gamma_1(C) > 0$ (recall Proposition 3.44) the hypothesis (b) of Theorem 14.55 is satisfied. Note that if Ω is bounded, this condition holds if

$$\gamma_2(F(C)) \ge (L + \varepsilon(C))\gamma_1(C)$$

for every countable $C \subseteq \Omega$ and $\varepsilon(C) > 0$.

Summarizing, if $\Omega \subseteq X$ is bounded, the most serious hypothesis of Theorem 14.55 is (a). Note that this condition holds with $\Omega = B_R(0)$ for $R > M$ if all solutions (14.36) with $F(x(0)) = \lambda G(x(T))$ for some $\lambda \in [0, 1]$ satisfy the *a priori* bound $\|x(t)\| \leq M$ for all $t \in [0, T]$. Hence, we obtain the well-known heuristic principle that *a priori* bounds lead to the existence of solutions. If T is sufficiently large, one can also attempt other techniques (Ljapunov functions, guiding functions etc.) to verify a "quantitative" form of dissipativity of (14.36) which can be used to verify (a) (with $\Omega = B_R(0)$ and sufficiently large R and T).

Of course, Theorem 14.55 presents only a simple and very straightforward application of the degree theory which we had developed. For the case $F = \mathrm{id}$, that is, if one is interested in periodic solutions of (14.36), results of the type of Theorem 14.55 had already been considered for quite a while by means of degree theory for multivalued maps; we refer to [120] and its references for more details. However, for the case that F^{-1} does not exist or is discontinuous, the equation does not have a structure for which classical degree theory for multivalued maps is applicable: In this case, the degree theory for function triples seems to be the simplest topological approach to study boundary value problems of type (14.36), (14.37). Note that in any case all three functions of the function triple have a very natural meaning given by the form of the problem, and so the approach by a function triple degree seems to be a very natural one.

Bibliography

[1] R. Abraham and J. Robin, *Transversal Mappings and Flows*, Benjamin, New York, 1967.

[2] R. R. Akhmerov, M. I. Kamenskiĭ, A. S. Potapov, A. E. Rodkina and B. N. Sadovskiĭ, *Measures of Noncompactness and Condensing Operators*, Birkhäuser, Basel, Boston, Berlin, 1992.

[3] J. Andres and R. Bader, Asymptotic Boundary Value Problems in Banach Spaces, *J. Math. Anal. Appl.* **274** (2002), 437–457.

[4] J. Andres, G. Gabor and L. Górniewicz, Boundary Value Problems on Infinite Intervals, *Trans. Amer. Math. Soc.* **351** (1999), 4861–4903.

[5] J. Andres and L. Górniewicz, *Topological Fixed Point Principles for Boundary Value Problems*, Kluwer, Dordrecht, 2003.

[6] J. Andres and M. Väth, Coincidence Index for Noncompact Mappings on Nonconvex Sets, *Nonlinear Funct. Anal. Appl.* **7** (2002), 619–658.

[7] J. Appell and M. Väth, *Elemente der Funktionalanalysis*, Vieweg & Sohn, Braunschweig, Wiesbaden, 2005.

[8] R. F. Arens, Extension of Functions on Fully Normal Spaces, *Pacific J. Math.* **2** (1952), 11–22.

[9] J.-P. Aubin and A. Cellina, *Differential Inclusions*, Springer, Berlin, Heidelberg, New York, 1984.

[10] J.-P. Aubin and H. Frankowska, *Set-Valued Analysis*, Birkhäuser, Boston, Basel, Berlin, 1990.

[11] R. Bader and W. Kryszewski, Fixed-Point Index for Compositions of Set-Valued Maps with Proximaly ∞-connected Values on Arbitrary ANR's, *Set-Valued Anal.* **2** (1994), 459–480.

[12] E. G. Begle, The Vietoris Theorem for Bicompact Spaces, *Ann. of Math.* **51** (1950), 534–543.

[13] E. G. Begle,The Vietoris Theorem for Bicompact Spaces II, *Michigan Math. J.* **3** (1955), 179–180.

[14] P. Benevieri and A. Calamai, A Borsuk-Type Theorem for some Classes of Perturbed Fredholm Maps, *Topol. Methods Nonlinear Anal.* **35** (2010), 379–394.

[15] P. Benevieri, A. Calamai and M. Furi, A Degree Theory for a Class of Perturbed Fredholm Maps, *Fixed Point Theory and Applications* **2005** (2005), 185–206, doi:10.1155/FPTA.2005.185.

[16] P. Benevieri and M. Furi, On the Uniqueness of the Degree for Nonlinear Fredholm Maps of Index Zero Between Banach Manifolds, *Comm. Appl. Nonlinear Anal.*, (to appear).

[17] P. Benevieri and M. Furi, A Simple Notion of Orientability for Fredholm Maps of Index Zero between Banach Manifolds and Degree Theory, *Ann. Sci. Math. Québec* **22** (1998), 131–148.

[18] P. Benevieri and M. Furi, On the Concept of Orientability for Fredholm Maps between Real Banach Manifolds, *Topol. Methods Nonlinear Anal.* **16** (2000), 279–306.

[19] P. Benevieri and M. Furi, A Degree for Locally Compact Perturbations of Fredholm Maps in Banach Spaces, *Abstr. Appl. Anal.* (2006), 1–20, doi:10.1155/AAA/2006/64764.

[20] C. E. Blair, The Baire Category Theorem Implies the Principle of Dependent Choices, *Bull. Acad. Polon. Sci. Sér. Sci. Math. Astronom. Phys.* **25** (1977), 933–934.

[21] Yu. G. Borisovich, B. D. Gel'man, A. D. Myshkis and V. V. Obukhovskiĭ, Topological Methods in the Fixed-Point Theory of Multi-Valued Maps (in Russian), *Uspekhi Mat. Nauk* **35** (1980), 59–126, Engl. transl.: Russian Math. Surveys **35** (1980), no. 1, 65–143.

[22] Yu. G. Borisovich, B. D. Gel'man, A. D. Myshkis and V. V. Obukhovskiĭ, Multivalued Maps (in Russian), *Itogi Nauki—Seriya "Matematika" (Progress in Science—Mathematical Series)* **19** (1982), 127–230, Engl. transl.: J. Soviet Math. **24** (1984), 719–791.

[23] Yu. G. Borisovich, B. D. Gel'man, A. D. Myshkis and V. V. Obukhovskiĭ, Multivalued Analysis and Operator Inclusions (in Russian), *Itogi Nauki—Seriya "Matematika" (Progress in Science—Mathematical Series)* **29** (1986), 151–211, Engl. transl.: J. Soviet Math. **39** (1987), 2772–2811.

[24] Yu. G. Borisovich, B. D. Gel'man, A. D. Myshkis and V. V. Obukhovskiĭ, *Introduction to the Theory of Multivalued Maps and Differential Inclusions* (in Russian), 2nd ed, Librokom, Moscow, 2011.

[25] Yu. G. Borisovich and Yu. E. Gliklikh, On the Lefschetz Number for a Class of Multi-Valued Maps, in: *7. Let. Mat. Shkola, Katsiveli 1969* (Yu. A. Mitropol'skij and Yu. Yu. Trokhimchuk, eds.), pp. 283–294, Akad. Nauk Ukrain SSR, Kiev, 1970.

[26] Yu. G. Borisovich, V. G. Zvyagin and Yu. I. Sapronov, Nonlinear Fredholm Mappings (in Russian), *Uspekhi Mat. Nauk* **32** (1977), 3–54, Engl. transl.: Russian Math. Surveys **32** (1977), no. 4, 1–54.

[27] K. Borsuk, Sur les Prolongements des Transformations Continues, *Fund. Math.* **28** (1937), 99–110.

[28] K. Borsuk, *Theory of Retracts*, Polish Scientific Publ., Warszawa, 1967.

[29] L. E. J. Brouwer, Über Abbildungen von Mannigfaltigkeiten, *Math. Ann.* **71** (1912), 97–115.

[30] F. E. Browder, *Nonlinear Operators and Nonlinear Equations of Evolution in Banach Spaces*, Proceedings of Symposia in Pure Mathematics XVIII, Part 2, Amer. Math. Soc., Providence, R. I., 1976.

[31] R. Caccioppoli, Sulle corrispondenze funzionali inverse diramate: teoria generale e applicazioni ad alcune equazioni funzionali non lineari e al problema di Plateau, *Atti Accad. Naz. Lincei Rend. Cl. Sci. Fis. Mat. Natur.* **24** (1936), 258–263, 416–421.

[32] A. Calamai, The invariance of Domain Theorem for Compact Perturbations of Nonlinear Fredholm Maps of Index Zero, *Nonlinear Funct. Anal. Appl.* **9** (2004), 185–194.

[33] S.-S. Chang and J.-W. Song, Coincidence Indices for Set-Valued Compact Mapping Pairs, *J. Math. Anal. Appl.* **148** (1990), 469–488.

[34] A. Cwiszewski and W. Kryszewski, *The Fixed Point Index for c-Admissible Set-Valued Maps on Arbitrary Absolute Neighborhood Retracts.*

[35] S. J. Daher, Fixed Point Theorems for Nonlinear Operators with a Sequential Condition, *Nonlinear Anal.* **3** (1979), 59–63.

[36] G. Darbo, Punti Uniti in Trasformazioni a Codominio non Compatto, *Rend. Sem. Mat. Univ. Padova* **24** (1955), 84–92.

[37] K. Deimling, *Nichtlineare Gleichungen und Abbildungsgrade*, Springer, Berlin, 1974.

[38] K. Deimling, *Nonlinear Functional Analysis*, Springer, Berlin, Heidelberg, 1985.

[39] A. Dold, *Lectures on Algebraic Topology*, Springer, Berlin, Heidelberg, New York, 1972.

[40] C. H. Dowker, On Countably Paracompact Spaces, *Canad. J. Math.* **3** (1951), 219–224.

[41] C. H. Dowker, On a Theorem of Hanner, *Ark. Mat.* **2** (1952), 307–313.

[42] C. H. Dowker, Inductive Dimension of Completely Normal Spaces, *Quart. J. Math. Oxford Ser.* **4** (1953), 267–281.

[43] J. Dugundji, An Extension of Tietze's Theorem, *Pacific J. Math.* **1** (1951), 353–367.

[44] J. Dugundji and A. Granas, *Fixed Point Theory*, Polish Scientific Publ., Warszawa, 1982.

[45] J. Dydak and G. Kozlowski, Vietoris-Begle Theorem and Spectra, *Proc. Amer. Math. Soc.* **113** (1991), 587–592.

[46] J. Dydak and J. Segal, *Shape Theory — an Introduction*, Lect. Notes Math. 688, Springer, Berlin, Heidelberg, New York, 1978.

[47] Z. Dzedzej, *Fixed Point Index Theory for a Class of Nonacyclic Multivalued Mappings*, Dissertationes Math. (Rozprawy Mat.) 253, Polish Scientific Publ., Warszawa, 1985.

[48] S. Eilenberg and D. Montgomery, Fixed Point Theorems for Multi-Valued Transformations, *Amer. J. Math.* **68** (1946), 214–222.

[49] S. Eilenberg and N. Steenrod, *Foundations of Algebraic Topology*, Princeton Univ. Press, Princeton, 1952.

[50] G. Eisenack and C. Fenske, *Fixpunkttheorie*, BI Wissenschaftsverlag, Mannheim, Wien, Zürich, 1978.

[51] K. D. Elworthy and A. J. Tromba, Degree Theory on Banach Manifolds, in: *Global Analysis. Proceedings of the Symposium in Pure Mathematics of the American Mathematical Society held at the University of California, Berkeley, California July 1-26, 1968* (S.-S. Chern and S. Smale, eds.), pp. 86–94, Amer. Math. Soc., Providence, R. I., 1970.

[52] R. Engelking, *Theory of Dimension—Finite and Infinite*, Heldermann, Lemgo, 1995.

[53] C. C. Fenske and H.-O. Peitgen, Repulsive Fixed Points of Multivalued Transformations and the Fixed Point Index, *Math. Ann.* **218** (1975), 9–18.

[54] P. M. Fitzpatrick, J. Pejsachowicz and P. J. Rabier, The degree of proper C^2 Fredholm mappings. I, *J. Reine Angew. Math.* **427** (1992), 1–33.

[55] P. M. Fitzpatrick and W. V. Petryshyn, Fixed Point Theorems for Multivalued Noncompact Acyclic Mappings, *Pacific J. Math.* **54** (1974), 17–23.

[56] P. M. Fitzpatrick and W. V. Petryshyn, Fixed Point Theorems and the Fixed Point Index for Multi-Valued Mappings in Cones, *J. London Math. Soc. (2)* **12** (1975), 75–85.

[57] I. Fonseca and W. Gangbo, *Degree Theory in Analysis and Applications*, Clarendon Press, Oxford, 1995.

[58] G. Fournier and M. Martelli, Eigenvectors for Nonlinear Maps, *Topol. Methods Nonlinear Anal.* **2** (1993), 203–224.

[59] G. Fournier and D. Violette, A Fixed Point Index for Compositions of Acyclic Multivalued Maps in Banach Spaces, in: *Operator Equations and Fixed Point Theorems* (S. P. Singh, V. M. Sehgal and J. H. W. Burry, eds.), 1, pp. 203–224, The MSRI-Korea Publications, 1986.

[60] M. Furi and M. P. Pera, A Continuation Method on Locally Convex Spaces and Applications to Ordinary Differential Equations on Noncompact Intervals, *Ann. Polon. Math.* **47** (1987), 331–346.

[61] M. Furi and M. P. Pera, On the Fixed Point Index in Locally Convex Spaces, *Proc. Roy. Soc. Edinburgh Sect. A* **106** (1987), 161–168.

[62] D. Gabor, The Coincidence Index for Fundamentally Contractible Multivalued Maps with Nonconvex Values, *Ann. Polon. Math.* **75** (2000), 143–166.

[63] D. Gabor and W. Kryszewski, A Coincidence Theory Involving Fredholm Operators of Nonnegative Index, *Topol. Methods Nonlinear Anal.* **15** (2000), 43–59.

[64] D. Gabor and W. Kryszewski, A Global Bifurcation Index for Set-Valued Perturbations of Fredholm Operators, *Nonlinear Anal.* **73** (2010), 2714–2736.

[65] D. Gabor and W. Kryszewski, Alexander Invariant for Perturbations of Fredholm Operators, *Nonlinear Anal.* **74** (2011), 6911–6932.

[66] R. E. Gaines and J. L. Mawhin, *Coincidence Degree, and Nonlinear Differential Equations*, Lect. Notes Math. 568, Springer, Berlin, Heidelberg, New York, 1977.

[67] H. G. Garnir, M. de Wilde and J. Schmets, *Analyse Fonctionnelle. Théorie Constructive des Espaces Linéaires à Semi-Normes, Tome I: Théorie Générale*, Birkhäuser, Basel, Stuttgart, 1968.

[68] J. Girolo, Approximating Compact Sets in Normed Linear Spaces, *Pacific J. Math.* **98** (1982), 81–89.

[69] C. Good, I. J. Tree and W. S. Watson, On Stone's Theorem and the Axiom of Choice, *Proc. Amer. Math. Soc.* **126** (1998), 1211–1218.

[70] L. Górniewicz, *Homological Methods in Fixed-Point Theory of Multi-Valued Maps*, Dissertationes Math. (Rozprawy Mat.) 129, Polish Scientific Publ., Warszawa, 1976.

[71] L. Górniewicz, *Topological Fixed Point Theory of Multivalued Mappings*, Kluwer, Dordrecht, 1999.

[72] L. Górniewicz, A. Granas and W. Kryszewski, On the Homotopy Method in the Fixed Point Index Theory of Multi-Valued Mappings of Compact Absolute Neighborhood Retracts, *J. Math. Anal. Appl.* **161** (1991), 457–473.

[73] L. Górniewicz and Z. Kucharski, Coincidence of k-Set Contraction Pairs, *J. Math. Anal. Appl.* **107** (1985), 1–15.

[74] L. Górniewicz and M. Lassonde, Approximation and Fixed Points for Compositions of R_δ-Maps, *Topology Appl.* **55** (1994), 239–250.

[75] A. Granas, The Leray-Schauder Index and the Fixed Point Theory for Arbitrary ANRs, *Bull. Soc. Math. France* **100** (1972), 209–228.

[76] H.-P. Heinz, On the Behaviour of Measures of Noncompactness with respect to Differentiation and Integration of Vector-Valued Functions, *Nonlinear Anal.* **7** (1983), 1351–1371.

[77] M. W. Hirsch, *Differential Topology*, Springer, New York, Heidelberg, Berlin, 1976.

[78] P. J. Huber, Homotopical Cohomology and Čech Cohomology, *Math. Ann.* **144** (1961), 73–76.

[79] D. M. Hyman, On Decreasing Sequences of Compact Absolute Retracts, *Fund. Math.* **64** (1969), 91–97.

[80] L. Janos and M. Martelli, Sequentially Condensing Maps, *Univ. u Novom Sadu Zb. Rad. Prirod.-Mat. Fak. Ser. Mat.* **16** (1986), 85–94.

[81] T. J. Jech, *The Axiom of Choice*, North-Holland Publ. Company, Amsterdam, 1973.

[82] D. S. Kahn, An Example in Čech Cohomology, *Proc. Amer. Math. Soc.* **16** (1969), 584.

[83] M. I. Kamenskiĭ, V. V. Obukhovskiĭ and P. Zecca, *Condensing Multivalued Maps and Semilinear Differential Inclusions in Banach Spaces*, de Gruyter, Berlin, 2001.

[84] T. Kato, *Perturbation Theory for Linear Operators*, Springer, New York, 1966.

[85] J. L. Kelley, The Tychonoff Product Theorem Implies the Axiom of Choice, *Fund. Math.* **37** (1950), 75–76.

[86] I.-S. Kim and M. Väth, Some Results on the Extension of Single- and Multivalued Maps, *Topol. Methods Nonlinear Anal.* **28** (2006), 133–153.

[87] I.-S. Kim and M. Väth, Some Remarks on Measures of Noncompactness and Retractions onto Spheres, *Topology Appl.* **154** (2007), 3056–3069.

[88] M. A. Krasnoselskiĭ and P. P. Zabreĭko, *Geometrical Methods of Nonlinear Analysis* (in Russian), Nauka, Moscow, 1975, Engl. transl.: Springer, New York 1984.

[89] W. Kryszewski, *Topological and Approximation Methods of Degree Theory of Set-Valued Maps*, Dissertationes Math. (Rozprawy Mat.) 336, Polish Scientific Publ., Warszawa, 1994.

[90] W. Kryszewski, Some Homotopy Classification and Extension Theorems for the Class of Compositions of Acyclic Set-Valued Maps, *Bull. Sci. Math.* **119** (1995), 21–48.

[91] W. Kryszewski, The Fixed-Point Index for the Class of Compositions of Acyclic Set-Valued Maps on ANR's, *Bull. Soc. Math. France* **120** (1996), 129–151.

[92] W. Kryszewski, Remarks to the Vietoris Theorem, *Topol. Methods Nonlinear Anal.* **8** (1996), 383–405.

[93] W. Kryszewski, *Homotopy Properties of Set-Valued Mappings*, Univ. N. Copernicus Publishing, Toruń, 1997.

[94] Z. Kucharski, A Coincidence Index, *Bull. Acad. Polon. Sci. Sér. Sci. Math. Astronom. Phys.* **24** (1976), 245–252.

[95] Z. Kucharski, Two Consequences of the Coincidence Index, *Bull. Acad. Polon. Sci. Sér. Sci. Math. Astronom. Phys.* **24** (1976), 437–443.

[96] K. Kuratowski, *Topology*, I, Academic Press, New York, San Franisco, London, 1966.

[97] J. Leray and J. Schauder, Topologie et équations fonctionnelles, *Ann. Sci. École Norm. Sup.* **51** (1934), 45–78.

[98] Y. C. Liou, J. C. Yao and V. V. Obukhovskiĭ, Application of the Coincidence Index to some Classes of Impulsive Control Systems, *Nonlinear Anal.* **69** (2008), 4392–4411.

[99] N. G. Lloyd, *Degree Theory*, Cambridge Univ. Press, Cambridge, 1978.

[100] J. Loś and C. Ryll-Nardzewski, On the Applications of Tychonoff's Theorem in Mathematical Proofs, *Fund. Math.* **38** (1951), 233–237.

[101] W. A. J. Luxemburg and M. Väth, The Existence of Non-Trivial Bounded Functionals Implies the Hahn-Banach Extension Theorem, *Z. Anal. Anwendungen* **20** (2001), 267–279.

[102] T.-W. Ma, *Topological Degrees of Set-Valued Compact Fields in Locally Convex Spaces*, Dissertationes Math. (Rozprawy Mat.) 92, Polish Scientific Publ., Warszawa, 1972.

[103] M. Mandelkern, A Short Proof of the Tietze-Urysohn Extension Theorem, *Arch. Math. (Basel)* **60** (1993), 364–366.

[104] J. L. Mawhin, *Topological Degree Methods in Non-Linear Boundary Value Problems*, Amer. Math. Soc., Providence, R. I., 1979.

[105] E. Michael, A Note on Paracompact Spaces, *Proc. Amer. Math. Soc.* **4** (1953), 831–838.

[106] E. Michael, Some Extension Theorems for Continuous Functions, *Pacific J. Math.* **3** (1953), 789–806.

[107] J. W. Milnor, *Topology from the Differential Viewpoint*, University Press of Virginia, Charlottesville, 1965.

[108] H. Mönch, Boundary Value Problems for Nonlinear Ordinary Differential Equations in Banach Spaces, *Nonlinear Anal.* **4** (1980), 985–999.

[109] H. Mönch and G.-F. von Harten, On the Cauchy Problem for Ordinary Differential Equations in Banach Spaces, *Arch. Math. (Basel)* **39** (1982), 153–160.

[110] K. Morita, Star-Finite Coverings and the Star-Finite Property, *Math. Japon.* **1** (1948), 60–68.

[111] J. Nagata, *Modern Dimension Theory*, Heldermann, Berlin, 1981.

[112] L. Nirenberg, *Topics in Nonlinear Functional Analysis*, Courant Institute of Mathematical Sciences, New York, 1974.

[113] L. Nirenberg, An Application of Generalized Degree to a Class of Nonlinear Problems, in: *3ieme Coll. sur l'Analyse fonction., Liege 1970*, pp. 57–74, Centre Belge de Recherches Matheématiques. Vander éditeur, Louvain, Belgique, 1971.

[114] L. Nirenberg, Generalized Degree and Nonlinear Problems, in: *Contributions to Nonlinear Analysis. Proc. Symp. Univ. Wisconsin, Madison, 1971* (E. H. Zarantonello, ed.), pp. 1–9, Academic Press, New York, London, 1971.

[115] R. D. Nussbaum, The Fixed Point Index and Asymptotic Fixed Point Theorems for k-set-contractions, *Bull. Amer. Math. Soc. (N.S.)* **75** (1969), 490–495.

[116] R. D. Nussbaum, The Fixed Point Index for Local Condensing Maps, *Ann. Mat. Pura Appl.* **89** (1971), 217–258.

[117] R. D. Nussbaum, Some Fixed Point Theorems, *Bull. Amer. Math. Soc.* **77** (1971), 360–365.

[118] R. D. Nussbaum, Degree Theory for Local Condensing Maps, *J. Math. Anal. Appl.* **37** (1972), 741–766.

[119] R. D. Nussbaum, Some Asymptotic Fixed Point Theorems, *Trans. Amer. Math. Soc.* **171** (1972), 349–375.

[120] V. V. Obukhovskiĭ, On the Topological Degree for a Class of Noncompact Multivalued Mappings (in Russian), *Funktsionalnyi Analiz, Ul'yanovsk* **23** (1984), 82–93.

[121] V. V. Obukhovskiĭ, P. Zecca and V. G. Zvyagin, On Coincidence Index for Multivalued Perturbations of Nonlinear Fredholm Maps and some Applications, *Abstr. Appl. Anal.* **7** (2002), 295–322.

[122] D. O'Regan, Y. J. Cho and Y.-Q. Chen, *Topological degree theory and applications*, Chapman & Hall/CRC, Boca Raton, London, Now York, Washington, 2006.

[123] E. Ossa, A Simple Proof of the Tietze-Urysohn Extension Theorem, *Arch. Math. (Basel)* **71** (1998), 331–332.

[124] E. Outerelo and J. Ruiz, *Mapping Degree Theory*, Amer. Math. Soc., Real Sociedad Matemática Española, Providence R. I., Madrid, 2009.

[125] J. Pejsachovich and P. J. Rabier, Degree Theory for C^1 Fredholm Mappings of Index 0, *J. Anal. Math.* **76** (1998), 289–319.

[126] P. J. Rabier and M. F. Salter, A Degree Theory for Compact Perturbations of Proper C^1 Fredholm Mappings of Index 0, *Abstr. Appl. Anal.* **2005** (2005), 707–731.

[127] M. E. Rudin, A New Proof that Metric Spaces are Paracompact, *Proc. Amer. Math. Soc.* **20** (1960), 603.

[128] M. E. Rudin, A Normal Space X for which $X \times I$ is not Normal, *Fund. Math.* **73** (1971), 179–186.

[129] B. N. Sadovskiĭ, Limit-Compact and Condensing Operators (in Russian), *Uspekhi Mat. Nauk* **27** (1972), 81–146, Engl. transl.: Russian Math. Surveys **27** (1972), no. 1, 85–155.

[130] H. W. Siegberg and G. Skordev, Fixed Point Index and Chain Approximations, *Pacific J. Math.* **102** (1982), 455–486.

[131] S. Smale, An Infinite-Dimensional Version of Sard's Theorem, *Amer. J. Math.* **87** (1965), 861–866.

[132] L. A. Steen and J. A. Seebach, Jr., *Counterexamples in Topology*, Springer, New York, 1978.

[133] D. K. Su, Fixed-Point Theorems for Set-Valued Mappings on Banach Spaces (in Chinese. English summary), *Zhejiang Daxue Xuebao* **22** (1988), 1–5.

[134] T. tom Dieck, *Topologie*, 2nd ed, de Gruyter, Berlin, New York, 2000.

[135] M. Väth, *Ideal Spaces*, Lect. Notes Math. 1664, Springer, Berlin, Heidelberg, 1997.

[136] M. Väth, Fixed Point Theorems and Fixed Point Index for Countably Condensing Maps, *Topol. Methods Nonlinear Anal.* **13** (1999), 341–363.

[137] M. Väth, An Axiomatic Approach to a Coincidence Index for Noncompact Function Pairs, *Topol. Methods Nonlinear Anal.* **16** (2000), 307–338.

[138] M. Väth, *Volterra and Integral Equations of Vector Functions*, Marcel Dekker, New York, Basel, 2000.

[139] M. Väth, Coincidence Points of Function Pairs Based on Compactness Properties, *Glasgow Math. J.* **44** (2002), 209–230.

[140] M. Väth, Merging of Degree and Index Theory, *Fixed Point Theory Appl.* **2006** (2006), Article ID 36361, 30 pages.

[141] M. Väth, Degree and Index Theories for Noncompact Function Triples, *Topol. Methods Nonlinear Anal.* **29** (2007), 79–118.

[142] M. Väth, A General Degree for Function Triples, *Topol. Methods Nonlinear Anal.* (2011), (to appear).

[143] L. Vietoris, Über den höheren Zusammenhang kompakter Räume und eine Klasse von zusammenhangstreuen Abbildungen, *Math. Ann.* **97** (1927), 454–472.

[144] D. Violette and G. Fournier, Un Indice de Point Fixe pour les Composées de Fonctions Multivoques Acycliques Dans des Espaces de Banach, *Ann. Sci. Math. Québec* **22** (1998), 225–244.

[145] J. D. M. Wright, Functional Analysis for the Practical Man, in: *Functional Analysis: Surveys and Recent Results (Proceedings of the Conference on Functional Analysis, Paderborn, Germany 1976)* (K.-D. Bierstedt and B. Fuchssteiner, eds.), pp. 283–290, North-Holland, Amsterdam, 1977.

[146] P. Zecca, V. G. Zvyagin and V. V. Obukhovskiĭ, On the Oriented Coincidence Index for Nonlinear Fredholm Inclusions (in Russian), *Dokl. Akad. Nauk* **406** (2006), 443–446, Engl. transl.: Dokl. Math. **73** (2006), no. 1, 63–66.

[147] E. Zeidler, *Nonlinear Functional Analysis and its Applications*, I, Springer, New York, Berlin, Heidelberg, 1986.

[148] V. V. Zelikova, On a Topological Degree for Vietoris Type Multimaps in Locally Convex Spaces (in Russian), *Sb. Stat. Asp. Stud. Mat. Fak. Vgu., Voronezh* (1999), 45–51.

[149] V. G. Zvyagin and N. M. Ratiner, *Oriented Degree of Fredholm Maps of Non-Negative Index and its Applications to Global Bifurcation of Solutions*, Global Analysis – Studies and Applications V (Yu. G. Borisovich and Yu. E. Gliklikh, eds.), Lect. Notes Math. 1520, Springer, Berlin, Heidelberg, New York, 1992, pp. 111–137.

Index of Symbols

Index

M

P

R

Bei Fragen zur Produktsicherheit wenden Sie sich bitte an:
If you have any questions regarding product safety,
please contact:

Walter de Gruyter GmbH
Genthiner Straße 13
10785 Berlin
productsafety@degruyterbrill.com